D0290611

High Temperature Vapors

MATERIALS SCIENCE AND TECHNOLOGY

A. S. Nowick and B. S. Berry, Anelastic Relaxation in Crystalline Solids, 1972

E. A. Nesbitt and J. H. Wernick, Rare Earth Permanent Magnets, 1973

W. E. Wallace, Rare Earth Intermetallics, 1973

J. C. Phillips, Bonds and Bands in Semiconductors, 1973

H. Schmalzried, Solid State Reactions, 1974

J. H. Richardson and R. V. Peterson (editors), Systematic Materials Analysis, Volumes I, II, and III, 1974

A. J. Freeman and J. B. Darby, Jr. (editors), The Actinides: Electronic Structure and Related Properties, Volumes I and II, 1974

A. S. Nowick and J. J. Burton (editors), Diffusion in Solids: Recent Developments, 1975

J. W. Matthews (editor), Epitaxial Growth, Parts A and B, 1975

J. M. Blakely (editor), Surface Physics of Materials, Volumes I and II, 1975

G. A. Chadwick and D. A. Smith (editors), Grain Boundary Structure and Properties, 1975

John W. Hastie, High Temperature Vapors: Science and Technology, 1975

In preparation

Gerald Burns, Introduction to Group Theory with Applications

High Temperature Vapors

Science and Technology

JOHN W. HASTIE

Inorganic Materials Division
National Bureau of Standards
Washington, D. C.

ACADEMIC PRESS New York San Francisco London 1975

A Subsidiary of Harcourt Brace Jovanovich, Publishers

To Duncan

ACADEMIC PRESS, INC.
111 Fifth Avenue, New York, New York 10003

United Kingdom Edition published by
ACADEMIC PRESS, INC. (LONDON) LTD.
24/28 Oval Road, London NW1

Library of Congress Cataloging in Publication Data

Hastie, John W
High temperature vapors.

(Materials science and technology series)
Bibliography: p.
Includes index.
1. High temperatures. 2. Vapors. 3. Chemistry, Technical. I. Title.
QD515.H33 536'.57 75-3585
ISBN 0-12-331950-1

PRINTED IN THE UNITED STATES OF AMERICA

Contents

7 Chemistry of High Temperature Species in Space

Foreword

The time is ripe for a book summarizing the past 25 years of successful efforts in characterizing and understanding high temperature phenomena. In the early studies of bulk vaporization processes and phase diagrams as well as the current sophisticated efforts to relate crystal surfaces to evaporation rates and to induce nuclear fusion at 10^8 K by laser irradiation of hydride pellets, one finds a strong basic requirement for thermodynamic data, crystal structures, molecular configurations, and chemical kinetics. High temperature scientists have pioneered the adaptation of nearly every kind of experimental apparatus for use in the high temperature regime. Mass spectroscopy, optical spectroscopy, microwave spectroscopy, electron diffraction, x-ray crystallography, and the somewhat paradoxical combination of low temperature matrix isolation of high temperature species have all been utilized, along with many other tools of chemistry and physics, to elucidate high temperature processes. Doing high temperature experiments has to take precedence over maintaining a highly sensitive machine in perfect condition, but the results are worth the trouble. Although the same principles of structure and bonding apply in high temperature as for more conventional systems, there is much more variety in terms of oxidation states, electron-deficient bonds, electron-excess bonds, and other special cases which cover nearly every category of the physical state. Thus, the high temperature scientist has to be an innovator and yet a practical person who can combine the instrumental techniques and theoretical approaches with the practical viewpoints and questions which come up when one wishes to design a flame-retardant fabric, a container for liquid metal at 2000 K, or the skin of a high speed airplane.

Over the world there have evolved a number of centers of high temperature research, and from the nearly 50 or so such laboratories, there has emanated a group of young scientist-engineers trained in the breadth and depth required for successful high temperature research. Dr. John W. Hastie, a product of two such laboratories, is now established at still a third where the techniques of chemistry, physics, and engineering are all required to produce a successful approach to this particular area of chemistry under one of the most important of the extreme conditions available—high temperatures.

"High Temperature Vapors" should serve as a valuable reference for scientists in chemistry, physics, materials science, and metallurgy, as well as in areas of engineering where high temperatures are regularly applied but not always studied in fundamental detail. Dr. Hastie has combined a deep appreciation of basic science with the practical viewpoint of one concerned intimately with flame retardancy, combustion, and related practical developments. The book should be useful at an advanced undergraduate or beginning graduate level as a reference, as a text, or simply for providing a broad background of appreciation and understanding for this important field of science and technology which is especially involved with materials and their behavior under extreme conditions.

JOHN L. MARGRAVE

Preface

The subject of high temperature chemistry, with emphasis on the vapor state, has developed over the past three decades primarily as a result of modern technology's needs for new materials. To a large extent, these requirements arose initially in connection with the atomic energy program. More recently, new technologies, particularly those of the aerospace and military establishments, have added to the need for continued development of high temperature materials and new high temperature chemical processes.

This period of rapid development has seen the appearance of a number of books dealing with fundamental aspects of high temperatures, such as temperature generation and measurement, and also property characterization techniques for high temperature systems. However, very little attention, in the form of readily available monographs, for example, has been given to the relationship and application of this fundamental knowledge to the various problem areas of high temperature science and technology. Thus the primary intent of this monograph is to direct the readers' attention to the relationship of the basic science of high temperature vapors to some areas of discernible practical importance in modern science and technology. It is therefore not the task of this book to provide a detailed discussion of experimental techniques nor to derive relevant mathematical relationships. There is, however, a sufficiently broad citation of literature so that the interested reader may locate such information.

The major high temperature problem areas selected for discussion include: chemical vapor transport and deposition; the vapor phase aspects of corrosion, combustion, and energy systems; and a brief discussion of extraterrestrial high temperature species. I have given extra emphasis to the chapter on combustion for several reasons. First, I am not aware of any other account where the high temperature chemistry aspects of combustion are considered at reasonable length; and second, the variety of existing combustion problems or applications that derive from the presence of high temperature inorganic species provides many examples of the general practical utility of high temperature vapor phase chemistry.

In order to demonstrate the existence of a common thread between these apparently diverse subject areas, the discussion has been reduced to a fundamental, i.e., molecular, level involving basic thermodynamic and kinetic arguments. The material presented should therefore be of interest to students with a basic background in physical chemistry as well as researchers with an interest in high temperature chemistry, combustion chemistry, and pyrometallurgy. The areas covered here certainly do not represent the totality of applications for high temperature vapor phase science, but it is hoped that they will serve as representative examples, and perhaps form a basis for the realization of new areas of application. I am also hopeful that the need for continued research on the basics of high temperature chemistry will be apparent in the examples considered.

Many of the scientific details of applied technology are either proprietary or not readily available to the general reader. The technological aspects considered are therefore only those that are common knowledge in the open literature. In order to aid the reader in obtaining literature material, I have limited the citation of U.S. Government reports (National Technical Information Service) that are not routinely available in libraries. The subject matter dealt with here is therefore intermediate between applied technology, as practiced, and treatises on the fundamental aspects of high temperature chemistry.

Readers who are expert in any of the areas dealt with, and who may take umbrage at a necessarily oversimplified account of their subject of special interest, should keep in mind that their subject is being viewed through a filter, so to speak, which passes information that is primarily in the realm of high temperature vapors. A definition of what we consider as a high temperature vapor state has been given in the introductory chapter. This chapter also serves the purpose of providing some historical perspective and indicates the diverse role of high temperature vapors in materials science.

In the preparation of this manuscript, the primary literature sources used were those listed as an Appendix to Chapter 1. A scanning of *Current Contents* and *Chemical Abstracts* was also made. The literature survey emphasized mainly the period 1966–1973. Of the titles surveyed, about 8000 papers were considered to be relevant to the objectives of the monograph. However, it was not possible to include all of this material in a monograph of this type, and a screening was necessary to reduce the amount of material to a manageable level. Many of the smaller sections in this book, i.e., those of about 20 pages in length, are themselves major topics, each worthy of many hundreds of pages of discussion. Therefore, in order to rationalize the scope of their highly contracted discussion, to indicate the relevance of high temperature vapors to each topic, and to mention more extensive literature sources, these sections have been prefaced by an introduction.

My wife, Hilary, deserves a very special acknowledgment, both for her direct secretarial assistance and for assuming the life of a "book widow" during the past several hundred weekends. The expert typing of the final manuscript by Mrs. Ellen Ring is also greatly appreciated.

JOHN W. HASTIE

Terminology

Throughout the chapters to follow the reader will be confronted with the use of either more than one symbol for a particular quantity, e.g., J or F for flux, or the same symbol for different quantities, e.g., σ for symmetry number or collision diameter. This results, in part, from the use of the same basic concepts and mathematical expressions by different scientific or engineering disciplines. Rather than attempting to reduce this "entropy effect" by introducing a systematic terminology, with the resultant loss of free energy (by the author), I have elected to adopt the terminology as used by the authors of the various pieces of cited literature. This creates a small amount of repetition but should aid the reader when referring to the original literature sources. It is expected that such referrals will be made by those readers requiring additional detail, such as descriptions of experimental and computational procedures which, for the most part, are to be intentionally avoided in the discussion to follow. Also, in order to remain consistent with the cited literature, no attempt to use SI units is made here. For a similar reason, F is used to denote free energy rather than the recently recommended IUPAC terminology of G for Gibbs free energy. (In current application-oriented literature, e.g., journals of the American Institute of Aeronautics and Astronautics, preference is still given to the use of F.) A listing of IUPAC recommended symbols and terminology, including SI units, may be found in *Pure and Applied Chemistry* **21,** 1–44 (1970).

1

Introduction

I. The Nature of High Temperature Vapors

The chemical and physical nature of high temperature vapors can be described in terms of the identity of the various species present and their thermodynamic, kinetic, and transport properties.†

Application of the basic principles of thermodynamics, kinetics, and gas dynamics permits a rigorous connection to be made between the micro—i.e., molecular—structure of high temperature vapors and macroscopic observables such as temperature, density, viscosity, and thermal or electrical conductance. The following examples, which are typical of the many others considered in later chapters, serve to emphasize the importance of this connection, which is virtually unique for the gaseous state of matter.

The various molecular degrees of freedom such as rotation, vibration,

† The main relationships of thermodynamics, kinetics, and transport phenomena are summarized in Section V.

dissociation, and ionization can have an appreciable influence on properties of practical importance, such as the kinetic energy of a propellant gas, because of the heat-sink effect which results from the distribution of energy among the various internal degrees of freedom, e.g., see Holzmann (1969).

Another example, from the area of combustion, is given by the relationship of basic reaction rate data for the elementary molecular processes to ignition temperature, flame speed, rate of heat release, and other macroscopic observables. Knowledge of species identity is crucial to the micro–macro link. Existing knowledge is often inadequate in this regard—as demonstrated by the example of SO_2, where its predicted role in combustion systems containing sulfur is currently in doubt owing to conflicting reports on the existence of a new high temperature isomer of SO_2, or alternatively a dimer, e.g., see Cullis and Mulcahy (1972), Hellner and Keller (1972), and Basco and Morse (1971).

The characterization of high temperature vapors, both at the micro and macro levels, is a very extensive subject and much progress has been made in this area in recent years, e.g., see the reviews of Hastie *et al.* (1970), Thorn (1966), and Eyring (1967, 1969, 1971, 1972). Only brief mention of this fundamental characterization aspect will be given in the chapters which follow, where the main emphasis will be on areas of application. This restriction will be relaxed where previous treatment of the fundamentals, in the more classical high temperature literature sources, appears to have been limited. Fortunately, there are several literature sources available where extensive discussions of the characterization of high temperature vapors appear and the interested reader is referred to the book on "Characterization of High Temperature Vapors" (Margrave, 1967a) and the series of "Advances in High Temperature Chemistry" (Eyring, 1967, 1969, 1971, 1972), which provide useful background to the ensuing discussion.

Defining a High Temperature Vapor

The subject of high temperature vapors is, to a large extent, synonymous with high temperature chemistry. Indeed, some authors refer only to the vapor phase when discussing high temperature chemistry, e.g., see Novikov and Gavryuchenkov (1967). However, in practice, chemistry at high temperatures can involve solid-, liquid-, and vapor- or gas-phase interactions—including the critical state. Studies of high temperature liquid-phase chemistry are usually discussed under the synonyms of molten- or fused-salt chemistry, as many high temperature materials are in salt form (Blander, 1964; Sundheim, 1964; and Braunstein *et al.* 1971). Discussions of molten oxide chemistry are also usually made in the same context as molten salts. Solids at high temperatures are often considered in conjunction with the other phases. The frequent coexistence of several phases in high temperature chemical systems points to the restrictive nature of subject area definitions such as high temperature vapors, molten-salt chemistry, or high temperature solid-state chemistry.

The terminology "high temperature vapor" is without rigid definition. Many definitions of what constitutes "high temperature chemistry," or "high temperature species," have been given (Drowart *et al.*, 1969; Thorn, 1966; Margrave, 1962; Brewer and Searcy, 1956). Perhaps the most practical definition is that offered privately, by workers in the field, to the effect that it is the subject matter of interest to researchers who prefer to be described as high temperature chemists. Thus, from a survey of the various bibliographies, journals, monographs, and conferences that have dealt with high temperature chemistry, one can develop a feeling for what this interdisciplinary subject consists of. Given this approach, it is apparent that the interest and effort devoted to various temperature regimes are currently at a maximum for the 1000–2000°C temperature range. This is followed by a lesser—but still significant—effort for the 200–1000 °C and 2000–3000 °C temperature intervals, and a growing interest in the classical plasma region of 3000–10,000 °C and the nuclear fusion plasma region of $\sim 10^7$ °C.

With this degree of flexibility in what constitutes high temperature, it follows that the definition of a "vapor" at high temperatures is similarly broad. The conventional definition of vapor as a gas which may be condensed by an increase in pressure at constant temperature, is too rigid for a discussion of the practical aspects of gas- or vapor-phase phenomena at high temperatures. In the ensuing discussion, a high temperature vapor may have the following characteristics. It may—

be an *inorganic* vapor either in or out of equilibrium with its condensed state;

be an *unsaturated* inorganic vapor with no condensed state present, such as may occur, for example, in hydrocarbon flames containing metals in trace amounts;

contain a *permanent or common gas*, such as H_2O, O_2, or H_2 which, for example, may be interacting with an inorganic system to yield new vapor species;

contain *radicals* and charged species, such as positive *ions*, negative ions, and electrons, which result from thermal—i.e., equilibrium—or kinetic processes at high temperature;

TABLE 1.1

Representative Examples of High Temperature Species—
in the Presence of a Condensed Phase

Vapor species	Condensed phase	Vapor species	Condensed phase
Li_2O	$Li_2O(c)$	B_2O_3	$B_2O_3(c)$
$(WO_3)_n$, $n = 1$–5	$WO_3(c)$	$Be(OH)_2$	$BeO(c) + H_2O$
$Ni + O_2$	$NiO(c)$	$(NaOH)_2$	$NaOH(c)$
C_n, $n = 1$–5	$C(c)$	$SbS + S_2$	$Sb_2S_3(c)$
CeC_2, CeC_4	$CeC_2(c)$	Sb_4O_6	$Sb_2O_3(c)$
$Al + N_2$	$AlN(c)$	$NaAlF_4$	$NaAlF_4(c)$
Si_2N	$Si(c) + N_2$	$NdAl_3Cl_{12}$ }	$NdCl_3(c) + Al_2Cl_6$
$Ti + B$	$TiB(c)$	$NdAl_4Cl_{15}$ }	
$NaCl$, $(NaCl)_2$	$NaCl(c)$	$LiGaO$	$Li_2O(c) + Ga_2O(c)$
$(CuCl)_3$	$CuCl(c)$	$SnPbS_2$	$SnS(c) + PbS(c)$
BaF_2	$BaF_2(c)$	$CsRbCl_2$	$CsCl(c) + RbCl(c)$
$Al_2O + O_2$	$Al_2O_3(c)$	$AuPd$	$Au(c) + Pd(c)$

in a majority of practical instances, have a pressure in the range of 10^{-7}–10^3 atm;

have a temperature in the range of 200–2000 °C, though a few departures from this range will be considered.

The classical description of a high temperature vapor usually also involves the presence of a condensed phase. Some representative examples, indicating the variety of species that can be present under this vapor-saturated condition, are given in Table 1.1. Many hundreds of such species have now been identified. The recent review of De Maria (1970) gives emphasis to this varied character of high temperature vapors.

As most of the ensuing discussion deals with vapor-phase species, only the condensed-phase systems are explicitly identified, e.g., as $Al_2O_3(s)$ and $SiO_2(l)$ or $Al(c)$—where solid or liquid is present; gas-phase species are indicated simply as SiO, for example.

II. The Scope and Literature of High Temperature Vapor-Phase Chemistry

The changing emphasis, over a period of several decades, and the present-day scope of the vapor phase in high temperature chemistry, is revealed by the contents of the various reviews and monographs—as outlined chronologically below.

1950

The early treatise edited by Lebeau and Trombe (1950) dealt primarily with the production of high temperatures by gas combustion, resistance, induction, and solar furnaces, H-atom torches, and electric discharges.

At about the same time, the results of the Plutonium Metallurgical Project, which was part of the Manhattan Project, appeared in the open literature (Quill, 1950). This work dealt with the high temperature thermodynamic properties of many materials, such as the metals and their halides, nitrides, carbides, sulfides, silicides, and phosphides, and the common gases. The sparseness of experimental data in this area required the development of estimation procedures. In attempting to discern systematic trends in the limited data available, Brewer and co-workers noted several anomalies, particularly between the estimated and "experimental" entropy of vaporization for cuprous chloride (e.g., see the review of Brewer, 1972). Experimental work by Brewer and Lofgren (1948) revealed that CuCl(s) did not vaporize as a monomeric CuCl species, but as the more complex and unexpected trimeric form, $(CuCl)_3$. The importance of observations of this type is that, quoting from Brewer's recent reflections on this early work (Brewer, 1972),

> thermodynamic calculations no matter how precisely done, are worthless if the major species are not recognized and the correct net reactions are not considered.

As will become evident in the chapters to follow, severe limitations still exist in applying high temperature science to technological problems, as a result of uncertainties in species identity.

1956–1957

Later reviews of high temperature chemistry emphasized the complexity and unusual oxidation states associated with high temperature vapors (Brewer and Searcy, 1956; Brewer, 1957). Several rules were proposed by Brewer to provide a basis for predicting the nature of high temperature vaporization.† Essentially these are

- that the minor species will increase in relative importance with increasing temperature, and
- that, from empirically based observations, an almost constant entropy

† A more detailed discussion of these rules may be found in Section V.A.

change results for processes where the same difference occurs between the number of moles of reactant and product *free-molecular species.*

The ability to estimate entropies allows the main temperature-dependent contribution to free energy changes to be predicted. The usefulness of this entropy rule will become evident in later chapters (e.g., see Chapter 2, Section VIII.E, and Chapter 3, Section IV.D).

In 1957, Margrave outlined the following areas where the nature and properties of high temperature species are of importance:

- understanding high temperature processes involving vaporization and dissociation,
- the testing of various bonding theories,
- gas–solid interactions,
- combustion processes,
- reactions in arcs and discharges,
- chemical synthesis,
- chemical kinetics.

These areas have received considerable attention in subsequent years and will be discussed further in later chapters.

1959–1964

The fundamental physicochemical aspects of high temperature were reviewed in the book edited by Bockris *et al.* (1959).

A review by Searcy (1962) reemphasized the systematics in thermodynamic properties of the type discussed earlier by Brewer.

In 1961 a small introductory monograph on high temperature chemistry appeared and emphasis was given to the use of combustion and electric arcs for temperature generation (Lachnitt, 1961). A very brief mention was also given, at this time, to the role of high temperatures in areas of materials technology—including metallurgy, nuclear energy, chemical synthesis, and refractory materials.

Further evidence of an increased interest in high temperature technology was provided by the setting up of periodic High Temperature Technology International Conferences (see Asilomar, 1960, 1964, 1969). At the first meeting in 1959 (Asilomar, 1960), only a single presentation—albeit an important one—dealt specifically with high temperature vapors, namely, that of Inghram and Drowart (1960) on "Mass Spectrometry Applied to High Temperature Chemistry." A more significant fraction of the second meeting in 1963 (Asilomar, 1964) dealt with various scientific and technological aspects of high temperature vapors, involving thermodynamics, kinetics, chemical bonding, gas–metal reactions, and materials requirements for space propulsion.

Also during this period, a chapter on high temperature technology appeared (Ferguson and Phillips, 1962). In addition to the usual discussion of temperature, its measurement, and methods of attainment, mention was given to rocket chemistry and plasma processes.

Drowart and Goldfinger (1962) reviewed the gas-phase aspects of high temperature chemistry, acknowledging that the chemistry of the other states of matter should be considered in a more comprehensive treatment of the subject. Likewise, Gilles (1961) reviewed the field of high temperature chemistry giving emphasis to the techniques and results of high temperature vaporization studies.

In 1962 Margrave suggested the following goals for the future development of high temperature chemistry:

- development of accurate temperature standards,
- development of new techniques for attaining temperatures of greater than 4000 K,
- development of apparatus to withstand oxidizing conditions, particularly for the temperature range of 2000–3000 K,
- development of better techniques for structure determination of high temperature species,
- the determination of accurate thermodynamic data.

Partial attainment of these goals has been achieved in subsequent years.

With the accumulation of experimental data, particularly that developed by mass spectrometric analysis, the complexity of high temperature vapors was further demonstrated, and in 1962 Brewer again emphasized this aspect of high temperature chemistry.

A number of underdeveloped areas in high temperature chemistry were defined by Searcy (1964). These included the need for determining entropies of both complex vapor species and high temperature solids. Kinetic data relating to molecular reaction rates, vaporization coefficients, gas–solid reactions, and disproportionation reactions were also indicated to be lacking.

1966–1968

The current and future problems of high temperature chemistry were discussed in 1966 by a committee on high temperature chemical phenomena (see Margrave, 1967b). Among the problems considered were the application of optical, electron spin resonance, and molecular beam spectroscopic techniques to the characterization of high temperature species. The general need for basic thermodynamic, kinetic, and transport data was again emphasized. Problems mentioned in connection with rocket technology included the determination of spectroscopic opacities of high temperature gases, the synthesis of high energy propellants, characterization of the ablation rocket reentry process, and the description of mass and energy transport. The need for intermolecular force data for calculations of transport parameters at high temperatures and high pressures was indicated.† The lack of understanding of elementary plasma processes was also stressed. Several other problem areas were mentioned in passing. These include geochemistry, nuclear fission and fusion power systems, and chemical rockets. Also, the wide scatter of papers and data relating to high temperature chemistry was of some concern. This difficulty was also evident to Thorn (1966) in his review of the field. The committee (Margrave, 1967b) therefore suggested that a systematic study of technical requirements for basic data in applied fields should be made. Unfortunately, this suggestion does not appear to have come to fruition, and it is hoped that the discussion of problem areas given in the chapters to follow will provide some general insight into the basic data requirements.

Weller and Bagby (1967), in considering the present and future demands of high temperature technology, predicted an increased use of higher process temperatures. This trend results naturally from the potential increased efficiency in energy utilization at high temperatures. The fundamental problems associated with this use of increased temperatures result primarily from heat dissipation and thermal capacity limitations. Natural heat sinks such as fusion, vaporization, dissociation, and ionization develop with increasing temperature. Thus the ability of chemical systems to exhibit heat capacity is limited to temperatures below about 10,000–20,000 K, and hence higher temperature systems need to convert energy to other forms, such as electrical. Some examples of high temperature applications were given by Weller and Bagby (1967) as: the processing of metals and carbides; the generation and control of rocket turbine, MHD, and nuclear-power systems, and the use of combustion-based devices in weaponry. The authors were of the opinion that dramatic advances in the technology of high temperature materials should not be expected.

Many of the modern techniques developed for the characterization of high temperature vapors were described in the book edited by Margrave (1967a). These included classical vapor pressure techniques, mass spec-

† In the author's opinion this is still a very much neglected area of high temperature vapors.

trometry, optical spectroscopy, microwave spectroscopy, electron diffraction, and the statistical calculation of thermodynamic data from basic molecular parameters.

The expanding interest in high temperature vapors, in connection with high temperature technology, was revealed at the third Asilomar meeting in 1967 (Asilomar, 1969), where the papers presented dealt predominantly with the vapor phase. In contrast, the review of Kubaschewski (1968) on experimental thermochemistry at high temperatures dealt substantially with the condensed-phase aspects of solution thermodynamics, solid–liquid-phase equilibria, nonstoichiometry, and vapor pressures. An important role of high temperatures in thermodynamic systems was considered to be the increased rate of approach to equilibrium and the increased average phase width; that is, a wider range of homogeneity exists at high temperatures.

1970–1973

A further attempt to systematize the high temperature thermodynamic data for inorganic materials was made by Searcy in 1970 (Searcy *et al.*, 1970).

More recently, a two-volume treatise on the utilization of high temperatures in physics and chemistry has appeared (Chaudron and Trombe, 1973a, b). In addition to the usual discussion of characterization techniques and methods of attaining high temperatures, some coverage is given to materials preparation by vapor deposition techniques and combustion applications such as magnetohydrodynamic energy converters.

III. Thermodynamic Relationships between High Temperature Vapors and the Other States of Matter†

In general, a saturated equilibrium vapor phase controls the nature of the condensed phase through the basic principles of phase equilibria, e.g., see Reisman (1970). The variation of vapor pressure with temperature, or with condensed-phase composition, can in fact prove to be useful in defining phase boundaries—as was emphasized recently by Wynnyckyj (1972). Conversely, one can determine the vapor pressure or rate of vaporization for a multicomponent liquid phase using existing phase diagrams.

† A summary of basic thermodynamic relationships and concepts is given in Section V.

An important aspect of vapor–solid interdependences concerns the sensitivity of solid-state composition—and related properties—to the mode of vaporization. In particular, vaporization from nonstoichiometric oxides such as $Fe_3O_4(s)$ and $WO_3(s)$ can lead to pronounced changes in their chemical and physical nature, especially when the vaporization is noncongruent. This problem has been alluded to briefly by Anderson (1963) and by Ackermann *et al.* (1961), for example.

The question of the chemical stability of refractory materials at high temperatures may be answered in part by considering reaction free energies between the various system components, e.g., see Livey and Murray (1959). The following cases may then be defined where the vapor-phase influences the chemical stability of the solid phase:

$$\text{solid I} + \text{solid II} = \text{vapor } [+ \text{ solid (or liquid) III}]$$

$$C(s) + TiN(s) = \tfrac{1}{2}N_2 + TiC(s) \quad \text{and}$$

$$Si(s) + SiO_2(s) = 2SiO$$

$$\text{solid I} + \text{gas I} = \text{solid II} + \text{gas II}$$

$$2C(s) + SiO = SiC(s) + CO$$

$$\text{solid I} + \text{gas I} = \text{gas II}$$

$$BeO(s) + H_2O = Be(OH)_2$$

$$\tfrac{1}{2}xAl_2O_3(s) + \tfrac{1}{2}yS_2 = Al_xS_y + \tfrac{3}{4}xO_2 \qquad (x, y = 1 \text{ or } 2)$$

In these cases a precise knowledge of the vapor species identity is essential to the prediction of solid-state thermochemical stability.

An enhanced stability of a particular valence state, usually a lower oxidation state, may occur in the vapor phase at high temperature. Thus, on condensation a disporportionation may occur; such is the case for SiO, i.e.,

$$2SiO \rightarrow SiO_2(s) + Si(s).$$

This limits the use of silica refractories as containers under high temperature reducing conditions owing to the contamination of the sample with solid Si and SiO_2.

The energetic characteristics of a high temperature vapor may be related to those of its condensed solid state via thermodynamic relations such as the Born–Haber, e.g., see Gutmann and Mayer (1972), and related cycles, e.g., see Oshcherin (1971). For example, with the defect structure model of zirconium trihalide disproportionation to the tetrahalide and dihalide, the difference between the heat of disproportionation to the tetrahalide and the heat of sublimation of the tetrahalide has been suggested as being a measure of the energy required by the electron transfer and creation of lattice vacancies (Shelton *et al.*, 1969).

Many refractory nonmetallic solids are ionic in their bonding character and their stability can be expressed in terms of a crystal or lattice energy, i.e., for an Na^+Cl^- type solid:

$$U = N\left(-\frac{Ag_1g_2e^2}{r} + \frac{B}{r^n}\right),$$

where

N is Avogadro's number,
A the Madelung constant which represents the summation of ionic interactions over the whole crystal,
g_1, g_2 are charges of cation and anion,
e is the unit of electrostatic charge,
r the internuclear separation of ions, and
B, n are empirical constants representing repulsive interactions.

For NaCl the calculated and experimental crystal energy is 184 kcal mol^{-1}. This is the energy to produce in free space the ions Na^+ and Cl^- from the solid. Now NaCl vaporizes as molecular NaCl (with minor amounts of dimers and trimers), and the energy to produce Na^+ and Cl^- from a NaCl molecule is 147 kcal mol^{-1}. Thus in this case the crystal energy is not very much greater than the corresponding dissociation energy (i.e., to the ions) for the vapor molecule. However, for the case of ZnO, the crystal energy is 964 kcal mol^{-1}, whereas the corresponding energy for the ZnO vapor molecule is only 471 kcal mol^{-1}. In terms of dissociation to the elements, the molecule has a dissociation energy of less than 66 kcal mol^{-1}; that is, ZnO is a relatively unstable vapor species. The crystalline solid, on the other hand, is relatively stable. Thus, the preferred vaporization process is the dissociative reaction:

$$ZnO(c) = Zn + \tfrac{1}{2}O_2,$$

rather than

$$ZnO(c) = ZnO.$$

For metallic phases the structures and stabilities can be related to the electronic structure of the atoms, i.e., the number of s, p, and d electrons in the valence state, as shown by the Brewer–Engel theory for transition elements (Brewer, 1965, 1968). Conversely, one can infer the electronic structure of certain atoms from condensed-phase thermodynamic properties, as was shown by Brewer (1971) for the lanthanide and actinide metals.

The nature of the chemical bond in polyatomic vapor-phase species should also be relatable to the condensed-phase stability and structure.

However, very little attention appears to have been given to this question and the relationships are not immediately obvious, as evidenced by the following example. Consider the vaporization of Sb_2O_3(s) and Sb_2S_3(s) which have a similar structure in the solid state. In the former case, Sb_4O_6 is the main equilibrium vapor species (Hastie, 1972), whereas for the latter case the vapor species are predominantly SbS and S_2 (Faure *et al.*, 1972). A fundamental difficulty in relating the free molecular state to the condensed phase is the lack of applicability of *localized* molecular orbital treatments in the latter case. That is, atomic orbitals overlap far beyond nearest neighbors in the condensed phase, by virtue of the high density of nuclei. Discussions of bonding in solids must therefore be based on semi-empirical arguments such as those given by Phillips (1973).

The statistical free energy calculation of atomic clusters (up to 70 atoms) provides a link between the gas and solid phase, e.g., see Hoare and Pal (1972). It is reasonable to expect that the molecular nature of high temperature vapors may eventually be linked to that of the corresponding solids by a similar statistical approach, but giving due consideration to electrostatic interactions.

A fundamental thermodynamic link between the condensed and vapor phase is given by activities and related quantities. For example, activities of condensed-phase components can be expressed in terms of partial vapor pressures with reference to the standard state pressure. Hence, a measurement on the vapor phase can provide activity data for the condensed phase and vice versa. Thermodynamic activities may be used to *suggest* structural models in the condensed phase, particularly for molten-salt mixtures, e.g., see Sundheim (1964) and Lumsden (1966).

IV. The Role of High Temperature Vapors in Materials Science

The need for high temperature materials in modern technology is well recognized, and no detailed discussion of this point is required here. However, the important role of high temperature vapors in materials science and technology does not appear to be widely appreciated. This is particularly evident from a survey of the contents of symposia and textbooks on materials science, even for those where the high temperature aspects are of specific interest.

In considering research prospects on inorganic materials, Tananaev (1971) emphasized the chemical synthesis and physical properties of new or improved materials. Vapor-phase technology was considered only in connection with the synthetic aspects. A recent treatise on materials science and technology focuses its attention on the solid state with the exception of a chapter on chemical vapor deposition (Herman, 1972). Similarly, a text on metallurgy at high pressures and high temperatures deals entirely with solids (Gschneider, 1964). The treatise on tools and techniques in physical metallurgy considers solid-state measurement techniques, together with a section on vapor deposition in connection with crystal growth and alloy preparation (Weinberg, 1970). A chapter on emission and atomic absorption spectrophotometry deals, in passing, with some of the elements of vapor-phase high temperature chemistry.

The reason for such a limited consideration of the vapor phase in connection with materials science and technology may be as follows. Physical characteristics of materials at high temperatures, in contrast to the chemical properties, are more readily identifiable with the practical usage and limitations.† On the other hand, chemical phenomena involve molecular processes which tend to be more abstract, and therefore less readily identifiable as important factors in the ability of a material to function at high temperature. The recent rapid development of high temperature chemistry, with an emphasis on the gas or vapor phase, can contribute much to the appreciation of the role of chemistry in materials applications at high temperatures.

Areas of Application

A number of technologically important areas where the chemical aspects of high temperature vapors are of fundamental importance are indicated in Table 1.2. The contents of later chapters deal with these and other areas of application. Furthermore, the discussion to follow is designed to be generally in keeping with the objectives of materials science and engineering according to the definition:

> Materials science and engineering are concerned with the generation and application of knowledge relating the composition, structure, and processing of materials to their properties and uses.

From a consideration of the so-called materials cycle of:

- obtaining the raw material—e.g., mining of ores;
- metals extraction and refinement—i.e., metallurgy;
- conversion to basic product materials—e.g., crystals, alloys, ceramics, coatings;
- disposal of used products—e.g., dumping, incineration and recycling,

† Some recent examples of physical processes that lead to materials limitations at high temperatures may be found in the articles edited by Kennedy (1968).

TABLE 1.2

High Temperature Vapor-Phase Systems in Materials Science and Technology[a]

Energy systems			Chemical systems	
Heat sinks	Combustion	Nuclear fission and fusion	Material transport	Chemical modification
Vaporization Dissociation Ionization	Flames Explosions Chemical rockets MHD Gas turbines Corrosion	Corrosion Heat transfer Fuel processing	VD CVD Flame spraying Ablation Lamps Ore genesis Extractive metallurgy	Corrosion inhibition Fire resistance Smoke control Combustion-knock control

[a] VD, vapor deposition; CVD, chemical vapor deposition; MHD, magnetohydrodynamics.

it becomes apparent that high temperature vapors can play a significant role at each stage of the materials cycle. For instance, the separation and transportation of ores in the period of orogenesis may have involved high temperature vapors, as discussed in Chapter 3. The metallurgical extraction and refinement process can utilize high temperature vapors to advantage, as is indicated in Chapter 3. Similarly, the preparation of product materials can be achieved by vapor solvation and transport processes, as shown in Chapter 3. And finally, the disposal or recycling of these materials can utilize combustion or pyrometallurgical procedures, based on the chemistry discussed in Chapters 3 and 5. Corrosion may also enter into various stages of the materials cycle, as discussed briefly in Chapter 4.

Another aspect of the materials cycle involves the utilization of energy to effect the various materials transformations. Thus the efficient or controlled use of energy is linked to a proper handling of the materials cycle, which would result from a thorough basic understanding and optimization of the various materials- and energy-transformation processes. Likewise, the role of materials as fuels and containers in the energy producing systems themselves is of significance. Hence we find that high temperature vapors can also play a role in energy systems, as discussed in Chapter 6.

High temperature species also serve to indicate the material composition of stellar bodies such as hot stars, as indicated in Chapter 7, and thereby provide clues as to the primogenetic state of inorganic materials.

In concluding these remarks on areas of application it may be argued that, given our broad definition of what constitutes a high temperature vapor (see Section I), we have artificially enhanced the importance of high temperature vapors in materials science. However, from the subsequent discussion dealing with applications of the science of high temperature vapors to apparently diverse materials problems, it is hoped that the general role of high temperature vapors will be demonstrated.

V. Basic Concepts and Controlling Factors in High Temperature Vapor-Phase Processes

In practice, many high temperature vapor-phase processes are controlled by a combination of physical and chemical effects. The physical influence usually takes the form of a gas-dynamic effect, such as forced convection, and this movement of gas can lead to gradients of material concentration and temperature. These gradients serve as a force for mass and energy transport and thereby influence the overall progress of any chemical process that may be occurring. One can readily conceptualize the interdependence of physical and chemical effects by recalling the limited range of gas flows that may be used to produce a visually steady flame on a laboratory torch. In this case the chemical kinetic energy of the combustion process exerts a balancing influence on the forced flow of gas entering the flame, and a steady flame zone results.

Even though many practical systems are not in a state of general equilibrium there are instances, especially at high temperatures, where local equilibrium is achieved. For the example of the laboratory torch flame, there are regions in space where a local equilibrium exists but also other areas where not all reactions have reached equilibrium. Hence both thermodynamic and chemical kinetic factors, together with mass and energy transport, may control a particular high temperature gas- or vapor-phase system.

Many examples where one or more of these controlling factors operate concurrently will emerge in the chapters to follow. The pertinent thermodynamic, kinetic, or transport expressions will usually be considered as the practical examples arise. However, for the reader's convenience, a brief outline is given here of the more important basic concepts underlying the chemical and physical process-controlling phenomena. The object will be either to refresh the reader's memory or to indicate literature sources where a detailed discussion of these basic concepts may be found.

The point should be made that thermodynamics is the underlying basis for all high temperature processes regardless of whether the system is rate-limited or not. It can also be stated that only the thermodynamic properties of high temperature vapors are reasonably well established. Relatively little basic work has been done in the areas of chemical kinetics and transport processes. In view of this restriction, the discussion in the chapters to follow is largely based on thermodynamic arguments but, where possible, rate-limited phenomena will also be considered.

A. Equilibrium Processes

An equilibrium process is really a special case of a rate process, the rate conditions being such that the overall chemical reaction is sufficiently fast in both directions that the reaction free energy is minimized. We may therefore consider that equilibrium "lies in the eye of the beholder." That is, if observations are made on the proper characteristic time scale, as determined by the process kinetics, then one may appear to have an equilibrium system. This time scale can vary from $\sim 10^{-6}$ sec, e.g., for bimolecular flame processes, to many years for mineralogical processes; for examples of the latter case see Figure 2 given by Barton *et al.* (1963). We can also expect to observe systems where some—but not all—chemical reactions are at equilibrium; that is, certain species may be in a state of so-called local equilibrium.

1. Basic Relationships

Consider a general equilibrium vapor-phase reaction involving ideal gases:

$$\mathrm{A} + \mathrm{B} = \mathrm{C} + \mathrm{D}.$$

From thermodynamics, the extent to which the reactants A and B interact to yield products C and D can be defined by the difference in standard free energies† between the reactants and products, i.e.,

$$\Delta F_{\mathrm{r}} = F_{\mathrm{C}} + F_{\mathrm{D}} - F_{\mathrm{A}} - F_{\mathrm{B}}.$$

The concentrations, or partial pressures, p, of the reaction components are then interrelated via ΔF_{r} according to the relation

$$\Delta F_{\mathrm{r}} = -RT \ln K_{\mathrm{p}},$$

where K_{p} is the temperature-dependent equilibrium constant, R the

† The common use of F° to define the standard state property is neglected here in favor of F since all thermodynamic functions considered in the discussion to follow are referenced to the standard state. Note also, F is the Gibbs free energy [see p.140, Pitzer and Brewer (1961)].

universal gas constant, and T the temperature; also,

$$K_p = p_C \cdot p_D / p_A \cdot p_B.$$

For the case where no additional species are present, the sum of these partial pressures must be equivalent to the total pressure P, i.e.,

$$p_A + p_B + p_C + p_D = P.$$

Also, the total pressure is related to the other degrees of freedom by the ideal gas relation

$$P = (n/V)\,RT,$$

where n is the total number of moles and V is the volume.

The enthalpy, or heat at constant volume, released or absorbed (from the surroundings) by the reaction is given by

$$\Delta H_r = \Delta H_{f\,C} + \Delta H_{f\,D} - \Delta H_{f\,A} - \Delta H_{f\,B},$$

where ΔH_f represents the standard heat of formation for the various components. Hence, if the products have more negative (or less positive) heats of formation than the reactants, then ΔH_r is negative, i.e., exothermic, and heat is released by the reaction. ΔH_r is also related to ΔF_r by the relation

$$\Delta F_r = \Delta H_r - T\,\Delta S_r,$$

where ΔS_r is the reaction entropy and is given by

$$\Delta S_r = S_C + S_D - S_A - S_B,$$

where S_C, etc. are the entropies of the reaction components.

Empirically, we find that ΔH_r and ΔS_r are almost invariant with temperature, and the standard values at 298 K can often be used to a reasonable approximation for high temperature conditions. As a typical case consider the reaction

$$2AlCl_3 = Al_2Cl_6,$$

where at 298 K

$$\Delta H = -30.2 \text{ kcal mol}^{-1},$$

$$\Delta S = -36.6 \text{ cal deg}^{-1} \text{ mol}^{-1},$$

and at 1000 K

$$\Delta H = -27.5 \text{ kcal mol}^{-1},$$

$$\Delta S = -31.9 \text{ cal deg}^{-1} \text{ mol}^{-1}.$$

One should note from the relationships among K_p, ΔF_r, ΔH_r, and

ΔS_r that the most favorable reaction conditions are determined by

$$\Delta F_r - ve, \qquad \Delta H_r - ve, \qquad \Delta S_r + ve.$$

Negative values of ΔH_r can result from the production of either additional or relatively stronger bonds in the reaction products. Positive values of ΔS_r can result primarily from an increased number of molecular species produced by chemical reaction. Gas molecules inherently have high entropies, and this makes them increasingly important chemical entities as the temperature is raised.

It is useful to recognize that the entropy of a gaseous species can be calculated from basic molecular parameters, by statistical thermodynamics, and is therefore not reliant on experimental thermodynamics, e.g., see Margrave (1967a). As an example, for a vapor comprised of nonlinear polyatomic species the entropy, in units of cal deg^{-1} mol^{-1}, is given by

$$S_T = 6.8634 \log M + 18.3025 \log T + 2.2878 \log(I \times 10^{117}) - 4.5756 \log \sigma - 2.3493 + S_{\text{vib}} + S_{\text{elec}},$$

where

M is the molecular weight,
I the product of the moments of inertia—as determined from the molecular geometry and the masses of the component atoms,
σ the symmetry number,
S_{vib} the vibrational entropy contribution—as determined from the fundamental molecular frequencies, and
S_{elec} the electronic entropy contribution—as determined from a knowledge of the ground and low-lying molecular electronic states.

Thermodynamically, the entropy is related to the heat capacity C_p by

$$S_T = \int_0^T (C_p/T)\, dT.$$

These, then, are the main thermodynamic relationships by which the composition and thermal properties of an equilibrium system may be determined from basic data. Thermodynamic tables of free energies, enthalpies, and entropies are available for many *but not all* reactants and products, and the following data sources are most useful for high temperature vapor systems: The Joint Army Navy Air Force Thermochemical Tables (JANAF, 1971); "Thermodynamic Properties of the Elements," Stull and Sinke (1956); Kelley and King (1961); Kelley (1960); Wagman *et al.* (1969), and earlier issues in the series; Schick (1966); Kubaschewski and Evans (1956); and Brewer and Rosenblatt (1969), for diatomic metal oxides.

Basic application-oriented discussions of equilibrium processes and thermodynamics may be found in the following recent sources: Balzhiser *et al.* (1972), Henley and Rosen (1969), Gordon and McBride (1971), Eriksson (1971), and Holzman (1969).

2. *The Multicomponent Equilibrium Problem*

In many practical systems the reactants and products may be involved in more than one reaction. The equilibrium composition is then the result of a number of competing processes and can only be defined from the solution of a set of simultaneous equations. As an example, consider the $mH_2 + O_2$ combustion system where m is the number of moles of H_2 per mole of O_2. The composition will be determined by the equilibrium constants for the reactions

$$2H + O = H_2O, \qquad K_1 = p_1/p_5^2p_6, \tag{1}$$

$$H + O = OH, \qquad K_2 = p_2/p_5p_6, \tag{2}$$

$$2H = H_2, \qquad K_3 = p_3/p_5^2, \tag{3}$$

$$2O = O_2, \qquad K_4 = p_4/p_6^2, \tag{4}$$

where $p_1 - p_6$ are the partial pressures of H_2O, OH, H_2, O_2, H, and O, respectively. These are the minimum set of independent reactions needed to specify the system; also, as with most equilibrium reactions, as written, they do not necessarily represent the elementary molecular steps but the overall result of a number of such steps. Additional system constraints are given by the mass conservation relations:

$$P = \sum_{i=1}^{6} p_i, \tag{5}$$

$$a = (1/A)(2p_1 + p_2 + 2p_3 + p_5) = 2m, \tag{6}$$

$$b = (1/A)(p_1 + p_2 + 2p_4 + p_6) = 2, \tag{7}$$

where

$A(mH_2 + O_2) = P$,
a is the number of moles of H per mole of $(mH_2 + O_2)$, and
b the number of moles of O per mole of $(mH_2 + O_2)$.

Given these seven simultaneous nonlinear equations, it is possible to solve for the seven unknowns of A and p_1–p_6.

It is clear from this example, where only two elements H and O are involved, that for practical systems where many elements are commonly present, the solution of a large number of simultaneous equations is required to specify the composition. A number of computer-aided computational procedures has been developed for this purpose—see Gordon and McBride (1971), Henley and Rosen (1969), Balzhiser *et al.* (1972), and Eriksson (1971) for recent examples.

3. *The Phase Rule*

We have seen that an equilibrium vapor-phase system is constrained by certain thermodynamic relationships between the variables of pressure, composition, volume, and temperature. In many instances the vapor will be in equilibrium with another phase, i.e., a solid or a liquid. The number of phases which can coexist at equilibrium is given by the Phase Rule:

$$P_{\mathrm{h}} = C + 2 - F,$$

where

P_{h} is the number of phases,
F the number of degrees of freedom, which may be pressure, temperature, and composition, and
C the number of components which is usually the number of chemical elements.

As an example, consider the system of components

$$\mathrm{FeO(c)} + \mathrm{Fe(c)} + \mathrm{CO} + \mathrm{CO_2},$$

where $C = 3$ and $P_{\mathrm{h}} = 3$ (2 condensed + 1 gas), and, from the Phase Rule, $F = 2$. Hence, if the temperature and composition are specified, for example, the pressure does not affect the system.

A recent comprehensive account of the Phase Rule and phase equilibria has been given by Reisman (1970).

4. *Stability*

The term stability is often used in discussions of high temperature species, and this monograph is no exception. We can consider stability in either a physical or a chemical sense, as demonstrated by the following examples.

Consider the relative stabilities of the species CuO and CuH. A reasonable measure of their stabilities would be the energy needed to dissociate the molecular bond which for CuO is 81 kcal mol^{-1} and for CuH is about 66 kcal mol^{-1}. Hence, if a system containing just these species was subjected to a dissociating force, such as may be provided by high energy electron impact or by a shock wave, one could expect CuH to dissociate more readily than CuO. However, suppose that H_2 is introduced to the system and the temperature is sufficiently high for chemical reaction to proceed, i.e.:

$$CuO + H_2 = H_2O + Cu \qquad \Delta H_r = -48 \text{ kcal mol}^{-1},$$

and

$$CuH + H_2 = CuH + H_2 \qquad \Delta H_r = 0 \text{ kcal mol}^{-1}.$$

As ΔS_r can be neglected in these cases, then ΔF_r is mainly determined by the magnitude of ΔH_r from which it follows that the equilibrium constant for the first reaction is large. That is, CuO is rendered unstable by the presence of H_2, and in *this* chemical system CuH has greater stability than CuO.

Chemical stability is then a property of the reactions, their free energies, and the concentrations of the reactants, and any reference to chemical stability should include these controlling conditions. Discussions of practical chemical systems are often conveniently made with the use of so-called stability diagrams, which are also known as Ellingham or Kellogg diagrams. These diagrams are essentially plots of ΔF_r vs T or ΔF_r vs p_A and their utility will be demonstrated in later chapters.

5. *Brewer's Rule*

It follows from the above arguments that the relative chemical stability of two species may change with temperature. This was recognized by Brewer (1950) in his rationalization of the unexpectedly complex molecular character of high temperature vapors. Brewer's rule essentially predicts that if a high temperature vapor system—present in equilibrium with a condensed phase—contains several molecular species, then the lower concentration species will increase in relative importance as the temperature is increased.

The thermodynamic basis of this prediction is as follows. Let M be a major species and D a minor one. The concentrations, or partial pressures, of these species will then be given by

$$-RT \ln p_{\mathrm{M}} = \Delta H_{\mathrm{M}} - T\,\Delta S_{\mathrm{M}}, \qquad -RT \ln p_{\mathrm{D}} = \Delta H_{\mathrm{D}} - T\,\Delta S_{\mathrm{D}},$$

where the ΔH and ΔS terms are the reaction enthalpy and entropy, respectively, for the vaporization process. Now a major factor in the production of entropy is the number of moles of free molecular components produced by chemical reaction. This means that endothermic vaporization processes such as

$$A(s) = \mathrm{M}, \qquad A(s) = \mathrm{D}$$

will have similar entropies (per mole of vapor species) and this has been verified empirically; so that we have

$$\Delta S_M \simeq \Delta S_D.$$

Therefore, since

$$\ln p_D < \ln p_M,$$

then,

$$\Delta H_D > \Delta H_M,$$

and as ΔH_D and ΔH_M are the respective slopes of the curves for ln p_D and ln p_M vs T^{-1}, it is obvious that p_D increases more rapidly with temperature than does p_M. That is, the minor component increases in relative importance with increasing temperature and may, in fact, become the major species at high temperature.

This behavior is characteristic of vapor-phase systems in equilibrium with a condensed phase. For the case where no equilibrium condensed phase is present it is important to recognize that an increase in temperature leads to a decrease in the proportion of complex or unusual molecular species, and for the following reason. Consider the case where D is actually a dimer of M, i.e., D is $(M)_2$, and the reaction

$$(M)_2 = 2M$$

controls the relative concentrations of $(M)_2$ and M. This dissociation reaction has a positive entropy which therefore contributes favorably to the reaction free energy. Also, since this contribution is in the form of a $T\ \Delta S$ term, then it increases with increasing temperature. That is, $(M)_2$ or D becomes less important with increasing temperature for an unsaturated vapor-phase system.

B. Rate Processes

The progress of a given chemical reaction, for a particular time scale, may be determined by rate processes of a chemical kinetic or mass and thermal transport nature. These various rate processes may act simultaneously on a system. However, their dependence on system variables, such as pressure and temperature, is usually sufficiently different that conditions can be found where either chemical kinetic or transport factors are rate controlling. For example, many metallurgical processes involving gas–solid (or liquid) reactions exhibit a tendency for chemical kinetic control at low temperatures ($\lesssim$500°C) and mass-transport or gaseous-diffusion control at high temperatures ($\gtrsim$1000 °C).

1. Chemical Kinetics

Chemical kinetics can be considered either on a phenomenological or on a molecular level. With the phenomenological approach one deals with macroscopic observables such as composition of reactants and products and time. At the molecular level, one considers also the reaction intermediates and their *fundamental* molecular processes. In principle, both approaches should lead to the same final result. However, the results of a phenomenological observation, in contrast to those at the molecular level, are not sufficiently basic to be transferable to different chemical systems. Indeed, in later chapters we shall see that phenomenological kinetic observations on the same chemical system can differ widely between laboratories. This is particularly noticeable for gas–solid reactions, e.g., see Bradshaw (1970). Unfortunately, with a few exceptions, there is insufficient basic kinetic information about fundamental molecular processes involving high temperature vapors to allow a molecular level approach to practical problems.

At any level of observation, the rate at which a reaction proceeds is proportional to the activities or partial pressures of the reactants. This is the law of mass action which is supported by both empirical observation and gas kinetic collision theory. Thus, for a reaction

$$\mathrm{A + B \rightleftarrows C + D}$$

the rate of production of C is given by

$$\frac{d[\mathrm{C}]}{dt} = k_{\mathrm{f}}[\mathrm{A}][\mathrm{B}],$$

where k_{f} is the forward-direction reaction rate constant and [C], etc., denote concentrations. The reaction may also proceed in the backward direction, with rate constant k_{r}, and this leads to a loss of C according to

$$\frac{-d[\mathrm{C}]}{dt} = k_{\mathrm{r}}[\mathrm{C}][\mathrm{D}].$$

Hence, the net rate of production of C is given by

$$\frac{d[\mathrm{C}]_{\mathrm{net}}}{dt} = k_{\mathrm{f}}[\mathrm{A}][\mathrm{B}] - k_{\mathrm{r}}[\mathrm{C}][\mathrm{D}].$$

This type of equation is essentially a species continuity relation. As the rate constants are related to the equilibrium constant, K, by

$$K = k_{\mathrm{f}}/k_{\mathrm{r}},$$

then,

$$\frac{d[\mathrm{C}]_{\mathrm{net}}}{dt} = k_{\mathrm{f}}([\mathrm{A}][\mathrm{B}] - [\mathrm{C}][\mathrm{D}]/K).$$

Clearly, the net rate of production of component C is favored by large values of k_{f}, K, [A], and [B]. These rate expressions can be integrated to yield solutions describing the time dependence of the various species concentrations, e.g., see Pratt (1969).

In many practical systems a steady state is achieved; that is, the species concentrations do not vary with time. Such is the case, for example, with a laminar flow Bunsen hydrocarbon flame where the luminous cone appears stationary to the eye and hence so are the concentrations of the species responsible for the luminosity. This condition of

$$\frac{d[\mathrm{C}]}{dt} = 0$$

simplifies the solution of rate expressions and for the above example the steady-state concentration of [C] is given by

$$[\mathrm{C}] = k_{\mathrm{f}}K[\mathrm{A}][\mathrm{B}]/[\mathrm{D}].$$

These arguments concerning net species production rates can be generalized to cases where many elementary reactions are involved. In practice, species continuity equations, expressing the net rate of production of each species in terms of elementary rate constants and species concentrations, are constructed. These simultaneous differential equations may then be integrated, subject to certain initial and final boundary conditions, to give the gas composition at any time. Bittker and Scullin (1972), for instance, have developed a computer program to deal with multireaction chemical kinetic calculations.

As an example of a complex kinetic system, consider the well-known H_2–Br_2 reaction

$$H_2 + Br_2 = 2HBr,$$

which is controlled by the following elementary reactions:

$$Br_2 + M \rightarrow 2Br + M, \tag{1}$$

$$Br + H_2 \rightarrow HBr + H, \tag{2}$$

$$H + Br_2 \rightarrow HBr + Br, \tag{3}$$

$$2Br + M \rightarrow Br_2 + M, \tag{4}$$

$$H + HBr \rightarrow H_2 + Br, \tag{5}$$

where M is a nonreactive third-body species. The net rate of production of Br and H is then given by

$$\frac{d[\mathrm{Br}]}{dt} = 2r_1 - r_2 + r_3 - 2r_4 + r_5$$

and

$$\frac{d[\mathrm{H}]}{dt} = r_2 - r_3 - r_5,$$

where

$$r_1 = k_1[\mathrm{Br_2}][\mathrm{M}], \text{ etc.},$$

are the individual reaction rates. For a steady-state condition, these rate expressions for radical intermediates may be solved with the elimination of the radical concentration terms, and the net rate of HBr production is then given by

$$\frac{d[\mathrm{HBr}]}{dt} = \frac{2k_2(k_1/k_4)^{1/2}[\mathrm{Br_2}]^{1/2}[\mathrm{H_2}]}{1 + (k_5/k_3)[\mathrm{HBr}]/[\mathrm{Br_2}]},$$

which is of the same form as that obtained empirically (i.e., the phenomenological rate expression) by monitoring the reactants and product concentration as a function of time. This agreement tends to verify the selection of elementary reactions.

It follows from the above discussion that the basic data for chemical kinetics are the rate constants and equilibrium constants. The rate constants are temperature dependent and quite often extremely so. It is not uncommon for the rate constant to increase by ten orders of magnitude over a temperature interval of 300 to 1000 K. The temperature dependence of the rate constant k is usually expressed in the approximate form:

$$k = Ae^{-E/RT},$$

where A is normally a constant, or a slowly varying function of temperature, which is a measure of the number of active molecular collisions between the reactants; E is a constant and is considered to be the energy required to activate the reaction. For an endothermic process, E is usually equal to—or slightly greater than—the equilibrium reaction enthalpy. Activation energies of about 20–30 kcal mol^{-1} are quite common and, in this case, temperatures of about 1000 K are sufficiently high to give reasonably rapid reaction rates. It is interesting to note that both equilibrium constants and rate constants vary exponentially with temperature (i.e., for ΔH and $E \neq 0$).

Some basic rate data pertinent to high temperature vapors may be found in the compilations of Kondratiev (1972) and Baulch *et al.* (1972). Fristrom (1972) has listed some additional less comprehensive data sources. However, for most inorganic high temperature species, no rate data exist. Very approximate data can be obtained from a consideration of potential energy functions, e.g., see Mayer *et al.* (1967), but this depends on a knowledge of molecular parameters which are also usually unavailable.

2. *Mass and Heat Transport*

Just as an equilibrium chemical reaction, which when disturbed by an external influence such as the removal of a product, readjusts in such a way as to oppose the effect of the disturbance (i.e., in accordance with the Le Chatelier principle), so too do nonequilibrium systems tend to readjust to the action of forces. These forces may derive from gradients of concentration, pressure, or temperature, and when acting result, respectively, in a transport flux of mass, momentum, and energy in the form of heat.

In many practical high temperature vapor systems these transport effects are superimposed on those derived from chemical reaction and may even control the overall process. As an example, the rate of combustion of ammonium perchlorate, as a solid propellant, is controlled by diffusion effects both at the surface and in the bulk gas (Jacobs and Russell-Jones, 1968; Jacobs and Powling, 1969). Also, many metallurgical processes are rate limited by mass diffusion effects, e.g., see Bradshaw (1970). The discussion of the chapters to follow will reveal many examples where chemical reaction and transport processes are intricately combined.

For the sake of simplicity the following discussion deals with one-dimensional processes though the arguments may readily be extended to the more general two- and three-dimensional cases, as may be seen by reference to the cited literature.

a. *Mass Transport*

At high temperatures, chemical reactions are sufficiently fast that they are rarely the rate-limiting factor. However, mass transport to and from the reaction zone may be relatively slow and hence, rate limiting. In a fluid system mass transport may occur by diffusion or by convection. Concentration and, to a lesser degree, temperature gradients provide the driving force for diffusion, and convection is controlled by density gradients.

The basic transport law for concentration diffusion is Fick's law:

$$J_i = -D_i \frac{\partial C_i}{\partial x},$$

where

J_i is the flux of species i, i.e., mass transported per unit area per unit time,
D_i the transport coefficient of diffusivity for species i in the medium, and
$\partial C_i/\partial x$ the gradient of concentration (moles per unit volume) in the x direction.

Note that the negative sign defines the direction of transport as being down the concentration gradient. The diffusional flux is also sometimes expressed in terms of the mass fraction gradient, i.e.,

$$J_i = -\rho D_i \frac{\partial w_i}{\partial x},$$

where w_i is the species mass fraction and ρ the gas density. Alternatively, the flux may be expressed in terms of the species partial pressure gradient, i.e.,

$$J_i = -\frac{D_i}{RT}\frac{\partial p_i}{\partial x}.$$

For the common problem of steady-state diffusion of a single species along a tube the mass transport is given by

$$J = DP(N_a - N_b)/RTL,$$

where L is the length of tube and N_a and N_b are the species mole fractions at each end of the tube.

With multicomponent mixtures (n species), the laminar flow flux is usually expressed in terms of the Stefan–Maxwell equation:

$$\frac{\partial N_i}{\partial x} = \sum_{j=1}^{n} \frac{1}{CD_{ij}} (N_i J_j - N_j J_i),$$

where N represents mole fraction. The problem of dealing with numerous D_{ij} binary diffusion coefficient terms is simplified by defining D_{im} as an "effective binary diffusivity" of component i in the mixture such that

$$J_i = -CD_{im}\frac{\partial N_i}{\partial x} + N_i \sum_{j=1}^{n} J_j;$$

combining this with the Stefan–Maxwell equation gives

$$\frac{1}{D_{im}} = \frac{\sum_{j=1}^{n} (1/D_{ij})(N_j J_i - N_i J_j)}{J_i - N_i \sum_{j=1}^{n} J_j}.$$

For the particular case of a binary gas mixture (species 1 and 2) the Stefan–Maxwell equation simplifies to give

$$J_1 = -CD_{12}\frac{\partial N_1}{\partial x} + N_1(J_1 + J_2).$$

In the presence of bulk gas flow processes, such as convection and turbulence, the general mass transport expression is

$$J_i = -D_i\frac{\partial C_i}{\partial x} + vC_i - E\frac{\partial C_i}{\partial x},$$

where v is the laminar flow velocity of fluid motion, i.e., convection in the x-direction, and E is the coefficient of eddy diffusion to account for turbulent flow.

Mass transport computations must satisfy a condition of overall mass continuity which is given by the continuity equation

$$\frac{\partial \rho}{\partial t} + \frac{\partial \rho v}{\partial x} = 0.$$

Thus no generation or destruction of mass occurs in the transport process. Also, for an n-component mixture, there are n equations of this type expressing individual species continuity, i.e.,

$$\rho D\frac{\partial w_i}{\partial t} + \frac{\partial (j_i)}{\partial x} = \psi_i,$$

where

$j_i = \rho_i(v_i - v)$ is the specific mass flow rate, and $\psi_i + ve$ or $- ve$ indicates the generation or loss of species i, respectively.

b. Momentum Transfer

Considerations similar to those given for mass transport apply to the transport of momentum, that is mv. The basic transport law is Newton's law:

$$J_{mv} = -\eta \frac{\partial v}{\partial x},$$

where the transport coefficient η is the gas viscosity.

c. Mass Transport by Free Vaporization

In the absence of diffusion—or chemical reaction—limitations the transport of material by vaporization is given by the molar flux relation:

$$J = P(2\pi RT/M)^{-1/2},$$

where M is the molecular weight. Where chemical reaction rate limitations are present, this flux is reduced by a factor α known as the vaporization coefficient. Many inorganic materials, such as metals and salts, have unit values of α and are therefore not rate limited. However, there are instances where large reductions in the vaporization flux can occur, e.g.,

$$4As(s) \rightarrow As_4, \quad 505\text{–}580 \text{ K}, \quad \alpha = 7.1 \times 10^{-5};$$

$$4Sb(s) \rightarrow Sb_4, \quad 606\text{–}697 \text{ K}, \quad \alpha = 0.17$$

(Rosenblatt, 1970). A general discussion of vaporization processes may be found in the chapters edited by Margrave (1967a).

d. The Boundary Layer

At a gas–solid (or liquid) interface it is difficult to define adequately the mass flow conditions, and this has led to the concept of a boundary layer. With this concept, the flux relationships may be expressed in terms of an empirical factor δ_i known as an effective boundary layer thickness. The mass transport flux, for species i, is then given by

$$J_i = (D/\delta_i)(C_{i,\text{ at } x=0} - C_{i,\text{bulk}}),$$

where the concentration terms $C_{i,\text{ at } x=0}$ and $C_{i,\text{bulk}}$ define the concentrations at the boundary layer solid–gas and the boundary layer bulk–gas interfaces, respectively. The empirical factor δ_i tends to decrease with increasing bulk–gas flow rate. This concept of a boundary layer can also be applied to momentum and heat transport although the magnitudes of δ_i are not necessarily the same.

An example of the effect of a stagnant gas boundary layer on the rate of vaporization of a solid, such as tungsten metal, has been given by Harvey (1972). From stagnant film mass transfer theory, the metal vaporization rate has the following proportional dependence on temperature and total gas pressure:

$$J_W \propto \exp(-\Delta H/RT)/(P)^{0.9},$$

where ΔH is the enthalpy of metal vaporization.

For practical processes, such as the heterogeneous oxidation of carbon in an open hearth furnace, effective boundary layers of the order of 10^{-3} cm are derived—e.g., see Elliott *et al.* (1963). Similarly, a mass transport description of the chemical vapor deposition of tungsten by a surface reaction of WF_6 and H_2 indicated a value of

$$\delta \sim 4.5 \times 10^{-4} \text{ cm}$$

(Brecher, 1970).

e. Diffusion Coefficients

From the foregoing discussion it is clear that the basic parameter in mass transport theory is the diffusion coefficient, D. Experimental values for D are available for many of the common permanent gases, such as CO_2, H_2, etc. However, for metal-containing high temperature species such data are virtually nonexistent and estimation procedures are required. The gas kinetic Chapman–Enskog theory, or some modification thereof, is usually used as the basis for estimating transport coefficients; e.g., see Hirschfelder *et al.* (1954) and also Schwerdtfeger and Turkdogan (1970). The theory is based on binary collision considerations, and is valid for near equilibrium conditions, but strictly applies only to species with no internal degrees of freedom as elastic adiabatic collisions are assumed.

For polyatomic molecules the theory is extended by considering approximate potential energy functions for the essentially nonelastic molecular interactions. The coefficient for self-diffusivity is then given by

$$D = 1.8583 \times 10^{-3}\, T^{3/2}(2/M)^{1/2}/P\sigma^2\Omega^{(1,1)*} \quad \text{cm}^2\,\text{sec}^{-1},$$

where P is in atmospheres; σ (Å) is the collision diameter for low-energy collisions as obtained from the Lennard–Jones 6-12 potential function; and $\Omega^{(1,1)*}$ is the reduced collision integral, which for the case of rigid elastic molecules would be unity. A table of $\Omega^{(1,1)*}$ values for various values of kT/ϵ and σ may be found in the chapter of Schwerdtfeger and Turkdogan (1970); k is the Boltzmann constant and ϵ is the maximum energy of attraction in the Lennard–Jones potential function. This function takes the form

$$\phi(r) = 4\epsilon[(\sigma/r)^{12} - (\sigma/r)^6],$$

where r is the internuclear separation. For bimolecular collisions between unlike molecules the following approximations are used

$$\epsilon_{12} = (\epsilon_1\epsilon_2)^{1/2} \qquad \text{and} \qquad \sigma_{12} = \tfrac{1}{2}(\sigma_1 + \sigma_2).$$

The diffusivity function holds for nonpolar molecules and is a satisfactory approximation for polar gas molecules in high temperature rate processes, *but* for the highly polar metal-containing vapor species little is known about the validity of these estimation procedures.

Where the potential function data are unavailable the following empirical relations are used:

$$\epsilon/k = 0.77T_c = 1.15T_b = 1.92T_m$$

and

$$\sigma = 0.841\,(V_c)^{1/3} = 1.166\,(V_b)^{1/3} = 1.221\,(V_m)^{1/3},$$

where ϵ/k is in units of K, and V is the molar volume at the critical point (c), the boiling point (b), or the melting point (m).

Binary diffusion coefficients are similarly given by the kinetic theory expression:

$$D_{12} = 1.8583 \times 10^{-3} \frac{T^{3/2}(1/M_1 + 1/M_2)^{1/2}}{P\sigma_{12}{}^2\Omega^{(1,1)*}} \quad \text{cm}^2\ \text{sec}^{-1}.$$

Note that the diffusivity is predicted to be inversely proportional to pressure; this is found experimentally to hold over a wide pressure range for pressures of greater than about 10^{-3} atm. Also, the diffusivity is predicted to be proportional to $T^{1.5}$. However, empirically one finds

$$D/D_0 = (T/T_0)^{1.8},$$

where D_0 and T_0 are reference conditions.

Comprehensive discussions of mass transport processes may be found in the texts of Eckert and Drake (1972), Jost (1960), and Crank (1956).

f. Heat Transport

If a system is perturbed from equilibrium by the effect of a temperature gradient, $\partial T/\partial x$, the perturbation tends to be eliminated by a flux of heat, J_q, down the gradient. This gives rise to the phenomenon of thermal conduction. As may be predicted from kinetic theory, thermal conduction is an increasing function of temperature.

The basic law of thermal conductivity is Fourier's law:

$$J_q = -\lambda \frac{\partial T}{\partial x},$$

where λ is the coefficient of thermal conductivity.

In a multicomponent system with mass transport present, additional heat flux components can occur in the form of thermal diffusion and convection. The heat flux is then given by

$$J_q = -\lambda \frac{\partial T}{\partial x} + \rho \sum_{i=1}^{n} h_i N_i V_i + C_p \rho v T,$$

where h_i are component specific enthalpies, N_i is the mole fraction, V_i is the diffusion velocity, and the last term expresses the contribution of convection. A small and usually negligible thermal diffusion effect, known as the Dufour effect, has been omitted from this heat flux expression. The second term represents the total enthalpy, per unit area per second, flowing relative to the mass average motion of the mixture. The *specific* enthalpy is obtained from

$$h_i = (m_i)^{-1}(U_i + kT),$$

where m_i is the mass of species i, U_i the internal molecular energy, and k the Boltzmann constant. In most thermochemical tables the enthalpy is given in terms of molar quantities. Usually, standard heat of formation and enthalpy increment data are given and H (molar) is then obtained from the relation

$$H = \Delta H_{f,298} + (H_T - H_{298}).$$

Energy conservation can be expressed in terms corresponding to a zero divergence of the total flux, i.e.,

$$\frac{\partial}{\partial x}(J_q) = 0.$$

The thermal self-conductivity transport coefficient is, from the gas kinetic theory, given by

$$\lambda = \frac{10^{-7} \times 1989.1\,(T/M)^{1/2}}{\sigma^2 \Omega^{(2,2)*}(T^*)},$$

where $\Omega^{(2,2)*}(T^*)$ is a collision integral obtained from an assumed molecular potential function such as the Lennard-Jones potential; e.g., see Taylor and Petrozzi (1968). These authors also give thermal conductivity expressions for multicomponent systems and estimated collision-integral data for a number of high temperature species, such as LiOH, Li_2O, HBO_2, and AlOCl.

The coefficients of thermal conductivity and viscosity are related by the expressions

$$\lambda = (15/4)\ (R/M)\eta,$$

for monatomic gases, and

$$\lambda = (C_p + 5R/4M)\eta,$$

for polyatomic gases.

g. Transport Coefficient Interrelationships and Dimensionless Numbers

There are a number of useful interrelationships between the various transport coefficients and these are often expressed, and utilized, in the form of dimensionless numbers. When the transport coefficients are redefined such that they each have units of $cm^2\ sec^{-1}$, it becomes apparent that their magnitudes are also similar (Frank-Kamenetskii, 1969). For laminar flow conditions we define

$$a = \lambda / C_p \rho$$

as the coefficient of thermal diffusivity, and

$$\nu = \eta / \rho,$$

as the coefficient of kinematic viscosity. Then,

$$a \sim \nu \sim D \sim \Lambda U,$$

where Λ is the mean free path and U is the mean thermal molecular velocity.

Dimensionless numbers are commonly used in transport theory, as they allow the results of different experimental conditions to be more readily related to each other and extended to an engineering scale. For high temperature vapor processes the more commonly used dimensionless numbers include

Nusselt	$\mathrm{Nu} = \beta \mathrm{d}/D$	(for diffusion),
	$\mathrm{Nu} = \alpha d/\lambda$	(for heat transfer);
Prandtl	$\mathrm{Pr} = \nu/a = C_p \eta/\lambda$;	
Schmidt	$\mathrm{Sc} = \nu/D = \eta/\rho D$;	
Reynolds	$\mathrm{Re} = Vd/\nu = \rho V d/\eta$	(V is velocity);
Lewis	$\mathrm{Le} = \mathrm{Sc}/\mathrm{Pr} \simeq 1$	(for many cases);

where β is defined by

$$J = \beta\, \Delta C$$

and α is defined by:

$$J_q = \alpha\, \Delta T;$$

d is a characteristic dimension such as length of a flat plate or diameter of a cylinder, and is related to δ by

$$\delta = d/\mathrm{Nu}.$$

The Prandtl number is a measure of the relative importance of momentum and heat transfer, Sc indicates the relative importance of momentum and mass transfer, Le indicates the ratio of energy transported by conduction to that by diffusion, and the magnitude of Re determines the nature of the gas motion as being laminar flow or turbulent flow. For the case of a straight tube, the transition between laminar and turbulent flow occurs for Re values in the approximate range of 2100–3200.

3. *Effects of Chemical Reaction on Mass and Heat Transport*

Under the combined effects of chemical reaction, mass diffusion, and convection the time dependence of species concentration is given by

$$\frac{\partial C}{\partial t} = f(c) + D\frac{\partial^2 C}{\partial x^2} - v\frac{\partial C}{\partial x},$$

where $f(c)$ is the law of chemical reaction rate, e.g., for a first-order reaction

$$f(c) = -kC,$$

where k is the reaction rate constant. Combined rate equations of this type can be integrated by separation of the variables and solutions obtained for particular boundary conditions, as shown, for example, by Jost (1960) and by Crank (1956). The solution for a steady-state condition is

$$C = C_0 \exp\{(vx/2D)[1 - (1 + 4kD/v^2)^{1/2}]\},$$

where $C = C_0$ at $t = 0$.

Species conservation requires that the divergence of species flux be equal to K_i—the net rate of species appearance, or disappearance, due to chemical reaction. That is,

$$\frac{\partial}{\partial x}[N_i(v + V_i)] = K_i.$$

The conditions whereby either mass transport or chemical reaction become rate controlling can be demonstrated by the following example. Consider a gas–solid steady-state reaction, where C is the bulk concentration of the reacting gas species, C_s is its concentration at the surface, and δ is the gas boundary layer thickness. Then, the diffusion rate of C is given by

$$\text{diffusion rate} = (D/\delta)(C - C_s).$$

Suppose that the species reacts with the surface according to the first-order expression

$$\text{reaction rate} = kC_s.$$

Then, since the gas is at a steady state, the rate of species diffusion and rate of species consumption by reaction must be equal; it therefore follows from the above rate expressions that

$$C_s = \left(\frac{D}{\delta}\right)\frac{C}{(k + D/\delta)}.$$

Thus, where $D/\delta \gg k$ we have

$$\text{overall rate} \sim kC,$$

and the process is chemically controlled. For the reverse condition of $k \gg D/\delta$:

$$\text{overall rate} \sim (D/\delta)C$$

and the process is diffusion controlled.

The effect of chemical reaction on heat flux can be expressed by

$$J_q = -(\lambda_f + \lambda_r)\frac{\partial T}{\partial x},$$

where λ_f and λ_r are the thermal conductivities due to molecular collisions and to chemical reaction, respectively (Butler and Brokaw, 1957). To demonstrate the effect of chemical reaction on the thermal conductivity consider a dissociation reaction

$$A = nB,$$

with an enthalpy of ΔH. The thermal conductivity increment arising from this reaction is then,

$$\lambda_r = \frac{D_{AB}P}{RT}\frac{\Delta H^2}{RT^2}\frac{N_A N_B}{(nN_A + N_B)^2}.$$

In this instance the heat released by the reverse recombination process, which is favored at relatively low temperatures, results in heat transport from a hot to a cool region.

Energy conservation in a chemically reacting system can be expressed by

$$h + \tfrac{1}{2}v^2 = H_s,$$

where

$h = \sum_{i=1}^{n} \sigma_i H_i$ is the static enthalpy per unit mass;
H_i is the molar enthalpy of species i,
σ_i is the number of moles of species i per unit mass of mixture, and
H_s is the total specific enthalpy, including that provided by chemical reaction.

The effect of a chemically reacting system is manifested as a redistribution of species, leading to changes in σ_i and H_i as well as in the specific kinetic energy expressed by $\frac{1}{2}v^2$.

A detailed treatment of transport phenomena, including cases where chemical reaction is present, may be found in the text book of Geiger and Poirer (1973). An example of practical significance, which indicates the combined effects of mass and heat transfer in the presence of chemical reaction, is given by the ablative combustion of a carbon surface in a laminar air stream, as has been discussed in detail by Eckert and Drake (1972, p. 734). Further discussion of the effects of chemical reactions on heat transport will be given later in connection with energy systems (Chapter 6, Section II).

Appendix. Major Literature Sources for High Temperature Chemistry

High Temperature (1963–). A. E. Sheindlin (ed.), translation of *Teplofiz. Vysok. Temp.* by Consultants Bureau, New York.

High Temperature Science (1969–). J. L. Margrave (ed.), Academic Press, New York.

High Temperatures-High Pressures (1969–). E. Fitzer and J. Lees (eds.), Pion Ltd., London.

High Temperature Bulletin. Information Centre on High Temperature Processes, Dept. of Fuel Sci., The University, Leeds, England.

Bibliography on the High Temperature Chemistry and Physics of Materials. Int. Union Pure Appl. Chem., M. G. Hocking (ed.), Metallurgy Dept., Imperial College, London, England.

2

Gas–Solid Reactions with Vapor Products

I. Introduction

Many high temperature systems of practical concern are heterogeneous, involving gas–solid reactions. As examples of such reactions will appear in later chapters dealing with vapo- or pyrometallurgy, corrosion, energy systems, and combustion-related problems, it is convenient to consider in this chapter some of the more general aspects of gas–solid reactions. Particular emphasis is given to the case of water vapor or steam–solid interactions as many practical systems contain H_2O as an atmospheric component. Also, the basic data for such systems is relatively sparse, and the discussion to follow is intended to reveal areas where further study appears necessary.

A practical interest in gas–solid reactions with vapor products, and at temperatures typically greater than 1000 °C, results from problems associated with high temperature oxidation and corrosion of refractory materials, combustion of metals in reactive gases, erosion or ablation of rocket components, and the reduction of iron oxide ore to the metal. Examples of recent studies in these areas may be found in the AGARD Conference Proceedings on "Reactions Between Gases and Solids" (AGARD, 1970); in the discussions of the scientific aspects of blast furnace technology (Szekely, 1972); and elsewhere in this book in chapters on corrosion (Chapter 4), energy systems (Chapter 6), and combustion (Chapter 5).

Another practical aspect of high temperature (>1000 °C) gas–solid interactions, but one which is not widely recognized, is that in the firing of ceramics, atmospheric influences are as important as time and temperature in determining the physical and chemical character of the ceramic product. In the common preparation of ceramics such as porcelain, glass, and cement, where combustion of hydrocarbon fuel is used as a direct-contact heat source, water vapor is inevitably present in the atmosphere. This is desirable in that it provides for improved heat transfer relative to air. However, an undesirable aspect of water vapor is its possible chemical interaction with the ceramic, e.g., see the review of Koenig and Green ($\sim$1967). For instance, the solubility of H_2O in glass is believed to result in the loss of Si–O–Si linkages with the formation of Si–OH bonds. In some cases a change in ceramic composition can result owing to the formation—and vapor transport—of new hydrated oxide species. Boron can be lost from ceramics in the form of boric acid vapor (Miwa *et al.*, 1964). Similarly, the kiln heating of the raw cement materials of feldspar, mica, and illite, in the presence of moisture, leads to the volatilization of alkalis, particularly potassium (Goes and Kul, 1960). This is probably due to the stability of the KOH vapor species.

At moderately high temperatures, i.e., $\sim$500–1000 °C, the interaction of gases with solids has practical consequences in connection with catalysis of gas-phase processes,† metallurgical processing, ore genesis, and other chemical transport phenomena—as discussed in detail in Chapter 3. A potential application of metal–gas systems as reduced pressure standards has also been considered recently, e.g., see Lundin (1969).

A little recognized aspect of gas–solid reactions is their role in what appears to be solid–solid reactions. For instance, the formally written solid–solid interaction of

$$\mathrm{ZnO(s) + C(s) = Zn + CO,}$$

which forms the basis of the Imperial Smelting Process for Zn production, actually *proceeds* via two gas–solid reactions, i.e.,

$$\mathrm{ZnO(s) + CO = Zn + CO_2}$$

† The catalytic effect of solids on gas phase reactions is of great practical importance, e.g., see Banerjee (1968), but is outside the scope of this monograph.

and

$$CO_2 + C(s) = 2CO,$$

with the latter reaction being rate limiting, e.g., see Lumsden (1972).

In describing gas–solid interactions the following reaction steps may be defined:

- diffusion of reactant gas to the surface,
- adsorption of gas on the surface,
- formation of product molecules,
- desorption of products, and
- diffusion of products away from the surface.

The first and last steps are often neglected and are not usually the rate-determining steps, except for high gas pressure and relatively high temperature conditions. Important factors in gas–solid reactions, then, are: the identity of the intermediates and products formed, their state of aggregation, their rates of formation, and their influence on further reaction.

The application of laboratory rate data to "real" conditions may be complicated by aerodynamic effects, such as boundary layer formation and convection. These effects are difficult to define in "real" systems; indeed even for model-system laboratory studies, basic data on mass transport coefficients for high temperature species are usually not known.

In many instances the heterogeneous chemical reactions are sufficiently rapid that the gas–solid interface can be described by equilibrium thermodynamics. Even for the case where material transport in the gas phase is rate determining, it is still possible to qualitatively compare reaction rates on the basis of equilibrium constant and activity data, as shown, for example, by Whiteway (1971) for reactions between gases and molten Fe–C alloys.

Heterogeneous reaction mechanisms and, in particular, gas–solid reactions, have been discussed recently by Wagner (1970a). A discussion of the catalytic aspects of gas–solid reactions at the molecular level may be found in the review of Wagner (1968). However, it is usually the case that very little basic chemical insight is available when considering catalytic phenomena, e.g., see Myers (1971). The role of chemisorption, solid-state diffusion, and other nonvapor phase processes in gas–solid reactions is not to be considered here. Kofstad (1964, 1966) deals with these aspects, particularly in connection with reactions between gases and refractory metals.

II. Gas–Solid Reaction Types

A somewhat artificial but informative breakdown of the subject of gas–solid interactions, in terms of reaction types, is given as follows—where the reaction may proceed to yield the following final states:

(a) a scattering of the gas without modification of the surface;
(b) a physically adsorbed gas layer at the surface;
(c) chemisorption, i.e., a strong, but reversible, gas adsorption;
(d) dissociation of the incident gas without modification of the surface;
(e) chemical interaction to yield a chemically modified solid surface such as a solid solution or a new solid phase;
(f) chemical interaction with the surface to yield liquid products;
(g) chemical interaction with the surface material to yield vapor-phase products.

Some examples of these reaction types are as follows:

Type a: Scattering of rare gases from crystal faces of metals. Using supersonic molecular beams, with a translational energy of several tenths of an electron volt, it is possible, by mapping the scattering trajectory of such beams, to infer something about the surface structure, e.g., see Bishara *et al.* (1968).

Type b: This type of gas–solid process is of interest in connection with the fundamentals of crystal growth and the reverse process of evaporation. In particular, the growth of semiconductor materials is of practical interest and adsorption-desorption studies of As_2, P_2, and Bi species at GaAs and GaP (111) surfaces have been made (Arthur, 1968). In some cases the extent of adsorption can be quite high. For instance, adsorption probabilities for oxygen on metals such as Ta, Nb, and V ($T > 1000$ °C) are in the range of 0.2–0.8 (e.g., see Fig. 6 of Rosner, 1972a). It is of note that the theoretical description of adsorption and desorption kinetics contains partition functions for both free and adsorbed molecular species; that is, the vibrational and rotational structure of such species is required as basic input. However, basic data of this type is virtually nonexistent for adsorbed high temperature species.

Type c: Some gas–solid interactions are accompanied by high heats of adsorption, e.g., more than 100 kcal mol^{-1} for H_2 on Pt, and this is indicative of electron transfer and chemical bond formation at the surface. However, the adsorption process is still reversible. This type of interaction can be rationalized in terms of a molecular orbital electron exchange model (Jansen, 1968).

Type d: The dissociation of H_2 on metals is a well-known example of this reaction type, e.g.,

$$H_2 + Ta(s) \rightarrow 2H + Ta(s), \qquad T \sim 1400\text{–}2500\ K$$

(Krakowski and Olander, 1968).

Type e: Several examples of this reaction type are as follows:

$$CO + FeO(s) \rightarrow Fe(s) + CO_2, \qquad T \sim 1000\ °C,$$

$$yH_2O + xTa(s) \rightarrow Ta_xO_y(s) + yH_2, \qquad T \sim 1600\ °C,$$

$$\tfrac{1}{2}yO_2 + xW(s) \rightarrow W_xO_y(s), \qquad T \sim 600\ °C,$$

and

$$2NaCl(s) + ZrCl_4 \rightarrow Na_2ZrCl_6(s), \qquad T \sim 485\ °C$$

(Pint and Flengas, 1971). These reactions tend to be diffusion limited and mass transport phenomena, both in the gas and solid phase, are of primary importance. A detailed discussion of mechanisms for this reaction type may be found in the review of Hills (1970), see also Szekely and Evans (1972). For these systems the identity of the surface species is often of interest and flash desorption, with mass spectrometric analysis of the desorbed species, has been used for this purpose, e.g., see Hudson (1970) and McCarroll (1967a,b).

Type f: The following interaction:

$$\tfrac{7}{2}O_2 + SO_2 + 2V\ (\text{alloy}) + 2Na\ (\text{as NaOH or NaCl}) \rightarrow V_2O_5(l) + Na_2SO_4(l)$$

can be an important source of corrosion (see Chapter 4, Section II).

Type g: The following reactions:

$$\tfrac{3}{2}O_2 + W(s) \rightarrow WO_3$$

and

$$H_2O + WO_3(s) \rightarrow WO_2(OH)_2$$

are important in connection with the moderately high temperature corrosion of tungsten, e.g., in incandescent lamps (see Chapter 3, Section IX). Other important examples of this type are

$$Ni(s) + 4CO \rightarrow Ni(CO)_4, \qquad 70\text{–}180\ °C,$$

$$UF_4(s) + F_2 \rightarrow UF_6, \qquad 160\text{–}353\ °C,$$

$$Si(s) + 2Cl_2 \rightarrow SiCl_4, \qquad 375\text{–}404\ °C,$$

$$Fe_2O_3(s) + 6HCl \rightarrow 2FeCl_3 + 3H_2O, \qquad 450\text{–}600\ °C,$$

and

$$2Re(s) + \tfrac{7}{2}O_2 \rightarrow Re_2O_7, \qquad 600\text{–}1500\ °C.$$

III. Reaction Mechanisms

A. Phenomenological Level

For reactions of the type g, i.e.,

$$\text{solid} + \text{gas I} \rightarrow \text{gas II},$$

the isothermal reaction rates are found to be of the form

$$\text{isothermal rate} = kp^n,$$

where k is an overall rate constant, p is the pressure of gas I, and n usually has values in the range 0.5–1.0, e.g., see the review of Habashi (1969). For those cases where $n = 1$, such as for the chlorination of Si, the observed proportionality of reaction rate to reactant gas partial pressure *suggests* mechanisms such as

$$Si(s) + Cl_2 \rightarrow SiCl_2 \quad \text{and} \quad SiCl_2 + Cl_2 \rightarrow SiCl_4,$$

with the first step being the rate-determining step (Landsberg and Block, 1965). For cases where n is less than unity, such as the chlorination of tungsten at about 700°C, the rate-determining step is thought to be the dissociation of the reactant gas at the surface, i.e.,

$$Cl_2(ads) \rightarrow 2Cl(ads).$$

This type of gas dissociation rate-limited process is also believed to apply to the chlorination of metal oxides (see Chapter 3, Section VII); and for reactions (at ~1000 °C) between CO_2 and FeO, Fe_3O_4, graphite, and various metals; and reactions of CH_4 or NH_3 with Fe; as well as for reactions of O_2 with Cu_2O or NiO, e.g., see the review of Worrell (1971).

For nonisothermal conditions the combined effects of reaction area A, pressure p, and surface temperature T on the reaction rate may be described by the expression

$$\text{rate} = Akp^n e^{-E/RT},$$

where E is an overall reaction activation energy and R is the gas constant. Activation energies covering a range of about 1–80 kcal mol^{-1} are found but many of the reactions studied have values in the region of 20 kcal mol^{-1}. Table 2.1 gives examples of activation energies for the carbothermic chlorination of metal oxides by reactions such as,

$$ZrO_2(s) + 2Cl_2 + 2C(s) = ZrCl_4 + 2CO,$$

which are of metallurgical importance (see Chapter 3, Section VII).

The inadequacy of the phenomenological approach to defining rate

TABLE 2.1

Phenomenological Activation Energies for Metal Oxide Chlorinations[a]

Oxide	Activation energy (kcal mol^{-1})	Oxide	Activation energy (kcal mol^{-1})
Fe_2O_3	59.1	Fe_2O_3	26.0
ZrO_2	55.1	$ZrSiO_4$	23.3
SnO_2	45.0	TiO_2	21.7
Nb_2O_5	44.0	$ZrSiO_4$	18.0
TiO_2	36.4	TiO_2	16.6
Ta_2O_5	31.0		

[a] See O'Reilly *et al.* (1972) for the original literature citations.

processes is revealed by the wide variation of activation energies determined for the same oxide system—as shown in Table 2.1 for Fe_2O_3 and TiO_2 (see also the Fig. 1 given by Bradshaw, 1970).

A similar rate expression may be derived from gas kinetic theory, for an adsorption rate-limited process, i.e.,

$$\text{rate} = A(q/\lambda)e^{-E/RT},$$

where E is the adsorption process activation energy, λ is the stoichiometric factor and q is the collision number of reactant gas with unit surface area (cm^2) and is related to pressure by the equality

$$q = 3.52 \times 10^{22} p/(MT)^{1/2},$$

where p is in Torr, T is in degrees Kelvin, and M is the molecular weight of reactant gas. Comparison of experimental reaction rates with the gas kinetic theory prediction then provides an indication of the overall reaction efficiency. For instance, the oxidation of Re with O_2, at 1507 °C and at a gas flow rate of 3.6×10^{18} molecules sec^{-1} and a pressure of 2 Torr, occurs about once every two-hundred collisions (Gulbransen *et al.*, 1969).

At high temperatures, the reaction rates are usually sufficiently rapid that the rate of gas diffusion can become the rate-limiting step and the rate of reaction is then given by

$$\text{rate} = A \frac{D}{RT}\left(\frac{dp}{dx}\right),$$

where dp/dx is the pressure gradient at distance x from the surface, and D is the diffusivity of the reactant gas above the surface—see Chapter 1, Section V for additional discussion of gas diffusivity. Mass transport limitations of this kind are particularly common in metallurgical processes (Bradshaw, 1970).

For reactions of the type e, i.e.,

$$\text{solid I} + \text{gas I} \rightarrow \text{solid II} + \text{gas II},$$

the reaction rate may be limited by the rates of diffusion of reactants and products through a "shell" of solid II. An interesting example of this effect is given by the initial reduction of iron oxide ore to a lower valence oxide form. In such cases first-order overall reaction kinetics may usually be assumed. However, in some reaction regimes, chemical kinetics may play a secondary role to physical processes, such as diffusion in porous material and sintering. Szekely and Evans (1972) have recently proposed a model which incorporates these effects.

Another class of phenomenologically based mechanistic studies is represented by the observations of Rosner and co-workers on the effects of dissociated gases, i.e., atoms of O, F, N, etc, on refractory solids, e.g., see Rosner and Allendorf (1970, 1971) and earlier cited work. In this instance the surface gas dissociation step, which is often rate limiting, is eliminated and rapid reaction rates, relative to the corresponding undissociated case, are found.

B. Molecular Level

The phenomenological reaction rates may be the result of a number of individual reaction steps, each operating at the molecular level. In order to rationalize phenomenological rate data and extrapolate it to "real-life" or nonlaboratory conditions, the ultimate goal of the gas–solid reaction mechanist must be to identify and characterize the fundamental individual reaction steps. However, relatively little is known about the molecular level details of gas–solid reactions, e.g., see the recent review of Rosner (1972). In this connection Kubaschewski (Kubaschewski *et al.*, 1970, p. 3) stated that:

> Neither the causes of phase stability nor the mechanisms of oxidation and diffusion are sufficiently well established, and the field is wide open to intelligent research and thought.

A recent review of the dynamics of gas–solid interactions, emphasizing microscopic aspects such as gas scattering trajectories and their relation to surface crystallographic structure and the extent of adsorption and reaction, has been made by Saltsburg (1973). This type of work represents a more advanced level of sophistication than the common phenomenological kinetic studies which contain the more fundamental dynamic information but in an irretrievable form.

An example of the number and type of elementary reaction steps that may be present in a single process is given by the relatively simple case of a metal-surface dissociation of molecular CH_4, i.e.,

$$CH_4 = 2H_2 + C(\text{surface})$$

The most likely individual steps for this process are considered to be

$$\begin{aligned} CH_4 + \square &= CH_4(\text{ad}), \\ CH_4(\text{ad}) &= CH_3(\text{ad}) + H(\text{ad}), \\ CH_3(\text{ad}) &= CH_2(\text{ad}) + H(\text{ad}), \\ CH_2(\text{ad}) &= CH(\text{ad}) + H(\text{ad}), \\ CH(\text{ad}) &= C(\text{ad}) + H(\text{ad}), \\ 2H(\text{ad}) &= H_2, \\ C(\text{ad}) &= \square + C(\text{in metal}), \end{aligned}$$

where $\square$ represents a surface active site as discussed below (see also Worrell, 1971). Thus, when gas–solid reactions are considered at this elementary level, the importance of high temperature radical species, such as CH_3 or CH_2, to gas–solid processes becomes more apparent.

The concept of a surface active site has been of great significance to the development of molecular level models of gas–solid reactions. One of the more generally useful descriptions of surface sites is the terrace–ledge–kink model, where the terraces are low-index crystallographic planes. A kink is an energetically preferred site for adsorption, but adsorption at a terrace or ledge site is also possible. It is possible to describe a chemical adsorption process in terms of an equilibrium reaction:

$$AB + \square \rightleftarrows AB(\text{ad}),$$

where AB is a gas species, $\square$ is a vacant surface site, and AB(ad) is the chemisorbed molecule. An equilibrium constant can be written for such a reaction from which the form of an adsorption isotherm may be predicted; this is known as the Langmuir adsorption isotherm. Discussions of the applicability or limitations of this and other adsorption isotherm models may be found in the review of Worrell (1971) and earlier cited literature.

1. *Application of Time-Resolved Mass Spectrometry*

One of the key problems in gas–solid reaction theory concerns the identity of the molecular intermediates, such as CH_3 in the above example.

The recent development of molecular beam techniques with mass spectrometric detection has now provided the means for identifying at least some of the participating species. Also, the use of beam modulation techniques provides a time base for kinetic studies on some of the individual reaction steps at the surface.

With the modulated incident beam mass spectrometric approach, the experimental variables are the composition, pressure, and temperature of the scattered or desorbed molecular beam, the surface temperature, and the modulation frequency of the incident molecular beam, e.g., see Olander (1968). The experimental information output is provided by the signal amplitude of the desorbed beam and its phase lag $\Delta\phi$ relative to a reference signal. The signal amplitude S of the lock-in detector output is given by the proportionality

$$S \propto \sin(\omega t - \Delta\phi),$$

where $\omega = 2\pi f$, f is the modulation frequency, and t is time. Exact relationships between S and $\Delta\phi$ have been given by Harrison *et al.* (1964). Given a model of the surface processes, $\Delta\phi_i$ and S_i (species i) may be related to the periodic behavior of the surface concentration—see Olander (1968) for the mathematical details of various kinetic cases.

The surface lifetime τ and desorption activation energy E_d for a particular species, may be determined as follows (Arthur, 1968). Consider a first-order process. Let F be the incident flux at time $t = 0$ and α is the fraction of molecules adsorbed with a mean lifetime τ prior to desorption; then $1 - \alpha$ is a measure of the flux immediately reflected. The flux of species leaving the surface at a time $t \geq 0$ is then given by

$$\frac{dN}{dt} = \dot{N} = (1 - \alpha)F + \alpha F(1 - e^{-t/\tau}),$$

and τ may be determined from the time dependence of

$$\dot{N}_{(t)} - \dot{N}_{(t=0)}.$$

Also,

$$\tau = \tau_0 \exp(E_d/RT).$$

Hence, for conditions where E_d is constant, e.g., low surface coverage, E_d may be obtained from the temperature dependence of τ. As the sensitivity of the mass spectrometric method is better than 10^{10} atoms cm^{-2} sec^{-1}, measurements can be made under low surface coverage conditions where the desorption kinetics may be more readily defined. Results for τ and E_d have been given by Arthur (1968) for adsorption–desorption of Bi, Bi_2, As_2, and P_2 from GaAs and GaP surfaces. The atomic species Bi was found to have a much higher sticking probability than the molecular species Bi_2, and this was attributed to the surface active sites being separated by too great a distance to allow adsorption of both ends of a Bi–Bi molecular species.

The application of similar modulated molecular beam techniques to kinetic studies of high temperature gas–solid reactions, as distinct from adsorption–desorption processes, has been demonstrated, for example, by Schwarz and Madix (1968). In this case the product desorption rate constant, k_d is related to the experimental observables of signal amplitude and phase lag by

$$S = (\text{const})k_d/(k_d^2 + \omega^2)^{1/2},$$

and

$$1/k_d = \tau = (1/\omega)[\cot \Delta\phi - (S_2/S_1) \csc \Delta\phi],$$

where τ is the surface residence time, $\Delta\phi = \phi_2 - \phi_1$, and the subscripts 1 and 2 refer to different temperatures. For the more general case, where more than one intermediate i is involved,

$$S = \text{const} \prod_i k_i/(k_i^2 + \omega^2)^{1/2},$$

where k_i is the rate constant for the depletion of the ith intermediate in the sequence.

The time-dependent interaction of Cl_2, at pressures of $\sim 10^{-5}$–10^{-6} Torr, with Si and Ge surfaces to yield the volatile products $SiCl_2$ and $GeCl_2$, respectively, has been monitored by Madix and Schwarz (1971). From the dependence of product species signal amplitude and phase shift on the Cl_2 modulation frequency and the Ge, or Si, surface temperature, i.e., 900–1100 and 770–1500 K range, respectively, the following reaction mechanism is suggested:

$$M(s) + Cl_2 \xrightarrow{s^*} (M\text{—}Cl_2) \xrightarrow{k_c} (=MCl_2)_a \xrightarrow{k_1} (=MCl_2)_a{}^+ \xrightarrow{k_d} MCl_2,$$

where

M is Si or Ge,
s^* is the sticking coefficient into the precursor state (M–Cl_2),
k_c the inverse of the characteristic time for conversion of the precursor to the stable surface species $(=MCl_2)_a$,
k_1 the inverse of surface diffusion lifetime for $(=MCl_2)_a$,
k_d the first-order rate constant for desorption of MCl_2.

The overall rate constants (units of sec^{-1}) for the first-order surface reactions were determined to be

$$2 \times 10^7 e^{-22700/RT} \qquad \text{for Ge(111)},$$

$$6 \times 10^7 e^{-25600/RT} \qquad \text{for Ge(100)},$$

$$3 \times 10^8 e^{-40000/RT} \qquad \text{for Si(111)},$$

where the surface temperature is greater than 800 K for Ge and 1050 K for Si.

Similar studies have been made for the overall reaction

$$Ge(s) + O_3 \rightarrow GeO + O_2$$

(Madix *et al.*, 1971).

Additional examples of reaction mechanisms are considered in the discussion to follow.

IV. Oxidation Reactions

A. General Considerations

Oxide volatility is particularly important in the gas–solid corrosion of W, Mo, and the platinum metals (see Chapter 4). Reviews concerning the fundamentals of the oxidation and hydroxation of metals have been given by Speiser and St. Pierre (1964) and also by Kofstad (1966). A literature review covering the period 1969–1971 has been made by Wadsworth (1972). Also, the vaporization aspects of the gas–solid interactions involving the oxidation of C, Si, Cr, Mo, and Nb have been summarized recently by Gulbransen and Jansson (1970).

The principal volatile species involved in metal oxidation reactions are listed in Table 2.2. This listing is somewhat of an oversimplification as higher temperature and lower oxygen chemical-potential conditions may lead to a predominance of lower valence species, such as WO (e.g., see Fig. 2.3 and also Rovner *et al.*, 1968). For the purpose of examining the likely importance of these volatile oxide species, under equilibrium conditions, it is convenient to express the thermochemical data in terms of stability diagrams or plots of $\log p$ (oxide species) vs $\log pO_2$ (e.g., see Fig. 4.3). The effect of temperature is usually expressed in terms of $\log p$ (oxide species) vs T^{-1} plots (e.g., see Fig. 2.1). These plots may be used to discuss and predict the role of thermochemical effects in metal oxidation at high temperatures.

TABLE 2.2

Some Important Volatile Species in Metal Oxidation Reactions[a]

Metal	Volatile oxide	Temperature (°C) (approximate)
Ir	Ir_2O_3 or IrO_3	1200
Ru	RuO	1280
Rh	RhO_2	1400
Pt	PtO_2	1400
Cr	CrO_3	1200
Mo	MoO_3 + polymers[b]	1000
W	WO_3, WO_2 + polymers[b]	1100
Re	ReO_3	1200
	Re_2O_7	<1000
Ta	TaO and TaO_2	2000
Zr	ZrO and ZrO_2	1800
Nb	NbO and NbO_2	1400
Si	SiO	1200
V	VO	1300

[a] Taken in part from Argent and Birks (1968) and based on literature data; see also Rovner *et al.* (1968) for direct evidence of species identity.
[b] e.g., $(MO_3)_{n=2,3}$.

Under isothermal equilibrium conditions, the loss of metal is determined by the stoichiometry of the volatile oxide product, i.e., for a reaction

$$x\mathrm{M(s)} + \tfrac{1}{2}y\mathrm{O_2} = \mathrm{M}_x\mathrm{O}_y, \qquad p(\mathrm{M}_x\mathrm{O}_y) = K_p p\mathrm{O_2}^{y/2}.$$

For instance, the isothermal oxidation of Pt is found to be directly proportional to pO_2 and the reaction is therefore believed to be

$$\mathrm{Pt(s)} + \mathrm{O_2} = \mathrm{PtO_2}$$

(Alcock and Hooper, 1960). If the partial pressures of reactant and product species can be individually monitored it is then possible to test the constancy of K_p with pressure and hence verify the establishment of an equilibrium condition. However, it is often difficult to achieve equilibrium in gas–solid reactions (see Section VI) and various rate processes usually determine the extent of reaction rather than the thermodynamic stabilities of reactants and products.

Experimentally, it is found that volatile metal oxide species can affect the overall oxidation kinetics when their partial pressure is greater than about 10^{-9} atm. For most practical purposes the range of oxygen pressures

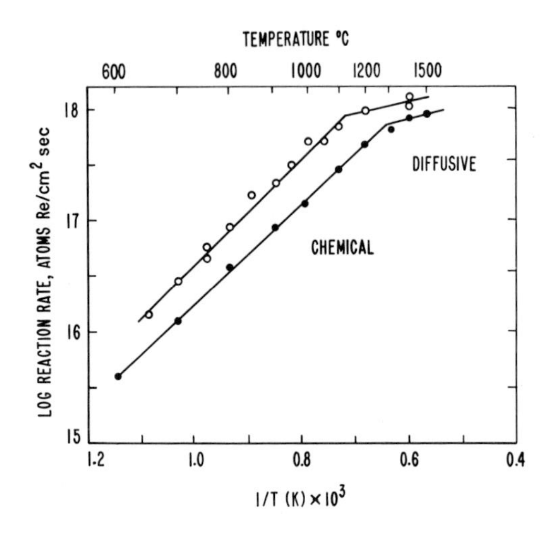

Fig. 2.1 Rate of reactive volatilization of Re in O_2, at 5 Torr (open circles) and 2 Torr (closed circles) pressure, as a function of temperature, and with O_2 flows of $\sim 10^{18}$–10^{19} molecules sec^{-1}; between 600–1100 °C the rate is controlled by chemical reaction (activation energy $\simeq$ 21 kcal mol^{-1}), above about 1100 °C for 5 Torr and 1300 °C for 2 Torr O_2 pressure the rate is gas-diffusion limited (Gulbransen *et al.*, 1969).

of interest may be taken as $10^{-6} - 1$ atm. With oxide species partial pressures of $\sim$1% of the total gas pressure, the degree of volatilization is sufficiently high that a layer of gaseous or condensed (fume) products may limit the rate of oxygen diffusion to the surface. As an example, for Re the onset of a diffusion-limited oxidation occurs when the temperature and pressure conditions are such that overall reaction rates of about 10^{18} atoms of metal cm^{-2} sec^{-1} occur (Gulbransen *et al.*, 1969). The onset of this diffusion-limited condition is rather sudden, as shown in Fig. 2.1. This type of behavior is also found with Mo and graphite oxidation and generally results from the reduced temperature dependence of gas diffusion as compared with that for chemical reaction. A quantitative assessment of the degree to which gas diffusion controls the overall oxidation process can be made using a rotating disk method (Olander and Schofill, 1970; Olander, 1967).

The rate-limiting reaction steps in the oxidation of metals vary considerably with conditions of temperature and pressure, as discussed in detail by Argent and Birks (1968). For example, at relatively low pressures, i.e., 10^{-2}–10 Torr, and temperatures, i.e., 300–600 °C, where the products are nonvolatile, the oxidation rate of Nb is proportional to one-half power of the oxygen pressure. In this case the initial reaction steps are believed to be

$$O_2 \rightarrow O_2(\text{adsorbed})$$

$$O_2(\text{adsorbed}) \rightarrow 2O(\text{chemisorbed})$$

and

$$2O(\text{chemisorbed}) \rightarrow 2O(\text{dissolved}),$$

with the last step being rate limiting. At higher pressures the adsorption sites become saturated, and the oxidation rate is then insensitive to pressure.

If diffusion of O_2, or O, into the metal is rate controlling, then parabolic oxidation is observed, e.g., as for Ta at 250–450 °C and 76 Torr O_2 pressure. The so-called Wagner theory satisfactorily accounts for this type of oxidation, and a mathematical description may be found in Argent and Birks (1968).

Concerning the mechanism of oxidation at the molecular level, the initial step in the majority of cases is believed to be the dissociative adsorption of oxygen by the metal. This type of process should lead to a scrambling of the oxygen atoms such that if an isotopic mixture of $^{16}O_2$ and $^{18}O_2$ is used, then the products would be $M_x{}^{16}O^{16}O$, $M_x{}^{16}O^{18}O$, and $M_x{}^{18}O^{18}O$. Using matrix isolation infrared spectroscopy as an analytical technique for the isotopic species, Ogden *et al.* (1970) showed that the oxidation of Tl occurs as a nondissociative process, i.e., no $Tl_x{}^{16}O^{18}O$ was observed, whereas for Ge, Sn, Ga or In, the more common dissociation process is operative, i.e., $M_x{}^{16}O^{18}O$ was observed. These authors suggested that the two types of oxidation might be correlated approximately with the heats of formation of the oxides. Thus Tl_2O has a heat of formation of -42 and Ga_2O of -82 kcal mol^{-1}, and in the former case the heat associated with the formation of the oxide may be too low to favor the dissociation of molecular O_2. As most oxides have higher heats of formation than Tl_2O, then we can expect that the dissociative process is relatively common.

B. Oxidation of Tungsten

A classical example of metal oxidation with vapor-phase products is given by the W(s)–O_2 reaction system which has received extensive study over a wide range of temperature and pressure. This extended range over which rate of W losses can be monitored allows for a more stringent test of proposed theories of adsorption–oxidation–vaporization processes. Tungsten oxidation under conditions of high temperature and relatively high O_2 pressure has also been of concern in connection with the corrosion of aerospace components. A discussion of W oxidation for these particular temperature–pressure conditions is deferred until Chapter 4, Section III. Our immediate attention is directed to oxidation under relatively low pressure conditions where mass transport effects are not likely to be rate limiting.

At relatively low temperatures, that is, between 800 and 1300 K, tungsten oxidizes to form a nonvolatile scale, i.e.,

$$x\mathrm{W(s)} + \tfrac{1}{2}y\mathrm{O_2} \rightarrow \mathrm{W}_x\mathrm{O}_y\mathrm{(s)}.$$

Between 1000 and 1300 K, this scale consists of two visibly distinct oxide layers. The formation of one of these layers is characterized by the normal parabolic growth time dependence and the other layer by a linear rate law (Webb *et al.*, 1956). Flash desorption of the oxide species WO, WO_2, and WO_3, with mass spectrometric detection, has been used to infer the nature of the low temperature chemisorption and oxidation of tungsten (McCarroll, 1967b).

At high temperatures, i.e., $1400 < T < 3500$ K, volatile oxides are

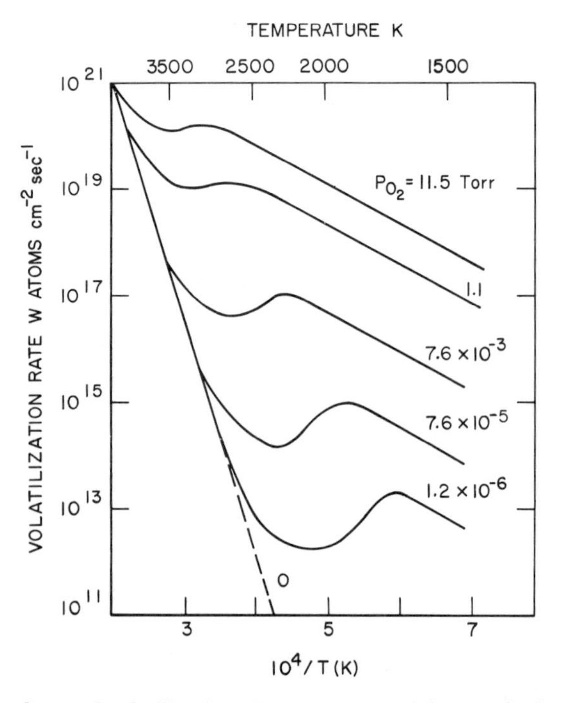

Fig. 2.2 Total rate of volatilization of tungsten-containing species in oxygen at the indicated O_2 pressures, and as a function of temperature; these curves are based both on experimental data and the semiempirical quasi-equilibrium model. (Reprinted with permission from J. C. Batty and R. E. Stickney, *Oxid. Metals* **3**, 331. Copyright ©1971 by Plenum Publ. Corp., New York.)

produced during oxidation and a net weight loss of W results. Weight-loss data for W, in the presence of relatively low pressures of O_2, are given in Fig. 2.2. As is suggested by the nonmonotonic dependence of the vaporization rate on the reciprocal temperature shown in the Fig. 2.2, the high temperature low pressure oxidation of W is complex. This complexity results from the presence of a variety of vapor species, i.e., W_3O_9, W_2O_6, WO_3, WO_2, and O—as identified mass spectrometrically by Schissel and Trulson (1965). The dip in the vaporization-rate curves, shown in Fig. 2.2, is apparently a consequence of the decreasing stability of WO_3, relative to less volatile and less oxygen-consuming species, such as WO, at the higher temperature regions.

1. *Reaction Models*

A kinetic model, based on mass spectrometric studies of species flux as a function of tungsten temperature (~1400–3100 K) and oxygen pressure (~10^{-4} Torr), involves dissociative adsorption of O_2 as the initial step (Schissel and Trulson, 1965). At temperatures of about 2600 K and greater, an almost complete conversion of O_2 to desorbed O atoms is found i.e., the O_2 sticking coëfficient is near unity.

These tungsten-oxidation results have more recently been analyzed in terms of a quasi-equilibrium surface model, e.g., see Batty and Stickney (1971) and also the review of Rosner (1972a). The model proposed by Batty and Stickney (1971) considers the various tungsten–oxygen species to be in local chemical equilibrium with each other at the surface but not necessarily in general equilibrium with the total amount of impinging O_2 reactant gas. The rate-determining step in the overall oxidation process is believed to involve the initial adsorption of O_2. Only a fraction of the impinging O_2 molecules are adsorbed *and equilibrated* at the surface and the empirically determined probability, $G'O_2$, of this adsorption increases exponentially with temperature according to the relation

$$G'O_2 = 1.16 \times 10^8 (Z'O_2)^{-0.3} 10^{-(0.8\times 10^4/T)},$$

where $Z'O_2$ is the total oxygen impingement rate, which typically has values of about 10^{17} cm^{-2} sec^{-1} at 10^{-4} Torr oxygen pressure.

According to this model, the Langmuir vaporization rates $R_{W_xO_y}$ of the various W_xO_y species are related by the expression

$$2G'O_2 Z'O_2 = R_O + 2R_{O_2} + R_{WO} + 2R_{WO_2} + \cdots,$$

i.e., oxygen is conserved. The rates $R_{W_xO_y}$ may be determined from the partial pressures pW_xO_y which, in turn, are derived from the known equilibrium constants $K_{W_xO_y}$. That is,

$$R_{W_xO_y} = pW_xO_y/(2\pi M_{W_xO_y}kT)^{1/2},$$

where

k is the Boltzmann constant,
p is partial pressure, and
M is molecular weight,

and for the general reaction

$$xW(s) + \tfrac{1}{2}yO_2 \rightleftarrows W_xO_y,$$

$$pW_xO_y = K_{W_xO_y} \cdot pO_2^{y/2}.$$

The various important species are O, WO, WO_2, WO_3, W_2O_6, W_3O_8, W_3O_9, and W_4O_{12} and the model allows predictions to be made for the contribution

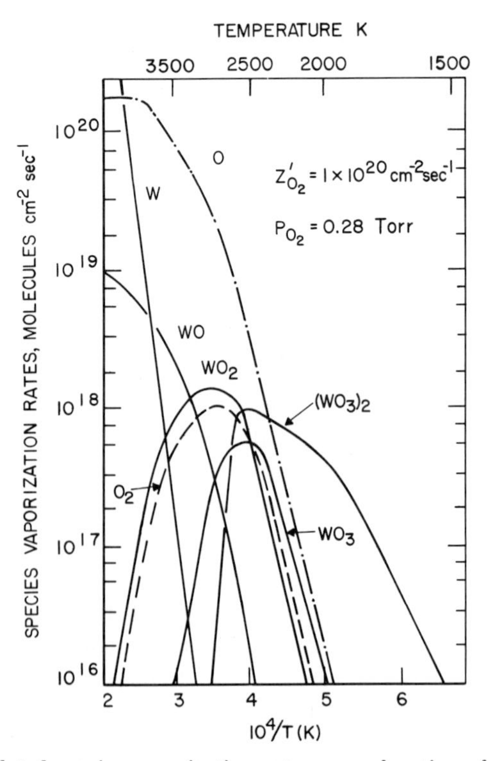

Fig. 2.3 Calculated species vaporization rates, as a function of temperature, resulting from tungsten óxidation and using the quasi-equilibrium model (Reprinted with permission from J. C. Batty and R. E. Stickney, *Oxid. Metals* **3,** 331. Copyright ©1971 by Plenum Publ. Corp., New York.)

of these various species to the total vaporization rate, as shown by the curves of Fig. 2.3. Also, the total vaporization rate data of Fig. 2.2 are consistent with the model.

This is an empirical model in that the determination of $G'O_2$ is based on a self-consistent treatment of experimental vaporization rate data, i.e., see Batty and Stickney (1971) for details. However, once an expression relating $G'O_2$ to $Z'O_2$ and T is derived, then the model allows vaporization rate calculations to be made for conditions not covered, or perhaps accessible, by experiment.

The assumption of equilibrium between species at the surface has also been used by Schafer (1970) to explain the gross phenomena of W oxidation and transport along a temperature gradient.

It is of note that the oxidation of W by a molecular beam of Ga_2O to yield the species WO_2, WO_3, and W_2O_6, as reported by Burns and Blaz (1971), does not appear to support the quasi-equilibrium theory, though no details of the extent of deviation from theory are given by the authors. The conditions under which this theory may be applied generally need to be established.

V. Halogenation Reactions

A. General Considerations

Halogenation reactions of the gas–solid type are of considerable practical importance in connection with the conversion of ores to metals (see Chapter 3), nuclear fuel processing (see Chapter 6), corrosion (see Chapter 4), metal halide lamps (see Chapter 3), the in situ generation of volatile metal halide flame inhibitors (see Chapter 5), and other vapor transport applications as discussed throughout Chapter 3. The following discussion serves to illustrate what is known at a basic level (see also the examples given earlier in Section III).

Halogenation reactions with metal surfaces, particularly where the halogen is Cl_2, frequently lead to the formation of volatile metal halide species. For example, the chlorination of W and Mo leads to volatile product species, and a clean metal surface is retained. In this case, reaction rates are found to depend on the 0.6 power of Cl_2 pressure, which suggests that dissociation of Cl_2 at the surface is the rate-limiting step in the reaction mechanism and atomic Cl is the reactant species (Landsberg *et al.*, 1971). The identity of the metal halide species formed is not known but is most likely MCl_n, where n varies from 6 to 1, depending on the conditions of surface temperature and halogen pressure.

In some metal–Cl_2 reactions, the formation of relatively complex vapor species may occur. For example, Bell *et al.* (1961), using a combination of static and dynamic vaporization techniques, infer that the transport of solid Pd by Cl_2 results from a reaction giving $PdCl_2$ and Pd_5Cl_{10} as vapor-phase products. At a Cl_2 pressure of about 1 atm and a temperature of 859 °C, the partial pressure of the predominant species is $p(Pd_5Cl_{10}) = 1.2 \times 10^{-2}$ atm, and at 1506 °C, $PdCl_2$ becomes the predominant species and $p(PdCl_2) = 0.1$ atm.

In other cases stable solid halide phases may form. For example, at relatively low temperatures, i.e., ~100–500 °C, the formation of the stable solid phases $Au_2Cl_2(s)$ and $Au_2Cl_6(s)$ results in a nonmonotonic dependence of chlorination rate on temperature (see Fig. 3.10). Thus from about 100 to 200 °C, depending on the pressure of Cl_2, the rate increases with temperature but from about 200–500 °C, a rate decrease results due to the stability of $Au_2Cl_2(s)$ over this temperature interval. At higher temperatures of 500–800 °C this solid phase dissociates and an increasing rate is again observed (Landsberg and Hoatson, 1970). Similarly, with the $VOCl_2(s)$ and $SbOCl(s)$ systems (see Knacke, 1972; Hastie, 1973a, respectively), the formation of a series of intermediate condensed oxide and oxychloride phases can thermodynamically limit the rate of production of the volatile metal chlorides during chlorination reactions.

The rates of chlorination can be further limited by chemical reaction kinetics and by mass transport effects. Conditions of temperature, pressure, and gas flow can usually be adjusted to show the conditions where either of these limitations is predominant. Fruehan (1972), for example, has found that for iron at >680 K, for Ni at >993 K, and for Sn at 340–396 K, the transport of Cl_2 through a metal halide vapor boundary layer is rate controlling. At lower temperatures the transport of $NiCl_2$ vapor is rate limiting for Ni, and for Fe a slow surface reaction of Cl_2 and $FeCl_2(s)$ to yield $(FeCl_3)_2$ is rate limiting.

Another related case where mass transport, in the vapor phase, can be rate limiting is for the H_2 reduction of metal chlorides, i.e.:

$$MCl_2 + H_2 \rightleftarrows M(s) + 2HCl.$$

In this instance the reaction rate depends on the vaporization of condensed MCl_2 and removal of the HCl product. The rate-limiting effect of an HCl boundary layer at the $MCl_2(s) + M(s)$ surface has been defined by Rigg (1973) in terms of the expression

$$\text{reaction rate} = \frac{D\gamma P_e}{RTBx_0}\left(\frac{P_e - P}{P_e}\right) \text{mol cm}^{-2}\,\text{sec}^{-1},$$

where

D is the HCl diffusivity,
γ represents the porosity of the product layer [M(s)],
P_e is the final HCl pressure, i.e., that external to tne reduction zone,
P the HCl pressure at the surface,
x_0 a distance, defining the zone where MCl_2 vapor diffusion is significant, and
$B \simeq P_e/P_{MCl_2}$ is taken as a measure of diffusional resistance to MCl_2.

This expression serves to indicate the factors that enter into a practical gas–solid reaction system.

B. Nickel–Halogen System

Relatively few gas–solid kinetic studies have been made with a direct monitoring of reactant, intermediate, and product species. However, McKinley (1966, 1969) has made several studies of this type using mass spectrometric line-of-sight detection techniques.

For the case of the F_2–Ni(s) system, with F_2 pressures of 10^{-4}–10^{-7} Torr, a strong adsorption of F_2 was found at surface temperatures of 1400–1600 K. At lower temperatures, i.e., ~950 K, a much reduced degree of adsorption occurred, and this is probably due to the relative involatility of NiF_2 and NiF surface layers which protect the Ni metal from further reaction with the impinging F_2 gas. At higher temperatures the surface product species are desorbed, as shown in Fig. 2.4. From the temperature dependence of these species desorption rates, the activation energies for desorption processes are determined to be 28 ± 2 kcal mol^{-1} for NiF and 39 kcal mol^{-1} for NiF_2. These are much lower values than those for bulk sublimation where the heat of sublimation of NiF_2, for instance, is about 78 kcal mol^{-1}. Analogous results were obtained for the Br_2–Ni system.

The surface reaction mechanism that is most consistent with these observations is as follows (S denotes surface):

$$\mathrm{Ni(S)} + \tfrac{1}{2}\mathrm{F_2} \rightarrow \mathrm{NiF(S)}, \qquad (1)$$

$$\mathrm{NiF(S)} + \tfrac{1}{2}\mathrm{F_2} \rightarrow \mathrm{NiF(S)\cdots F}, \qquad (2)$$

(i.e., a mobile layer of F atoms forms on a NiF surface)

$$\mathrm{NiF(S)\cdots F} \rightarrow \mathrm{NiF_2(ads)}, \qquad (3)$$

$$\mathrm{NiF_2(ads)} \rightarrow \mathrm{NiF_2(g)}, \qquad (4)$$

$$\mathrm{NiF(S)} \rightarrow \mathrm{NiF(g)}, \qquad T > 1100\ \mathrm{K}, \qquad (5)$$

$$\mathrm{NiF(S)\cdots F} \rightarrow \mathrm{NiF(S)} + \mathrm{F(g)}, \qquad (6)$$

and

$$\mathrm{NiF(S)} \rightarrow \mathrm{Ni(S)} + \mathrm{F}. \qquad (7)$$

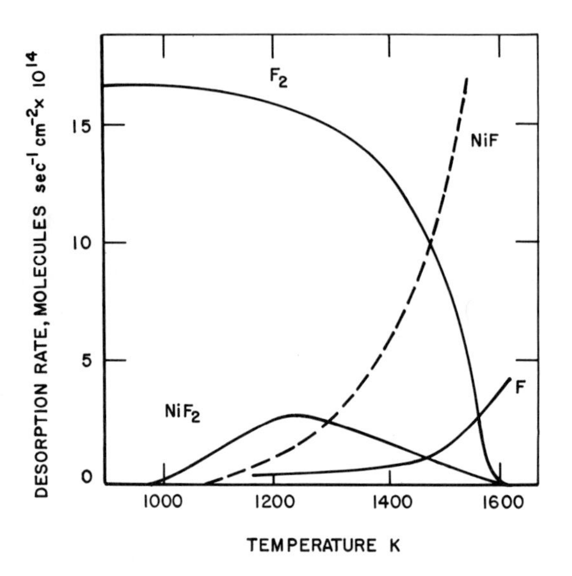

Fig. 2.4 Desorption rates for species in the Ni(s)–F_2 reaction system versus Ni surface temperature at 10^{-5} Torr F_2 pressure (McKinley, 1966).

The energetics of step (5) is also similar for the bromine and chlorine analogs, and it is suggested that a Ni substrate, NiX, bond-breaking is occurring (X = halogen). The rate of NiF_2 desorption is limited by the dissociative adsorption of F_2 on the NiF surface, i.e., reaction (2).

It should be noted that studies of this type rely on an assumed temperature-independent electron impact ionization process for mass spectrometric detection. For a low bond energy species such as F_2, it is conceivable that the F^+ to F_2^+ ion intensity ratio from F_2 could be sensitive to the temperature and internal energy of F_2. This would affect the form of the desorption rate versus temperature profiles, such as those given in Fig. 2.4, for F and F_2.

Henderson *et al.* (1964) have used a similar mass spectrometric technique to show that the reaction of F_2 with a boron surface at temperatures of 1000–1640 K leads to the formation of BF and BF_3 as the major vapor species.

VI. Synergistic Reactions

Synergistic, i.e., nonadditive, effects may result from the interaction of gas mixtures with solids. As a case in point, for the reaction

$$Cl_2 + O_2 + W(s) \rightarrow WO_2Cl_2,$$

the loss of W in this gas mixture is greater than the sum of the separate effects of Cl_2 and O_2, for pressures of 5×10^{-4}–10^{-6} Torr and over a temperature range of 1200–2400 K (McKinley, 1969). The relative abundance of the major species WO_2Cl_2, WO_2, and WO_3 correlates with their free energies of formation, and this is consistent with the quasi-equilibrium model discussed earlier for tungsten oxidation.

A related study on the W(s) + O_2 + Br_2 system has been made more recently under both chemical kinetic and equilibrium conditions (Gupta, 1971). These systems are of particular interest in lamp chemistry (see Chapter 3, Section IX). It was found in this study that the reaction

$$WO_2(s) + Br_2 = WO_2Br_2$$

does not attain equilibrium in a Knudsen cell over the temperature range 430–880 K. However, close to equilibrium conditions were found in the higher temperature range of 900–1200 K. A similar departure from equilibrium in Knudsen cells has been noted for the overall process

$$\tfrac{1}{3}Ti(s) + TiCl_4 \rightarrow \tfrac{4}{3}TiCl_3,$$

(Hastie *et al.*, 1971).

VII. Graphite–Gas Reactions

A. General Considerations

Apart from the prominant role played by carbon oxidation in steel making (Ward, 1962), one of the main areas of practical interest in carbon or graphite–gas reactions occurs in nuclear reactor technology (Eatherly and Piper, 1962). In this connection it is found that neutron irradiated graphite crystals oxidize at a higher rate than unirradiated crystals owing to the formation of lattice defects. Also, the rate of reaction of graphite with H_2 at 500–600 °C increases in the presence of neutron irradiation (Riley, 1968).

Coal gasification technology, e.g., see Chapter 6, Section V, is another area where graphite–gas-type reactions are of practical significance though it cannot be generally claimed that coal, which has an approximate composition of $CH_{0.7}$, and graphite will necessarily follow the same reaction pathways.

B. Graphite Oxidation

There exists a considerable body of fundamental literature on the oxidation of graphite, e.g., see the recent review of Rosner (1972a) and the articles edited by Belton and Worrell (1970), and no attempt to correlate or rationalize this information is made here, except to note that the oxidation is reaction-rate controlled at temperatures of less than about 1100 °C, diffusion controlled between about 1400 °C and 2500 °C, and sublimation controlled at temperatures of more than 2500 °C. Also, the increased rate of oxidation of porous graphite points to the importance of condensed state morphology on the course of gas–solid reactions.

Several molecular level studies, using the modulated incident beam mass spectrometric detection technique, have been carried out recently by Olander *et al.* (1972a,b). Reaction mechanisms are suggested, from these studies, for the oxidation of the basal and prism planes of pyrolytic graphite.

C. Graphite–H_2 Reactions

The reaction of H_2 with graphite is thermodynamically not very favorable for high temperature and low H_2 pressure conditions. However, in practice, hydrocarbon products such as CH_4, C_2H_4, C_2H_2, C_6H_6, and higher molecular weight species, can form from such a reaction under nonequilibrium conditions, e.g., see Steck *et al.* (1969) and Gulbransen (1966). There is a current interest in high pressure H_2–coal reactions for the generation of CH_4 as a substitute for natural gas. In fact, nonequilibrium reaction conditions may have existed in the primordial atmosphere and the formation of natural hydrocarbon deposits, e.g., oil, could have resulted in part from graphite–hydrogen reactions under gas flow conditions typical of volcanic activity (Gulbransen, 1966).

The interaction of H_2, and also N_2, with graphite in Knudsen cells has been utilized as a means of determining heats of formation for high temperature species such as C_2H (Wyatt and Stafford, 1972), CH_2 (Chupka *et al.*, 1963), and CN (Berkowitz, 1962). Difficulty in achieving gas–solid equilibrium was recognized in these studies, and in some cases a comparison of second and third law data was used to support the equilibrium assumptions. In passing, one might consider the applicability of the balanced reaction concept, as used in flame studies, e.g., see Chapter 5, Section II.I, to gas–solid systems where only a local equilibrium may exist.

VIII. Water Vapor–Solid Reactions

A. General Considerations

The interaction of water vapor with solids can lead to the formation of new condensed phases, such as metal hydroxides or oxides, or to the formation of corresponding vapor species, depending to a large extent on the conditions of temperature and pressure. Water vapor can also have important catalytic effects on condensed-phase phenomena involving ceramics, such as crack growth (Lawn, 1974) and devitrification (Meek and Braun, 1972). In such cases it is thought that the formation of hydroxide surface intermediates occurs.

Many metal oxides form solid hydroxide phases that are stable at room temperature but decompose to yield the oxide and water vapor at only moderately elevated temperatures (i.e., ~100–500 °C), e.g., see Yamaoka *et al.* (1970), MacKenzie (1970), and the JANAF Thermochemical Tables (JANAF, 1971). In some cases a stepwise decomposition is observed for, example,

$$\tfrac{1}{2}La_2O_3(s) + \tfrac{3}{2}H_2O \rightarrow La(OH)_3(s), \qquad 90\ °C,$$

$$La(OH)_3(s) \rightarrow LaO(OH)(s) + H_2O, \qquad 425\ °C,$$

and

$$2LaO(OH)(s) \rightarrow La_2O_3(s) + H_2O, \qquad 570\ °C$$

(Rybakov *et al.*, 1969). The alkali metal–hydroxide systems are notable exceptions to this tendency for dehydration. In this case one finds the following reactions:

$$M_2O(s) + H_2O = 2MOH(s) \qquad \text{and} \qquad MOH(s) = MOH.$$

At relatively high temperatures, i.e., $T \gtrsim 1000$ °C, where the condensed hydroxide phases are unstable, there is a tendency for metals, or their oxides, to form volatile hydroxide species in the presence of water vapor. It is informative to consider the thermodynamic basis for this relative instability of condensed-phase hydroxides, as compared with that of the corresponding vapor-phase species at high temperatures.

Consider the following reactions which are typical for hydroxide systems:

$$Fe(OH)_2(c) = Fe(OH)_2, \tag{1}$$

$$Fe(OH)_2(c) = FeO(c) + H_2O, \tag{2}$$

and

$$FeO + H_2O = Fe(OH)_2. \tag{3}$$

At a temperature of 1500 K, the free energies for these reactions are

$$\Delta F_1 = -5.1\ \text{kcal mol}^{-1}, \qquad \Delta F_2 = -38.2\ \text{kcal mol}^{-1},$$

and

$$\Delta F_3 = -32.9\ \text{kcal mol}^{-1}.$$

Hence the decomposition of the condensed hydroxide [reaction (2)] is much more favorable than its nondissociative vaporization [reaction (1)].

It also follows that the reaction:

$$FeO(c) + H_2O = Fe(OH)_2$$

is not a very favorable process for forming the hydroxide vapor species. However, for the all-gas-phase process of reaction (3) the free energy is very favorable. Thus, hydroxide species formation appears to be particularly favored under conditions where the oxide is volatile. This point is further demonstrated by the following example.

As a demonstration of the interplay between the thermodynamic factors that may, or may not, lead to the production of volatile metal–hydroxide species, consider the comparison systems:

$$WO_3(s) + H_2O = WO_2(OH)_2, \qquad (4)$$

and

$$\text{“}TaO_2(s)\text{”} + H_2O = TaO(OH)_2, \qquad (5)$$

where “$TaO_2(s)$” is, in fact, the mixture $Ta_2O_5(s) + Ta(s)$. Reaction (4) is known to occur readily at about 1300 K, where $\Delta H_4 = 39.9$ kcal mol^{-1} and $\Delta S_4 = 18.8$ cal deg^{-1} mol^{-1}. On the other hand, the analogous reaction (5) does not occur under similar conditions of temperature and pressure. The thermodynamic rationale for this difference in behavior is as follows.

Both WO_3 and TaO_2 are known vapor-phase species, and their *homogeneous* interaction with H_2O to yield the hydroxide species would be very similar. This follows from the similar reaction entropies and enthalpies that may be estimated for reactions (4) and (5). In both cases the reaction involves the conversion of a metal–oxygen double bond to two metal–hydroxyl bonds. Hence, under conditions where the oxide vapor species are present, one could expect species such as $TaO(OH)_2$ to form—as was shown above to be the case for the $Fe(OH)_2$ system. However, in practice, TaO_2 vapor is not observed below temperatures of about 2000 K, which is 700 K higher than the corresponding vaporization temperature for WO_3. Thus any production of $TaO(OH)_2$ at the lower temperature region of about 1300 K would have to result from a gas–solid reaction such as (5). The main thermodynamic difference between the processes (4) and (5) is in the heats required to project the oxide into a gas-phase form. Experimentally, one finds that the sublimation enthalpy of “TaO_2” is more endothermic than for WO_3 by about 24 kcal mol^{-1}. Hence, for reaction (5):

$$\Delta H_5 \sim \Delta H_4 + 24 \sim 64 \text{ kcal mol}^{-1}.$$

The entropy change would be similar for both reactions (4) and (5). It, therefore, follows from the above differences in ΔH_5 and ΔH_4 that the concentration ratio of $TaO(OH)_2/WO_2(OH)_2$, under similar conditions of temperature and H_2O pressure, would be of the order of 10^{-5}, and hence one would not expect to find $TaO(OH)_2$ in the presence of only "$TaO_2(s)$" and H_2O.

From this example it is apparent that, given sufficient basic thermodynamic data and the "rules" of analogy, it is possible to rationalize and predict the conditions under which water vapor–solid oxide reactions can lead to the formation of hydroxide vapor species. Jackson (1971) has recently made such predictions (e.g., see Fig. 2.6). Further discussion along these lines will be given in the sections to follow.

B. Water Vapor–Solid Reactions with Oxide Products

1. H_2O as an Oxidant

Many metals react with oxygen-containing gases such as H_2O or O_2 to form solid or, to a lesser extent, vapor-phase oxide products. As an example of the latter case, the gross evaporation of Ta is enhanced in the presence of water vapor, as shown by the results of Fig. 2.5. The principal reaction is the reduction of water to form tantalum oxides and H_2, although small

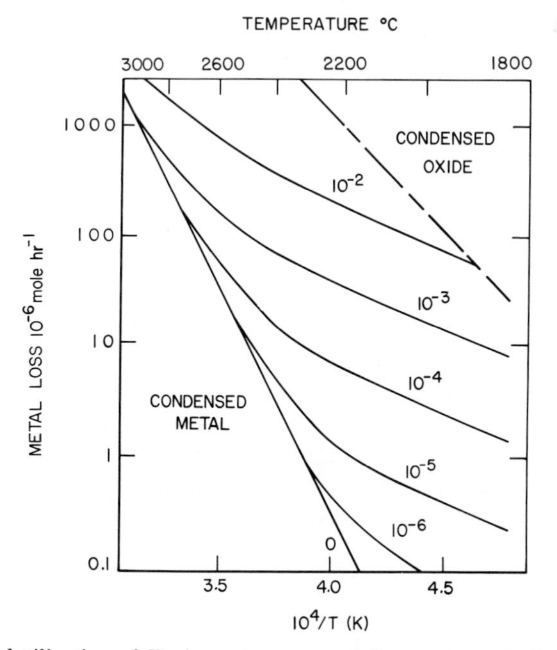

Fig. 2.5 Volatilization of Ta in water vapor at the pressures indicated (pH_2O in Torr) and as a function of temperature. (Reprinted with permission from E. Fromm and H. Jehn, *Vacuum* **19,** 191. Copyright ©1969 by Pergamon Press Ltd.)

contributions of unknown metal hydroxides to the metal volatility should not be discounted. Similar data have been summarized by Fromm and Jehn (1969) for the reactions of Nb metal with H_2O and other gases.

Water vapor may also be used as an oxidant in chemical transport processes, e.g.,

$$2GaP(s) + H_2O = Ga_2O + P_2 + H_2, \qquad T \sim 1000\ °C.$$

2. H_2O as a Catalyst

The interaction of water vapor with metals at elevated temperatures, i.e., ~500–1000 °C, can result in a catalytic reduction of metal oxides, in the vicinity of another metal, by a gas–solid process of the type

$$M(s) + H_2O = MO(s) + H_2,$$

$$H_2 + M'O(s) = M'(s) + H_2O,$$

where M and M′ are different metals (Kononyuk and Kulikovskaya, 1970). For example, NiO and Fe_2O_3 can be reduced in the presence of Ta, Ti, or Fe metal together with trace impurity levels of H_2O at temperatures above 600 °C and vacuum conditions of 10^{-1}–10^{-3} Torr. Thus, the role of diffusion in apparently solid–solid reactions may be obscured by catalytic gas–solid reactions of this type.

C. Water Vapor–Solid Reactions with Hydroxide Vapor Products

1. Introduction

A review of early evidence for molecular species resulting from water vapor–solid interactions was made by Elliott (1952). At that time he concluded that there did not appear to be any periodicity, or systematics, in the ability of metals or their oxides to form hydrated species in the vapor phase. However, in part, this was the result of insufficient or inadequate experiments and a general lack of basic data on high temperature species. As is discussed elsewhere (Chapter 5, Section II.K), there appears to be a valid correlation between the thermodynamic functions for halides and hydroxides, and on this basis a periodicity in the stability of metal mono- and dihydroxides can be predicted, as shown for example in Fig. 2.6 for the metal dihydroxides.

Referring to the Fig. 2.6, the broken horizontal line in the region of $\log K \simeq -4$ may be taken as an approximate boundary above—or about—which the presence of vapor-phase hydroxide species will probably be important in water vapor–metal oxide interactions. The formation of $Fe(OH)_2$, which is known from the transport of iron oxide (FeO, *l*) in the

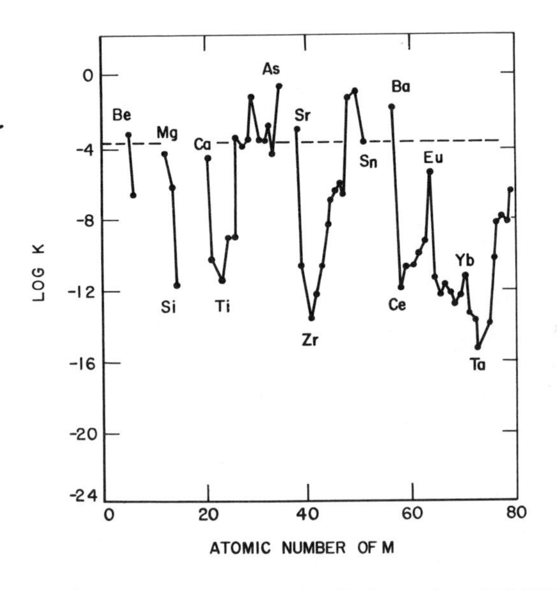

Fig. 2.6 Calculated equilibrium constants for the formation of $M(OH)_2$ at a temperature of 2000 K and based on the general reaction

$$MO_x(s, l) + H_2O = M(OH)_2 + \tfrac{1}{2}(x - 1)O_2,$$

where the H_2O and O_2 pressures are each set equal to 1 atm; the broken horizontal line is explained in the main text (Jackson, 1971).

presence of H_2O vapor, has a value of $\log K = -3.4$. Hence the dihydroxides with K values greater than this should be readily detectable. These gas–solid processes are generally endothermic, and K will therefore increase with increasing temperature. That is, the production of metal dihydroxide vapor is favored by high temperatures. Most of the dihydroxide species indicated in the Fig. 2.6 have yet to be identified experimentally.

2. *Identity of the Vapor Products*

Several relatively recent reviews have been made on the evidence for vapor-phase hydroxide formation, i.e., see Glemser and Wendlandt (1963), Greene *et al.* (1959), and Jackson (1971); see also Chapter 5, Section II.K. for a discussion of metal hydroxides in flames. It is apparent that much of the evidence for hydroxide species is of an indirect, and sometimes ambiguous, nature. Usually, in gas–solid systems, an enhanced rate of transport of an oxide in the presence of water vapor is taken as evidence for the formation of volatile hydroxide species. A systematic variation in the experimental conditions of gas composition and temperature can then lead to an inferred formula for the transport species, e.g., see Glemser and Wendlandt (1963).

Many of the known water vapor–solid interactions, leading to volatile products, are of the type

$$n\mathrm{A(s)} + q\mathrm{H_2O} = r[\mathrm{A}_{n/r}\cdot(\mathrm{H_2O})_{q/r}],$$

where the equilibrium constant is given by

$$K_p = p[\mathrm{A}_{n/r}\cdot(\mathrm{H_2O})_{q/r}]^r / p[\mathrm{H_2O}]^q,$$

and p represents partial pressure. If the system is at thermodynamic equilibrium, and constant temperature conditions are maintained, then K_p is constant for variable pressure. It follows that a plot of $\log p[\mathrm{A}_{n/r}\cdot(\mathrm{H_2O})_{q/r}]$ vs $\log p(\mathrm{H_2O})$ should yield a line with slope equal to q/r, i.e., the number of moles of H_2O associated with a mole of complex product. As the solid A is usually present at constant activity it is not possible to determine n/r, i.e., the number of moles of A incorporated into the product, and a value of n equal to unity is usually assumed.

The exact identity of the hydroxide species is important in determining and extrapolating the pressure dependence of transport and deposition processes. This extrapolation of laboratory data can be related to practical questions, such as when and where solid deposits will form in steam turbines and in mineral veins of geological structures (Elliott, 1952). However, very few verifications of species identity by spectroscopic techniques have been made to date. The hydroxides of the alkali metals, and also Ba and B, have been characterized by both mass and optical spectroscopy, and they are probably the best established metal–hydroxide systems at present.

The transport technique provides an indirect measure of the degree of importance of hydroxide formation. For example, in the presence of about 1-atm H_2O vapor, which is flowing over the condensed oxide under saturated equilibrium conditions, an "apparent" vapor pressure of oxide of about 1 Torr would result for BeO at $\sim$1850 K, WO_3 at $\sim$1300 K, MoO_3 at $\sim$930 K, and TeO_2 at $\sim$950 K. The oxide vapor pressures in the absence of H_2O are negligible at these temperatures. However, at lower pressures of H_2O the oxide vapor species may become more significant than the hydroxides. This would appear to be the case for Mo vapor transport. The observation of a greater degree of Mo transport in O_2 as opposed to H_2O vapor, at pressures of around 10^{-6} Torr and temperatures of 1300–2200 K, suggests the metal oxides to be more significant than the hydroxides in this case (Wahl and Batzies, 1970).

An example of the effect of water vapor pressure and temperature on the rate of transport, or apparent vapor pressure, of an oxide, is shown in Fig. 2.7. These are the basic data from which inferences as to the molecular

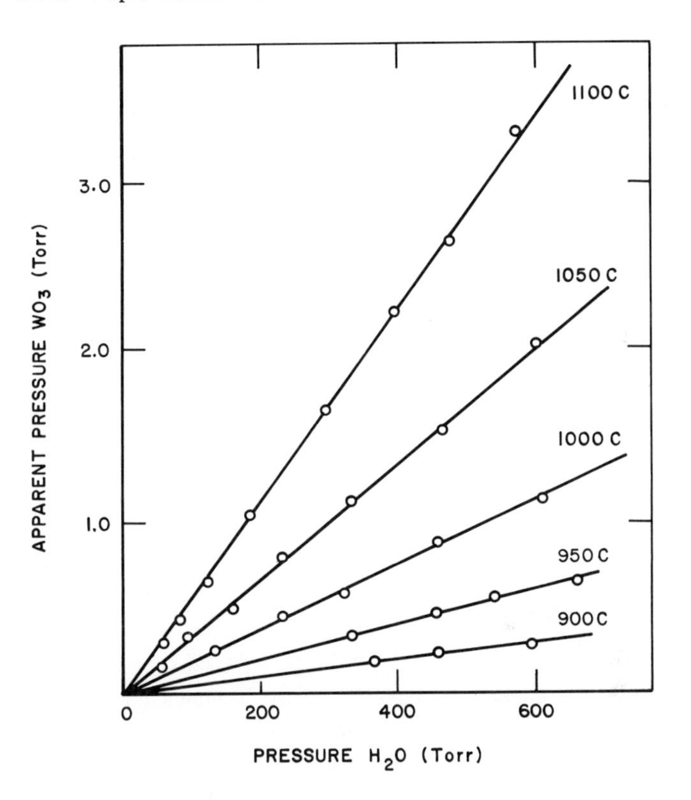

Fig. 2.7 Isotherms of apparent vapor pressure of $WO_3(c)$ as a function of water vapor partial pressure (Glemser and Wendlandt, 1963).

formula of transport species and the heat and entropy of reaction are usually made. For the reaction

$$x WO_3(s) + H_2O = W_xO_{3x-1}(OH)_2,$$

the number of water molecules associated with the transport species can be deduced from the isothermal dependence of WO_3 transport on H_2O partial pressure. The linearity of the plots shown in the Fig. 2.7, for example, indicate the presence of a single H_2O molecule in the transport species. Thus, if $x = 1$, then the reaction is

$$WO_3(s) + H_2O = WO_2(OH)_2 \qquad (\text{or } WO_3 \cdot H_2O).$$

The identity of the hydroxide product given in this reaction has been verified by an unpublished mass spectrometric analysis, cited by Glemser and Wendlandt (1963). This hydroxide species is believed to be important in the water vapor catalyzed degradation of incandescent W-filament lamps (see Chapter 3, Section IX).

From the temperature dependence of the transport data, the enthalpy for the above reaction is about 40 kcal mol^{-1} ($T \sim 1000$ °C). The endothermic character of this process indicates that the formation of the hydroxide is favored by an increase in temperature, as shown in Fig. 2.7. If one considers the enthalpy and entropy associated with introducing the condensed oxide into the vapor phase, i.e.,

$$WO_3(s) \rightarrow WO_3,$$

then the homogeneous reaction

$$WO_3 + H_2O = WO_2(OH)_2 \qquad (\text{or } WO_3 \cdot H_2O)$$

has an enthalpy change of about -87 kcal mol^{-1}. Such a high heat of reaction, and stability of the hydroxide product species, excludes a weak hydration of the type $WO_3 \cdot H_2O$ as the bonding mechanism; the formation of pseudohalide–metal-hydroxyl type bonds is the only reasonable explanation in this case.

3. *Summary of "Known" Hydroxide Species Important in Water Vapor–Solid Reactions*

Table 2.3 summarizes "known" hydroxide species which are believed to be important in gas–solid reactions. Several species of expected but unproven existence, as given by the JANAF Thermochemical Tables (JANAF, 1971), are also listed. Also the known nonmetallic, relatively low temperature, hydroxides are indicated for comparison. Most of the observations were made under conditions of relatively low water vapor pressure, i.e., 1 atm and lower.

In addition to the species listed in Table 2.3, there is spectroscopic, and other, evidence for the presence of metal hydroxide species in flames, as summarized in Chapter 5, Section II.K.

D. Hydroxide–Halide Conversion

Previous discussions, e.g., see Section VIII.C (and also Chapter 5, Sections II.J–L; Section V.D; Section VI.C; Section VII), have indicated the importance of metal–hydroxide vapor species in the presence of water vapor. However, many practical or experimental systems may also contain a halogen source which is usually present in the form of the hydrogen halide, HF or HCl. These halides can noticeably influence the formation of metal hydroxides, and it is important to recognize the thermodynamic conditions under which this occurs.

TABLE 2.3

Hydroxide Species Important in Water Vapor–Solid Interactions at High Temperature

Species	Comments and references
AlOH	[a], [e]
AlOOH	[a], [e]
$B(OH)_2$	[a], [b]
$B(OH)_3$	[a]
$(HBO_2)_3$	[a]
$[B(OH)_2]_2$	[a]
BaOH, $Ba(OH)_2$	[e], Mass spectrometric thermodynamic equilibrium study (Stafford and Berkowitz, 1964); $D[Ba—(OH)_2] = 206$ kcal mol^{-1}, and $D[Ba—OH] = 107$ kcal mol^{-1},
BeOH	[a]
$Be(OH)_2$	[a], [c], Mass spectrometric negative result (Hildenbrand, 1963)
CaOH	[a], [e]
$Ca(OH)_2$	[a], [e]
$Co(OH)_2$	[c]
$CrO_2(OH)$	[g]
$CrO_2(OH)_2$	[g]
CsOH, $(CsOH)_2$	Schoonmaker and Porter (1959)—mass spectrometry
$Cu(OH)_2$	Suggested from transport of Cu oxides in stream (Tskhvirashvili and Vasadze, 1966)
$Fe(OH)_2$	[a], [e]
KOH	[a], [e], mass spectrometry (Gusarov and Gorokhov, 1968), $D[K—OH] = 84.7 \pm 2$ kcal mol^{-1}
$(KOH)_2$	[a]
LiOH, $(LiOH)_2$	[a] Berkowitz *et al.* (1960), $Li_2O + H_2O$ at 1100 K
$Mg(OH)_2$	[a]
$MoO_2(OH)_2$	[a], [e]
NaOH	[a]
$(NaOH)_2$	[a]
$Ni(OH)_2$	[c]
RbOH, $(RbOH)_2$	Schoonmaker and Porter (1959)—mass spectrometry
ReO_3OH	$Re_2O_7 + H_2O$, $T \sim 700$ K, Weber *et al.* (1971a), Semenov and Shalkova (1969)
$TeO(OH)_2$	[c], [d], [f]
$WO_2(OH)_2$	[a], $W + 4H_2O = WO_2(OH)_2 + H_2$, $\Delta F_T = 26{,}700 - 5.56\,T$ kcal mol^{-1}; $T = 1200$–$1500°C$, S_{1600} $WO_2(OH)_2 = 136.7 \pm 1$ cal deg^{-1} mol^{-1} (Belton and McCarron, 1964); see also Speiser and St. Pierre (1964)
$Zn(OH)_2$	[c], [d]

TABLE 2.3 (*Continued*)

Species	Comments and references
Known nonmetallic hydroxides	[d]
NO_2OH, NOOH	
$PO(OH)_3$, P_3O_7OH	
$SO_2(OH)_2$	
ClOH, ClO_3OH	
$CO(OH)_2$	
$SeO_2(OH)_2$	

[a] See the JANAF Thermochemical Tables (JANAF, 1971), for thermodynamic data and original literature citations.

[b] Existence predicted from thermodynamic analogies, experimental verification not yet made.

[c] Species identity inferred from transport experiments only.

[d] See the review of Glemser and Wendlandt (1963).

[e] See Chapter 5, Section II.K on hydroxides in flames.

[f] This observation has been further verified by a recent transpiration study (Malinauskas *et al.*, 1970); also $\Delta H_{1000\text{ K}} = -30.2$ kcal mol^{-1} and $K_{p,970} = 27.9 \times 10^{-5}$ atm^{-1} for the reaction

$$TeO_2 + H_2O = TeO(OH)_2.$$

[g] See Graham and Davis (1971), Caplan and Cohen (1961), Glemser and Mueller (1962), and Table 5.13.

A common metal valence state for high temperature species is two, and we shall therefore consider, as an example, the competition between $M(OH)_2$ and MX_2 species where X may be F or Cl and M is a metal. The thermodynamic arguments used may, however, be applied equally well to other valence state systems. In a high temperature system containing water vapor and a source of chloride or fluoride, the predominant halogen species will be HCl or HF. The competition between MX_2 and $M(OH)_2$ is then determined by the free energies of reactions of the type

$$MO(c, g) + H_2O = M(OH)_2, \tag{1}$$

$$MO(c, g) + 2HX = MX_2 + H_2O, \tag{2}$$

and the difference between (1) and (2) is given by the reaction

$$MX_2 + 2H_2O = M(OH)_2 + 2HX. \tag{3}$$

From the free energy change of reaction (3) we may obtain the relative concentrations of $M(OH)_2$ and MX_2 for various conditions of H_2O and HX concentration. To determine this free energy change, ΔF_3, it is necessary to consider the related enthalpy and entropy quantities of ΔH_3 and

ΔS_3, respectively. Now the entropy of reaction (3) is given by

$$\Delta S_3 = (S_{M(OH)_2} - S_{MX_2}) + (2S_{HX} - 2S_{H_2O}).$$

By using the entropy data given in the JANAF Thermochemical Tables (JANAF, 1971), and also calculations of $S_{M(OH)_2}$ derived from estimated molecular and spectroscopic parameters, we can show that ΔS_3 generally has values of

$$\Delta S_3 = -4 \pm 2 \text{ cal deg}^{-1} \text{ mol}^{-1}, \qquad T \sim 1000\text{–}2000 \text{ K}.$$

This small calculated entropy change is also in accord with the entropy "rules" given in Chapter 3, Section IV. For the analogous cases involving $M(OH)_3$ and $M(OH)_4$, ΔS_3 values of -3 and -2 cal deg^{-1} mol^{-1}, respectively, are similarly determined. This approximately constant reaction entropy for various M results from the striking similarity between hydroxide and halide species—OH behaving as a pseudohalogen.

The heat of reaction (3) is given by

$$\Delta H_3 = \Delta H_f[M(OH)_2 - MX_2] + \Delta H_f[2HX - 2H_2O].$$

It is convenient to express this heat in terms of bond dissociation energies, i.e.,

$$\Delta H_3 = -2(\bar{D}_{M-OH} - \bar{D}_{M-X}) + (2D_{H_2O} - 2D_{O-H} - 2D_{H-X}),$$

where $\bar{D}_{M-OH}$ is the average M–OH bond dissociation energy in $M(OH)_2$; that is, it represents one-half of the heat for the process

$$M(OH)_2 = M + 2OH.$$

Similarly, $\bar{D}_{M-X}$ is the average bond dissociation energy for the M–X bond in MX_2; D_{H_2O} represents the energy to dissociate H_2O to the elements; D_{O-H} and D_{H-X} are the bond dissociation energies for the species OH and HX, respectively.

Consider the case where X = F. From the correlation between metal hydroxide and halide bond dissociation energies, i.e., see Chapter 5, Section II.K.5,

$$\bar{D}_{M-OH} \simeq 0.8\bar{D}_{M-F}.$$

Therefore, the reaction enthalpy is

$$\Delta H_3 = 0.4\bar{D}_{M-F} - 34 \text{ kcal mol}^{-1}.$$

Obviously, the production of $M(OH)_2$ relative to MX_2 is favored by a low value of $\bar{D}_{M-F}$. For strong M–F bonded species, such as SiF_2 where $\bar{D}_{Si-F} =$ 142 kcal mol^{-1} and $\Delta H_3 = +23$ kcal mol^{-1}, the high endothermicity of reaction (3) strongly favors the presence of MF_2 over $M(OH)_2$.

Consider the case where X = Cl. In this case as $\bar{D}_{\mathrm{M-Cl}} \sim \bar{D}_{\mathrm{M-OH}}$, then ΔH_3 is independent of M and has a general value of

$$\Delta H_3 \simeq +28 \text{ kcal mol}^{-1}$$

for MCl_2 systems. Therefore,

$$\Delta F_3 = \Delta H_3 - T\,\Delta S_3 = 32 \text{ kcal mol}^{-1}, \qquad T = 1000 \text{ K}.$$

Hence the equilibrium constant is $K_3 \simeq 10^{-7}$ and

$$p\mathrm{MCl_2}/p\mathrm{M(OH)_2} \simeq 10^7 (p\mathrm{HCl}/p\mathrm{H_2O})^2, \qquad T = 1000 \text{ K},$$

where p represents species partial pressure. It is apparent from this relationship that in order for the hydroxide to be present in similar amount to the corresponding chloride, the concentration of H_2O relative to HCl should be in the range of 10^3–10^4. A similar excess concentration of H_2O would be required for fluoride systems.

A few representative examples where both $M(OH)_2$ and MCl_2 species may be present, are considered as follows.

Example A. Following the discussion of Chapter 3, Section VIII, at a temperature of about 1000 K, H_2O pressures of about 10^3 atm and HCl pressures of 10 atm are typical conditions expected for ore genesis. Hence, from the equilibrium constant of $K_3 = 10^{-7}$, as derived above, it follows that

$$p\mathrm{M(OH)_2}/p\mathrm{MCl_2} \sim 10^{-3}.$$

Given an uncertainty in K of about a factor of 10, it is still apparent that the chloride is the more predominant metal-containing vapor species in the vapor transport model of ore formation—but see Section E.

Example B. Consider a fossil-fueled combustion furnace, such as is used in a steam generator power plant, and where $p\mathrm{H_2O} \sim 0.1$ atm and $p\mathrm{HCl} \sim 10^{-3}$ atm are typical concentrations; temperatures can be expected to fall in the range of 1000–2000 K depending on the location within the furnace. Under these conditions the predicted relative hydroxide concentration is given by

$$p\mathrm{M(OH)_2}/p\mathrm{MCl_2} \sim 10^{-3}, \qquad T \sim 1000 \text{ K}$$
$$\sim 1.0, \qquad T \sim 2000 \text{ K}.$$

Therefore, the hydroxides are relatively insignificant species only for the cooler regime of this combustion process.

Example C. The thermodynamic characterization of high temperature species is most frequently made in Knudsen cells where the total pressure is necessarily restricted to an upper limit of about 10^{-4} atm. The study of metal–hydroxide species in such cells would usually involve the interaction of H_2O with a condensed metal oxide. Depending on the method of preparation, the metal oxide sample could contain a halogen impurity such that HCl pressures of about 10^{-6} atm would be generated. Hence,

$$pM(OH)_2/pMCl_2 \sim 10^{-3}, \qquad T = 1000 \text{ K},$$

and the hydroxide species concentration could be reduced by as much as three orders of magnitude, owing to the presence of the halogen impurity—depending on the extent of the relative participation of these species in other reactions. However, at temperatures of greater than about 2000 K, the hydroxide concentration would not be appreciably affected by the presence of halogen.

E. High Pressure (>1 atm) Water Vapor–Solid Reactions with Vapor Products

1. Introduction

The solubility of inorganic solids in water vapor (steam) at high pressures but not excessively high temperatures, i.e., <1000 °C, has been known for many years; for instance, see the reviews of Glemser and Wendlandt (1963), Morey (1957), and Elliott (1952). Typical solubility data for inorganic solids in steam are given in Table 2.4.

TABLE 2.4

Solubility of Inorganic Solids in Water Vapor (Mainly at 500°C and 1050 atm, in ppm by Weight)[a]

GeO_2	8700	Nb_2O_5	28
Na_2SO_4	4300	NiO	20 (at 2100 atm)
SiO_2	2600	$CaSO_4$	20
KCl	278 (at 300 atm)	NaCl[b]	11.5 (at 590 atm); 11.82 (at 1051 atm and 650°C)
ZnS	204		
BeO	120		
$PbSO_4$	110	SnO_2	3
Fe_2O_3	90	Al_2O_3	1.8
$BaSO_4$	40	UO_2	0.2
Ta_2O_5	30		

[a] From the reviews of Morey (1957) and Glemser and Wendlandt (1963), unless indicated otherwise.

[b] Sourirajan and Kennedy (1962).

It is apparent that this solubility phenomenon can enhance the presence of a metal oxide or salt in the vapor phase by many orders of magnitude over that expected by normal unreactive vaporization. In the case of NaCl and H_2O this interaction could, for example, limit the effectiveness of vaporization methods for desalination. In fact, the original transport of NaCl from subterranean magma could well have occurred by a steam–NaCl vapor transport process (Sourirajan and Kennedy, 1962). The solubility of metal oxides in steam may have as yet unrecognized consequences, particularly in connection with the possible inability to maintain a protective oxide surface for metals in the combined presence of steam and other corrosive agents. Silica-containing refractory ceramics are known to be degraded by high pressure steam, and this was a problem in the development of the commercial naphtha-reforming process (Huggett and Piper, 1966).

A few attempts have been made to rationalize the observed solubility data in terms of gas-phase thermodynamics involving formation of discrete molecular complexes, such as metal hydroxides or hydrates, e.g., see Glemser and Wendlandt (1963). In view of our expanded but still inadequate knowledge regarding the thermodynamic properties of hydroxide and hydrated species, it is desirable to reconsider these early ideas. In general terms, and as a first-order approximation, it is thermodynamically reasonable to expect solids having a low-energy requirement for vaporization to exhibit the greatest solubility. Thus the high heat of vaporization for Al_2O_3 (369 kcal mol^{-1}), as compared with SiO_2 ($\sim$145 kcal mol^{-1}), would be consistent with the observed relative solubilities given in Table 2.4. Similarly, BeO, ZnS, GeO_2, and KCl each have relatively low vaporization energies and high solubilities. However, the heat of vaporization for UO_2 ($\sim$157 kcal mol^{-1}) does not differ appreciably from that of SiO_2, and the low UO_2 solubility therefore seems anomalous. It is therefore apparent that a more detailed thermodynamic analysis of each solubility system is necessary. A few specific cases are considered in later sections.

Solid solubility data are usually given in terms of the weight fraction L of dissolved solid relative to H_2O. The pressure of H_2O is conveniently represented by its density D. The exact pressure–density relationship is given by the expression

$$P(\text{atm}) = \text{density}(\text{gm cm}^{-3}) \times 82.06 \times T(\text{K})/18$$

for the case where the molecular weight is given by that of monomeric H_2O. One can determine, to a good approximation, the number of moles of H_2O associated with a hypothetical complex vapor species from the slope of $\log L$ versus $\log D$ curves. This is an analogous procedure to that used for low-pressure H_2O–solid reactions, as discussed in Section VIII.C.

A more rigorous determination of the H_2O content of the product species can be made if the molar volume of the solid is taken into account. The nonideal gas nature of H_2O is implicitly taken into account by the use of experimental density data. In this case it can be shown, e.g., see Glemser and Wendlandt (1963), that for a process involving a solid, S:

$$x\mathrm{S} + m(\mathrm{H_2O}) = \mathrm{S}_x\cdot(\mathrm{H_2O})_m$$

$$\log X = (m - 1) \log \rho(\mathrm{H_2O}) + (\log K_c + XV_sP/4.574T),$$

where X is the mole fraction of the complex product, $\rho(H_2O)$ is the density of H_2O vapor, V_s is the molar volume of solid, P is the total pressure, and the equilibrium constant K_c is given by

$$K_c = X/\rho(\mathrm{H_2O})^{m-1}a_s{}^x,$$

where a_s is the activity of solid S. In practice, the V_s term has only a small effect on the slopes of the log L–log D solubility versus density curves and the slopes have values close to $(m - 1)$. For large values of m, i.e., $\gtrsim 5$, the uncertainty in the solubility data usually does not allow one to distinguish between slopes of m and $(m - 1)$.

2. *Nonideal Gas Properties of H_2O*

For practical purposes, hot gases at pressures of 1 atm and lower obey the ideal gas law, i.e.,

$$PV = nRT.$$

Essentially there is no strong chemical interaction between the molecular species which in the case of steam are H_2O molecules. Fugacity and pressure are then equivalent under these ideal gas conditions. At pressures in excess of 1 atm or so, a significant departure from the ideal interdependence of P, V, and T is found experimentally, and this is usually expressed in a series form such as

$$PV = RT + B(T)P + C(T)P^2 + \cdots,$$

where $B(T)$, $C(T)$, etc., are temperature-dependent coefficients known as the second, third, etc. virial coefficients. These coefficients represent the degree of departure from ideality. They are also related to the fugacity f or escaping tendency. For instance, where $C(T)$, etc., are small relative to $B(T)$ we have

$$f/P = 1 + B(T)P/RT.$$

In considering chemical reactions for "real" gases, it is convenient to utilize standard state thermodynamic properties which are based on ideal gas behavior and then, where necessary, correct for the nonequivalence of fugacity with pressure, i.e.:

$$F_{\text{real}} - F_{\text{stand}} = RT \ln f.$$

The ratio f/P can be estimated from expressions such as the Berthelot equation (e.g., see Pitzer and Brewer, 1961, p. 187), provided the critical pressure and temperature are known. For the case of H_2O at 1300 K and 10,000 atm, the fugacity correction to standard state free energy amounts to only about 10%.† This correction becomes less at lower pressures and at higher temperatures. Hence, for purposes of the discussion to follow, standard state thermodynamic data may be used as a *satisfactory approximation* (± 2 kcal mol^{-1} in ΔF) for the description of high pressure steam reactions.

a. Molecular Composition of Steam

In discussing the chemistry of steam–solid solubility processes, it is more informative to consider explicitly the molecular character of steam rather than bypass its probable complexity by using gas density as an experimental parameter. Both the virial coefficients and recent experimental measurements of the molecular structure of steam may be used for this purpose.

We may consider the real gas to be an equilibrium mixture of readily interconvertible single, double, triple, etc. molecules (Hirschfelder *et al.*, 1942). The second virial coefficient can be expressed in terms of contributions both from stable bound double molecules and interacting, but separate, molecules (Stogryn and Hirschfelder, 1959). It has been shown for gases such as H_2O, CO_2, CH_4, and O_2, that stable bound double molecules provide the main contribution to the nonideality of these gases. For the case of H_2O, Hirschfelder *et al.* (1942) have determined from the second virial coefficient that, at 1 atm and 298 K, water vapor contains 0.045 mole fraction of double molecules, i.e., $(H_2O)_2$. Also $(H_2O)_2$ was calculated to be about 3 kcal mol^{-1} more stable in its enthalpy of formation than the separated monomers. It follows from these data that the standard free energy change associated with dimer formation is

$$\Delta F_{298} = 1.6 \text{ kcal mol}^{-1}.$$

This indirect determination of the composition of water vapor is supported by more recent theoretical and experimental work.

From ab initio molecular orbital calculations on $(H_2O)_2$ Morokuma and Pederson (1968) determined an upper-limit stability, relative to the

† A table of f/P data for steam may be found in the chapter of Schwerdtfeger and Turkdogan (1970).

monomers, of 12.6 kcal mol^{-1}. A more recent, and presumably more accurate, calculation by Morokuma and Winick (1970) gives a dimer stability of 6.55 kcal mol^{-1}.

Using molecular beam sampling of steam, with mass spectrometric identification, Greene *et al.* (1969) have observed directly the species $(H_2O)_2$, $(H_2O)_3$, and $(H_2O)_4$ present in equilibrium with H_2O as the major component. From a second law thermodynamic treatment of their data, an energy for dimerization of about -6.5 kcal mol^{-1} was determined. An extrapolation of their data, taken at 373 K, to a temperature of 298 K and a total pressure of 1 atm indicates a dimer content of about 0.0174 mole fraction. Considering the uncertainties in both techniques, this is considered to be in satisfactory agreement with the value of 0.045 mole fraction derived from the second virial coefficient.

Consider the entropy change resulting from dimer formation. The second law data of Greene *et al.* (1969) suggest a value of about -30 cal deg^{-1} mol^{-1}, whereas their third law value, based on a structure similar to that determined by the ab initio calculations and estimated vibration frequencies is -31 ± 2 cal deg^{-1} mol^{-1}. These values are consistent with the average entropy change of -30 ± 4 cal deg^{-1} mol^{-1} found for a large number of dimerization, or similar, reactions (see Chapter 3, Section IV). There is a tendency, however, for association reactions with low negative ΔH values, i.e., less than about -10 kcal mol^{-1}, to have ΔS values at the high end of this range. The most probable values of ΔS and ΔH for water dimerization are considered to be

$$\Delta S_{\text{dim}} = -32 \pm 2 \text{ cal deg}^{-1}\text{ mol}^{-1}$$

and

$$\Delta H_{\text{dim}} = -7.4 \pm 2 \text{ kcal mol}^{-1}.$$

These values are consistent with the data of Greene *et al.* (1969) and, as will be shown subsequently, with the molar entropy properties of water vapor.

For the formation of trimer $(H_2O)_3$ from monomer, the data of Greene *et al.* (1969) indicate

$$\Delta F_{373\text{ trimer}} = 5.4 \text{ kcal mol}^{-1}.$$

We estimate

$$\Delta S_{373\text{ trimer}} = -56 \text{ cal deg}^{-1}\text{ mol}^{-1},$$

from analogy with similar reactions, and hence,

$$\Delta H_{373\text{ trimer}} = -15.4 \text{ kcal mol}^{-1}.$$

This value for the enthalpy of trimer formation indicates a binding energy per H_2O molecule similar to that found for the dimer, which is not at all unreasonable.

From these measured and estimated data it can be shown for temperatures of about 600 °C that, at a pressure of 1 atm, water vapor is primarily monomeric H_2O, at pressures of the order of 100 atm the dimer is relatively more significant than the monomer, and at 1000 atm the trimer is relatively more significant than the dimer.

Independent support for this model of an increased formation of molecular aggregates with increased water vapor pressure is provided by the known molar entropy values of water vapor at various pressures (Pistorius and Sharp, 1960). These data have been plotted as a function of pressure for fixed temperatures of 600 and 500 °C, as shown in Fig. 2.8. The decreasing molar entropy with increasing pressure provides indirect evidence of the nonideality of high pressure water vapor. As was indicated above, this nonideality can be considered in terms of an equilibrium mixture of associated species. Also, standard state thermodynamic functions can be used for the description of this equilibrium, at least to a reasonable approximation. Therefore, the molar entropy may be expressed in terms of a summation of entropy contributions from the individual molecular components, i.e.,

$$S_{\text{molar}} = \sum_i (1/n) X_i S_i$$

where n is the number of molecules of H_2O in species i, i.e., 2 for dimer, 3 for trimer, 4 for tetramer, etc., S_i is the entropy of species i, and X_i is the mole fraction of species i.

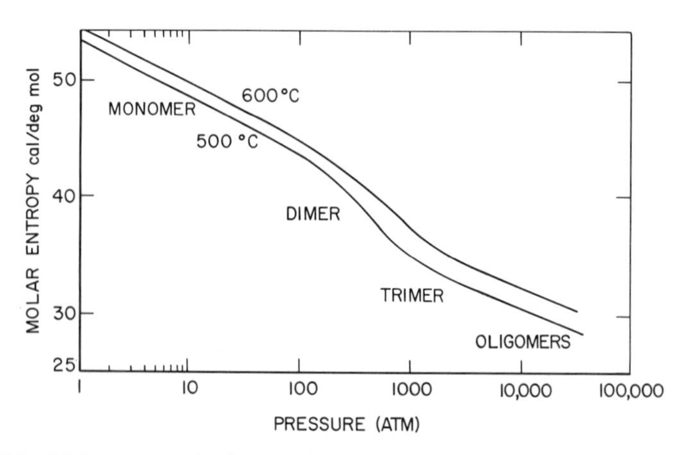

Fig. 2.8 Molar entropy isotherms of water vapor as a function of pressure; data of Pistorius and Sharp (1960).

From earlier discussion, it has already been established that H_2O is the predominant species at 1 atm pressure and ~600 °C. This is further supported by the observation that $S_{molar} \simeq S_1$ at 1 atm. However, at 100 atm and 600 °C, for example, $S_{molar} = 44.87$ cal deg^{-1} mol^{-1} whereas $S_1 = 54.3$ cal deg^{-1} mol^{-1}; this rapid decrease in molar entropy with increasing pressure is attributed to molecular aggregation. In fact, at pressures of around 100,000 atm, the molar entropy of steam does not differ appreciably from that expected of the liquid phase. Given any reasonable estimate of the entropies of the individual polymeric species, i.e., $(H_2O)_2$, $(H_2O)_3$, and $(H_2O)_4$, it is possible to show, from the molar entropy data, that the contribution of monomer to high pressure water vapor must be negligible at pressures of the order of several hundred atm. It can also be shown that the dimer $(H_2O)_2$ is a major component of the vapor at pressures of several hundred atm, and that $(H_2O)_3$ and $(H_2O)_4$ become the predominant species at higher pressures, as indicated in Fig. 2.8.

These indirect arguments concerning the molecular character of steam are further supported by a summary report on spectroscopic evidence for the polymeric nature of water vapor at elevated temperatures, i.e., ~350 °C, and pressures (High Temperature Institute, 1971). At pressures of ~28 atm, and lower, monomers are indicated as the predominant species, at ~85 atm dimers are present to the extent of about 15–20% together with monomers, and at ~284 atm the dimer content is about 80 ± 5%. Using this composition at 284 atm, the molar entropy is calculated to be 41.4 ± 1 cal deg^{-1} mol^{-1}, and this agrees very well with the experimental result of 42 cal deg^{-1} mol^{-1} (i.e., see Fig. 2.8). This spectroscopic study indicated an open structure for the dimer in agreement with the theoretical result.

One can conclude from this discussion that solid solubility chemistry at steam pressures in the region of 1000 atm necessarily involves reactions of molecular aggregates such as $(H_2O)_2$, $(H_2O)_3$, or $(H_2O)_4$ and not the standard state form of molecular H_2O.

3. *SiO_2–H_2O Reactions*

The solubility of quartz in steam has been frequently studied and can be as high as 61.8 wt % at 9,500 atm and about 1050 °C (Kennedy *et al.*, 1962; Semenova and Tsiklis, 1970).

It is of interest to attempt to provide a thermodynamic rationale for the observed solubility properties of quartz as this gives information about the vapor species formed. From the data on the dependence of solubility on pressure, as reviewed by Brady (1953) and by Glemser and Wendlandt (1963), the following reactions have been suggested, by the latter authors, for a temperature of 1000 K:

$$SiO_2(s) + 2H_2O = Si(OH)_4, \qquad \text{density} \leq 0.05 \text{ gm cm}^{-3} \text{ (i.e., } \lesssim 228 \text{ atm);†} \tag{a}$$

$$2SiO_2(s) + 3H_2O = Si_2O(OH)_6, \qquad \text{density up to} \sim 0.45 \text{ gm cm}^{-3} \text{ (i.e., } \sim 2000 \text{ atm);} \tag{b}$$

$$SiO_2(s) + H_2O = SiO(OH)_2, \qquad \text{density} \geq 0.65 \text{ gm cm}^{-3} \text{ (i.e., } \gtrsim 3000 \text{ atm).} \tag{c}$$

From the discussion of the previous section, concerning the molecular constitution of water vapor, it follows that for the upper density range considered for the reaction (a), the major water species is $(H_2O)_2$. For the conditions of reaction (b) the species $(H_2O)_3$ is probably predominant, and for reaction (c) it is likely that $(H_2O)_3$ and $(H_2O)_4$ are the major species. Hence the above reactions, as written, most likely do not represent the actual molecular processes.

Reaction (a). For the lowest density regime, i.e., that for reaction (a), the isothermal pressure dependence of $SiO_2(s)$ solubility in H_2O suggests that two H_2O molecules are present in the vapor complex. Since the partial pressure of $(H_2O)_2$ varies as the square of the monomer pressure, then it follows that, provided the monomer is still present as a major species, the observed isothermal solubility dependence on pressure is also consistent with $(H_2O)_2$ as a reactant, in addition—or alternative—to $2(H_2O)$.

Consider first the case of H_2O monomer as the solvating species in reaction (a). The temperature-dependent solubility data indicate a reaction enthalpy of $\Delta H = 9.5$ kcal mol^{-1}, for temperatures of 200–600 °C and pressures of up to 2000 atm. As $\Delta H_T = 144$ kcal mol^{-1} for the reaction

$$SiO_2(s) = SiO_2,$$

then the reaction enthalpy for the homogeneous gas-phase reaction,

$$SiO_2 + 2H_2O = Si(OH)_4,$$

is $\Delta H_T = -134.5$ kcal mol^{-1}. It follows from this enthalpy change and the known bond energies in SiO_2 and H_2O that the average bond energy, $\bar{D}[\text{Si–OH}]$, in $Si(OH)_4$, equals 117 ± 5 kcal mol^{-1}.

As an alternative test of the reasonableness of reaction (a), but modified by taking $(H_2O)_2$ as the reactant species, consider the following thermodynamic arguments. Using the solubility data listed by Morey (1957), an *apparent* partial pressure for SiO_2 in H_2O at 333 atm and 600 °C is

$$p(SiO_2) = 0.12 \text{ atm}.$$

† The conversion of density to pressure is very approximate since the gas molecular weight is not accurately known.

This corresponds to a free energy for reaction (a) of

$$\Delta F_{873} = 15.0 \text{ kcal mol}^{-1},$$

assuming $(H_2O)_2$ to be the predominant species. One can calculate the entropy for $Si(OH)_4$ as 108 ± 4 cal deg^{-1} mol^{-1}, by using the statistical method and estimated molecular constants. Hence, from the known entropy of SiO_2(s) (JANAF, 1971) and the previously estimated entropy value of about 76 cal deg^{-1} mol^{-1} for $(H_2O)_2$, (see Section E.2), it follows that the entropy change for the modified reaction (a) is

$$\Delta S_{873} = 7 \pm 5 \text{ cal deg}^{-1} \text{ mol}^{-1}.$$

Therefore, from these ΔS and ΔF values, the enthalpy change is

$$\Delta H_{873} \sim 21 \text{ kcal mol}^{-1}.$$

This may be combined with the known heats of formation for SiO_2(s) and $(H_2O)_2$ [see Section E.2] to indicate that the average bond energy, $\bar{D}$[Si–OH], in $Si(OH)_4$ is about 115 ± 6 kcal mol^{-1}. The pseudohalogen model for estimating bond energies, as discussed elsewhere (Chapter 5, Section II.K), indicates a value of 104 ± 5 kcal mol^{-1} which is in satisfactory agreement with the two results derived from the solubility data. Thus the model proposed for the solubility of quartz, at pressures of the order of several hundred atmospheres, is thermodynamically reasonable.

Reactions (b) and (c). Consider the higher pressure conditions as covered by the proposed reactions (b) and (c). The solubility of SiO_2 varies with steam pressure, at $T \sim 1050$–1100 °C, according to the curve given in Fig. 2.9 (Kennedy *et al.*, 1962). It is seen that at a pressure of about 2000 atm, the curve slope is approximately unity. From the molar entropy, i.e., see Fig. 2.8, and related discussion given earlier, the major water species for these conditions is most probably $(H_2O)_3$. Hence, reaction (b) could be rewritten as

$$y\text{SiO}_2(\text{s}) + (\text{H}_2\text{O})_3 = (\text{SiO}_2)_y \cdot (\text{H}_2\text{O})_3, \tag{b'}$$

and for $y = 2$ the most likely reaction product would be $Si_2O(OH)_6$—an analog of disiloxane. At higher pressures the slope of the curve contained in Fig. 2.9 increases, and one can speculate as to the formation of species such as $\{SiO(OH)_2\}_n$, which would be analogs of known cyclic oxohalides.

4. *NaCl–H_2O Reactions*

The solubility of NaCl in high pressure high temperature steam can enhance the "vaporization" of NaCl by more than four orders of magnitude. For example, at 200 atm H_2O pressure and 725 °C the steam con-

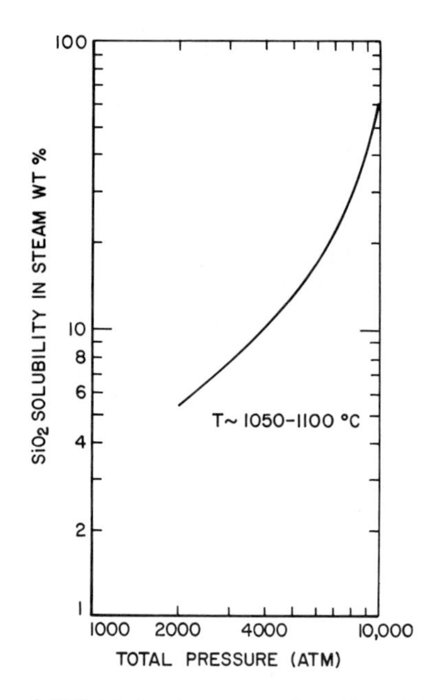

Fig. 2.9 Solubility of SiO_2(s) in steam as a function of total pressure at 1050–1100 °C; data of Kennedy *et al.* (1962).

tains about 3% NaCl as compared with about 10^{-4}% provided by non-reactive free molecular vaporization.

Sourirajan and Kennedy (1962) have determined the solubility of NaCl in steam over a considerable range of temperature (i.e., 250–750 °C) and H_2O pressure (i.e., ~75–1240 atm). From these data it is possible to suggest reasonable chemical reactions to account for the observed degree of solubility and its dependence on temperature and pressure.

For the pressure interval of about 20–400 atm, the isothermal dependence of NaCl solubility, or apparent partial pressure, on steam pressure indicates that for each mole of NaCl present in the vapor, two moles of H_2O are involved in the "solvation process," i.e.,

$$NaCl(s) + 2H_2O = NaCl \cdot 2H_2O. \qquad (1)$$

From earlier discussion on the molecular composition of steam it was indicated that for this pressure interval both monomer and dimer are major species. Hence, one could also consider the solvation process as

$$NaCl(s) + (H_2O)_2 = NaCl \cdot (H_2O)_2. \qquad (2)$$

Since $(H_2O)_2$ varies as the square of the H_2O pressure, then this process would also be in accord with the observed dependence of solubility on pressure. Alternatively, one could consider the hydrolysis reaction

$$NaCl(s) + 2H_2O\ [or\ (H_2O)_2] = NaOH \cdot H_2O + HCl. \tag{3}$$

However, for relatively low temperature conditions, i.e., <500 °C, the neutral pH of the solution indicates the absence of a hydroxide-forming process such as reaction (3), e.g., see Elliott (1952).

The variation of NaCl solubility, or apparent partial pressure, with temperature at a constant pressure of 100 atm is shown in Fig. 2.10. If a single reaction were present one would expect a linear dependence of log pNaCl on T^{-1} and the slope of the curve would indicate the heat for the solvation reaction. From the curve of Fig. 2.10, it is apparent that at least two processes are present, one over the temperature interval of ~400–600 °C, and a second over the temperature interval of ~600–750 °C. From the slope of the curves in these regions, approximate reaction heats of +5 kcal mol^{-1} and +80 kcal mol^{-1} are indicated, respectively. Either of the processes (1) and (2) would be consistent with a reaction enthalpy of +5 kcal mol^{-1} if the binding energy of $2H_2O$, or $(H_2O)_2$, to an NaCl molecule is about −50 kcal mol^{-1}. As will be shown by later discussion, this is a reasonable binding energy for a "hydrated" NaCl molecule.

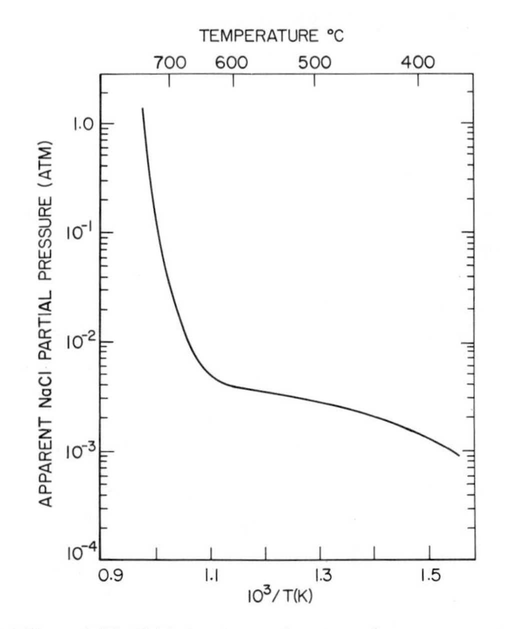

Fig. 2.10 Solubility of NaCl(s) in steam (expressed as apparent partial pressure) as a function of temperature and at a total pressure of 100 atm; data of Sourirajan and Kennedy (1962).

Additional evidence for reactions (1) or (2) as the solvation processes, at 400–600 °C, is provided by a comparison of measured and expected reaction entropy. If reaction (1) was considered as a homogeneous gas-phase process, i.e., NaCl(g) rather than NaCl(s), then from the reaction entropy generalizations discussed in Chapter 3, Section IV, one would expect $\Delta S_1 \simeq -55$ cal deg^{-1} mol^{-1}. Taking into account the known entropy of sublimation for NaCl, i.e., $\sim+32$ cal deg^{-1} mol^{-1}, the entropy change of the heterogeneous reaction (1), as written, would then be

$$\Delta S_1 \simeq -23 \text{ cal deg}^{-1} \text{ mol}^{-1},$$

for the temperature interval of interest. From the solubility data, the apparent partial pressure of NaCl is $\sim 4 \times 10^{-3}$ atm at 100 atm steam pressure and 873 K temperature. This leads to $\Delta F_1 = 25.5$ kcal mol^{-1} and, as $\Delta H_1 \simeq +5$ kcal mol^{-1}, it follows that $\Delta S_1 = -23.5$ cal deg^{-1} mol^{-1} which is basically in agreement with the predicted value. Hence, the observed pressure dependence and temperature dependence for the interval $\sim$400–600 °C, together with the actual magnitude of the NaCl solubility, are each consistent with reactions (1) and (2) as the solvating chemical processes.

Consider the higher temperature interval of $\sim$600–750 °C where a reaction enthalpy of about +80 kcal mol^{-1} is indicated. This temperature interval also leads to a marked increase in the extent of NaCl solubility, i.e., see Fig. 2.10. It is not possible to write "hydration" reactions, such as reaction (1) or (2), having such a highly endothermic reaction enthalpy. The only reactions that can conceivably account for a reaction enthalpy of about +80 kcal mol^{-1} involve the hydrolysis of NaCl(s) to form NaOH and HCl. At temperatures approaching the melting point of NaCl(s) this type of hydrolysis is well known. The reaction

$$\text{NaCl(s)} + \text{H}_2\text{O} = \text{NaOH} + \text{HCl}$$

has an enthalpy change of about 84 kcal mol^{-1} which is close to the observed value. However this reaction does not account for the observed square dependence of NaCl solubility with H_2O pressure. The reaction

$$\text{NaCl(s)} + (\text{H}_2\text{O})_2 = \text{NaOH}\cdot\text{H}_2\text{O} + \text{HCl} \tag{4}$$

would be more consistent with the observed pressure dependence and would have a reaction enthalpy of +80 kcal mol^{-1} if the binding energy of NaOH and H_2O in the species $NaOH \cdot H_2O$ was -11 kcal mol^{-1}. This is not an unreasonable value since $H_2O \cdot H_2O$ is about -7 kcal mol^{-1} and, as was suggested earlier, $NaCl \cdot H_2O$ is probably about -25 kcal mol^{-1}. Also, the solubility data for NaOH in steam suggest that the species $NaOH \cdot H_2O$ may exist, e.g., see Elliott (1952) and cited literature.

It appears, then, that the enhanced vaporization of NaCl in steam results from hydration at relatively low temperatures and hydrolysis at elevated temperatures.

5. *KCl–H_2O Reactions*

It is possible to make an analysis of KCl solubility data similar to that for NaCl. Morey (1957) has summarized the solubility data for KCl in steam. At a pressure of 250 atm and a temperature interval of 400–500 °C the solubility reaction is exothermic by about -15 kcal mol^{-1}. Also, the solubility decreases from about 800 to about 130 ppm over this increasing temperature interval. The only reaction type that is consistent with these observations is one involving hydration, and from entropy and enthalpy arguments analogous to those considered for the NaCl–H_2O case the most consistent reaction is

$$KCl(s) + 3(H_2O)_n = KCl \cdot 3(H_2O)_n,$$

where n is probably 2 or 3. This reaction is also consistent with the pressure dependence of the observed solubility where approximately $7H_2O$ per KCl are indicated.

6. *MCl·H_2O Binding Energies*

The above interpretation of salt solubility, in terms of hydrated species formation, has provided some indication of the stability of these species. For the NaCl–H_2O case, the species $NaCl \cdot 2H_2O$ or $NaCl \cdot (H_2O)_2$ have hydration binding energies of about -50 kcal mol^{-1}, and for the KCl–H_2O case the postulated species $KCl \cdot 3(H_2O)_n$ has a binding energy of about -66 kcal mol^{-1}.

The only similar vapor species for which hydration binding energies are known are ions of the type $M^+ \cdot (H_2O)_n$, where M is an alkali metal, e.g., see Dzidic and Kebarle (1970), and $Cl^-(H_2O)_n$ (Arshadi *et al.*, 1970). For the reaction

$$M^+ + H_2O = M^+ \cdot H_2O,$$

ΔH has values of -18 kcal mol^{-1} for K, -23 kcal mol^{-1} for Na, and -35 kcal mol^{-1} for Li, i.e., the hydration energy increases with increasing charge density of M^+. For the reaction

$$M^+ \cdot (H_2O)_5 + H_2O = M^+ \cdot (H_2O)_6,$$

ΔH has values ranging from -10 to -15 kcal mol^{-1} for various M. Similarly, for the reaction

$$Cl^- + H_2O = Cl^- \cdot H_2O,$$

$\Delta H = -13$ kcal mol^{-1}, and for the reaction

$$Cl^- \cdot (H_2O)_5 + H_2O = Cl^- \cdot (H_2O)_6$$

$\Delta H = -4$ kcal mol^{-1}.

Now it is well known that species such as NaCl and KCl are comprised of M^+Cl^- and, as the binding energy of $Na^+ \cdot H_2O$ is -23 kcal mol^{-1} and of $Cl^- \cdot H_2O$ is -13 kcal mol^{-1}, one can reasonably expect $Na^+Cl^- \cdot H_2O$ to exhibit a binding energy of at least -23 kcal mol^{-1} and possibly as high as -36 kcal mol^{-1}. These estimates compare favorably with the -50 kcal mol^{-1} found for $NaCl \cdot 2H_2O$ or $NaCl \cdot (H_2O)_2$, i.e., about -25 kcal mol^{-1} for each H_2O molecule associated with NaCl. Similarly, the $KCl \cdot 3(H_2O)_n$ binding energy of about -66 kcal mol^{-1} is in accord with the hydration energies for K^+ and Cl^-.

7. *Metal Oxide–H_2O Reactions*

Metal oxides such as BeO and Al_2O_3 are considered to be refractory materials, their vapor pressures being insignificantly low at temperatures of less than about 2000 K. However, in the presence of high pressure steam these and other metal oxides show an appreciable vapor-phase solubility as indicated in Table 2.4.

The interaction of water vapor with BeO to form $Be(OH)_2$ is well known, and from the available thermodynamic data for $Be(OH)_2$ it is possible to calculate the solubility of BeO due to formation of this hydroxide. This calculated solubility is many orders of magnitude less than that observed at 500 °C. The only reaction that satisfies the observed solubility data is a hydration process

$$BeO(s) + (H_2O)_n = BeO \cdot (H_2O)_n,$$

where n is probably 3 or 4. However, for $Al_2O_3(s)$ the observed solubility at 500 °C could be attributed to the hydroxide-forming process,

$$Al_2O_3(s) + (H_2O)_3 = 2Al(OH)_3,$$

in addition to a hydrate-forming process. This suggestion is based on an estimated heat of formation for $Al(OH)_3$ using the Al–OH bond energy data for the species Al–OH and Al–$(OH)_2$, e.g., see Chapter 5, Section II.K. The possible existence of the species $Al(OH)_3$ has been suggested previously from transport experiments on $Al_2O_3(s)$ in low pressure water vapor (Von Wartenberg, 1951).

The stability of hydroxides of molybdenum is well known and the solubility of MoO_3 in steam has been analyzed in terms of hydroxide-forming processes, e.g., see Glemser and Wendlandt (1963). Species as complex as $Mo_7O_{18}(OH)_6$ have been suggested from analogies with liquid solution behavior.

From these examples of solid solubility phenomena in steam, it is apparent that both hydrate and hydroxide species may be involved, depending to a large extent on the temperature. An uncertainty in the species identity limits the extrapolation of solubility data to temperature and pressure conditions that may be of practical significance. A better understanding of the general high pressure steam–solid solubility phenomenon clearly requires the identification and thermodynamic characterization of the inferred species.

IX. Miscellaneous Examples of Practical Interest

A. Molten Iron Oxidation and Steelmaking

Modern steelmaking processes, which utilize selective oxidation of impurities, involve movement of molten iron through an oxidizing environment, as for example in spray steelmaking and pouring of deoxidized streams (Ward, 1962). Under these conditions of relatively short gas–solid interaction times, i.e., ~1 sec or less, kinetic rather than thermodynamic factors are likely to be important, e.g., see Baker (1967), Baker and Ward (1967), and cited references. Hence, a number of kinetic studies have been made on molten iron oxidation. For example, Robertson and Jenkins (1970) have studied the levitation of molten Fe in O_2 with millisecond time scale photographic observations of oxidation, slag formation, and fuming processes.

The results of recent work by Vig and Lu (1971), on the oxidation rates of free falling iron droplets through oxygen atmospheres similar to that of air, are consistent with a forced convection diffusion-limited rate model. The rate of transfer of O_2 to the surface of the falling droplet is considered as the rate-limiting step, and the chemical reaction rate is believed to be relatively fast. At relatively high melt temperatures, e.g., 1755 °C, an explosion of the droplet occurs during the free fall. From earlier work, e.g., see Baker and Ward (1967) and cited references, it is believed that the explosion results from a rapid generation of a CO bubble within the molten droplet, followed by mechanical disintegration of the drop as the CO moves through the liquid surface.

At lower O_2 partial pressures ($\sim 10^{-5}$–10^{-2} atm) than those used by Vig and Lu (1971), the vaporization rate of Fe, at 1750 °C, is sufficiently high that gas-phase oxidation with subsequent production of fume occurs, e.g., see Distin and Whiteway (1970). The overall chemical processes are believed to be

$$Fe + \tfrac{1}{2}O_2 = FeO(c),$$

followed by further reaction of FeO(c) to yield Fe_3O_4 and Fe_2O_3 condensed fume. The production of this fume leads to an enhanced apparent vaporization rate for Fe, as discussed elsewhere (Chapter 3, Section IV.F). However, the observed rates are still less than one-tenth of the maximum, i.e., Langmuir, free vaporization rates.

The mechanism of decarburization of thin iron foil, and also molten steel, is considered to be

$$CO_2 \rightleftarrows CO + O(ad),$$

$$C\ (in\ Fe) + O(ad) \rightleftarrows CO,$$

with the former step being rate controlling, e.g., see Wagner (1970a). Gas diffusion may become rate controlling for thicker samples.

The decarburization of iron by hydrogen, i.e.,

$$C\ (in\ Fe) + 2H_2 = CH_4$$

is believed to be controlled by the following mechanism:

$$H_2(g) = 2H(ad),$$

$$C\ (in\ Fe) + H(ad) = CH(ad),$$

$$CH(ad) + H(ad) = CH_2(ad),$$

$$CH_2(ad) + H(ad) = CH_3(ad),$$

$$CH_3(ad) + H(ad) = CH_4(g),$$

with the second to last step being rate controlling, e.g., see the review of Wagner (1970a). However, the equilibrium concentration of CH_4 is low and, in practice, the overall rate may be determined by the transport of CH_4 away from the surface.

B. Welding Chemistry

Arc-welding is an example of a practical system where gas–metal reactions occur, and the chemical processes may be rationalized in terms of thermodynamic equilibria, i.e., see Pollard and Milner (1971). For example, the ability of additions of Si and Cr to raise the carbon level of the steel melt in CO_2 arc welding may be thermodynamically explained by the competition of Si and Cr for oxygen which would otherwise lead to CO production. The production of CO can be deleterious to welds, because of bubble formation in the melt and subsequent porosity in the solid weld. A similar problem arises in connection with the oxyacetylene torch welding of tough-pitch copper (i.e., contains oxygen) where a too-reducing flame results in water vapor bubble formation according to the process

$$Cu_2O(c) + H_2 \text{ (from flame)} = 2Cu(c) + H_2O.$$

A practically oriented account of various welding processes may be found in Houldcraft (1967) and also Kubaschewski *et al.* (1970, p. 63).

Appendix. Some Additional Recent Literature on High Temperature Gas–Solid Reactions

Banerjee, N. N., Murty, G. S., Rao, H. S., and Lahiri, A. (1973). *Fuel* **52,** 168. Flash pyrolysis of coal: effect of nitrogen, argon and other atmospheres in increasing olefin concentration and its significance on the mechanism of coal pyrolysis.

Bittel, J. T., Sjodahl, L. H., and White, J. F. (1969). *J. Amer. Ceram. Soc.* **52,** 446. Steam oxidation kinetics and oxygen diffusion in UO_2 at high temperatures.

Feininger, T. (1969). *Amer. J. Sci.* **267,** 1011. Solubility of albite in supercritical water in the range 400 to 600° and 750 to 3500 bars. Comment.

Grabke, H. J. (1972). *Carbon* **10,** 587. Oxygen transfer and carbon gasification in the reaction of different carbons with carbon dioxide.

Hennecke, J. F. A., and Scherff, H. L. (1971). *J. Nucl. Mater.* **38,** 285. Carbon monoxide equilibrium pressures and phase relations during the carbothermic reduction of uranium dioxide.

Iguchi, Y., Ban-Ya, S., and Fuwa, T. (1969). *Trans. Iron Steel Inst. Japan* **9,** 189. Solubility of water in liquid $CaO–SiO_2$ with Al_2O_3, TiO_2, and FeO at 1550°.

Kieffer, R., Nowotny, H., Ettmayer, P., and Freudhofmeier, M. (1970). *Monatsh. Chem.* **101,** 65. On the stability of transition element carbides in N_2 up to pressures of 300 atm.

Leitnaker, J. M., Lindemer, T. B., and Fitzpatrick, C. M. (1970). *J. Amer. Ceram. Soc.* **53,** 479. Reaction of UC with nitrogen from 1475° to 1700°C.

Lintz, H. G., Pentenero, A., LeGoff, P. (1969). *J. Chim. Phys.* **66,** 692. Study of gas–solid interactions by means of a nozzle-beam.

Loginov, M. V., and Mittsev, M. A. (1971). *Zh. Tekh. Fiz.* **41,** 709. Thermal dissociation of $SrCl_2$ molecule on a tungsten surface.

Long, L. C., Jr., and Joiner, J. R., Jr. (1967). Refractories in a Reactive Atmosphere. U. S. Govt. R and D Rep. No. AD 651 435.

Merrill, L. S., Jr. (1973). *Fuel* **52,** 61. The coal–oxygen reaction near the ignition temperature.

Nemkin, V. M., and Morev, I. I. (1970). *Zh. Fiz. Khim.* **44,** 208. Thermodynamic analysis in reduction of semi-dispersed iron-ore concentrates at temperatures above 1400°C.

Nishida, S., and Berkowitz, N. (1973). *Fuel* **52,** 262. Reactions of coal with discharge-generated (excited) nitrogen species.

Nishida, S., and Berkowitz, N. (1973). *Fuel* **52,** 267. Reactions of coal with nitrogen/hydrogen mixtures in a discharge.

Ollis, D. G., Lintz, H. G., Pentenero, A., and Cassuto, A. (1971). *Surface Sci.* **26,** 21.

Study of gas–solid chemical interactions by noble gas molecular beams. IV. Adsorption states and oxidation of molybdenum and rhenium.

Orcutt, R. H. (1970). *J. Res. Nat. Bur. Std.* **A74,** 45. Generation of controlled low pressures of nitrogen by means of dissociation equilibriums.

Rao, Y. K., and Jalan, B. P. (1972). *Met. Trans.* **3,** 2465. Rates of carbon–carbon dioxide reaction in the temperature range 839 to 1050°.

Secco, E. A., and Yeo, R. S. C. (1971). *Can. J. Chem.* **49,** 1953. Gas solid exchange reactions; zinc vapor and monocrystalline zinc telluride.

Sell, N. J., Marx, W. F., and Lester, J. E. (1972). *High Temp. Sci.* **4,** 222. The effect of a manganous chloride beam on the vaporization rate of a KCl crystal.

Weigel, F., and Wishnevsky, V. (1970). *Chem. Ber.* **103,** 193. Vapor-phase hydrolysis of lanthanide(III) chlorides. III. Heat and Gibbs free energy of the reaction $MCl_3(s) + H_2O(g) = MOCl(s) + 2HCl(g)$ (M = Er, Tm).

3

Chemical Vapor Transport and Deposition

I. Introduction

A. Scope of Application

The transport and deposition of inorganic materials via a high temperature vapor phase can be accomplished either by a vaporization-condensation process, or by a reversible gas–solid reaction whereby the solid of interest is, in effect, solvated by the reactive transport gas. Solid deposits can also be formed by the coreaction of gases or vapors at a heated surface—such a process is known as chemical vapor deposition. An introductory discussion and some applications of these various types of material transport is given in this chapter.

Transport processes involving high temperature vapors have practical utility in the preparation of a wide variety of materials. These include single crystals and thin films for electronic and optical components, refractory coatings for rocket engines and nuclear reactor components, metals—as extracted from their ores—and endothermic or metastable high temperature reaction products. A knowledge of such transport processes also finds application in theories of ore genesis and in the development of metal halide lamps.

One can define a variety of subject areas that fall in the category of chemical vapor transport and deposition. These are listed in Table 3.1 where some interrelationships between the various areas and high temperature species are given, together with a general indication of practical applications.

TABLE 3.1

Relationships and Scope of Application for the Various Subject Areas of Chemical Vapor Transport and Deposition

Subject area	Relationships	Scope of application
II. Chemical and physical vapor deposition processes	High temperature species control composition and production rate	Crystals, coatings
III. Chemical transport along a temperature gradient	A special case of II	Crystals, metal purification, tungsten-lamp regeneration
IV. Enhanced volatility processes and complex vapors	Broadens scope of application of II, III, VII, and VIII	Crystals, metal extraction
V. Chemical synthesis with high temperature species	A special case of II and III	Synthesis of unusual compounds
VI. Plasma chemistry	Special case of II and V, has application to VII, IX	Chemical synthesis, metal extraction, preparation of ceramics
VII. Vapor phase metallurgy	Incorporates II, III, IV, VI	Pyrometallurgical processing, metal purification
VIII. Ore genesis	Incorporates II, III, VII	Basis for mineral exploration and synthesis
IX. Metal halide lamps	Incorporates some elements of II, III, VI	Improved light sources, better electrical efficiency

B. Process Controlling Factors

It is useful at this point to indicate briefly the major controlling factors in vapor transport processes. The primary process parameters are mass and energy transport. Detailed mathematical accounts of the theory of mass and energy transfer, as they relate to molecular processes, may be found in the books of Luikov and Mikhailov (1965) and Eckert and Drake (1972), in addition to the specific references cited throughout this chapter (see also Chapter 1, Section V).

The management of materials and heat, which are interrelated through thermodynamics, is of basic importance to industrial chemical processing since the recycling of materials and heat or energy is often economically advantageous. This may prove to be of even greater significance in the future, owing to the predicted shortages and higher cost of both minerals and energy. It appears likely that, given an increased need for materials recycling and low-grade ore processing, the development of new processes, and perhaps some involving vapor transport at elevated temperatures, will be required.

Plant design is usually centered around the performance of a chemical reactor which may involve both heat and materials input and output. A discussion of the basic principles underlying the choice of reactor type for a chemical process may be found in Meissner (1971). The necessity for management of heat energy usually arises from the fact that chemical reactions, particularly those involving high temperature vapors, are often carried out at higher temperatures than the surroundings. The initial reactants and final products are also managed at lower temperatures than the reactor. Heat must therefore be supplied to, or removed from, the reactor, depending on the endo- or exothermic nature of the reaction, respectively. The efficiency of heat exchangers in chemical processing is therefore of economic significance. Also, the location and temperature of the exchanger must be such that no chemical reactions, such as condensation or vaporization, occur within the exchanger. From these few considerations it is apparent that a knowledge of the thermodynamic properties of the reactants and products is an essential part of process and reactor design.

In addition to a thermodynamic system characterization, it is usually necessary to consider the rate-controlling factors of mass transport and chemical kinetics. For nonequilibrium processes, such as those which utilize flow or stirred reactors or plasma quenching, the degree of chemical conversion is related to the residence time of reactants within the reactor. Frequently, however, the course and degree of a chemical reaction can be described in terms of local chemical equilibrium, but where the overall process may involve transport of material or energy along concentration and temperature gradients. Thus the evaluation of transport fluxes under various conditions is of practical concern.

A recent account of transport phenomena involving gas mixtures and condensed phases has been given by Schwerdtfeger and Turkdogan (1970). These authors have also described experimental techniques for handling gas mixtures and measuring their equilibrium and transport properties. The theory and application of transport processes has also been discussed by Oxley (1966)—see also Chapter 1 Section V. Penner (1955) has elaborated on the kinetics of chemical reactions under flow conditions. In particular, the case of chemical reactions occurring during flow through a De Laval nozzle, i.e., an adiabatic expansion process, is considered in connection with rocket motor operation.

II. Chemical and Physical Vapor Deposition Processes

A. Introduction

Vapor deposition processes involving physical or chemical techniques form the basis of many modern preparative methods for inorganic materials, particularly as used by the electronics industry. Physical vapor deposition usually involves vaporization of the material of interest and its condensation without chemical transformation; chemical vapor deposition (CVD) differs in that the material is transported in a vapor or gas state to a substrate where chemical reaction occurs to form a new condensed phase material, such as a crystal or thin film. This chemical transformation may generally be achieved by processes of the following type†:

- disproportionation of high temperature species, e.g.:

$$2GeI_2 \xrightarrow{\text{cool}} Ge(s) + GeI_4;$$

- reduction, e.g.:

$$WF_6 + 3H_2 \xrightarrow{\text{heat}} W(s) + 6HF;$$

- pyrolysis, e.g.:

$$GeH_4 \xrightarrow{\text{heat}} Ge(s) + 2H_2.$$

† In addition, processes involving chemical transport along a temperature gradient are also sometimes considered in the broad context of CVD, but a detailed discussion of this process type is deferred to Section III.

For at least the latter two process types, the deposition substrate must be kept at a relatively high temperature. With the first process type a high temperature vapor phase is usually required to generate the reduced valence species necessary to disproportionation reactions. High temperatures may also be involved in physical vapor deposition, both in generating the vapor and in the heat evolved by the exothermic condensation process. High temperature species are therefore key intermediates in the overall process of material transformation, and it is this aspect of vapor deposition processes that is to be given emphasis in the discussion to follow.

A comprehensive and more general account of vapor deposition processes may be found in the book of Powell *et al.* (1966) or the review of Haskell and Byrne (1972). Holzl (1968) has elaborated on the general techniques for chemical vapor deposition. Also, some specific aspects of vapor deposition have received separate discussion elsewhere in this monograph. In particular, the reader should refer to the discussions of chemical transport along a temperature gradient (Section III), of gas–solid reactions (Chapter 2), and of enhanced volatility processes (Section IV), to gain a more fundamental insight into CVD processes. Additional examples of application for vapor deposition are provided in the discussions of vapor-phase metallurgy (Section VII), ore genesis (Section VIII), and metal halide lamps (Section IX).

B. Scope of Application

1. General Applications

Vapor deposition, in general, competes with electrodeposition as a commercial molecular-forming process. The former process is more versatile in that it is not restricted to the production of metallic coatings; that is, it can also produce refractory materials such as oxides, sulfides, carbides, nitrides, borides, silicides, and fluorides. Thin films of metastable materials can also be produced by vapor deposition. Another advantage is that vapor deposition rates are usually greater than for electrodeposition. Contrary to condensed phase multicomponent systems, the vapor phase has no miscibility restrictions and, in principle, chemically complex uniform deposits can be obtained by vapor deposition.

For the particular case of chemical vapor deposition, which is also sometimes known as reactive vapor deposition, there are advantages over other coating techniques in the ability to coat complex geometrical shapes and to produce high density, high strength, and high purity materials. As vapor deposition processes are atomic processes where the substrate is built up atom by atom, or conceivably molecule by molecule, this provides for a high density material. A possible limitation with exothermic CVD processes results from finite heat transfer rates from the surface through the substrate. This may necessitate a relatively slow deposition rate. However, for many applications the "throwing power" of CVD is relatively high.

A technique for the preparation of alloys, ceramics, and mixed metal–ceramic materials with high deposition rates ($\sim 10^{-3}$ gm min^{-1}) has been described by Bunshah and Douglass (1971a, b). The technique involves electron beam heating of metals and ceramics contained in water-cooled copper crucibles. High power electron beams ($\sim$10 kV) provide high evaporation rates. The vapors are usually trapped on heated metal surfaces, since by controlling the surface temperature one can control grain size. A major advantage of this technique is the high purity capability. Also, multiple vaporization sources may be used for the production of alloys. Reactive evaporation between metal vapor and oxygen may be used to produce oxide layers such as Y_2O_3. Since the grain size—and hence mechanical properties—of the condensate is critically dependent on the collector temperature, the authors have calculated the deposit temperature which is the result of an energy balance in the substrate–condensate system. The energy terms considered involve mainly radiation, condensation, and reaction enthalpy (for reactive evaporation), and good agreement between calculated and measured temperature was found.

a. CVD—Examples.

Some typical examples of CVD reactions are given in Table 3.2. Additional examples of chemical transport-type reactions may be found in Section III. Practical processes are not always confined to these particular reaction types. For example, polymer films can be prepared by polymerization of siloxanes, such as hexamethyldisiloxane, in the presence of a glow discharge (Tkachuk *et al.*, 1966). Electron or ion bombardment, and electromagnetic radiation in the visible, x-ray, or γ-ray regions, can also be used for polymerization. Methods of generating polymer films are discussed in the book edited by Hass and Thun (1966).

In its most common form, a CVD process involves vapor phase transport of reactants to the surface of interest under conditions whereby very little gas phase reaction occurs except at, or in the vicinity of, the surface. The surface reaction may have an activation energy, and additional energy can be provided to the reaction either by surface heating or by irradiation with light, x-rays, an electron beam, or by the production of an rf or dc plasma. Bunshah and Douglass (1971b) emphasize the importance of carrying out this activation in the reaction zone itself, rather than externally, as is often done The prospect of activating chemical reactions by

TABLE 3.2

Examples of CVD Reactions

Reaction	Temperature (°C)
Disproportionation	
$Si(s) + 2I_2 \rightarrow SiI_4$	1000
$Si(s) + SiI_4 \rightarrow 2SiI_2$	1000
$2SiI_2 \rightarrow Si(s) + SiI_4$	900
$Si(s) + SiO_2(s) \rightarrow 2SiO$	1300
$2SiO + O_2 \rightarrow 2SiO_2(s)$	1200
Reduction	
$SiHCl_3 + H_2 \rightarrow Si(s) + 3HCl$	1100
$SiCl_4 + C_6H_6 \rightarrow SiC(s) + \cdots$	1700
$SiCl_4 + H_2 + C_7H_8 \rightarrow SiC(s) + \cdots$	1250
$BCl_3 + NH_3 \rightarrow BN(s) + 3HCl$	1300–1900
Chemical transport	
$MgCl_2 + 2FeCl_2 + 3H_2O + \frac{1}{2}O_2 \rightleftarrows MgFe_2O_4(s) + 6HCl$	800–1200
$Ge(s) + H_2O \rightleftarrows GeO + H_2$	
$4GaAs(s) + 2GaX_3 \rightleftarrows 6GaX + As_4$	
$4GaCl_3 + As_4 + 6H_2 \rightleftarrows 4GaAs(s) + 12HCl$	~670
Pyrolysis	
$Ni(CO)_4 \rightarrow Ni(s) + 4CO$	200–300
$SiH_4 \rightarrow Si(s) + 2H_2$	800–1300
$Si(OC_2H_5)_4 \rightarrow SiO_2(s) + 4C_2H_4 + 2H_2O$	750
$3SiH_4 + 4NH_3 \rightarrow Si_3N_4(s) + 12H_2$	900

other than thermal means is important as it would broaden the scope of application to less refractory deposits. As an example, the epitaxial deposition of Si can be achieved at relatively low temperatures in the presence of uv radiation, according to the process

$$Si_2Cl_6 + 3H_2 \rightarrow 2Si(s) + 6HCl, \qquad 700\ ^\circ C$$

(Frieser, 1968). The recent availability of lasers for activation of specific reaction paths would seem to be an interesting prospect for future applications in activated CVD processes.

b. Metal Coatings.

Many of the modern metal coating techniques involve vapor phase processes of the following types:

- pack cementation and the allied halogen streaming process;
- chemical vapor deposition;
- distillation—i.e., physical vapor deposition;

- plasma, arc, or flame spraying;
- fluidized bed vapor deposition.

The first two process types often involve metal halide intermediates. Some of the practical aspects of these coating processes have been discussed by Wachtell and Jefferys (1968).†

CVD-produced metal coatings are frequently formed by the hydrogen reduction of volatile metal halides, or by high temperature decomposition of the halides (particularly iodides) or carbonyls, as suggested by the examples given in Table 3.2. Deposition rates of the order of 10^{-3} gm min^{-1}, pressures in the range of 10^{-3}–1 atm and temperatures between 100–2000 °C are common operating conditions.

Aluminum coatings can be achieved by H_2 reduction of $AlCl_3$ but a higher purity product can be obtained by the disproportionation of AlCl, i.e.,

$$3AlCl \rightarrow AlCl_3 + 2Al(l).$$

At 1 atm and 1200 °C, the reaction is almost completely to the left, but at 1000 °C about 50% disproportionation occurs. A much lower temperature production of Al(s) can be achieved by the pyrolysis of tri-isobutylaluminum at 250 °C and about 5 Torr pressure.

Chromium plating can be achieved by high temperature decomposition of the iodides or chlorides. For example, the chromizing process of Ni-base alloys is believed to occur according to the reactions

$$CrX_2 + H_2 = 2HX + Cr \text{ (deposit)}$$

and

$$CrX_2 = X_2 + Cr \text{ (deposit)},$$

where X is a halogen (see Samuel, 1958).

Copper deposition has been observed when solid cuprous oxide, contaminated with chloride, is heated in a stream of CO, and it is thought that a hypothetical volatile species $CuCl \cdot CO$ is the transport agent.

The formation of $MoSi_2$ coatings by a pack cementation process is considered to involve vapor deposition reactions such as

$$2NH_4Cl(s) \rightarrow N_2 + 4H_2 + Cl_2,$$

$$Si(s) + 2Cl_2 \rightarrow SiCl_4,$$

$$Si(s) + Cl_2 \rightarrow SiCl_2,$$

$$Mo(s) + 2SiCl_4 \rightarrow MoSi_2(s) + 4Cl_2,$$

$$Mo(s) + 2SiCl_2 \rightarrow MoSi_2(s) + 2Cl_2,$$

† The National Materials Advisory Board has also reported on the status and problems for coatings on superalloys, refractory metals, and graphite (NMAB-263, PB 193400; 1970).

though this scheme is largely speculation and reactions involving HCl also seem likely (see Leeds, 1968). It is noteworthy that the presence of volatile impurities such as H_2O can markedly affect pack cementation processes.

c. *Refractory Products*

Titania or silica, in the form of condensed particles, may be made by the gas-phase hydrolysis or oxidation of the tetrachlorides. Sometimes additives, such as $AlCl_3$ or Al_2O_3, are used as nucleating agents. Control of particle size is particularly important for pigment applications, and the conditions of nucleation and particle formation are therefore of special interest (Mezey, 1966).

The processing of nuclear fuel elements, in the form of carbides, for high temperature gas-cooled reactors represents a recent application of CVD to refractory material preparation.

Chemical vapor deposition processes also have an important role in the preparation of fibers to be used in lightweight composites. For example, the process

$$2BCl_3 + 3H_2 \rightarrow 2B(s) + 6HCl, \qquad 1070\ °C$$

allows boron to be coated onto a heated W or C wire. CVD has also recently been used to prepare fibers of refractory materials, such as BN, TiN, NbN, B_4C, Mo_2C, and NbC. Except for BN, where B_2O_3 was reacted with NH_3, the reactants consisted of volatile metal chlorides and H_2 as a reductant.

Titanium carbide has been prepared by the reaction

$$TiCl_4 + CCl_4 + 4H_2 \rightarrow TiC(s) + 8HCl, \qquad 1200\ °C$$

(Nickl and Reichle, 1971), and the growth of TiC crystals has been achieved by processes of the type

$$TiCl_4 + C_xH_y + H_2 \rightarrow TiC(s) + HCl + \cdots, \qquad 1200\text{–}1350\ °C$$

(Takahashi *et al.*, 1970).

Epitaxial growth of Al_2O_3 has been achieved at 1400–1600 °C using the reaction

$$2AlCl_3 + 3H_2O = Al_2O_3(s) + 6HCl$$

(Messier and Wong, 1971).

2. *Crystal Growth*

A relatively recent review of the theoretical and practical aspects of crystal growth from the vapor phase has been given by Ellis (1968). Both physical and chemical vapor deposition processes are used for crystal growth and the latter technique is particularly useful for epitaxial growth applications.

The main advantages of crystal growth from the vapor phase are greater chemical purity, fewer crystal defects, and more diversity in shape—such as for thin film and epitaxial layer production. In the case of MgF_2, for example, it is possible to grow single crystals by physical vapor deposition which have far fewer lattice defects than crystals grown from melts (Recker and Leckebusch, 1969).

In order to grow large single crystals from the vapor phase, it is necessary to control the rate of sublimation and the temperature gradient. Diffusion may also be a controlling factor. For the case of a cylindrical sample container and 1 atm total pressure, the transport of material along the tube is diffusion controlled, and Wirtz and Siebert (1971) have outlined the theory of this transport and applied it to the growth of Tl_2O_3 single crystals. Further discussion of diffusion controlled transport is given elsewhere (see Section III.C).

Passing mention should also be made of the use of flames as a vapor transport medium for crystal growth, e.g., see Khambatta *et al.* (1972) and Chapter 5, Section II. A.5.

3. *Thin Films*

The primary applications of thin films are in the construction of integrated electronic components, optical surfaces, and magnetic information-storage devices. During the past decade, the commercial production of thin films by vapor deposition techniques has become competitive with more conventional plating methods. One of the advantages over other plating techniques is that greater film thicknesses can be produced; relatively high purity is also possible (Maissel and Glang, 1970). The use of an electron beam as a heat source, rather than the use of resistance or induction heating, has broadened the scope of vapor deposition and reduced container-interaction problems. Laser radiation has also found use as a heat source.

It is now possible to plate expensive refractory metals on less expensive mild- or stainless-steel substrates for the corrosion protection of jet engine components. Also, it is possible to coat ceramics with metal, or vice versa. Silicon dioxide coatings may be formed by a silicon monoxide vaporization-oxidation process or by direct electron beam heating of SiO_2, which is a higher temperature process. Such coatings may be formed on plastics, metals, and ceramics. An enhanced adherence of the coating to the substrate can sometimes be achieved by applying a negative voltage bias to the substrate and allowing the vapor to pass through a glow discharge of opposite sign; thus the gas is accelerated, with much greater than thermal energy, at the substrate.

In-line coaters allow deposition of multiple layers of different materials without loss of vacuum. Such coatings find application for camera lenses, aluminum-coated mirrors, laser reflectors, heat reflection coatings for architectural glass, and titanium–palladium–gold coatings for communications equipment. As an example of the application of multiple simultaneous vaporization sources, epitaxial magnetic garnet films, such as $Gd_3Fe_5O_{12}$, have been produced by separate vaporization of $FeCl_2$ and $GdCl_3$ followed by reaction with $H_2O + O_2$ vapor at the substrate, e.g., see Kay (1971).

Using a pulsed ruby laser as a heat source, optical quality films of the materials Sb_2S_3, As_2S_3, Se, ZnTe, Te, MoO_3, PbTe, and Ge have been obtained (Smith and Turner, 1965). However, it is difficult to control the stoichiometry of the deposit for materials such as Cu_2O, ZnO, and InSb, which decompose on heating. Pulsed lasers exhibit a tendency to eject condensed particles and this can be effectively reduced by the use of a continuous CO_2 gas laser (25W). Films of SiO, ZnS, ZnSe, PbF_2, Na_3AlF_6, SiO_2, MgF_2, Si_3N_4, $LaAlO_3$, TiO_2, and Al_2O_3 have been produced with such a laser (Groh, 1968).

A comprehensive account of the various evaporation techniques used for the deposition of thin films has been given by Behrndt (1968). More basic aspects of thin films have been discussed by Chopra (1969) and reviewed by Kay (1971). The various processes used to produce the electronically important GaAs films have been discussed by Von Munch (1971). The chemical vapor deposition technique has been considered in some detail by Feist *et al.* (1969)—see also Powell *et al.*, (1966).

C. Process Fundamentals

Physical vapor deposition is a relatively straightforward process, and the substrate composition and rate of deposition can often be predicted from basic thermodynamic data. Also, the nucleation and growth of a physically deposited thin film, for example, may be described by a statistical path probability method (Braunstein and Kikuchi, 1969). Chemical vapor deposition, on the other hand, is at the present time mainly an empirical science with thermodynamic guidelines. The main controlling factors in CVD are the reaction free energies, surface reaction rates, rate of gas diffusion, rate of heat transfer through the substrate, and homogeneous nucleation effects. The relative roles played by these factors can vary according to the conditions of temperature, pressure, and composition.

At elevated temperatures, particularly those greater than 1000 °C, the reduction of surface reaction activation energy barriers by thermal excitation is often sufficient to allow an equilibrium thermodynamic description of vapor deposition to be made. Reduced temperature nonequilibrium conditions may, however, sometimes be necessary—as, for example, with the deposition of Ni metal from $Ni(CO)_4$ where the oxide is the preferred equilibrium phase.

1. CVD Mechanisms—Homogeneous Reaction

Although very little is known about CVD mechanisms at the molecular level, the view is widely held that the reactions are heterogeneous, involving adsorption of gas molecules at the surface prior to reaction. It is considered likely that the surface acts as a catalyst for the molecular interactions. A discussion of such heterogeneous reactions is given elsewhere in connection with gas–solid reactions (Chapter 2, Section III).

Much of the evidence for mechanisms is speculative and there are those who argue that many CVD processes could be homogeneous, i.e., gas phase, in nature. Gretz (1966) states that

> vapor phase nucleation may be much more probable than has been generally considered to be the case.

In favor of this argument is the observation of particle formation above the substrate at high pressures—which necessarily must involve molecular intermediates in the gas phase. Also, recent mass spectrometric experiments have shown that, at a sufficiently low partial pressure of the reactants, NH_3 and HCl, molecular NH_4Cl appears in the vapor phase prior to condensation and in the absence of a surface (Hastie and Kaldor, 1972). This presence in the vapor of polyatomic species can influence the condensation process. Such is the case for condensation of metals from their vapors, and an extensive review of this phenomenon has been given by Gen and Petrov (1969).

A mathematical model of homogeneous CVD processes has been described by Sladek (1971). The model considers that a gas-phase reaction occurs with the generation of product molecules which can either diffuse to the substrate where they condense to form a film, or they may undergo self-collisions with the production of dimers, which very rapidly leads to the formation of particles, i.e., powder. With some simplifying assumptions, it can be shown that the rate of such particle production can be expressed in terms of the relation

$$k_{\mathrm{prod}}\, e^{-E/RT} = -D \frac{d^2C}{dx^2} + k_{\mathrm{dim}}\, C^2,$$

where

k_{prod} is the experimentally determined preexponential term analogous to a collision factor in gas-phase kinetics,
E the overall process activation energy,
x the distance from the substrate,
C the concentration of the gas-phase monomer product,
D the diffusion coefficient of the monomer, and
k_{dim} the rate constant for dimerization.

The dimerization process is assumed to be the rate-limiting step for the production of particles above the substrate. Note that the dimerization contribution is assumed to be a second-order process.

From the above relation it can be shown that the importance of powder formation in the deposition process can be represented by the dimensionless quantity

$$M \equiv k'\left(\frac{k_{\text{dim}}}{D^2}\right)\left(\frac{E}{RT_0}\right)^4\left(\frac{T_0'}{T_0}\right)^4,$$

where,

$k' = k_{\text{prod}}\, e^{-E/RT_0}$,
T_0 is the temperature at $x = 0$, i.e., at the gas–substrate interface, and
$T_0' = (dT/dx)_{x=0}$ is the temperature gradient at the surface.

An approximate solution of this expression indicates that M is of the order of unity at the critical condition where a transition from film to particle formation occurs. At low temperatures, where M is small and hence powder formation is negligible, the film growth rate should increase exponentially with T_0, but above a critical temperature where powder forms, an exponential decrease with increasing T_0 would result.

This model leads to calculated growth rates that agree with experiment to about an order of magnitude accuracy for solid Si_3N_4 deposition. In this instance the model assumes the formation of a hypothetical Si_3N_4 vapor species.

For more detail on nucleation theory reference should be made to Hirth and Pound (1963), Hirth (1972), and Kay (1971).

2. *Equilibrium Processes*

The mechanisms cited for CVD and other vapor transport processes are most frequently the result of thermodynamic reasoning with assumptions made regarding the existence of equilibrium and the mode of mass transport. From thermodynamic first principles, a prediction of the experimental conditions of composition, temperature, and pressure required for an optimum growth rate, can be made provided the species identities, heat of formation, and free energy function data are known, e.g., see Bleicher (1972) or Silvestri (1972). The possible significance of equilibrium processes involving reduced valence species is indicated by the following examples.

Reactions involving the production of a lower valence vapor species such as

$$SiCl_4 + Si(s) = 2SiCl_2$$

are usually endothermic, due to the high endothermicity associated with the introduction of an extra metal atom into the vapor phase. Hence, such reactions become more favorable with increasing temperature. Conversely, the reverse disproportionation reaction is favored at lower temperatures. Thus, from these thermodynamic considerations, there is an optimum upper limit temperature for the deposition of the metal in systems where reduced valence species are the key intermediates.

The formation of reduced valence vapor species during a thermal-reduction vapor deposition process can have an adverse effect on the rate and efficiency of deposition. For example, consider the deposition of Nb metal by the thermal decomposition of $NbCl_5$ vapor, i.e., by the process

$$NbCl_5 \rightarrow Nb(s) + 5Cl.$$

The pressure conditions required for an optimum isothermal deposition of Nb can be influenced by the formation of stable lower valent niobium halides, i.e.,

$$2NbCl_5 + 3Nb(s) \rightarrow 5NbCl_2$$

or

$$2NbCl_5 + \tfrac{4}{3}Nb(s) \rightarrow \tfrac{10}{3}NbCl_3.$$

For a particular pressure regime, these reactions will outweigh the Nb producing reaction and a loss of Nb-deposit will result.†

It is clear, then, that for practical CVD systems the deposition process may be controlled by more than a single reaction. Furthermore, the fact that both endothermic and exothermic reactions may be operating simultaneously can lead to a nonmonotonic variation of deposition rate with temperature. A thermodynamic analysis of such a system requires the solution of a set of simultaneous equations incorporating equilibrium constant data and species materials conservation restraints—that is, a multicomponent equilibria calculation of the conditions for an overall minimum free energy (see Chapter 1, Section V. A). With this approach,

† Figure 3.1 demonstrates this type of effect, e.g., at pressures of ≤ 113 Torr and for $1180 > T > 1440$ K.

thermodynamics can be used to predict the optimum gas solubility conditions of a CVD process. Usually, the partial pressures of the vapor species are determined as a function of overall stoichiometry and temperature. The solubility of a particular component is then obtained from a summation of the appropriate species partial pressures.

As an example of the utility of this type of calculation one may consider the pyrolytic deposition of carbon from CH_4 (Lydtin, 1970). In this system the pyrolysis yields intermediate species such as C_2H, C_2H_2, H, C_3, CH_2, and CH_3. Their predicted relative importance varies considerably with temperature, as a consequence of which the gas solubility of carbon is at a minimum at about 1400 K for a total pressure of 10^{-2} atm. In other words, a maximum solubility occurs at both higher and lower temperatures than the minimum. Hence, the thermodynamic calculations predict that carbon deposition can result by gas transport from a cold to hot or, alternatively, a hot to cold surface (see also Section III). This prediction has been verified experimentally by Lydtin (1970).

A similar thermodynamic analysis has been made concerning CVD conditions in the Si–H–Cl vapor deposition system (Hunt and Sirtl, 1970). Using the minimization of a free energy computational procedure, the composition profiles for the major species $SiCl_4$, $SiHCl_3$, HCl, SiH_2Cl_2, $SiCl_2$, and SiH_3Cl were calculated as a function of temperature. From these calculations it was possible to deduce the solubility of Si in the vapor phase and the weight of Si deposited as a function of temperature and initial reactant pressure. For the case where $SiCl_4$ is the sole initial reactant, the predicted deposition conditions are represented by the curves of Fig. 3.1.

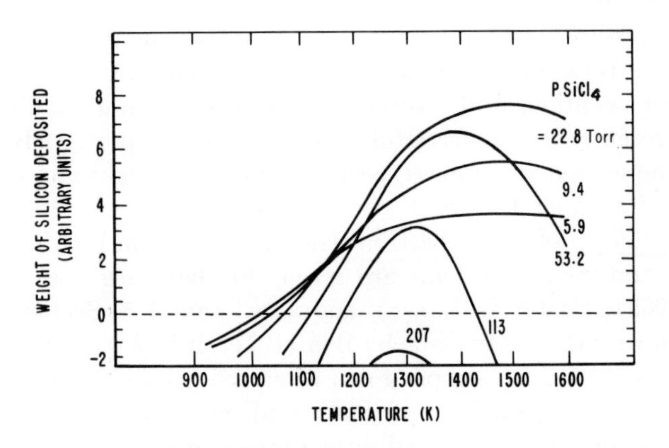

Fig. 3.1 Calculated weight of Si deposited from $SiCl_4$ as a function of temperature at the various initial pressures indicated; note that conditions of Si loss from the deposit occur below the broken horizontal line (Hunt and Sirtl, 1970).

For additional examples, where thermodynamic calculations have been applied to the predictive analysis of CVD systems, see: Reisman and Alyanakyan (1964), Ryabchenko *et al.* (1969), Bougnot *et al.* (1971), Sandulova and Ostrovskii (1971), Hunt and Sirtl (1972), Minagawa and Seki (1972), and Seki and Minagawa (1972). As is generally the case, the results are very dependent on the accuracy of the thermodynamic data and the choice of significant species.

3. *Nonequilibrium Processes*

Observations of apparent activation energies, which differ from reaction enthalpies, and other related kinetic phenomena, indicate that equilibrium thermodynamics provides only a crude description of many chemical transport processes. In practice, temperature and pressure regimes are observed where either mass transport—or chemical reaction—kinetics are rate determining. For instance, the transport of Ge (from a $\langle 111 \rangle$ crystal surface) due to the reaction

$$Ge(s) + 2HI = GeI_2 + H_2$$

is limited by surface reaction kinetics at temperatures below about 650 °C, and for an HI pressure of 3.8 Torr (Reisman and Berkenblit, 1966). At higher temperatures, or higher HI pressures, the surface reaction is sufficiently fast that gas transport kinetics become rate limiting. In this example it is thought that the rate-limiting kinetic step is the surface desorption of GeI_2. Another recent example of a surface-related kinetic limitation occurs in connection with the deposition of $GaAs_{1-x}P_x$ from the $PH_3 + AsH_3 + HCl$ system (Belouet, 1972).

For a precise description of CVD processes under nonequilibrium conditions, it is essential that the vapor-phase composition be known under the various conditions of deposition. This is experimentally a difficult task and only recently has a successful experiment been reported (Ban, 1971). The technique used was a mass spectrometric line-of-sight molecular beam sampling of a vapor transport system at 1 atm pressure. Details of the general technique of mass spectrometric sampling from high pressure and high temperature gas systems are given elsewhere, e.g., see Milne and Green (1969), Hastie (1973b, c), and also Chapter 5, Section II. F. In connection with the study made by Ban (1971), it is pertinent to note that a capillary was used for sampling and beam collimation purposes and that under the resulting gradual pressure gradient some shift in gas-phase composition could have occurred during sample extraction. However, it is unlikely that the species identity and general conclusions of the analysis are affected.

From a mass spectrometric monitoring of the concentration dependence of product on reactant species, the following reactions were established by Ban, for the formation of either GaAs or GaP solids using Ga (l), HCl, AsH_3 or PH_3, and excess H_2 as reactants

$$\mathrm{Ga(l) + HCl = GaCl + \tfrac{1}{2}H_2},$$

$$\mathrm{GaCl + \tfrac{1}{4}As_4\ [or\ P_4] + \tfrac{1}{2}H_2 = GaAs(s)\ [or\ GaP] + HCl},$$

and

$$\mathrm{GaCl + \tfrac{1}{2}As_2\ [or\ P_2] + \tfrac{1}{2}H_2 = GaAs(s)\ [or\ GaP] + HCl}.$$

This is a significantly different result from the GaAs (s)-forming reaction most frequently suggested, but from less-direct evidence, namely,

$$\mathrm{3GaCl + \tfrac{1}{2}As_4 = 2GaAs(s) + GaCl_3}.$$

No $GaCl_3$ was found in the mass spectrometric experiments.

The relative reactivity of the various pnictide species with GaCl was also determined, for mixed $AsH_3 + PH_3$ systems, to be

$$\mathrm{PAs_3 > P_3As \sim As_4 > P_2As_2 > As_2 \sim AsH_3 \sim P_4 > P_2 > PH_3}.$$

This order has no relation to what one would predict from thermodynamic data and is indicative of the importance of reaction kinetics to these deposition processes. The reaction of HCl with Ga (l) was also less complete than one would expect from the thermodynamics and the extent of departure of the actual, versus a predicted, degree of reaction decreased with increasing temperature over the range 400–800 °C and was minimal at 800 °C. Similarly, for the flow system of

$$\mathrm{GaCl + NH_3 \rightarrow GaN(s) + HCl + H_2},$$

only about 70 % of the equilibrium condition was achieved, with respect to HCl, owing to the limited residence time of HCl in the reaction zone (Ban, 1972a). It is interesting to note that the product GaN (s) is actually metastable under the conditions used.

Studies of this type have recently been extended and summarized by Ban (1972b). Also, the application of a similar mass spectrometric technique to the study of Au (s)–Cl_2 interactions is pertinent here, as in this case the ability to reproduce known thermodynamic equilibrium data was convincingly demonstrated (Hager and Hill, 1970). In this instance the sampling was also carried out through a heated molecular leak and involved a sampling pressure reduction of from $\sim$1 to 10^{-5} atm across the leak. The flow condition on the low pressure side of the leak was assumed to be isothermal hydrodynamic flow rather than the more usual adiabatic free jet expansion case.

4. Transport Factors†

Chemical vapor deposition reactions yielding some of the products in the gas phase may require rapid removal of the gaseous constituents to prevent back reaction—as is the case for the equilibrium process:

$$2Ti(s) + C_2H_2 \rightleftarrows 2TiC(s) + H_2.$$

Even for the case where the forward reaction is highly favorable, the rate of deposition is often limited by gas flow processes such as diffusion or convection. Lorel (1970) has derived theoretical expressions relating deposition rates with experimental parameters for conditions where diffusion and convection may be rate controlling. Diffusion and convection calculations are also described by Carlton and Oxley (1967).

As a particular example, Fujii *et al.* (1972) have modeled the epitaxial growth of Si in the flow reaction system

$$SiCl_4 + 2H_2 = Si(s) + 4HCl$$

under conditions where vapor-phase mass transport is rate limiting. In practice, a considerable enhancement of the deposition rate can be achieved if the gas boundary layer thickness is reduced by allowing a faster gas flow or, alternatively, a mechanical motion of the substrate through the reactant gases.

Other rate-limiting factors include heat transfer through the substrate and cluster formation in the region of the interface. The frequency of cluster formation can be described by the Becker–Doring equation, e.g., see Lydtin (1970). It is considered that the formation of clusters reduces both the deposition rate and chemical efficiency, owing to the thermodiffusion of clusters along the temperature gradient to cold regions with the resultant loss of reactants from the hot reaction zone.

5. Examples of Theoretical Modeling

By application of a theory combining the elements of thermodynamics, heterogeneous reaction kinetics, diffusion and mass transfer, and adsorption, Carlton *et al.* (1970) have predicted the deposition rate of boron from the BCl_3–H_2 system over a two order-of-magnitude range of rates with an accuracy of about 23%. Details of the general computational procedure

† Additional discussion of gas transport may be found in Section III and also Chapter 1, Section V.

have been given by Carlton and Oxley (1967). The model assumes that $HBCl_2$ is a significant reaction intermediate and this is supported by thermodynamic calculations. Also, the rate-determining step in the conversion of BCl_3 to elemental boron is considered to be the surface reaction

$$HBCl_2(ad) \rightarrow BCl(ad) + HCl.$$

Note that if this were a gas-phase process, the heat for the reaction would be about +80 kcal mol^{-1}, whereas the experimental apparent activation energy is only about 29 kcal mol^{-1}. Hence, the adsorption state has considerably reduced the energy requirements for the process.

The addition of small amounts of HCl to the gas phase greatly reduces the deposition rate, and this is apparently due to a reversal of the initial gas phase reaction

$$BCl_3 + H_2 = HBCl_2 + HCl$$

which, in turn, results in a loss of $HBCl_2$ (ad). Note that the reactants and intermediate products, i.e., HCl and $HBCl_2$, are believed to be in a state of local thermodynamic equilibrium at the deposition surface.

The deposition rate is also very dependent on the surface temperature but is remarkably insensitive to the total pressure and also the relative concentration of the initial reactants H_2 and BCl_3. Experimental deposition rates are typically of the order of 0.1 mg sec^{-1} cm^{-2}.

An analysis by Gruber (1970) suggests that the deposition of B on a hot filament by the reaction

$$2BCl_3 + 3H_2 \rightarrow 2B(s) + 6HCl$$

can be described in terms of a Langmuir–Hinshelwood adsorption model (Hinshelwood, 1926), where both BCl_3 and H_2 are adsorbed at the substrate surface and the H_2 is dissociated on adsorption.

A recent discussion of the basic phenomena in chemical vapor deposition, with particular emphasis on the $H_2 + WF_6$ system for the deposition of W(s), has been given by Haskell and Byrne (1972). The process of gas diffusion to and from the substrate is amenable to theoretical treatment, using the Stefan–Maxwell multicomponent diffusion equations (Haskell and Byrne, 1972; Brecher, 1970). The deposition reaction mechanism is believed to involve an initial adsorption of H_2 and WF_6 at the surface and this step is likely to be rapid and not rate controlling. This is followed by diffusion to growth steps where the reduction of WF_6 occurs. If too large an excess of WF_6 is present, then the growth is retarded, and this is apparently due to the formation of a reduced tungsten fluoride film which poisons the surface.

In conclusion, it is apparent that the nature of the intermediate species needs to be better established for the further development of CVD modeling.

6. Factors Affecting Structure of Deposits

As summarized by Gretz (1966), the structure of vapor deposits may be influenced by the following factors:

- the degree of supersaturation of the vapor or rate of deposition,
- the substrate temperature,
- the surface environment, e.g., CVD may result in the production of undersirable intermediates which become occluded in the deposit,
- the substrate crystal orientation, and
- the deposition mechanism, which may be either heterogeneous—i.e., nucleation on the substrate; heterogeneous, but in the gas phase; or homogeneous—i.e., nucleation in the gas phase. Conditions may exist where both heterogeneous and homogeneous processes occur simultaneously in the gas phase. Sputtering may be considered as an example where vapor-phase nucleation is most likely and the resulting deposits are frequently highly disordered.

Two structural forms that may occur are (1) amorphous deposits, which result from high supersaturation and low temperature conditions and (2) epitaxy, which is favored by low supersaturation and high temperature conditions. In achieving epitaxy by CVD, a dependence on the foreign substrate can occur, e.g., gallium arsenide can form an epitaxial deposit on tungsten but not on molybdenum.

Since the general morphology of deposits depends on the deposition conditions, it follows that stresses may also occur. The presence of stresses of varying degree may result in cracking or spalling, warping, changes in chemical reactivity (e.g., corrosion resistance) and changes in electrical and optical properties. Stresses may result from thermal expansion differences, e.g., when CVD is carried out on a foreign substrate; or from occlusions of gaseous products. The following kinds of material may be present as occlusions in CVD processes: lower valent halides of either the deposition compound or the basic metal used in the reduction of metal halides; hydroxides or oxychlorides resulting from hydrolysis of the metal halides; and a variety of metal–organic derived products and metal hydrides. Some other defined stresses are diffusion stresses, conversion stresses, excess energy stresses, surface tension stresses, and lattice misfit stresses. These are discussed in detail by Powell (1966).

III. Chemical Transport along a Temperature Gradient

A. Introduction

During the past decade the importance and utility of material or chemical transport along a temperature gradient has been recognized, in particular by Schafer (1964, 1971, 1972a, b, 1973). The general principles of chemical transport have also been considered recently by Reisman (1970) and Spear (1972). Basically, the technique represents a special case of the more general area of CVD (see Section II).

An early example of chemical transport, as recognized by Bunsen (1852), was the observation of solid Fe_2O_3 transport in volcanic gases containing HCl. The transport reaction is considered as

$$Fe_2O_3(s) + 6HCl \underset{\text{cool}}{\overset{\text{hot}}{\rightleftarrows}} 2FeCl_3 + 3H_2O,$$

where the direction of transport is from a hot to a relatively cool region. In effect, the HCl transport gas is a catalyst as it is not consumed by the final product. This example is typical of many chemical transport processes in that a solid is "solubilized" by chemical reaction, usually of the gas–solid type; it is then transported in the vapor phase along a temperature gradient, and deposited under thermodynamic conditions that favor a reversal of the initial solid–solubility reaction. Only trace quantities of the transport gas are required, and some so-called solid-state processes may in fact be the result of impurity catalyzed chemical vapor transport in the presence of a temperature gradient.

The chemical transport technique is widely used in the preparation of crystals and doped solids. Several chemical transport reactions of industrial interest include

- the Mond process which is used commercially to extract pure Ni from crude metal containing other transition elements, i.e.,

$$Ni(s) + 4CO \underset{180\ °C}{\overset{70\ °C}{\rightleftarrows}} Ni(CO)_4;$$

- the reaction

$$2Al(1) + AlCl_3 \underset{1080\ °C}{\overset{1180\ °C}{\rightleftarrows}} 3AlCl$$

which has been seriously considered for Al purification, where Al(l) is present initially in a crude alloy form;

- the reaction

$$2\mathrm{GaAs(s)} + \mathrm{GaI_3} \underset{\text{cool}}{\overset{\text{hot}}{\rightleftarrows}} 3\mathrm{GaI} + \tfrac{1}{2}\mathrm{As_4}$$

which can be used to prepare GaAs(s), a compound difficult to prepare by sublimation or melt processes;

- the reaction

$$\mathrm{W(s)} + \mathrm{I_2} + \mathrm{O_2} \underset{\text{hot}}{\overset{\text{cool}}{\rightleftarrows}} \mathrm{WO_2I_2},$$

which forms the basis of control over W-filament transport in commercial lamps, e.g., see Section IX.

As chemical transport can frequently be described by thermodynamic and gas diffusion arguments, it is possible, in principle, to use transport measurements for the derivation of new thermodynamic or diffusion data. However, this is not often attempted. The observation of a chemical transport is, of itself, indicative of the formation of molecular vapor species containing components of the transported material. In a number of instances, the inference of the formation of new vapor-phase species from transport observations preceded their characterization by other more affirmative means, such as mass spectrometry. This was the case for the aluminum monohalides and the silicon dihalides. Other notable examples where the existence of a new vapor species is indicated by chemical transport, but not yet supported by other techniques, include a beryllium hydroxide species, a low valence gold iodide, a tantalum sulfide, a titanium oxychloride, and the various metal halide complexes of $AlCl_3$ (see Section IV).

Most chemical transport reactions are carried out in the total pressure regime of $10^{-3} - 3$ atm, and they therefore tend to be rate limited by gas diffusion. At higher pressures convection can occur and this greatly increases the rate of diffusion-limited transport. Basically the transport relies on a significant shift in the equilibrium composition along a temperature gradient and also on reaction reversibility. The resulting gradients in species partial pressures then allow concentration diffusion to occur.

Schafer (1964) has discussed the chemical transport of some 150 inorganic materials which primarily include metals, chalcogenides, halides, oxides, and oxyhalides. The transporting reagents used were mostly halogen-containing compounds, particularly iodine, and to a lesser extent O_2, H_2, and H_2O. Most of the studies to date have been in the temperature range of quartz reactors, i.e., $< 1200\ ^\circ\mathrm{C}$, but an extension of chemical transport techniques to temperatures in the region of 2000 °C has been made recently in connection with the preparation of EuTe, EuSe, Eu_2SiO_4, and Al_2O_3 (Kaldis, 1971).

B. Degree of Reaction

The degree of reaction, or partial pressure of the transport species, determines the practical limits of material transport on a reasonable time scale. The preparative conditions are usually such that local thermodynamic equilibrium conditions exist and hence a basic factor, in the transport process, is the magnitude of the equilibrium constant and its temperature dependence. The equilibrium constant K can be expressed in terms of a reaction enthalpy change ΔH, and an entropy change ΔS, i.e.,

$$\ln K = (-\Delta H/RT) + \Delta S/R.$$

The following general thermodynamic rules are then derived for chemical transport along a temperature gradient. Consider a general transport process

$$\mathrm{A(s)} + \mathrm{B} = \mathrm{C},$$

where A (s) is the material to be transported, B is the reactive gas, and C is the transporting species.

Rule 1. To obtain significant transport rates, the equilibrium constant should approach unity; it also follows that

$$\Delta H \simeq T\,\Delta S$$

for an effective transport reaction.

Rule 2. The direction of transport along a temperature gradient is determined by the sign of ΔH.

If ΔH is positive, i.e., an endothermic reaction, then the equilibrium constant increases with increasing temperature and the transport direction is from T_2 to T_1 where T_2 is, by convention, greater than T_1. Alternatively, if ΔH is negative, i.e., an exothermic reaction, then the transport direction is from T_1 to T_2. It should be noted that if ΔH is zero, then there is no partial pressure differential along the temperature gradient, that is between the source and the product substrate or crystal, and therefore no transport results.

Rule 3. Provided that the condition of Rule 1 can be satisfied, reactions having large values of ΔH are most effective—in other words, for those reactions where the equilibrium constant changes substantially over a relatively small temperature gradient.

A more detailed discussion of these, and similar rules, has been given by Schafer (1964).

In practice, more than one chemical reaction may be occurring in a transport system. If both exothermic and endothermic processes are occurring simultaneously, the transport process may reverse direction with certain temperature conditions. For instance, with the iodide filament process (see Section VII.C) the competing reactions

$$M(s) + nI_2 = MI_{2n}, \qquad T_1 \rightarrow T_2$$

and

$$M(s) = M, \qquad T_2 \rightarrow T_1$$

can lead to a reversal in the direction of metal deposition. Similar reversals can occur for Cu_2O transport

$$Cu_2O(s) + 2HCl = \tfrac{2}{3}Cu_3Cl_3 + H_2O,$$

$$\Delta H = -15 \text{ kcal mol}^{-1} \qquad \text{and} \qquad T = 600 \rightarrow 900\ ^\circ\text{C};$$

$$Cu_2O(s) + 2HCl = 2CuCl + H_2O$$

$$\Delta H = +68 \text{ kcal mol}^{-1} \qquad \text{and} \qquad T = 1100 \rightarrow 900\ ^\circ\text{C}$$

and for Au transport

$$Au(s) + \tfrac{3}{2}Cl_2 = \tfrac{1}{2}Au_2Cl_6,$$

$$\Delta H = -12 \text{ kcal mol}^{-1} \qquad \text{and} \qquad T = 320 \rightarrow 450\ ^\circ\text{C}$$

$$Au(s) + \tfrac{1}{2}Cl_2 = \tfrac{1}{2}Au_2Cl_2,$$

$$\Delta H = +11 \text{ kcal mol}^{-1} \qquad \text{and} \qquad T = 800 \rightarrow 500\ ^\circ\text{C}.$$

The combined effects of such competing processes result in an optimum temperature for efficient transport.

The undesirable transport of metal in tungsten-filament incandescent lamps, containing trace amounts of H_2O or O_2, can be satisfactorily accounted for using the above thermodynamic rules—see Campbell (1969) and also Section IX.

C. Rate of Transport and Diffusion†

In a static reactor the conditions for transport may be kinetic, diffusive, or convective, depending on the gas pressure and temperature. With the Ga–I_2 system, for example, Bol′shakov *et al.* (1971) found kinetic transport at less than 0.3 atm, diffusive at 0.3–1.5 atm, and convective at higher pressures. In most applications the rate-limiting step in chemical transport is the vapor-phase diffusion process. This is largely the

† A comprehensive account of transport phenomena may be found in the text of Geiger and Poirier (1973)—see also Chapter 1, Section V.

result of the use of gas pressures of less than the convective regime but greater than 10^{-2} atm where the non-free-molecular flow gas motion is usually slow as compared with the chemical reaction rate. It therefore follows that the transport rate is often directly proportional to the concentration gradient (from Ficks' law) of the transport species and inversely dependent upon the total gas pressure (from diffusivity theory—see Chapter 1, Section V).

A few exceptions are known where the chemical reaction is sufficiently slow in one direction to be rate limiting. For instance, with the transport process

$$TiN(s) + 2Cl_2 = TiCl_4 + \tfrac{1}{2}N_2,$$

the reverse reaction is believed to be the slow step, owing to the initial difficulty of breaking the strong bond in N_2—see Spear (1972) and also Schafer and Fuhr (1962). In cases where the solids have low vaporization coefficients, such as arsenic, the transport may be limited by the vaporization rate rather than by diffusion. At low pressures, where free molecular flow is the main transport mode, the heterogeneous reaction rates are likely to be rate determining.

1. Relatively Static Conditions with Boundary Layer

Under relatively static bulk gas flow conditions, one can define a boundary layer of thickness δ for the almost stagnant gas in the vicinity of the sample surface, and the vaporizing species transport flux J_i is then given by the relationship

$$J_i = \frac{D_i(P_i - P_i')}{\delta RT} \quad \text{mol cm}^{-2}\,\text{sec}^{-1},$$

where for species i,

P_i is the vapor pressure at the surface,
P_i' is the vapor pressure at distance δ from the surface, i.e., in the bulk gas; and
D_i is the species diffusivity in the total gas medium.

The boundary layer thickness can be estimated from the relation

$$\delta = 1.5 L N_{\mathrm{s}}^{-1/3} N_{\mathrm{R}}^{-1/2},$$

where

L is the sample length parallel to the direction of gas flow,
N_{s} and N_{R} are the Schmidt and Reynolds numbers, respectively, and are

given by

$$N_s = v/D \qquad \text{and} \qquad N_R = VL/v,$$

where

V is the gas flow rate, and
v the kinematic viscosity, i.e., the ratio of the coefficient of viscosity to density.

As an example of this rate-limiting effect, the transport flux of CrO_3 through 0.1 atm of O_2 at 1200 °C, and 100 cm sec^{-1} gas flow rate, is experimentally found to be only 0.0027 times that of the maximum Langmuir free molecular vaporization rate (Graham and Davis, 1971).

2. Gas-Drift Conditions

Gas flow can occur even in a closed tube owing to molar volume changes accompanying chemical reaction. Faktor and Garrett (1971) have considered the GaAs/Cl_2 transport system and applied a quantitative model for diffusion under these gas drift conditions. The model assumes that no reaction occurs in the gas phase during the diffusion process. The viscous component of the transport flux for component A is then given by the expression

$$J(\text{visc } A) = UP_A/RT,$$

where U is the drift velocity of the bulk gas, and P_A is the partial pressure of component A. Where the stoichiometry of the substrate differs from that of the source, an additional flux may result from chemical concentration gradients. This diffusional flux component is given by

$$J(\text{chem } A) = -\frac{D_A}{RT}\left(\frac{dP_A}{dx}\right),$$

where

D_A is the diffusion coefficient of component A,
dP_A/dx the partial pressure gradient along a tube, and
x a distance relative to the crystal substrate interface where $x = 0$.

From the model, the dependence of GaAs crystal growth rate on parameters such as temperature gradient and pressure may be theoretically defined. It can also be shown that there exists a critical value of the growth rate, above which the growing interface is no longer in a position of maximum thermodynamic activity.

Watanabe *et al.* (1972) have also developed predictive equations for a similar vapor transport case.

3. *Single Reaction Case*

For the common case of a single transport reaction which is diffusion controlled, the material transport rate along a tube may be approximately calculated as follows (Schafer, 1964, 1971). Consider a process

$$i\mathrm{A(s)} + k\mathrm{B} = j\mathrm{C};$$

the number of moles of A transported, n_A, is given by the relation

$$n_\mathrm{A}/t = (i/j)(\Delta P_\mathrm{C}/P)(T^{0.8})(q/x)1.8 \times 10^{-4} \quad \text{mole/hour},$$

where

t is the time in hours,
P the total pressure (atm),
ΔP_C the pressure differential (atm) between the hot and cold zone for the transport species C,
T the average temperature along the diffusion path in units of K,
q the geometrical area cross section of the diffusion path in cm^2, and
x the distance of the diffusion path in cm.

In the derivation of this expression the diffusion coefficient, which is presumed to be an average value for all high temperature transport species in the vapor phase, is taken to be

$$D_0(273\ \mathrm{K}, 1\ \mathrm{atm}) = 0.1\ \mathrm{cm^2\ sec^{-1}}.$$

This order-of-magnitude value has been verified experimentally for species such as $CrCl_3$ and is of suitable accuracy for hydrogen free systems.† The diffusion coefficient (D) at the experimental pressure (P) and temperature (T) conditions is given by the semiempirical relationship

$$D = D_0(P_0/P)(T/T_0)^{1.8},$$

as obtained from data on permanent gases. The temperature and pressure dependence of D, and the suggested average value, are already included in the general transport expression given above. From this expression, for typical transport conditions of

$$\text{molecular weight} = 100, \quad \Delta P_\mathrm{C}/P = 10^{-5}, \quad x = 10\ \mathrm{cm},$$

$$q = 3.1\ \mathrm{cm^2}, \quad T = 1000\ \mathrm{K}, \quad \text{and} \quad i = j,$$

the transport rate is calculated to be 1.4 mg/100 hr.

† Reference to Fig. 3.2 indicates that a relatively small decrease occurs in D with increasing molecular weight for species with molecular weight $\gtrsim 100$.

With the carrier gas or transpiration method, and in the absence of boundary layer or diffusion limitations, the molar transport of component A by the carrier gas B may be obtained from the expression

$$n_A = (i/j)\,\Delta P_C \cdot n_B/P_B,$$

assuming $P_B \gg P_C$, which is often the case.

Where the transport rate is determined by convection, the molar transport is given by the expression

$$n_A = \frac{i}{j}\,[\Delta P_C \cdot P_B]\left[\frac{4.7r^4 l_w M_B T}{RT_1 \eta l}\right]\left[\frac{1}{T_1} - \frac{1}{T_2}\right]$$

where

l is the length of gas travel at T_1 in cm,
l_w the length of gas travel at T_2 in cm,
r the tube radius in cm,
η the gas viscosity at T_1 in gm cm^{-1} sec^{-1},
M_B the molecular weight of B, and
R the gas constant.

Note that in this case the transport is proportional to pressure, whereas for the diffusion-limited case an inverse proportionality to pressure was indicated. In practice, this difference in pressure dependence serves to identify whether the transport is under diffusive or convective control.

4. *Multireaction Diffusion-Limited Case*

For systems where more than one reaction is affecting the transport process, the general treatment of Mandel and co-workers may be used, e.g., see Lever and Mandel (1962), Jona and Mandel (1963, 1964), and Mandel (1962). This approach also allows for the inclusion of individual binary diffusion coefficients, rather than an average species diffusion coefficient, and is similar to that derived earlier for thermal transport processes by Brokaw (1961a, b), see also Chapter 6, Section II.

For the most general multireaction case, the net flux of solid J_s is given by

$$J_s = \frac{\Delta H_1\,\Delta T}{R^2T^3L}\,\frac{(1 + \sum_{j=2}^{N} \alpha_j)}{(\phi_{11} + \alpha_2\phi_{12} + \cdots + \alpha_N\phi_{1N})},$$

where the coupling parameters $\alpha_2 \cdots \alpha_N$ are given by a ratio of $(N - 1)$ dimensional determinants, i.e.:

$$\alpha_t = \frac{|\,[-\phi_{11} + (\Delta H_1/\Delta H_n)\phi_{n1}]\,|}{|\,[\phi_{it} - (\Delta H_1/\Delta H_n)\phi_{nt}]\,|}$$

and ϕ_{nt} is defined as

$$\phi_{nt} = \sum_i (q_{in}/p_i)\{\sum_j [(q_{jt}p_i - q_{it}p_j)/PD_{ij}]\};$$

where

q is a reaction coefficient, i.e., q_{in} is for species i in reaction n and is positive for products and negative for reactants;
p_i are average (with respect to position) partial pressures;
P is the total pressure;
$\Delta H_1 \cdots \Delta H_n$ are the reaction enthalpies for reactions $1 \cdots n$;
ΔT is the temperature difference between the source and the deposit;
T the average temperature;
R the gas constant;
L the distance between source and deposit; and
D_{ij} are temperature averaged binary diffusivities between species i and j.

As an example, consider the system ZnS/HCl, where the following two reactions are believed to be responsible for the transport of ZnS (s) (Jona and Mandel, 1963):

$$\mathrm{ZnS(s)} + 2\mathrm{HCl} = \mathrm{ZnCl_2} + \mathrm{H_2S} \tag{1}$$

and

$$\mathrm{ZnS(s)} + 2\mathrm{HCl} = \mathrm{ZnCl_2} + \mathrm{H_2} + \tfrac{1}{2}\mathrm{S_2}. \tag{2}$$

The coupling parameter is then given by

$$\alpha_2 = \frac{-\phi_{11} + (\Delta H_1/\Delta H_2)\phi_{21}}{\phi_{12} + (\Delta H_1/\Delta H_2)\phi_{22}}$$

and therefore

$$J_s = \frac{\Delta T}{R^2T^3L}\left(\frac{\Delta H_1(\phi_{22} - \phi_{21}) + \Delta H_2(\phi_{11} - \phi_{12})}{(\phi_{11}\phi_{22} - \phi_{21}\phi_{12})}\right).$$

As the reaction enthalpies and partial pressures may be obtained from basic thermodynamic data, then J_s may be calculated. The predicted flux of ZnS is found to be only about 30–40 % lower than the observed rates over an HCl pressure range of from 1 to about 15 atm. In view of the combined uncertainties in the basic thermodynamic and diffusion data required, this agreement is very satisfactory. However, at HCl pressures of less than 1 atm, the observed rate is much less than the calculated rate, and this may be attributed to heterogeneous reaction rate limitations.

For the similar ZnS/I_2 transport system the agreement between theory and experiment is very good for the pressure regime of $10^{-1} - 2$ atm (Jona and Mandel, 1964). At higher pressures, the observed flux tends to be larger than calculated, owing to the increasing importance of convective transport. At pressures of less than 10^{-1} atm, the experimental rates are lower than calculated, owing presumably to surface reaction kinetic limitations. The transport equilibria were taken as

$$ZnS(s) + I_2 = ZnI_2 + \tfrac{1}{2}S_2, \qquad (1)$$

$$ZnS(s) + 2I = ZnI_2 + \tfrac{1}{2}S_2, \qquad (2)$$

$$ZnS(s) + I = ZnI + \tfrac{1}{2}S_2, \qquad (3)$$

and the flux is then given by

$$J_s = \frac{\Delta H_1 \, \Delta T}{R^2 T^3 L} \frac{(1 + \alpha_2 + \alpha_3)}{(\phi_{11} + \alpha_2\phi_{12} + \alpha_3\phi_{13})}.$$

Here the α's, which express the relative importance of reactions (2) and (3) with respect to (1), are given by

$$(\Delta H_1/\Delta H_2)\phi_{21} - \phi_{11} = \alpha_2[\phi_{12} - (\Delta H_1/\Delta H_2)\phi_{22}] + \alpha_3[\phi_{13} - (\Delta H_1/\Delta H_2)\phi_{23}]$$

and

$$(\Delta H_1/\Delta H_3)\phi_{31} - \phi_{11} = \alpha_2[\phi_{12} - (\Delta H_1/\Delta H_3)\phi_{32}] + \alpha_3[\phi_{13} - (\Delta H_1/\Delta H_3)\phi_{33}].$$

In both these ZnS (s) transport studies, the fluxes were of the order of 10^{-7}–10^{-8} mole cm^{-2} sec^{-1}.

5. *Diffusion Coefficients*

Basic data on transport properties of gases† at high temperatures is relatively sparse and practically nonexistent for high temperature vapors—hence the frequent use of a single average diffusion coefficient irrespective of the molecular weight for the various gas components. Semitheoretical or empirical models are often used for the estimation of transport coefficients, e.g., see Ferron (1968) and Jona and Mandel (1963). The interrelationships between the various transport properties are usually based on the Chapman–Enskog general kinetic theory, e.g., see Hirschfelder *et al.* (1954) and Chapter 1, Section V. Self-diffusivity data are usually fitted to the Lennard-Jones potential function where the force constants are estimated from empirical relationships involving known parameters such as

† A recent compilation of gaseous binary diffusion coefficient data has been given by Marrero and Mason (1972). Some additional discussion of diffusion coefficients may be found in Chapter 1, Section V.

critical, boiling, and melting temperatures and molar volumes, e.g., see Turkdogan *et al.* (1963), Fujii *et al.* (1972), and Chapter 1, Section V. In very simple cases, molecular orbital calculations can provide the necessary potential energy curves for the theoretical determination of the transport properties of diffusion, viscosity, and thermal conductivity—for instance with the systems Li + H and O + H (Krupenie *et al.*, 1963).

Binary diffusion coefficients, between species i and j, may be estimated from the expression

$$D_{ij} = \frac{3}{8n\sigma_{ij}^2}\left[\frac{kT(M_i + M_j)}{2\pi M_i M_j}\right]^{1/2},$$

where

k is the Boltzmann constant,
n is the total gas density—as calculated from the total pressure and an average temperature,

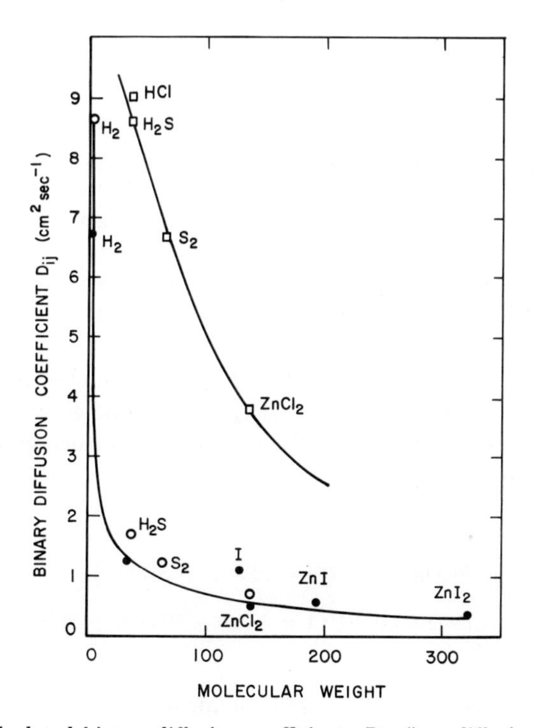

Fig. 3.2 Calculated binary diffusion coefficients D_{ij} (i.e., diffusion of i in j) as a function of component molecular weights and for 1268 K temperature and 1 atm pressure; open squares: $M_j = 2 = H_2$, open circles: $M_j = 34 = H_2S$, closed circles: $M_j = 64 = S_2$; data of Jona and Mandel (1963, 1964).

M_i, M_j are molecular weights, and
σ_{ij} are the "hard-sphere diameters" for the interacting molecular species,

$$\sigma_{ij} = \tfrac{1}{2}(\sigma_i + \sigma_j).$$

The σ's are usually unknown but may be estimated from the semiempirical formula:

$$\sigma_i = 2r_i + \delta_i,$$

where r_i is the radius of the molecule, as calculated from Pauling's covalent radii, and $\delta_i = 1.10$ Å is an empirical factor obtained from data on binary mixtures of permanent gases.† Several tables of gaseous binary diffusion coefficient data, together with a number of the relationships used to calculate diffusion coefficients, may be found in the book of Fristrom and Westenberg (1965, p. 259).

It is instructive to note the magnitudes of D_{ij} calculated for the various binary interactions in the ZnS transport systems, as shown in Fig. 3.2, where the importance of molecular weight differences is indicated.

D. Examples

A number of representative chemical transport processes are shown in Table 3.3. Unless indicated otherwise, the reactions given and their original literature source may be found in Schafer (1964). By convention, the transported solid is written on the left-hand side of the reaction. Both flow and closed tube (CT) conditions have been used for these transport preparations.

Multicomponent chemical systems are now being used to transport and prepare crystals of ferrites such as $NiFe_2O_4$. The solid components NiO and Fe_2O_3 may be transported by HCl along a temperature gradient (800 → 1000 °C) by the endothermic processes

$$NiO(s) + 2HCl = NiCl_2 + H_2O,$$

$$Fe_2O_3(s) + 6HCl = 2FeCl_3 + 3H_2O,$$

$$NiFe_2O_4(s) + 8HCl = NiCl_2 + 2FeCl_3 + 4H_2O$$

(Kleinert, 1969).

The utility of vapor complexes in chemical transport is beginning to be recognized, e.g., see the review of Emmenegger (1972). For example, the

† For an alternative approximation scheme see Chapter 6, Section II.B.

TABLE 3.3

Examples of Chemical Transport Reactions

Reaction	Conditions (T in °C)	Reference
$Ir(s) + \frac{3}{2} O_2 = IrO_3$	1325 → 1130 CT (closed tube)	
$Au(s) + \frac{1}{2} Cl_2 = \frac{1}{2} (AuCl)_2$	1000 → 700 flow, CT	
$Ta(s) + \frac{5}{2} I_2 = TaI_5$	580 → 1200	Jeffes and Marples (1972)
$Te(s) + I_2 = TeI_2$	440	Burmeister (1971)
$Mo(s) + 2MoO_3 = 3MoO_2$	1600 → T_1[a]	
$Al(s) + \frac{1}{4} Al_2S_3 = \frac{3}{4} Al_2S$	1300 → <1000 flow	
$Al(s) + \frac{1}{2} AlCl_3 = \frac{3}{2} AlCl$	1000 → 600 flow, CT	
$Si(s) + SiCl_4 = 2SiCl_2$	1300 → 1100 CT	
$Ti(s) + 2TiCl_3 = 3TiCl_2$	1200 → 1000 flow	
$Be(s) + BeCl_2 = 2BeCl$	1200 → 1000 CT	Gross and Lewin (1973)
$W(s) + xH_2O = WO_x + xH_2[x = 1 - 3]$	2400 → T_1 filament	
$Ni(s) + 2HCl = NiCl_2 + H_2$	1000 → 700 flow, CT	
$Cu(s) + HCl = \frac{1}{3} (CuCl)_3 + \frac{1}{2} H_2$	600 → 500 CT	
$Al_2O_3(s) + 2H_2 = Al_2O + 2H_2O$	2000 → T_1	
$Cu_2O(s) + 2HCl = \frac{2}{3} (CuCl)_3 + H_2O$	600 → 900 CT	
$BeO(s) + H_2O = Be(OH)_2$, or $(BeO)_n \cdot H_2O$ $[n = 1 - 3]$	1500 → T_1 flow	
$WO_3(s) + H_2O = WO_2(OH)_2$	1100 → T_1 flow	
$Li_2O(s) + H_2O = 2Li(OH)$	1000 → T_1 flow	
$SiO_2(s) + 4HF = SiF_4 + 2H_2O$	T_1 → T_2[b]	
$SiO_2(s) + H_2 + nH_2O = Si(OH)_2 \cdot nH_2O$	[c]	

TABLE 3.3 (*Continued*)

Reaction	Conditions (T in °C)	Reference
$MoO_2(s) + I_2 = MoO_2I_2$		Oppermann (1971)
$Al_2O_3(s) + 3CO + 6HCl = 2AlCl_3 + 3CO_2 + 3H_2$	1700	Schaffer (1965)
$V_2O_3(s) + \frac{3}{2} TeCl_4 = 2VCl_3 + \frac{3}{2} TeO_2$	$1050 \rightarrow 950$	Nagasawa *et al.* (1972)
$ZnO(s) + H_2 = Zn + H_2O$		Reisman and Landstein (1971)
$Cu(s) + \frac{1}{2} I_2 = CuI$ [or Cu_3I_3]	$T_1 \rightarrow T_2$	
$Pt(s) + COCl_2 + CO = Pt(CO)_2Cl_2$	$375 \rightarrow 475$	Schafer and Wiese (1971)
$MoCl_3(s) + MoCl_5 = 2MoCl_4$	$300 \rightarrow 250$ CT	
$PtCl_2(s) + 2CO = Pt(CO)_2Cl_2$	$375 \rightarrow 475$	
$Al_2S_3(s) + 3I_2 = 2AlI_3 + \frac{3}{2} S_2$	$T_2 \rightarrow T_1$	
$V_3Si(s) + 8VCl_3 = SiCl_2 + 11VCl_2$		Spear (1972)
$V_3Si(s) + 4SiCl_4 = 5SiCl_2 + 3VCl_2$		
$LnPO_4(s) + 3Cl_2 = LnCl_3 + POCl_3 + \frac{3}{2} O_2$		Orlovskii *et al.* (1971), Ln = lanthanide
$LnPO_4(s) + 5PCl_3 = LnCl_3 + 4POCl_3 + \frac{1}{2} P_4$		
$MgO + Al_2O_3(1) = MgAl_2O_4(1)$ ↓ $MgO(s)$	1900	Cockayne *et al.* (1971)

The following are considered plausible reactions (M = metal):

$MCO_3(s) + 2HCl = MCl_2 + H_2O + CO_2$

$MSiO_3(s) + 3SOCl_2 = MCl_2 + SiCl_4 + 3SO_2$

$MPO_4(s) + 3PCl_5 = MCl_3 + 4POCl_3$

$C(s) + 2Si = Si_2C$

$2C(s) + Si = SiC_2$

[a] T_1 represents the low temperature region.

[b] Observation of R. Gruehn, as cited by Schafer (1973).

[c] See Chapter 2, Section VIII.E for additional reactions involving silica and steam; note also the possibility of SiO_2 transport in N_2 atmospheres via the known vapor species Si_2N and Si_2NO (Muenow, 1973).

vapor growth of $Be_2SiO_4(s)$ using $Li_2BeF_4(s)$ as a transport agent probably involves the complex species $LiBeF_3$. Similarly, the transport of $Al_2SiO_5(s)$ by $Na_3AlF_6(s)$ most likely involves $NaAlF_4$ as a complex species. The transport of $CaNb_2O_6(s)$ with chlorine probably results from the formation of an unknown vapor complex containing Ca. Also, the chemical transport of EuS(s) is thought to involve a complex species of the type $Eu_xMo_yS_z$ (Kaldis, 1972). This particular example is interesting as surface reaction, rather than gas diffusion, is believed to be rate limiting.

For the particular case of $CrCl_3$ transport along a temperature gradient using Al_2Cl_6 ($P \sim 1$ atm) as a complexing reagent, the transport reaction is believed to be

$$CrCl_3(s) + \tfrac{3}{2}Al_2Cl_6 \underset{400\ ^\circ C}{\overset{500\ ^\circ C}{\rightleftarrows}} CrCl_3 \cdot 3AlCl_3$$

(Lascelles and Schafer, 1971). This stoichiometry for the complex vapor species was determined from the dependence of $CrCl_3$ transport on $P(Al_2Cl_6)$, as shown by Fig. 3.3. At 500 °C and $P(Al_2Cl_6) \simeq 1$ atm, the apparent partial pressure of the complex is about 0.2 atm.

A more detailed general discussion of complex vapor phenomena is given in the following section (Section IV).

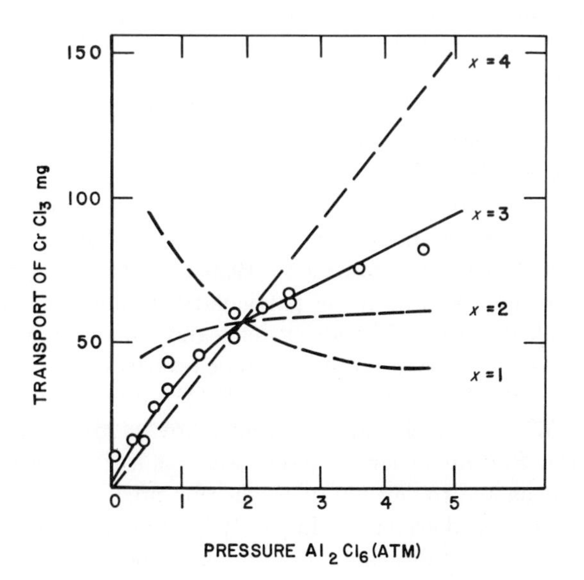

Fig. 3.3 Chemical vapor transport of $CrCl_3(s)$ as a function of the pressure of Al_2Cl_6 vapor; the open circles are experimental points; the curves denoted as $x = 1$–4 are calculated, where x is the assumed stoichiometric dependence of $CrCl_3$ on $AlCl_3$; the curve for $x = 3$ best fits the experimental points (Lascelles and Schafer, 1971).

IV. Enhanced Volatility Processes and Complex Vapors

A. Introduction

Most inorganic metal-containing compounds are relatively nonvolatile at conveniently accessible temperatures. The requirement of high temperatures to effect vapor-phase material transport of metal-containing compounds increases the problems associated with container corrosion and also leads to an uneconomical use of energy. Semiconductor materials technology is also limited by the high temperatures required for the more conventional CVD processes. Similarly, the efficiency of halide arc lamps (see Section IX) is limited by the need to maintain high wall temperatures for the volatilization of metal halides. Methods of obtaining an enhanced volatility, and thereby a lower vapor transport temperature, are therefore of current practical concern.

The most attractive technique for providing an increased "apparent volatility" is by addition of a complexing agent which has the property of generating volatile complex species.† At relatively low temperatures and high pressures, the formation of adducts with the complexing agent is possible. Thus, for example, the relatively volatile species $POCl_3$ may be used to form adducts with $AlCl_3$ or $ZrCl_4$, thereby increasing their volatility. For higher temperature–lower pressure conditions, adducts such as these tend to be thermodynamically unstable and $AlCl_3$ (or Al_2Cl_6) itself has been found to be a suitable, and particularly versatile, complexing agent for these more extreme conditions.

The discovery of complexing in high temperature vapors was somewhat unexpected and was manifested, initially, in the form of anomalously high partial pressures over multicomponent condensed halide systems, e.g., see Tarasenkov and Klyachko-Gurvich (1936). However, complex vapor phenomena are now known to be particularly prominent, and are relatively well studied, for metal halide systems and a number of reviews have appeared on the identity and thermodynamic properties of complex halides, e.g., see Hastie (1971), Sidorov *et al.* (1971), Buchler and Berkowitz-Mattuck (1967), Novikov and Gavryuchenkov (1967), and Bauer and Porter (1964). Several attempts have been made to correlate complex halide phenomena and to develop rules whereby predictions could be made for the many unstudied systems (Hastie, 1971; Novikov and Gavryuchenkov, 1967). The present discussion provides an opportunity for the further development of predictive schemes, and the great variety of possible com-

† Complexing has recently been utilized for chemical transport preparations of single crystals—e.g., see Section III.

plex interactions that can occur in high temperature systems indicates the need for such a systematic development and extension of the data, where possible. It should be pointed out that similar complex phenomena have been observed, but to a lesser extent, for mixed chalcogenide systems, and many of the generalizations given here for the halides may also be applicable to these systems, e.g., see Buchler and Berkowitz-Mattuck (1967).

Halide complexes are known to form in binary metal halide systems representative of Groups I–II, I–III, I–IV, I–transition element series, I–rare earth series, I–actinide series, II–III, III–transition element series, III–rare earth series, and III–IV of the periodic table. The thermodynamic data for these systems have been reviewed up to about 1970, e.g., see Hastie (1971) and cited literature, and emphasis is given to more recent work in the following discussion. Aluminum halide complexes are given particular attention as they have the ability to complex with a great variety of metal halides, and it is also believed that these complexes have the most potential for practical utility.

A brief discussion of $POCl_3$ complexing is warranted as it provides an enhanced metal halide volatility at relatively low temperatures and was actively considered as a means of separating Zr and Hf.

In addition to the importance of vapor-phase complex formation in a thermodynamic sense, complex species may also be significant as intermediates in reaction kinetics. For instance, the decomposition of the BrO radical proceeds via a dimer intermediate, i.e.,

$$2BrO \rightarrow (BrO)_2 \rightarrow 2Br + O_2$$

(Brown and Burns, 1970). Even though kinetic studies have yet to be made on complex metal halide vapors, the possible participation of such complex species as rate-controlling reaction intermediates should be considered. Indeed, these species could well serve as critical nuclei for nucleation in vapor deposition processes (see Section III).

B. $POCL_3$ Complexes

1. $AlCl_3 \cdot POCl_3$

The solid compound $AlCl_3 \cdot POCl_3$ melts congruently at 188° C and has a vapor density, over the temperature interval 300–400 °C, that indicates a molecular weight close to that for the complex $AlCl_3 \cdot POCl_3$ (Shubaev *et al.*, 1970). Mass spectrometric measurements, made under molecular flow conditions at 120–150 °C, also show the complex to be the major vapor species (Suvorov and Shubaev, 1971). For the dissociation reaction

$$AlCl_3 \cdot POCl_3 = AlCl_3 + POCl_3$$

the enthalpy change is $\Delta H_{298} = 28.4$ kcal mol^{-1}, which is similar to the heat for the reaction:

$$Al_2Cl_6 = 2AlCl_3,$$

and this is suggestive of similar bonding in both complexes—namely, halogen bridging.

2. $ZrCl_4 \cdot POCl_3$

Vapor density measurements indicate the formation of a vapor-phase complex species $ZrCl_4 \cdot POCl_3$ over mixtures of $ZrCl_4$ and $POCl_3$ (Suvorov and Krzhizhanovskaya, 1969b). Also, for the dissociation reaction

$$ZrCl_4 \cdot POCl_3 = ZrCl_4 + POCl_3,$$

$\Delta H_{691} = 24.3 \pm 1$ kcal mol^{-1}, and $\Delta S_{691} = 38.2 \pm 1$ cal deg^{-1} mol^{-1}.

Similar experiments on $ZrCl_4$–$PSCl_3$ mixtures did not indicate any complex vapor species. Also, the systems $TiCl_4$–$POCl_3$ and $SnCl_4$–$POCl_3$ do not lead to the formation of complex vapor species (Suvorov and Krzhizhanovskaya, 1969a).† However, the $HfCl_4$–$POCl_3$ system is known to form a complex vapor species. The dissociation thermodynamics of some of the condensed phase complexes has also been determined, e.g., see Krzhizhanovskaya and Suvorov (1971).

The separation of $ZrCl_4$ from $HfCl_4$ can be achieved by distillation of the vapor complexes $MCl_4 \cdot POCl_3$. The ratio of the volatilities at the boiling points, i.e., 360 and 355°C, for the Zr and Hf complexes, respectively, is 1.14 and the heat of vaporization for both complexes is 20.5 ± 0.5 kcal mol^{-1}. As a result of the importance of producing a Hf-free form of Zr as a nuclear reactor material, this process was developed on a pilot plant scale. However, the process apparently lacked commercial potential owing to the difficulty of retrieving $ZrCl_4$ from the solid complex $ZrCl_4 \cdot POCl_3$. It is now believed that the fractional distillation of the tetrachloride mixture is a better process (see Section VII. B.2). However, it would be useful if other volatile complexes could be found where the liberation of MCl_4 was more favorable than for the $POCl_3$ complex. Further discussion, and original literature citations, of these separation processes may be found in the articles reviewed by Thomas and Hayes (1960) and by Lustman and Kerze (1955).

† This does not preclude complex formation under different conditions of temperature and pressure, nor the presence of relatively small amounts that may be detectable by other means, such as mass spectrometry.

3. *$NbCl_5 \cdot POCl_3$*

Experiments similar to those reported for $ZrCl_4 \cdot POCl_3$ indicate the complex equilibrium

$$NbCl_5 \cdot POCl_3 = NbCl_5 + POCl_3,$$

where $\Delta H_{609} = 15.5 \pm 1$ kcal mol^{-1}, and $\Delta S_{609} = 29.5 \pm 1$ cal deg^{-1} mol^{-1}. Thus the stability of this complex, as represented by the enthalpy for dissociation, is less than that of the other lower valence metal chlorides. In fact, the dissociation reaction enthalpies are observed to decrease monotonically with increasing metal valence from 3 to 5. This trend can reasonably be expected to hold for other series of complexes.

C. Aluminum Halide Vapor Complexes

The ability of $AlCl_3$ to form a vapor complex has been known for the particular case of $NaAlCl_4$ for about 20 years. Similarly, the formation of $NaAlF_4$ over commercially used cryolite melts is well known. Recently, however, it has been found that the aluminum trihalides readily form vapor complexes with many other metal halides and over a wide range of metal valence, i.e., 1–5. In the presence of $AlCl_3$, or Al_2Cl_6, substantial increases in "apparent volatility" have been observed for metal chlorides representative of Groups I–V, and also the transition element, the rare earth element, and the transuranium element series of the periodic table. For the case of the rare earth and transuranium metal chlorides, the complexes with Al_2Cl_6 are sufficiently volatile for gas chromatographic separations to be carried out in a glass capillary column (Zvarova and Zvara, 1969, 1970a, b). Even the noble metal Au can be complexed with Al_2Cl_6 or Fe_2Cl_6. In the latter case, the complex vapor species has been mass spectrometrically identified as $AuFeCl_6$, and its formation enhances the volatility of $AuCl_3$ by at least a factor of 25 (Hager and Hill, 1970).

1. *$AlBr_3$–MBr_3 (M = P, As, Sb) Systems*

A complex vapor-phase interaction between $AlBr_3$ and the tribromides of P and Sb has been noted by Suvorov *et al.* (1969), using a vapor density technique. At temperatures of around 230 °C and total pressures of up to about 1 atm, the complex species $AlBr_3 \cdot PBr_3$ and $AlBr_3 \cdot SbBr_3$ are present at concentrations of about 10%.† Notably, no evidence for

† Values of ΔH and ΔS for the formation of these complex species may be found in Table 3.5.

complex formation was found for the corresponding arsenic system. This is indicative of the rather subtle interplay between the nature of M and the free energy required for vaporization of the complex species without dissociation and that required for dissociative vaporization.

Owing to the covalent bonding character of the component tribromides, the possibility of Al-pnictide bond formation arises. For the case of $AlBr_3 \cdot SbBr_3$ an electron diffraction study indicates Al–Sb bond formation rather than the alternative halogen bridging (Spiridonov and Malkova, 1969). This is a relatively weak bond, i.e., ~8 kcal mol^{-1}, and its formation would not compensate for the entropy decrease in the complex-forming reaction at temperatures of greater than ~300 °C. That is, this type of complex species is only thermodynamically important at relatively low temperatures and high pressures.

2. *The $NdCl_3$–$AlCl_3$ System*

A striking example of the degree of increased volatility to be gained from the formation of complex vapor species is given by the ability of aluminum trichloride, at pressures in the range of 1–7 atm, to increase the amount of $NdCl_3$ present in the vapor phase by a factor of 10^{13} at 330 °C (Øye and Gruen, 1969). This enhanced volatility is attributed to the complex forming process

$$NdCl_3(s) + nAl_2Cl_6 = NdAl_{2n}Cl_{(3+6n)}.$$

Using an absorption spectrophotometric technique to monitor the complex species, the coefficient n was found to fall in the range of 1.6 to 1.9, and two complex species were postulated, namely, $NdAl_3Cl_{12}$ (i.e., $n = 1.5$) and $NdAl_4Cl_{15}$ (i.e., $n = 2$).† Owing to the tendency of Al_2Cl_6 to dissociate to $AlCl_3$ at high temperature, the complex species vapor pressures reach a maximum at temperatures in the region of 700°C. This limiting-temperature phenomenon will receive further discussion later.

3. *Complexing between Metal Dichlorides and the Trichlorides of Aluminum and Iron*

Dewing (1967, 1970) has recently observed an enhanced "apparent volatility" for the dichlorides of Ca, Mg, Mn, Co, Pb, Zn, Cd, and for CuCl, in the presence of $AlCl_3$; and similarly for the dichlorides of Mn, Co, Ni, Zn, Cd, and Pb in the presence of $FeCl_3$. Volatility enhancement factors of up to 10^7 are found, as shown in Table 3.4, and complex species

† Values of ΔH and ΔS for the formation of these complex species may be found in Table 3.5.

TABLE 3.4

Enhanced Volatilities for Metal Dichlorides in the Presence of $AlCl_3$[a]

Dichloride	Temperature (°C)	Volatility enhancement factor[b]
$CaCl_2$	650	10^7
$MnCl_2$	600	1.7×10^3
$MgCl_2$	660	1.2×10^3
$CoCl_2$	600	8.6×10^2
$PbCl_2$	600	8

[a] From Dewing (1967, 1970).
[b] Defined as the ratio of apparent vapor pressure of dichloride (in complex form) to the vapor pressure of the pure dichloride phase.

partial pressures of greater than 10 Torr are readily attained for the typical complexing conditions of total pressures of about 1 atm and temperatures of 600–660°C.

The complexing processes appear to be

$$ACl_2(s) + 2MCl_3 = AM_2Cl_8 \tag{1}$$

and

$$ACl_2(s) + 3MCl_3 = AM_3Cl_{11}, \tag{2}$$

where A is the divalent metal and M is either Al or Fe. The former reaction is the more dominant process. For the transport of $MgCl_2(s)$ by $AlCl_3$, at 900 K:

$$\Delta H_1 = -13.8, \qquad \Delta H_2 = -42.6 \text{ kcal mol}^{-1}$$

and

$$\Delta S_1 = -17.9, \qquad \Delta S_2 = -55.4 \text{ cal deg}^{-1}\text{ mol}^{-1}.$$

Similar thermodynamic results have been obtained for the other $AlCl_3$–metal dichloride systems (Dewing, 1970).

The analogous $FeCl_3$ complex-forming reactions have enthalpies that are about 3 kcal mol^{-1} more negative than the $AlCl_3$ counterparts. This is consistent with a more negative dimerization enthalpy for $FeCl_3$ than $AlCl_3$, to the extent of 5 kcal mol^{-1}.

The stabilities of the AAl_2Cl_8 complexes, as defined by the free energies of formation from the component salts according to the above reaction (1), follow the sequence:

$$Ca > Ni > Mn \geq Co > Mg.$$

In this connection it is of note that the stability of $CaCl_2$ vapor is also greater than that of $MgCl_2$ vapor, as indicated by the respective heats of

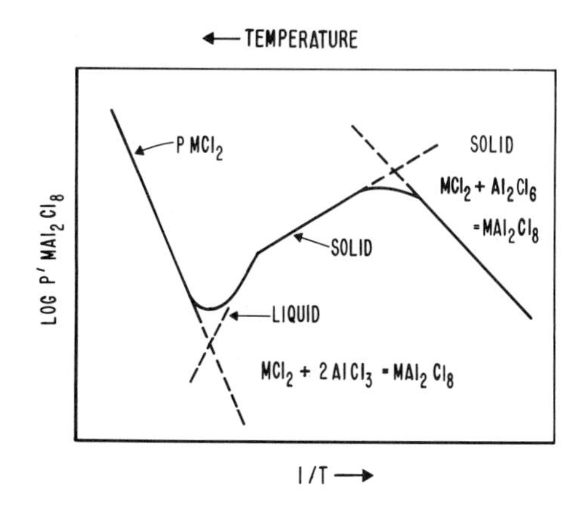

Fig. 3.4 Schematic of the apparent pressure of MAl_2Cl_8 versus reciprocal temperature (Dewing, 1970).

formation. More recent studies have shown that palladium dichloride solid also forms a complex species $PdAl_2Cl_8$, which has a stability similar to that found for the Ni-analog (Papatheodorou, 1973).

Owing to the thermodynamic interaction between the dimerization of MCl_3 and the complexing reactions of ACl_2, an upper temperature limit exists (~600–700 °C) for the optimum formation of the complex. This behavior, which also applies to the $NdCl_3$–$AlCl_3$ complexes, is demonstrated schematically in Fig. 3.4. Such a result clearly demonstrates the necessity of identifying—and thermodynamically characterizing—the competing processes involved in complex formation.

a. Metallurgical Significance

As was suggested by Dewing (1970, and US Patent No. 3425797, 1969), the complex interaction of $AlCl_3$, or $FeCl_3$, with metal dichlorides has potential as a novel means of distilling, and recovering, pure forms of the metal dichlorides. It also provides a means for extracting metals from their ores using a variation of the chlorination-type process (see Section VII).

In the normal ore-chlorination process, the oxide—or sulfide—ore is reacted with chlorine in the presence of a reducing agent such as carbon. Dewing has suggested the possibility of directly reacting chlorine and the complexing metal trichloride reagent with the ore, in a divalent form, to generate the dichloride–trichloride complex vapor species. In this manner, the recovery of $CoCl_2$ from a synthetic [$CoS + SiO_2$] ore was experimentally demonstrated using $AlCl_3$ or $FeCl_3$ at 1 atm and 600 °C. $FeCl_3$ was found to be more effective for this purpose than $AlCl_3$. The overall reaction is considered to be

$$2CoS(s) + 2Cl_2 + 4AlCl_3 = 2CoAl_2Cl_8 + S_2.$$

With an excess of Cl_2 present the formation of SCl_2 may also occur.

Some further insight into the proposed aluminum refining process based on aluminum monochloride disproportionation (see Section VII) is also provided by the evidence of stable complex formation between $AlCl_3$ and $CaCl_2$ or $MgCl_2$ which appear as troublesome impurities in the AlCl process.

The ability of $AlCl_3$ to complex with $CrCl_3$, as discussed elsewhere (Section III), may also have metallurgical implications in the extraction of Cr metal from the ore by chlorination (see Section VII.B).

4. *Systems UCl_5–Al_2Cl_6 and UCl_4–Al_2Cl_6*

The transport of UCl_5 and UCl_4 into the vapor phase, at temperatures of ~230–430 °C and using Al_2Cl_6 as a complexing agent, has been established by Gruen and McBeth (1969). For UCl_4, the volatility is enhanced by factors of 10^7 and 10^3 at ~230 °C and ~430 °C, respectively. At ~330 C and 1 atm pressure of Al_2Cl_6, the partial pressure of complexed UCl_4 is about 0.4 Torr. The principal reactions are believed to be†

$$UCl_4 + Al_2Cl_6 = UCl_4 \cdot Al_2Cl_6 \; \{\text{or } UAl_2Cl_{10}\}$$

and

$$(UCl_5)_2 + Al_2Cl_6 = 2(UCl_5 \cdot AlCl_3) \; \{\text{or } UAlCl_8\}.$$

5. *Summary of Thermodynamic Data*

A summary of known reactions, and their enthalpies and entropies, for aluminum halide complexing is given in Table 3.5. This data is typical of that for many other metal halide systems, as summarized elsewhere (Hastie, 1971).

D. Some Empirically Based Rules for Complex Halide Formation

1. *The Empirical Basis*

The rules to be outlined below rely on a certain degree of systematics in the structures and energetics of high temperature species and their complexes. A brief account of the evidence for such a systematic behavior, using metal halides as an example, is as follows. One should recognize at

† Values of ΔH and ΔS for the formation of these complex species may be found in Table 3.5.

TABLE 3.5

Thermodynamic Data for Complex Reactions Involving Aluminum Trihalides

Vapor equilibria	Temperature (K)	$-\Delta H_T$ (kcal/mole)	$-\Delta S_T$ (cal/deg-mole)	Reference
$LiF + AlF_3 \rightarrow LiAlF_4$	1000	73 ± 4	37	[a]
$Li_2F_2 + AlF_3 \rightarrow LiF + LiAlF_4$	1000	9.0 ± 2	4.6 ± 2	[a]
$NaF + AlF_3 \rightarrow NaAlF_4$	910	87.5 ± 3	31.5 ± 3	[b]
$2LiAlF_4 \rightarrow (LiAlF_4)_2$		43.3 ± 3	28.4 ± 3.5	[b]
$2NaAlF_4 \rightarrow (NaAlF_4)_2$	910	42.4 ± 2.5	27 ± 4	[b]
$2KAlF_4 \rightarrow (KAlF_4)_2$		34.0 ± 3	25.3 ± 3.5	[b]
$NaCl + AlCl_3 \rightarrow NaAlCl_4$		51		[c]
$BiCl + AlCl_3 \rightarrow BiAlCl_4$	1050	28.1 ± 0.5	42.8 ± 4	[d]
$BeCl_2 + AlCl_3 \rightarrow BeAlCl_5$				[e]
$FeCl_3 + AlCl_3 \rightarrow FeAlCl_6$				[f]
$UCl_4 + (AlCl_3)_2 \rightarrow UAl_2Cl_{10}$	600	32	32	[g]
$(UCl_5)_2 + (AlCl_3)_2 \rightarrow UAlCl_8$	600	−2.7	−6.1	[g]
$MCl_2(s) + 3\ AlCl_3 \rightarrow MAl_3Cl_{11}$	~1000			[h]
$MCl_2(s) + 2\ AlCl_3 \rightarrow MAl_2Cl_8$	~1000			[h]
(M = Ca, Mg, Mn, Co, Pb, Zn, Cd)				
$NdCl_3(s) + 1.5\ Al_2Cl_6 \rightarrow NdAl_3Cl_{12}$	815	−10.8	−2.5	[i]
$NdCl_3(s) + 2\ Al_2Cl_6 \rightarrow NdAl_4Cl_{15}$	815	−1.7	12.4	[i]
$CrCl_3(s) + 1.5\ Al_2Cl_6 \rightarrow CrAl_3Cl_{12}$				[j]
$PBr_3 + AlBr_3 \rightarrow AlBr_3 \cdot PBr_3$	~500	8.8 ± 1.5	18 ± 3	[k]
$SbBr_3 + AlBr_3 \rightarrow AlBr_3 \cdot SbBr_3$	~500	8.2 ± 1.5	15.1 ± 3	[k]

[a] Porter and Zeller (1960).
[b] Sidorov *et al.* (1971); see also Kolosov *et al.* (1971).
[c] See the review of Novikov and Gavryuchenkov (1967).
[d] Lynde and Corbett (1971).
[e] Semenenko *et al.* (1964a).
[f] Semenenko *et al.* (1964b).
[g] Gruen and McBeth (1969).
[h] Dewing (1970); see also discussion and thermodynamic data given in main text.
[i] Øye and Gruen (1969).
[j] Lascelles and Schaffer (1971); see also Section III.D.
[k] Suvorov *et al.* (1969).

the outset that for high temperature inorganic species, particularly metal halides, the bonding is largely ionic—e.g., see Hastie and Margrave (1969). That is, the stabilities and structures are determined mainly by coulombic interactions between ions such as Na^{+}–Cl^{-} or Al^{3+}–F^{-}.

With regard to the structure, or molecular geometry, of high temperature complex species, the evidence supports structures of the type represented in Fig. 3.5 (see Hastie, 1971; Hastie *et al.*, 1970). By edge- or face-sharing of tetrahedral structural units it is possible to extend the structures shown in Fig. 3.5 to include species of even greater complexity such as $PdAl_2Cl_8$, which, from its electronic absorption spectra, is believed to involve a square planar $PdCl_4$ group linked by Pd–Cl–Al bridges to $AlCl_4$ tetrahedra (Papatheodorou, 1973). Similarly, the structure of the species $CrAl_3Cl_{12}$ most likely involves a grouping of three $AlCl_4$ tetrahedra about a central Cr (Lascelles and Schafer, 1971).

Concerning the stabilities, or more specifically the bond dissociation energies, of high temperature complex metal halides, the most important factor is the stability gain accompanying metal halogen bridge formation. This is shown by the representative data given for molecular dimers in Table 3.6. As long as the ratio of bridge to terminal bond dissociation energies (as defined in Table 3.6) is greater than 0.5, the formation of dimers will be favorable enthalpy wise. That is, reactions of the type

$$2AX = (AX)_2$$

will be thermodynamically exothermic.

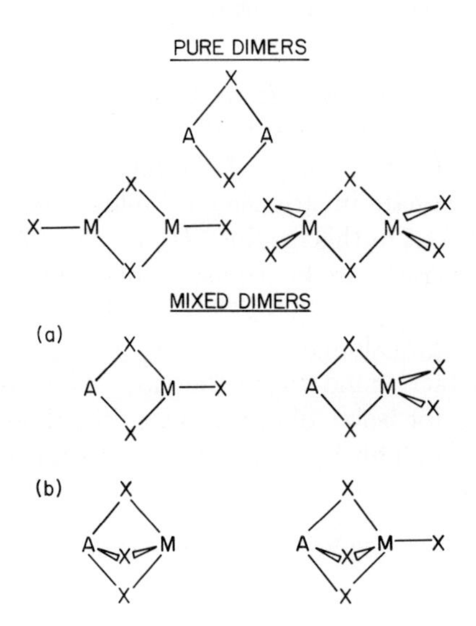

Fig. 3.5 Geometrical structures of dimeric metal halide species; X is halogen.

TABLE 3.6

Stability of Halogen Bridge Bond Formation[a]

Species	Bond energy (kcal mole^{-1})		
	Terminal[b] $\bar{D}_t$	Bridge[c] $\bar{D}_b$	Ratio $\bar{D}_b/\bar{D}_t$
$(LiF)_2$	138	84	0.61
$(NaF)_2$	115	72	0.63
$(NaCl)_2$	99	61	0.62
$(GeF_2)_2$	115	64	0.56
$(SnF_2)_2$	113	66	0.58
$(SnCl_2)_2$	93	51	0.55
$(AlF_3)_2$	139	81	0.58
$(AlCl_3)_2$	69	41	0.59
$(LaF_3)_2$	157	96	0.61
$(LaCl_3)_2$	123	73	0.59

[a] Calculations based on literature data—e.g., see the reviews cited in the main text and the JANAF Thermochemical Tables (JANAF, 1971), together with the structures given in Fig. 3.5.

[b] $\bar{D}_t$ in dimer is taken, by definition, as an average bond energy for the unbridged monomer, e.g., $\bar{D}_t = \frac{1}{2}\,\Delta H_{\text{atoms}}[GeF_2]$.

[c] $\bar{D}_b = (\Delta H_{\text{atoms}}[\text{dimer}] - m\bar{D}_t)/n$, where n is the number of bridge bonds in the dimer, and m is the number of terminal bonds.

It is particularly noteworthy from Table 3.6 that the ratio $\bar{D}_b/\bar{D}_t$ is approximately constant, and equal to 0.60 ± 0.04, irrespective of the absolute magnitude of $\bar{D}_t$, the charge of the metal ion, or the nature of the halogen. This approximate invariance also holds true for analogous metal chalcogenide species but in this instance $\bar{D}_b/\bar{D}_t \simeq 0.67$. It follows that one can expect bond energies to be transferrable between various species. For example, the cyclic species $CsLiCl_2$ could be expected to have Cs–Cl and Li–Cl bond energies similar to those in $(CsCl)_2$ and $(LiCl)_2$, respectively, which is in fact found to be the case. Thus very little enthalpy change should occur for isobonding reactions, i.e., those for which no new bond types are formed. This has been shown experimentally to be the case for reactions such as

$$PbCl_2 + PbBr_2 = 2PbClBr \qquad (\Delta H \sim 0)$$

and

$$Li_2O + Ga_2O = 2LiGaO \qquad (\Delta H \sim 0).$$

2. *The Rules*

From the experimental results of equilibrium studies on more than one-hundred complex halide vapor systems, it is now possible to make the following generalizations which are of predictive and heuristic value.

Rule 1. Reaction entropies are similar for those homogeneous reactions where the same change in the number of participating species occurs.

A corollary to this rule is that *reaction* entropies are usually not very temperature dependent, that is, they may vary by only several entropy units over a 1000 K temperature interval.

As an example of the effect of this rule, the entropy changes associated with the formation of metal halide species by vapor-phase reactions are found to have the following average values:

	Reaction	ΔS (cal deg^{-1} mol^{-1})
(a)	A + B = C + D	0 ± 3
(b)	A + B = C	−30 ± 5
(c)	2A + B = C	−60 ± 5
(d)	3A + B = C	−90 ± 10

The deviance limits represent standard deviations for an average of about 40 sets of data for reaction types (a) and (b). The data is more limited for reaction types (c) and (d), since the large entropy reduction for these processes tends to limit the extent to which the reaction may proceed.

The majority of reactions of type (b) actually have values within the limits of $\Delta S = -30 \pm 3$ cal deg^{-1} mol^{-1}. However, adduct complexes such as $BF_3 \cdot NH_3$ where the energy for dissociation, e.g., to BF_3 and NH_3, is relatively low, tend to be formed with an entropy decrease of about −40 cal deg^{-1} mol^{-1}. Such species tend to be unstable at high temperatures for normal pressures of 1 atm or less.

A few heavy metal halide complexes have reaction type (b) entropies as low as −20 cal deg^{-1} mol^{-1}, and this is indicative of very low fundamental frequencies for vibrations associated with ring deformation in the product species. Such frequencies of vibration are known to decrease according to the halogen sequence: F > Cl > Br > I and this causes a systematic decrease in, for example, the dimerization entropies of the tin dihalides which have values of −31, −26, −23, and −20 cal deg^{-1} mol^{-1}, respectively (Karpenko, 1969a). This effect should therefore be included in the entropy rule when systems with large differences in metal and halogen masses are being considered. In such cases it is preferable to consider isobonding reactions, if possible, where these ring effects tend to be offsetting for the reactants and products (e.g., see Table 3.7).

Examples of reaction type (a) include the mixed-dimer reactions of the alkali halides, i.e.,

$$(AX)_2 + (BX)_2 = 2(ABX_2),$$

where A and B are alkali metals and X is a halogen; and halogen exchange reactions such as,

$$PbCl_2 + PbBr_2 = 2PbClBr.$$

Examples of reaction type (b) are the simple dimerization reactions, e.g.,

$$2AlCl_3 = (AlCl_3)_2 \qquad (\Delta S_{800} = -32.9 \text{ cal deg}^{-1} \text{ mol}^{-1}),$$

$$2NaCl = (NaCl)_2,$$

and

$$2NaAlF_4 = (NaAlF_4)_2,$$

and reactions such as

$$NaF + AlF_3 = NaAlF_4 \qquad (\Delta S_{910} = -31.5 \text{ cal deg}^{-1} \text{ mol}^{-1}).$$

Examples of reaction type (c) are

$$MgCl_2 + 2AlCl_3 = MgAl_2Cl_8 \qquad (\Delta S_{800} = -58.5 \text{ cal deg}^{-1} \text{ mol}^{-1})$$

and

$$NdCl_3 + 2Al_2Cl_6 = NdAl_4Cl_{15}$$

or

$$NaF + 2VF_3 = NaV_2F_7$$

(Sholts *et al.*, 1970).

An example of reaction type (d) is

$$MgCl_2 + 3AlCl_3 = MgAl_3Cl_{11} \qquad (\Delta S_{800} = -96 \text{ cal deg}^{-1} \text{ mol}^{-1}).$$

While this rule has an empirical basis, it is possible to provide an independent rationalization for the rule in terms of structural arguments (Hastie, 1971). From the well-known statistical mechanical treatment, the entropy can be obtained as a summation of contributions from translation, rotation, vibration, and electronic degrees of freedom. We can show, for example, that for reactions of type (a), and where the bonding is similar for both sides of the reaction, e.g., the total number of M–X bridge or terminal bonds is unchanged, the main contribution to entropy-change results from changes in molecular symmetry. This contribution can vary from -0.2 cal deg^{-1} mol^{-1} to $+2.8$ cal deg^{-1} mol^{-1} for most of the likely symmetry changes. Such a variation is within the empirical range of $\Delta S = 0 \pm 3$ cal deg^{-1} mol^{-1}.

Rule 2. Bond dissociation energies are transferrable between species provided there is no change of valence or bond type, i.e., terminal or bridge bonds. It also follows that isobonding processes, where the number and type of bonds remain unchanged, should be essentially thermal neutral. The empirical basis for this rule has been indicated in the previous section.

Rule 3. The enthalpy change associated with complex formation, by the vapor-phase reaction of monomers, is always exothermic and usually by at least 30 kcal mol^{-1}.

This result is related to the tendency for complexing to occur by the formation of metal halogen bridge bonds. As was discussed in the previous section, bridge bonds are about 60% as strong as terminal bonds. Thus, for example, the dimerization of AX to yield $(AX)_2$, with a cyclic structure, would have an enthalpy change of

$$\Delta H \simeq 2D_{A-X} - 0.6 \times 4D_{A-X} \simeq -0.4D_{A-X},$$

where D_{A-X} is the A–X bond dissociation energy for the monomer. Since the bond dissociation energy for many metal halides is ~100 kcal mol^{-1}, then heats of dimerization of about −40 kcal mol^{-1} result.
Similarly, for a reaction

$$(A{-}X) + (X{-}B{-}X) = A\!\begin{array}{c} X \\ \diagup\ \diagdown \\ \diagdown\ \diagup \\ X \end{array}\!B{-}X,$$

using the bond additivity approximation of Rule 2:

$$\begin{aligned}\Delta H &\simeq (D_{A-X} + 2\bar{D}_{B-X}) - (2 \times 0.6D_{A-X} + 2 \times 0.6\bar{D}_{B-X} + \bar{D}_{B-X}) \\ &= -0.2(D_{A-X} + \bar{D}_{B-X}),\end{aligned}$$

which yields heats of similar magnitude to the heats of dimerization reactions.

Rule 4. Rules 2 and 3 provide a lower limit indication of the stability for complex species.

It follows from the discussion of Rules 2 and 3 that isobonding mixed-dimer reactions, such as

$$(AX)_2 + (BX)_2 = 2(ABX_2),$$

should occur with $\Delta H \sim 0$, since the number of bridge or terminal bonds remains the same on both sides of the reaction. However, in practice, we find that mixed-metal-dimer reactions have ΔH values ranging from about 0 to −81 kcal/2 mole complex, as shown in Table 3.7. As a consequence of this enhanced stability associated with mixed-dimer formation, these complexes may be observed under conditions where the component species dimers have a negligibly small concentration. For example, $CsPbCl_3$ is

TABLE 3.7

Enhanced Stabilities of Complexes Resulting from Isobonding Mixed Dimer Reactions[a]

Reaction	$-\Delta H$ (kcal/ 2 mole)	$+\Delta S$ (cal/deg- 2 mole)	Reference
$(CsCl)_2 + (LiCl)_2 = 2(CsLiCl_2)$	7	2±2	
$(CsCl)_2 + (NaCl)_2 = 2(CsNaCl_2)$	5.2	2±2	
$(CsCl)_2 + (KCl)_2 = 2(CsKCl_2)$	2.7	2±2	
$(CsCl)_2 + (RbCl)_2 = 2(CsRbCl_2)$	~1	2±2	
$(NaF)_2 + (BeF_2)_2 = 2NaBeF_3$	44	2	
$(KF)_2 + (SnF_2)_2 = 2KSnF_3$	23±10	—	o
$(NaF)_2 + (AlF_3)_2 = 2NaAlF_4$	71, 57	3, 3	b
$(KF)_2 + (AlF_3)_2 = 2KAlF_4$	81±10	3	b
$(LiF)_2 + (AlF_3)_2 = 2LiAlF_4$	38	4	m
$(NaCl)_2 + (AlCl_3)_2 = 2NaAlCl_4$	27	−6	l
$(NaBr)_2 + (AlBr_3)_2 = 2NaAlBr_4$	28	−10	c
$(NaCl)_2 + (LaCl_3)_2 = 2NaLaCl_4$	44	−6	
$(KCl)_2 + (MgCl_2)_2 = 2KMgCl_3$	34	−7	
$(NaCl)_2 + (BeCl_2)_2 = 2NaBeCl_3$	22	—	
$(NaBr)_2 + (ZnBr_2)_2 = 2NaZnBr_3$	−1	4	d
$(NaCl)_2 + (SnCl_2)_2 = 2NaSnCl_3$	29	0	
$(KCl)_2 + (SnCl_2)_2 = 2KSnCl_3$	34	−8	e
$(RbCl)_2 + (SnCl_2)_2 = 2RbSnCl_3$	31	2	f
$(CsCl)_2 + (SnCl_2)_2 = 2CsSnCl_3$	34	2	f
$(RbBr)_2 + (SnBr_2)_2 = 2RbSnBr_3$	16	−5	g
$(CsBr)_2 + (SnBr_2)_2 = 2CsSnBr_3$	20	6	g
$(KI)_2 + (SnI_2)_2 = 2KSnI_3$	16	−4	e
$(NaCl)_2 + (PbCl_2)_2 = 2NaPbCl_3$	14	$\lesssim 3$	j
$(UCl_5)_2 + (AlCl_3)_2 = 2UAlCl_8$	−3	6	h
$(GaCl)_2 + (GaCl_3)_2 = 2Ga_2Cl_4$	7.2	−3	i

TABLE 3.7 (*Continued*)

Reaction	$-\Delta H$ (kcal/2 mole)	$+\Delta S$ (cal/deg-2 mole)	Reference
$(InCl)_2 + (InCl_3)_2 = 2In_2Cl_4$	14	−0.6	[i]
$(TlCl)_2 + (InCl_3)_2 = 2TlInCl_4$	20	0	[k]
$(MgCl_2)_2 + (AlCl_3)_2 = MgAl_2Cl_8 + MgCl_2$	3.3	4	[n]
$(MgCl_2)_2 + 2(AlCl_3)_2 = MgAl_3Cl_{11} + MgCl_2 + AlCl_3$	4	0	[n]

[a] Unless indicated otherwise, the data and their original references are from Hastie (1971). The reaction heats and entropies are usually calculated, according to Hess's law, from the measured values for the simple dimerization reactions, e.g.,

$$2AX = (AX)_2 \quad \text{and} \quad 2BX_2 = (BX_2)_2,$$

and the mixed monomer reaction, e.g.,

$$AX + BX_2 = ABX_3.$$

Much of the thermodynamic data for the simple dimerization processes are given by the JANAF Thermochemical Tables (JANAF, 1971). The experimental uncertainties in ΔH and ΔS are typically ± 4 kcal mol^{-1} and ± 4 cal deg^{-1} mol^{-1}, respectively.

[b] See Sidorov *et al.* (1971).

[c] Suvorov and Malkova (1968).

[d] Schaaf and Gregory (1971); see Keneshea and Cubicciotti (1964) for $(ZnBr_2)_2$ data.

[e] Cited in Karpenko and Sevastyanova (1967).

[f] Karpenko (1970).

[g] Karpenko and Dogadina (1971), Karpenko (1969a, b).

[h] Gruen and McBeth (1969).

[i] Polyachenok and Komshilova (1972).

[j] More recent work on $PbCl_2$–alkali chloride systems has been reported by Belyaev *et al.* (1972).

[k] Polyachenok and Komshilova (1972); see also Rat'kovski (1972).

[l] Lynde and Corbett (1971) have reported on a similar system where for the reaction

$$BiCl + AlCl_3 = BiAlCl_4, \qquad \Delta H_{298} = -30 \text{ kcal mol}^{-1}.$$

[m] Rao (1970) has more recently published work on this system; his enthalpy data agree with the earlier work cited by Hastie (1971); his entropy data are used in this table.

[n] Calculated from the data of Dewing (1970); note for these isobonding reactions only one mole of complex product is formed.

[o] Data for $(SnF_2)_2$ taken from Dudash and Searcy (1969).

observed as a major species under conditions where $(PbCl_2)_2$ is undetectable and $(CsCl)_2$ is only a minor constituent.

It is apparent from Table 3.7 that where A and B are similar, for instance with the alkali halide mixed dimers, the reactions are close to being thermalneutral. However, as A and B become increasingly different in their size and formal charge character, i.e., valence, the value of ΔH tends to become more negative. This enhanced stability of the mixed-dimer complexes may be attributed to a relaxation of ionic repulsion in the ring of the mixed-dimer species.

According to the ionic model of bonding, which is known to apply reasonably well to alkali halide energetics, e.g., see the review of Bauer and Porter (1964), the cation–cation and anion–anion repulsion terms of the dimer potential energy are appreciable, being in the region of 100 kcal mol^{-1}. The observed trend in the heats of formation of the cyclic ring-structured mixed dimers can be explained in terms of a reduction in these repulsion effects, as follows.

In the absence of general information regarding the actual charges of the component ions, and the internuclear separations, the following approximation is suggested for correlating the effect of cation substitution with mixed-dimer stability. Consider the typical mixed-dimer reaction of

$$(AX)_2 + (BX_2)_2 = 2ABX_3,$$

where the species geometries are of the type indicated in Fig. 3.5. As a first approximation we consider only the cation coulombic repulsion terms in the potential energy of these species. Furthermore, for simplicity, a completely ionic character is assumed, i.e., the cations are A^{1+} and B^{2+}, and the separation of these cations in the molecule is taken as the sum of the ionic radii r. From Coulomb's law, the *change* in cation–cation repulsive potential energy that results from the formation of $2ABX_3$, by the above mixed-dimer reaction, is given by

$$\Delta U = [4/(r_A + r_B)] - [(1/2r_A) + (2/r_B)].$$

Analogous expressions may be readily derived for other ion combinations, such as A^{1+}–B^{3+} or A^{2+}–B^{3+}. For this approximation it is not necessary to consider explicitly the units of electronic charge, as only the relative magnitude of ΔU is of significance here. A value of $-\Delta U = 1.0$ Å^{-1} would be actually equivalent to an enthalpy difference of about 331 kcal mol^{-1}.

The extent to which ΔU correlates with the experimental excess stabilities, as given in Table 3.7, is indicated in Fig. 3.6. Despite the apparent scatter in the data there does appear to be an approximate one-to-one correspondence between $-\Delta U$ and $-\Delta H$, particularly for low values of $-\Delta U$. Much of the scatter can be attributed to departures from a completely ionic model. In fact, the repulsion-relaxation model would predict no stability enhancement, i.e., $\Delta H \sim 0$, for a completely covalent situation. Thus, for a fixed value of $-\Delta U$, the upper $-\Delta H$ boundary shown in Fig. 3.6 refers to mainly ionic-bonding systems and the lower $-\Delta H$ bound-

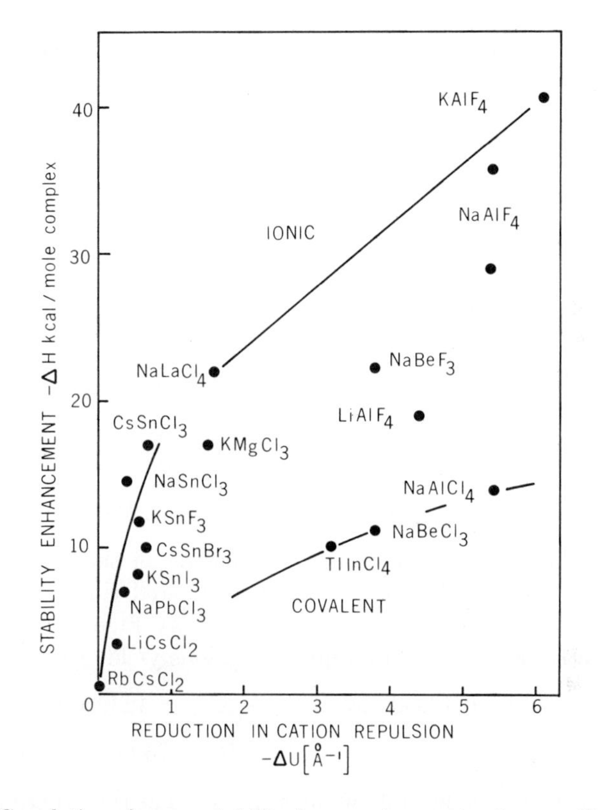

Fig. 3.6 Correlation of excess stability for complex species, i.e., $-\Delta H$ from Table 3.7, with decrease in cation–cation repulsive potential energy $-\Delta U$; the ionic radii used to calculate $-\Delta U$ were taken from Rich (1965).

ary to systems having a considerable degree of covalency. The observed reduction in the rate of change of stability enhancement with increasing $-\Delta U$ at values of greater than 1.0 Å^{-1} (see Fig. 3.6) probably results from effects of anion–anion interaction.

In applying this correlation to hypothetical systems, one can usually make an a priori estimate of whether the dimeric species should be classified as mainly ionic or mainly covalent. It then appears possible to estimate the unknown ΔH, from the correlation of Fig. 3.6, to an accuracy of better than ± 10 kcal mol^{-1}. This is about the same degree of uncertainty as some of the experimentally determined ΔH values.

3. *Examples of Application of the Rules*

Example 1. It has been noted by Novikov and Gavryuchenkov (1967) that the stabilities of the mixed complexes KCl·MCl$_n$ reach a maximum value in the region of $n = 2$, at least for the following examples:

	$-\Delta H$ (kcal mol^{-1})	
$KCl + KCl = (KCl)_2$,	44.5 ± 2	(1)
$KCl + SrCl_2 = KSrCl_3$,	67 ± 17	(2)
$KCl + LaCl_3 = KLaCl_4$,	61 ± 4	(3)
$KCl + ThCl_4 = KThCl_5$,	49 ± 3.	(4)

Consider the application of Rules 2, 3, and 4 to the above reactions. From the principles of bond additivity (Rule 2) and halogen bridge-bond stability (Rule 3), it follows that the reaction enthalpies for reactions (1)–(4) are given by

$$-\Delta H_1 = 0.4D_{KCl}, \qquad -\Delta H_2 = 0.2(D_{KCl} + \bar{D}_{SrCl}),$$

$$-\Delta H_3 = 0.2(D_{KCl} + \bar{D}_{LaCl}), \qquad -\Delta H_4 = 1.2D_{KCl} - 0.8\bar{D}_{ThCl}.$$

These derivations assume geometrical structures of the type indicated in Fig. 3.5; that is, K and M form a four-membered ring in each complex species. Substitution of the known bond or average bond dissociation energies into these reaction enthalpy expressions yields values of

$$-\Delta H_1 = 40.4 \text{ kcal mol}^{-1}, \qquad -\Delta H_2 = 42 \text{ kcal mol}^{-1},$$

$$-\Delta H_3 = 45 \text{ kcal mol}^{-1}, \qquad -\Delta H_4 = 26 \text{ kcal mol}^{-1}.$$

Clearly these values do not follow the same trend as the experimental data.

Consider now the enhanced stabilization, as derived from the correlation of Rule 4. By definition, no such stabilization occurs for reaction (1). From the calculated value of $-\Delta U = 0.5$ Å^{-1} for the complex of reaction (2), and the correlation of Fig. 3.6, a stabilization enthalpy of about -15 kcal mol^{-1} is estimated. Similarly for reactions (3) and (4), values of -20 and -15 kcal mol^{-1} are estimated, respectively. The total reaction enthalpies are then determined to be

$$-\Delta H_1 = 40.4 \pm 5 \text{ kcal mol}^{-1}, \qquad -\Delta H_2 = 57 \pm 10 \text{ kca' mol}^{-1},$$

$$-\Delta H_3 = 65 \pm 10 \text{ kcal mol}^{-1}, \qquad -\Delta H_4 = 41 \pm 10 \text{ kcal mol}^{-1}.$$

These predicted values are in good agreement with experiment, and the observed maximum reaction enthalpy in the region of $n = 2$ or 3 is, therefore, the result of the combined effects of bond dissociation energy trends and the enhanced stability of heterovalent cation rings.

Example 2. The transport of $NdCl_3$ with Al_2Cl_6 has been demonstrated at relatively low temperatures, i.e., 300–500 °C (Øye and Gruen, 1969). However, the proposed transport complexes, $NdAl_3Cl_{12}$ and $NdAl_4Cl_{15}$, are less favored species at higher temperatures where the formation of the complex $NdAlCl_6$ might be expected to occur. From the enhanced stability correlation given in Fig. 3.6, we can expect an excess stabilization of about -17 ± 5 kcal mol^{-1} complex, from which it follows that, for the reaction

$$NdCl_3 + AlCl_3 = NdAlCl_6,$$

$\Delta H = -52$ kcal mol^{-1}. From Rule 1 the entropy decrease is given by

$$\Delta S = -30 \text{ cal deg}^{-1} \text{ mol}^{-1}.$$

Hence, for conditions of 700 °C and 1 atm $AlCl_3$, it follows that the partial pressure of the complex $NdAlCl_6$ would be a factor of 10^4 greater than that of $NdCl_3$, i.e., about 10 Torr, and an appreciable enhanced transport of $NdCl_3$ into the vapor would result even at temperatures where the established higher molecular weight complex species are unstable.

A similar case can be made for the formation of a complex such as $CaAlCl_5$ at elevated temperatures where $CaAl_2Cl_8$ is relatively unstable and $AlCl_3$ is present rather than Al_2Cl_6. This would explain the unaccounted for excess transport of $CaCl_2$ found by Dewing (1970) at elevated temperatures.

Example 3. Consider the reason for the absence of any detectable complex formation in the NaF–MgF_2 and NaF–BaF_2 systems, whereas NaF–BeF_2 mixtures readily form complex vapor species such as $NaBeF_3$ (Belousov *et al.*, 1970). The substitution of Na^+ for Be^{2+} into a ring of $(BeF_2)_2$ results in a greater reduction in cation–cation repulsive potential energy than for either of the Mg^{2+} or Ba^{2+} cases, that is $-\Delta U$ equals 3.8 Å^{-1} for Be, 1.1 Å^{-1} for Mg, and 0.9 Å^{-1} for Ba. Hence, from the stabilization energy correlation derived from Rule 4, a greater stability would be expected for the Be system than for either of the Mg or Ba systems.

Example 4. Fluorination reactions for nuclear fuel processing can lead to the simultaneous formation of VOF_3 and UF_6, e.g., see Chapter 6, Section III.E. Some difficulty in separating vanadium from the desired uranium product can be expected if a complex such as $VOF_3 \cdot UF_6$ forms under the typical reaction conditions of 75°C temperature, and VOF_3 and UF_6 pressures of 150 and 1600 Torr, respectively.

Consider the hypothetical reaction

$$VOF_3 + UF_6 = VOF_3 \cdot UF_6.$$

From Rule 1, $\Delta S \simeq -30$ cal deg^{-1} mol^{-1}. If V and U are involved in halogen bridge bonding, then from Rule 3, $\Delta H \simeq -30$ kcal mol^{-1}. As the species involved in the reaction are most likely predominantly covalently bonded, no excess stability is likely for the complex species. Hence,

$$\Delta F = \Delta H - T\,\Delta S \simeq -20 \text{ kcal mol}^{-1},$$

from which it follows that the reaction equilibrium constant is about 10^{12} and $VOF_3 \cdot UF_6$ is virtually undissociated at the conditions of interest. However, in view of the relatively large number of anion groups involved in this complex species, one could argue for a lower stability than that indicated above. For instance, for the somewhat similar case of $NbCl_5 \cdot POCl_3$ discussed earlier, $\Delta H \simeq -15$ kcal mol^{-1}. If $VOF_3 \cdot UF_6$ is assigned a similar stability, the reaction equilibrium constant would be about 10^3. This would still favor formation of the complex adduct.

E. Dissociation of Complexes by Condensation

For practical utility, the advantage gained by the generation of a stable vapor phase complex may be offset by the difficulty of compound recovery from the condensate, as determined by the stability of the condensed phase complex. The already mentioned difficulty in separating $ZrCl_4$ from a $ZrCl_4 \cdot POCl_3$ condensate typifies this problem.

Phase diagrams may be consulted for an indication of the presence of a condensed phase complex. However, a quantitative assessment of the relative stabilities of condensed and vapor-phase species requires a thermodynamic description of the various vaporization processes, as given, for example, by Novikov and Gavryuchenkov (1967).

The relative stabilities between the condensed and vapor-phase species are determined basically by the competition between the volatilities of the complex and its dissociation products. From thermodynamic arguments it follows that the lower the volatility is for the dissociation products, the greater the stability of the vapor-phase complex becomes. It is difficult to develop predictive rules based on thermodynamics where a change of state, i.e., condensed to vapor phase, is involved. Hence the fate of complexes on condensation cannot, at present, be determined a priori.

Several examples of the various degrees of relative stabilities between phases are as follows. An extreme case of complex instability in the vapor phase is exemplified by the well-known example of NH_4Cl. In this case, the solid dissociates on vaporization to yield NH_3 and HCl as vapor products which recombine on condensation. The loss of stability in the vapor phase may be attributed to an increased covalency resulting from the decreased coordination number and loss of lattice energy generated by vaporization. The complexes of $HfCl_4 \cdot POCl_3$ and $SnCl_4 \cdot POCl_3$ also dissociate completely on vaporization, whereas $ZrCl_4 \cdot POCl_3$ does not.

Examples where stable complexes exist only in the vapor phase are given by $CoAl_2Cl_8$, which dissociates on condensation to yield $CoCl_2$(s) and $2AlCl_3$ (or Al_2Cl_6), and $NaPbCl_3$, which dissociates on condensation to yield NaCl and $PbCl_2$ in the condensed state. Notably, in these cases, the volatility of one of the dissociation products is much greater than that of the other.

F. Enhanced Vaporization in Diffusion-Limited Systems

In many practical systems, the rate of vaporization is diffusion limited and may be orders of magnitude less than the "*in vacuo*" maximum rate. However, in some pyrometallurgical processes, metal oxide fumes are evolved at a much greater rate than one would expect from a diffusion-limited system. This enhancement of diffusion-limited rates of vaporization for metals can be quantitatively accounted for using a model developed by Turkdogan *et al.* (1963). The model involves the chemical interaction of the metal vapor with a reactant gas such as O_2 or N_2 (i.e., for Si) at the boundary layer. The product species of this interaction rapidly condense to form a mist. This effectively removes metal atoms from the boundary layer, which results in a decrease in the layer thickness and an increased rate of diffusion for metal atoms through the boundary. In practice, the mist is continuously removed by forced convection. The model has recently been extended by Distin and Whiteway (1970) and by Rosner (1972b).

A similar vaporization enhancement can be found for systems where the metal reacts to form a volatile metal compound, e.g., SiO, $Fe(OH)_2$, and WO_3 (see Chapter 2). The observed enhancement can be critically dependent on the concentration of reactant gas because of the possible formation of a liquid oxide vaporization barrier, for example, at the metal surface.

Appendix. Molecular Distillation

Molecular distillation refers to the technique of low pressure vaporization and condensation of a substance where the composition of the condensate differs from that of the initial vaporant (Burrows, 1960; Krell and Lumb, 1963). The molecular vapor species usually do not undergo gas or surface collisions between the time of vaporization and condensation.

Under molecular vaporization conditions, the transport rate, as determined by basic molecular kinetics, is given by the relation

$$J = 5.83 \times 10^{-2} \times P(M/T)^{1/2} \quad \text{gm sec}^{-1}\ \text{cm}^{-2},$$

where P is the pressure in Torr, M is the gm molecular weight, and T is the temperature in K. For practical distillation conditions, the solid angle for vaporization, as subtended by the condenser, is less than ideal (i.e., $<2\pi$) and the distillation rate will be less than the kinetic prediction.

For so-called normal evaporative distillation, where the vapor passes through a high pressure atmosphere, the distillation rate J is a diffusion controlled process and is given by the relation

$$J = 1.60 \times 10^{-5} \times P(M/T)(D/L) \text{ gm sec}^{-1} \text{ cm}^{-2},$$

where D is the diffusion coefficient in cm^2 sec^{-1}, and L is distance.

For multistage stills, such as a rectifying column, the minimum number of theoretical plates required for a separation process may be calculated from the relationship

$$(\alpha_{av})^n = [x_d/(1 - x_d)][(1 - x_n)/x_n],$$

where x_d represents the mole fraction of the distillate that reaches the top of the column, x_n is the mole fraction of component x at the nth plate, as counted from the top of the column, and α_{av} is the average relative volatility per plate. The relative volatility is given by

$$\alpha = (\gamma_1 P_1/\gamma_2 P_2)(M_2/M_1)^{1/2}$$

where γ represents activity coefficient, P is the partial pressure and M is the molecular weight. A plate is defined such that the ascending vapor leaving the plate is in equilibrium with the liquid descending from the plate. A distillation efficiency may then be given in terms of the ratio of the calculated number of theoretical plates to the actual number of plates required.

V. Chemical Synthesis with Quenched High Temperature Species

High temperature species have recently found application as unique reactants for the synthesis of both known and novel chemical products. Such species are potentially more reactive than their condensed state, on account of the elimination of much of the self-binding energy by the vaporization process (see Timms, 1972). The endothermicity associated with the transformation of these species from a condensed to a vapor state can be as high as several hundred kcal mol^{-1}, e.g., with the production of metal atoms from their metallic state. For species where only moderate energy gains are attained by vaporization, such as the rare gas or alkali metal atoms with heats of vaporization of about 0 and 20 kcal mol^{-1}, respectively, very little reactivity enhancement is achieved in the vapor phase. As a rule of thumb, one can expect those species with highly endothermic heats of formation to be the most suitable as high temperature chemical reactants.

The application of high temperature species to preparative chemistry has benefited greatly from the earlier development of techniques for producing and characterizing high temperature species. In particular, the accumulation of thermodynamic data on high temperature vapors has stimulated the use of vaporization under equilibrium conditions as a means of producing high temperature species for synthetic use. This equilibrium approach has allowed more control over the synthetic process than that provided by alternative techniques for producing reactive species, such as: electric discharges, plasmas (see Section VI), and photochemistry.

The synthetic procedure, as utilized primarily by the research groups of Skell (1971), Margrave (Ezell *et al.*, 1967), and Timms (1972), involves the generation of a high temperature vapor under conditions that favor a single reactant species. An experimental arrangement usually consists of a vacuum chamber with a "hot zone," for generating high temperature species, and a line-of-sight "cold zone" for condensation of the reactants. Additional coreactants or diluents, which are usually gases, are also simultaneously directed at the cold trapping surface. The use of low pressures of reactants, i.e., about 1 Torr or less, reduces the likelihood of a gas phase reaction occurring prior to the cocondensation process. The reactants are rapidly quenched at the cold surface and chemical reaction may occur during the quenching process. These frozen reactants, diluents, and products are then allowed to warm to a higher temperature, usually room temperature, where the volatile products are distilled off and collected. Reactions may also occur during this warm-up procedure.

The initial trapping temperature is usually -196 °C, i.e., that allowed by the use of liquid nitrogen as a coolant. An optimum trapping temperature cannot be defined a priori, but it would represent a compromise between the need to rapidly trap reactants and quench products, and the thermal energy that may be required to surmount a kinetic reaction activation energy barrier. The observation of reactions in rare gas matrices at temperatures as low as 20 K is of interest in this connection. In fact, the future use of matrix isolation spectroscopy to monitor the reactants and products in the higher temperature, i.e., -196 °C, synthetic work would be an obvious adjunct to the chemical synthesis technique.

As a number of reviews have recently appeared on the subject of chemical synthesis with high temperature species (Timms, 1972, 1973a, b; Havel *et al.*, 1973; Ezell *et al.*, 1967), only a few representative examples will be given here. The reactant high temperature species are usually atoms such as C, Si, and a number of the transition elements, or reduced valence species such as SiF_2, $SiCl_2$, BF, BCl, PF_2, CF_2, CCl_2, SiO, SiS, and B_2O_2.

TABLE 3.8

Typical Reactions Involving Quenched High Temperature Species

$$SiF_2 + C_2H_2 \rightarrow HC{\equiv}C\ SiF_2\ SiF_2\ CH{=}CH_2$$

$$SiF_2 + C_2H_4 \rightarrow \begin{array}{ccc} & CH_2 & \\ / & & \backslash \\ H_2C & & SiF_2 \\ | & & | \\ H_2C & & SiF_2 \\ \backslash & & / \\ & CH_2 & \end{array}$$

$$SiF_2 + BF \rightarrow F_2Si(BF_2)_2$$

$$SiF_2 + CF_3I \rightarrow CF_3SiF_2I$$

$$SiCl_2 + C_2H_2 \rightarrow \begin{array}{ccc} Cl & & Cl \\ & \backslash\ / & \\ & Si & \\ / & & \backslash \\ \| & & \| \\ \backslash & & / \\ & Si & \\ & /\ \backslash & \\ Cl & & Cl \end{array}$$

$$Cr + PF_3 \rightarrow Cr(PF_3)_6$$

$$Ni + PF_3 \rightarrow Ni(PF_3)_4$$

$$Fe + PF_3 \rightarrow Fe(PF_3)_5 + (PF_3)_3Fe(PF_2)_2Fe(PF_3)_3$$

$$Fe + C_5H_6\ \text{(cyclopentadiene)} \rightarrow Fe(C_5H_5)_2 + H_2$$

$$Ni + CO_2 \rightarrow NiO + CO + Ni(CO)_4$$

$$C + BCl_3 \rightarrow Cl_2C(BCl_2)_2 + ClC(BCl_2)_3$$

$$C_2 + BCl_3 \rightarrow Cl_2BClC{=}CClBCl_2$$

$$Cr + PF_3 + C_6F_6 \rightarrow (C_6F_6)Cr(PF_3)_3$$

$$Co + NO + PF_3 \rightarrow Co(NO)(PF_3)_3$$

$$Fe(CO)_5 + KCN \rightarrow K^+[Fe(CO)_4CN]^-$$

Methods of generating these species, at rates of about 10–200 mmole hr^{-1}, include the following high temperature reactions:

Reaction	T (°C)
$Si(s) + SiF_4 \rightarrow SiF_2$	1200
$Si(s, l) + SiCl_4 \rightarrow SiCl_2$	1350
$Si(s) + SiO_2(s)$ or "SiO"(s) $\rightarrow SiO$	1400
$B(s) + B_2O_3(l) \rightarrow B_2O_2$	1300
$B(s) + BF_3 \rightarrow BF$	1800

and

$$P_2F_4 \rightarrow PF_2 \qquad 800.$$

Reactants such as GeF_2, $GeCl_2$, and SiS have been notably unproductive as compared with the similar species of SiF_2 and SiO. One of the difficulties associated with an apparent lack of reactivity may be the competing process of self-association. For example, matrix isolation studies on SiO reactions reveal that the self-association to form $(SiO)_2$, $(SiO)_3$, and higher polymers is very competitive with the reactions of SiO with BF_3 or SiF_4 (Hastie *et al.*, 1969a). Another difficulty is that some reactants, e.g., SiO, tend to favor the production of nonvolatile products which are difficult to characterize. Some representative reactions achieved to date are indicated in Table 3.8.

The present status of this synthetic technique can best be summarized by the recent comment of Timms (1973b) to the effect that, there is much to be understood about the interaction of high temperature species and other compounds on the cold surface.

VI. Plasma Chemistry

A. Introduction

A recently developed technique for effecting high temperature chemical transformation and vapor-phase material transport involves the use of non-self-supporting plasmas. As compared with alternative high temperature chemical processing techniques, plasmas have advantages as "containers" for corrosive gases, e.g., HCl, and also in the rapid heating rates and high product-quench rates that can be achieved.

The application of plasmas as a source of high temperature species for chemical synthesis has been amply demonstrated during the past decade—e.g., see the reviews of Gross *et al.* (1969), Beguin *et al.* (1964), Jolly (1969), Stokes (1969, 1971), Freeman (1969b), Ibberson (1969), Sayce (1972), and the book by McTaggart (1967). Some additional recent literature on plasma chemistry is indicated in the Appendix.

Plasmas have also been used extensively as sources of excitation for in situ basic spectroscopic studies (Burgess, 1972). The use of plasma-arcs for the production of excited atomic and ionic species is well known in connection with analytical spectrophotometry, e.g., see Boumans (1972) and Mavrodineanu and Hughes (1964). Many of the chemical factors involved in converting the sample analyte to the desired spectroscopic condition are similar to those where flames are used as the analytical medium, as discussed in Chapter 5, Section IV.

A technique peculiar to the analytical use of plasmas involves the measurement of spectral intensities as a function of time; that is, time-of-wait curves, e.g., see Rozsa (1972). The time resolution of such curves can be controlled or modified by a so-called carrier method where a carrier additive is added to the analyte. The function of the additive, apparently, is to modify the vaporization or excitation processes in the arc-plasma (Pszonicki and Minczewski, 1963). Typical carriers are AgCl, CuCl, $PbCl_2$ and Ga_2O_3. It seems likely that for the case where oxide samples are in contact with the carbon arc electrode, together with Cl_2 as produced from AgCl decomposition, a carbothermic chlorination process (see Section VII.B) is responsible for the enhanced volatility. The carrier technique is largely empirical, but it could well be given a more fundamental basis by the development of theoretical models based on the existing body of related information regarding volatility and flame excitation processes, e.g., see Chapter 5, Sections II and IV.

The general technological applications of plasmas, including nonchemical uses, have been discussed by Brown *et al.* (1968) and by Goldberger (1966). Kettani and Hoyaux (1973) have recently summarized the physics of plasmas and their engineering applications. A discussion of the chemical engineering aspects of various plasma types and processes may be found in the review of Ibberson and Thring (1969). Also, a state of the art review of plasma chemical technology and a discussion of some fundamental aspects of plasma kinetics and thermodynamics has been given by Vurzel and Polak (1970).

Prospects for commercial plasma chemical processing have been considered recently by Parsons (1970) and also by Sayce (1972). The main limitation for commercial processing is the low product yield. This is probably related to the large temperature gradients present in plasmas, and the difficulty in achieving a rapid quench on an industrial scale. These low yields lead to an inefficient and uneconomical use of electricity in sustaining the plasmas. One should note, in this connection, that electricity as a source of energy is about six times as expensive as that produced by fossil fuel combustion.

Commercial processes have been used for the production of NO, from which nitric acid may be formed. At temperatures in excess of 3000 K the endothermic reaction

$$N_2 + O_2 = 2NO$$

results in the production of several percent of NO. Kinetically, however, the reverse reaction is relatively fast, and hence collection of the NO produced at high temperature requires a very rapid quenching.

Quenching rates in excess of 10^7 K sec^{-1} are possible when the plasma products are directed onto a cooled surface. The use of Laval nozzles for the adiabatic expansive cooling of high pressure gases can provide cooling rates of up to 10^6 K sec^{-1} and this cooling technique has been suggested for future plasma chemical processes. However, this type of cooling is based on the conversion of thermal into translational energy, and, in order to prevent a subsequent reversal of this energy conversion process, it is necessary to remove the excess translational energy generated. This may be conveniently achieved by allowing the gases to drive an impulse power generating turbine.

Acetylene has been prepared commercially by a plasma decomposition of saturated hydrocarbons. Plasmas are also in commercial use for producing ceramic or metal spray coatings (Hecht, 1972) and ceramic crystals by the Verneuil technique, e.g., see Mahe (1973). On a laboratory scale, the preparation of refractory materials such as titanium nitride has been achieved with reasonable yields.

The chlorination of ores, as a first step in metal extraction, can be greatly enhanced under plasma conditions. Refractory oxides, such as alumina, titania, magnesia, and silica, which are difficult to reduce by conventional means, can be reduced to the metals under plasma conditions. Reoxidation during quenching is a problem, but yields of 40–70 % are possible. The reduction of metal halides, such as $TiCl_4$, $ZrCl_4$, and BCl_3 in the presence of H_2 is also of commercial interest for the preparation of the metals. Several other potential commercial applications are the oxidation of SO_2 to SO_3 leading to its removal from industrial effluent gases, and the production of expensive chlorofluorocarbons. Additional discussion of metallurgical and chemical applications of plasmas is given in Section C.

As is emphasized in most reviews of the current status of plasma chemistry, and its extension to technology, in order to fully realize the commercial potential of plasma chemistry there exists a need for thermodynamic and kinetic data, and also for basic molecular-level studies on well-defined plasma systems—see, for example, the review of Ibberson (1969).

B. The Fundamental Nature of Plasmas

1. Plasma Types

The fundamentals of plasmas have been discussed by Gross *et al.* (1969) and by Krempl (1971), among others. Low density glow discharges (i.e., $P = 0.1$–10 Torr) have been discussed by Kaufman (1969), and the basic theoretical and spectroscopic techniques available for characterizing high density (i.e., $P \sim 1$ atm) plasmas have been elaborated on by Freeman (1969a).

When gases or vapors are heated to very high temperatures, i.e., 2000–10,000 K, the degree of molecular dissociation is high and thermal ionization to yield positive ions, negative ions, and electrons also occurs. As the gas contains charged species and, in particular, highly mobile electrons, it can be affected by an external electric or magnetic field. These fields can be used to provide energy to the system and maintain the gas temperature. They can also affect the motion of the gas constituents such that the system can be isolated and contained between suitable electrodes. Gases subjected to these conditions are usually known as plasmas.

In practice, plasmas may be supported by energy sources such as arcs, sparks, rf-discharges, shocks, focused laser radiation, and adiabatic compression, e.g., see Burgess (1972).

Plasmas may be categorized according to their energies and particle densities. The temperature and electron density regimes of various classical so-called "cool" plasmas, i.e., $T < 10^5$ K, are shown in Fig. 3.7. It should be noted that plasma electron densities are usually many orders of magnitude lower than the total particle density. "Hot" plasmas are defined by $T > 10^7$ K where nuclear reactions are possible (Venugopalan, 1971). Plasmas supported by photoionization are typically low energy, low density systems (i.e., < 1 eV or $< 10^4$ K, and 10^8 electron cm^{-3}) and those supported by electron impact are high energy, low density systems (i.e., 10^3 eV, 10^8 electron cm^{-3}). Glow discharges usually result from energies of ~ 1 eV and densities as low as 10^2 electron cm^{-3}.

Flames may be considered as a special case of a low or medium temperature plasma, as discussed by Calcote and Miller (1971). However, most plasmas differ from combustion flames in that the higher temperatures of these plasmas ($\sim 10{,}000$ K), as compared with flames (< 3500 K), allow

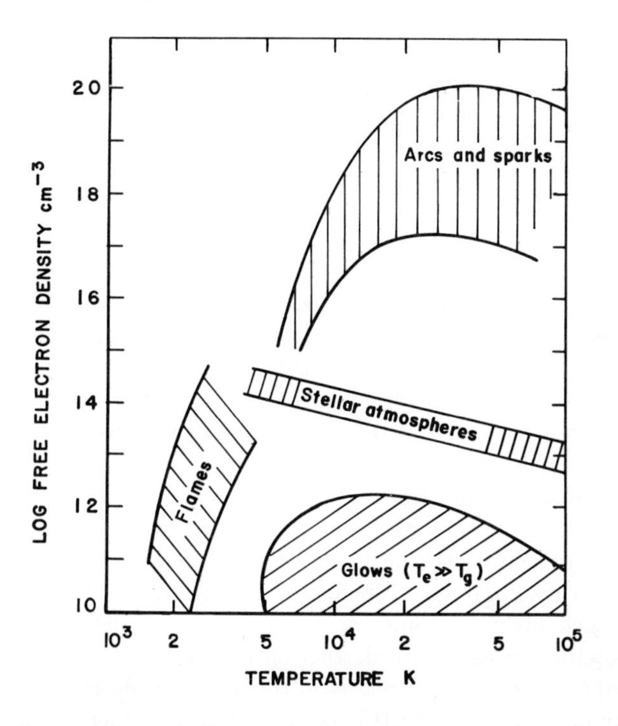

Fig. 3.7 Temperature and electron density regimes for various classical plasmas (adapted from Krempl, 1971).

a virtually complete dissociation of molecules to atomic and singly charged ionic species. The heat content of such highly dissociated gases is therefore much greater than for flames and a higher fraction of maximum available energy is present under plasma conditions (see Reed, 1967).

2. *Thermodynamic and Temperature Aspects*

At medium gas pressures, i.e., 10^{-1}–1 atm ($\sim10^{17}$–10^{18} cm^{-3}), and not excessively high temperatures, i.e., 10^3–10^4 K, the plasma species can be treated as being in local thermodynamic equilibrium. Hence, many of the plasma properties for arcs, flames, and plasma jets can be calculated from basic thermodynamic data. Thermonuclear plasmas, at energies of about 10^4 eV and particle densities of $\sim10^{14}$ cm^{-3}, are most likely in a state of complete thermodynamic equilibrium.

For plasmas having high electron and ion concentrations, that is about 10^{18} cm^{-3}, the coulombic interactions of these charged particles should be considered in conjunction with the thermodynamic calculations, and the Debye–Hückel treatment is usually used, e.g., see Gross *et al.* (1969).

The calculation of equilibrium properties for a plasma in which coulombic interactions are important has been described by Zeleznik and Gordon (1965). Essentially, the Debye–Hückel theory is used to correct the species partition functions for coulombic effects on translation and on the internal energy levels.

From thermodynamic calculations, common gases such as H_2, O_2, N_2, and CO at 1 atm pressure are completely dissociated to the elements at temperatures of approximately 6000, 7000, 11,000, and 12,000 K, respectively. Similarly, the degree of ionization for metals is almost 100% at temperatures of 15,000–20,000 K; at 6,000 K the degree of ionization is typically only about 10%. Calculated thermodynamic and transport data for simple multicomponent plasmas, such as for $Ar–O_2$ mixtures, have been given by Capitelli *et al.* (1970). Thermodynamics has also been used to predict chemical reactions in a dc carbon arc (Bril, 1968).

For the case of low pressure plasmas, where relatively few particle collisions occur, and also with other plasmas where processes such as mass and energy transport may be irreversible, a complete equilibrium is not reached. This results from an inefficient energy transfer between electrons and the much heavier atomic or molecular constituents because of the large mass differences involved. In this nonthermal situation, it is more meaningful to express temperatures of the different plasma species in terms of their kinetic energies. On this basis, electron temperatures are usually greater than, or equal to, the other species temperatures. The radiation properties of a plasma can be associated with an excitation temperature which is usually smaller than the particle kinetic temperature. This is due to the fact that radiation losses, in an essentially wall-less plasma, are more rapid than the collisional excitation process.

As energy transfer to rotational, vibrational, and electronic levels is primarily a collision process, it is also meaningful to define individual rotational, vibrational, and electronic temperatures for plasma species. The question of the effect of different "temperatures" for each degree of freedom on the composition behavior has been discussed by Manes (1969). He suggests that such systems may actually resemble equilibrium systems in their composition-dependent properties.

3. *Kinetic Aspects*

An extensive fundamental discussion of the kinetic aspects of plasmas has been given recently by Polak (1971).

One can best visualize why the various degrees of freedom for plasma species can have different kinetic temperatures—and hence different degrees of departure from equilibrium—in terms of the characteristic times for various interaction and relaxation processes. For plasma temperatures between 3000 and 15,000 K, and densities corresponding to approximately atmospheric pressure, the characteristic times are

- time between collisions $\sim 10^{-9}$ sec
- relaxation of molecular rotation $\sim 10^{-9}$ sec
- vibrational relaxation $\sim 10^{-7}$ sec
- time for dissociation, i.e., for O_2, $\sim 10^{-7}$ sec
- time to overcome a 60 kcal mol^{-1} activation energy $\sim 10^{-5}$ sec.

We should recognize that the classical rate expressions of gas kinetics may not necessarily apply to plasma processes. As a consequence of the significant population of various excited states, reactions may be multichannel processes, and consideration of an individual reaction path and rate for each quantum state may be more appropriate. Basic information of this kind is practically nonexistent, although the recent availability of lasers should permit future experimentation on energy-channeled rate processes.

a. Kinetic Models for NO Production

A kinetic description of the chemical transformations occurring during the quenching process is rarely possible, due to the lack of elementary rate data and experimental temperature–time histories. The constricted-arc-plasma production of NO in the N_2–O_2 system is one of the few examples where a kinetic model has been developed and compared with experiment (Timmins and Ammann, 1967). During the plasma quenching process, the important reactions affecting the steady-state production of NO are considered to be

$$2NO = N_2 + O_2, \tag{1}$$

$$O + O + M = O_2 + M, \tag{2}$$

$$N + O + M = M + NO, \tag{3}$$

$$N + N + M = N_2 + M, \tag{4}$$

$$NO + O = O_2 + N, \tag{5}$$

$$NO + N = O + N_2. \tag{6}$$

Reaction (1) is unimportant above 2500 K, where dissociation of the molecular species becomes significant. The atom-recombination processes (2), (3), and (4) are exothermic and are therefore approximately independent of temperature. Reactions (5) and (6) are fast in the reverse direction and lead to a rapid production of NO at relatively low temperatures.

From mass and energy transfer calculations, a gas quenching temperature–time history can be determined from experimental parameters. Quenching rates were calculated by Timmins and Ammann (1967) to be of the order of 10^5 K sec^{-1}. Combining this information with the known rates for the above reactions, leads to a knowledge of species concentrations at various temperatures during the quenching process—as shown, for example, in Fig. 3.8. The relative importance of the various reactions is revealed by these steady-state species concentration versus temperature profiles. Note that the formation of NO occurs mainly during the 4000 → 3000 K quench temperature interval. It is also apparent, from Fig. 3.8, that this NO production is associated with a rapid loss of N-atoms; that is, the reverse reaction (5) determines, to a large extent, the production of NO. When N atoms are present, the loss of NO due to the reaction (6) is also important.

As these calculations indicate about a 2% final production yield of NO, as compared with experimental values of about 6 %, it appears that additional NO-forming reactions should be included in the kinetic model. It is reasonable to assume that the reactor walls could enhance the recombin-

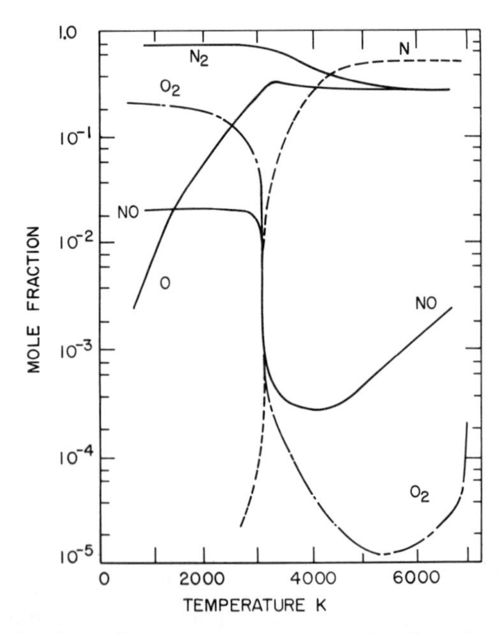

Fig. 3.8 Calculated species concentrations as a function of temperature in the N_2–O_2 plasma quenching model. (Reprinted from R. S. Timmins and P. R. Ammann, *in* "The Application of Plasmas to Chemical Processing" (R. F. Baddour and R. S. Timmins, eds.), by permission of the M.I.T. Press, Cambridge, Massachusetts. Copyright ©1967 by The M.I.T. Press.

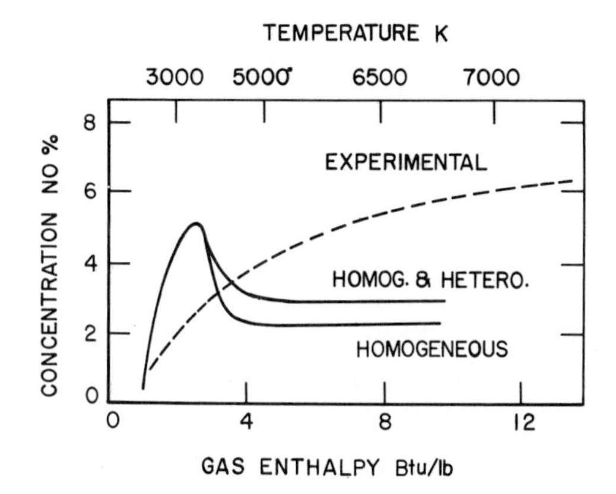

Fig. 3.9 Comparison of calculated and experimental NO concentrations in a quenched air plasma. (Reprinted from R. S. Timmins and P. R. Ammann, *in* "The Application of Plasmas to Chemical Processing" (R. F. Baddour and R. S. Timmins, eds.), by permission of The M.I.T. Press, Cambridge, Massachusetts. Copyright ©1967 by The M.I.T. Press.).

ation processes, i.e.,

$$N + N + \text{wall} \rightarrow N_2,$$

$$O + N + \text{wall} \rightarrow NO,$$

and

$$O + O + \text{wall} \rightarrow O_2.$$

Inclusion of these reactions into the kinetic calculations increases the NO production by less than 1%, to a level which is still several percent lower than the experimental result.

A comparison of the calculated and experimental NO yields, obtained under various plasma gas enthalpy conditions, is given by Fig. 3.9. The disparity between theory and experiment is probably due to the combined uncertaintities in the basic rate data, the derivation of the gas quenching time–temperature history, and experimental difficulties in determining gas temperatures and sample collection.

b. Heterogeneous Plasmas

The yields for heterogeneous plasma reactions tend to differ more significantly from the thermodynamic predictions than those for completely vapor-phase processes. This is attributable to both vaporization and condensation kinetic barriers. There is evidence of difficulty in achieving equilibrium between vapor and condensed species, and the low yield of Al from an Al_2O_3 plasma reduction may be due to such a vaporization-barrier effect. Alternatively, a condensation barrier may arise, owing to a relatively high number of collisions being required for nucleation. Thus, in the case of a fast quench, a homogeneous reaction which requires fewer collisions may be more favored than a heterogeneous reaction.

A probable example of the condensation barrier effect appears in the plasma production of acetylene, as discussed by Reed (1967). At temperatures of greater than about 5000 K the homogeneous reaction

$$2C + 2H = C_2H_2$$

is favorable and the equilibrium constant—and hence, production of acetylene—increases as the system is cooled. However, at about 4000 K, the heterogeneous reaction

$$C_2H_2 = 2C(s) + H_2$$

becomes more favorable and the formation of condensed carbon, at the expense of acetylene, should increase with decreasing temperature. In practice, for rapid quenching, this latter reaction does not reach equilibrium and an excess quantity of acetylene is thereby collected.

Another known rate limitation occurs with the decomposition of oxide particles of Al_2O_3, CuO, NiO, and TiO_2 in a 1-atm thermal argon plasma, supported by 4-MHz induction heating. In this case the rate-limiting step appears to be the rate of heat transfer across a vapor boundary layer surrounding the particles (Borgianni *et al.*, 1969).

4. *Plasma Composition*

The composition of plasmas is of primary importance to their utility as chemical reactors. In most plasmas, the chemistry is predominantly dissociative and the major chemical components are atoms, which may be electronically excited, and, to a lesser extent, singly charged positive ions. The regeneration of chemically bonded species then occurs as the plasma atoms are transported from the plasma and allowed to undergo recombination prior to and during quenching of the products. While thermodynamic and kinetic models provide some limited insight into the various stages of plasma chemical structure, it is also desirable to have direct experimental evidence.

Optical spectroscopic, e.g., see Kanaan *et al.* (1966), and mass spectrometric techniques, e.g., see Spokes and Evans (1965), are available experimental tools that are suited to the identification of plasma species. The latter technique potentially provides a greater diversity of analysis, but, in practice, a meaningful mass spectrometric sampling of high temperature and relatively high pressure gases is difficult (see Chapter 5, Section II.F).

O'Halloran *et al.* (1964) have described a time-of-flight mass spectrometric system for the analysis of argon plasma jets exhausting into the atmosphere. A total ion concentration of about 10^{13} ions cm^{-3} and the following individual ions and neutrals were determined: Ar^+, O_2^+, N_2^+, H_2O^+, O^+, N^+, Ar, O_2, N_2, and H_2O. From a later study the sensitivity of the analytical system to neutral plasma species was determined to be about 1 in 10^4 (O'Halloran and Walker, 1964). Mass separation effects and other perturbations on the sampling process were also identified in this work.

C. Chemical Applications

The major advantages to be gained from the use of plasmas as chemical reactors are

- ease of attainment and control of very high temperatures;
- the availability of clean "flames" which may be comprised of a single element, in contrast to the conventional multicomponent combustion flames;
- compact source of high energy;
- high heat transfer rates;
- high material flow rates; and
- high radical concentrations favoring the production of endothermic products.

The main areas of application for plasma chemical transformations are considered as follows.

1. Ore Transformations and Extractive Metallurgy

The potential applications of plasmas in extractive metallurgy have been recognized for at least a decade, e.g., see the review of Warren and Shimizu (1965), and the recent development of plasma torch devices on a pilot plant scale has generated a commercial interest in such applications (Thorpe, 1971, 1972; Sayce, 1972). For example, plasma devices and procedures have been tested for the production of ZrO_2 from $ZrO_2 \cdot SiO_2$ (zircon sand) and of TiO_2 from $TiCl_4$ stock. For relatively high cost materials such as these, the plasma techniques appear to be economically viable. Plasmas also appear to have ore beneficiation (breakdown) applications in separating MnO_2 from rhodonite ore (a manganese silicate), and in the generation of $BeCl_2$ from the chlorination of beryl ($3BeO \cdot Al_2O_3 \cdot 6SiO_2$).

Using an argon plasma jet reactor, Brown (1971) has obtained separations of iron and titanium from ilmenite ores ($FeTiO_3$) in the presence of

H_2 and NH_3. Reduction of iron ore to metallic iron has also been achieved in a dc plasma jet in the presence of hydrogen. Ore reductions of up to 64% were obtained. Similarly, ferric oxide has been reduced to the metal, but with 100% yield. Reductions of WO_3 and Ta_2O_5 to the metals have also been achieved but attempts to reduce TiO_2 and ZrO_2 were unsuccessful. For Al_2O_3, only a small amount ($\sim 1\%$) of reduction to the metal is normally obtained. However, by using very low feed rates, Rains and Kadlec (1970) have obtained a 20% reduction in an Ar–H_2–CO–CH_4 plasma. Apparently the oxide vaporization is the rate-limiting factor.

The carbothermic chlorination (see Section VII.B) of metal oxides has also been achieved in a plasma-arc.

Further details of these metallurgical processes may be found in the recent reviews of Stokes (1971) and Mahe (1973).

2. *Reduction of Halide to Metal*

The preparation of Zr metal from the tetrahalides, $ZrCl_4$ and ZrI_4, using a plasma-arc reduction process, has been demonstrated, e.g., see Rolsten (1961). Similar halide reductions have been used for the preparation of B, Nb, and Ti (see Sayce, 1972). From equilibrium thermodynamics, these halides should be completely dissociated at temperatures as low as 2100 K for the operating pressures of about 1 Torr. Arc temperatures are considerably greater than this, and hence, an effective reduction is possible, provided the quenching of reaction products from the arc is sufficiently rapid to prevent recombination of metal and halogen atoms. Fortunately, the production, stabilization, and collection of endothermic species, such as atoms, by plasma reduction is favored by the high entropy required for reassociation processes.

It has been demonstrated that various metal halides can be irreversibly dissociated to the elements in relatively low energy (<600 Watt) rf discharges (McTaggart, 1969). In these nonthermal discharges, it is believed that the dissociation results from impact by highly energetic electrons. Hence, it is not surprising that the metallic yields (70% or less) increase with a decrease in the bond dissociation energies of the parent alkali-halide molecules. For multivalent systems, such as the group II and group III halides, the products usually contained both metal and the initial halide. This observation suggested that the major plasma species were reduced halides, e.g., AlCl, which on condensation disproportionated, i.e.,

$$3AlCl \rightarrow 2Al(s) + AlCl_3.$$

The presence of AlCl, but not $AlCl_2$, in the plasma was established by mass spectrometric sampling experiments.

In connection with the chemical transformation of metal halides in plasmas, it should be mentioned that the presence of these halides can cause an unstable plasma operation. This is thought to be due to endothermic or electron capture processes affecting the plasma-coupling properties. It is, therefore, desirable to inject the halides into the tail of the plasmas (Audsley and Bayliss, 1969).

3. Chemical Synthesis

There have been numerous applications of electric discharges to laboratory scale synthetic chemistry. In most cases, the plasmas used were of low density and hence the reactions would not have been predictable from equilibrium thermodynamics. Only in a few instances have investigations of the reactive intermediates been made, e.g., see the mass spectrometric sampling study of Brinckman and Gordon (1967).

A number of reviews and detailed papers on the applications of plasmas to chemical processing may be found in Baddour and Timmins (1967). The applications described include nitrogen fixation, i.e., NO and HCN production (Timmins and Ammann, 1967); graphite–hydrogen reactions, e.g., acetylene production, (Clarke, 1967); and fluorine reactions, e.g., production of fluorocarbons (Bronfin, 1967).

It is interesting to note that the thermodynamics of CH_4–N_2 mixtures predicts the formation of about 10% of HCN and C_2H_2, together with H_2, as the major products of reaction for temperatures in the region of $\sim$3000 K, and that similar amounts of product are extracted from a rapidly quenched thermal plasma (Bronfin, 1969). This agreement between thermodynamic prediction and experiment is somewhat fortuitous, as the products are formed during a rapid quenching process where not all reaction paths necessarily lead to a local equilibrium state.

A number of reduced halides have been prepared by glow-discharge plasma reductions of gaseous halides. For example, BCl_3 yields B_2Cl_4, $SiCl_4$ yields Si_nCl_{2n+2}, $GeCl_4$ yields Ge_2Cl_6, $TiCl_4$ yields $TiCl_3$, and PCl_3 yields P_2Cl_4 (Jolly, 1969). Good yields (60–90%) of $TiCl_3$(s) were obtained by an argon thermal plasma-torch reaction of H_2 and $TiCl_4$. No reduction was found in the absence of H_2 (Miller and Ayen, 1969).

Some applications of plasmas to chemical synthesis utilize the plasma as a source of reactive gas which then impinges on a substrate outside the plasma region. For example, a plasma was used to generate active nitrogen, which, when directed onto an aluminum oxide pellet, produced the nitride AlN (Matsumoto *et al.*, 1968). Similarly, a plasma may be used to generate oxygen atoms which react with solid alkali hydroxides to yield superoxides (Hollahan and Wydaven, 1973). Superoxides are of interest as solid sources of molecular oxygen.

Recently, a centrifugal plasma furnace has been used for the preparation of surface active silica and alumina powders (Everest *et al.*, 1971; Barrett *et al.*, 1973). The surface activity of these powders relies on a hydroxylated surface condition, and it is perhaps pertinent to recall the comments made elsewhere concerning the formation of silicon and aluminum hydroxide vapor species—see Chapter 2, Section VIII.E. In fact, Barrett *et al.* (1973) note that the inability to achieve the same degree of hydroxylation for alumina as for silica may be related to a greater tendency for silica to form polymeric hydroxysilica species. This suggestion is in accord with the interpretations of oxide solubility in steam as given elsewhere (Chapter 2, Section VIII.E).

Plasma processes have also been used for producing coatings of zirconia and tungsten on rocket components (Mash, 1962).

A number of condensed metal nitrides, carbides, and cyanides have been produced from plasmas; these include: TaN, TiN, AlN, ZrN, HfN, Mo_2N, W_2N, Mn_3N_2, and Si_3N_4; TaC, Ta_2C, UC, WC, W_2C, Al_4C_3, VC_2, TiC, SiC, and VC; LiCN, $Ca(CN)_2$, $Mg(CN)_2$, $Zn(CN)_2$, and $Al(CN)_3$—as reviewed by Mahe (1973).

Plasmas may also be used to promote kinetically hindered chemical transport (see Section III) as shown recently by Veprek (1972).

Appendix. Some Additional Recent Literature on Plasma Chemistry

Anon. (1970). *Eng. Mater. Des.* **13,** 315. Metal spraying.

Audsley, A., and Bayliss, R. K. (1967). The Oxidation of Silicon Tetrachloride in an Induction Plasma Torch. Nat. Phys. Lab. IMU Rep. Ext. 5.

Blasingame, J. (1968). *Mater. Protection* **7,** 21. Metalizing takes back seat to flame spray.

Bourasseau, D., Cabannes, F., and Chapelle, Jr. (1970). *Astron. Astrophys.* **9,** 339. A study of local thermodynamic equilibrium in the argon plasma jets.

Dauvergne, J. P. (1967). *Rev. Hautes Temp.* **4,** 155. Homogeneous condensation of high melting oxide vapors in a plasma jet close to a cold surface.

Dauvergne, J. P., and LeGoff, P. (1967). *Rev. Hautes Temp.* **4,** 163. Simultaneous transfer of material and heat between a plasma and a cooled surface.

Dundas, P. H., and Thorpe, M. L. (1970). *Chem. Eng. Progr.* **66,** 66. Titanium dioxide production by plasma processing.

Dundas, P. H., and Thorpe, M. L. (1969). *Chem. Eng.* **76,** 123. Economics and technology of chemical processing with electric-field plasmas.

Feldman, H. F., Simons, W. H., and Bienstock, D. (1969). Calculating Equilibrium Compositions of Multiconstituent, Multiphase, Chemical Reacting Systems. PB-184316 (RJ-7257).

Gussein, M. A., Kir'yanov, Y. G., Rabotnikov, N. S., Sakhiev, A. S., and Syrkin, V. G. (1970). *Sov. Powder Met. Metal. Ceram.*, No. 12, 966. Manufacture of ultrafine iron powder by the plasma decomposition of iron pentacarbonyl.

Kubanek, G. R., and Gauvin, W. H. (1967). *Can. J. Chem. Eng.* **45,** 251. Recent developments in plasma jet technology.
Luzhnova, M. A., and Raikhbaum, Y. D. (1969). *High Temp.* (*USSR*) **7,** 283. Kinetics of metal particle evaporation in an arc plasma.
Nikitin, I. V., and Rosolovskii, V. Y. (1970). *Usp. Khim.* **39,** 1161. Reactions of fluorine and fluorides of non-metals in electric discharge.
Nishimura, Y., Takeshita, K., Adachi, Y., Nakashio, F., and Sakai, W. (1970). *Int. Chem. Eng.* **10,** 133. Pyrolysis of propane in an induction-coupled argon plasma jet.
Rains, R. K., and Kadlec, R. H. (1969). *Can. Chem. Eng. Conf., 19th and Symp. Catal., 3rd, Edmonton, 1969.* The reduction of Al_2O_3 to Al in a plasma.
Rautschke, R. (1967). *Proc. Int. Colloq. Spectrosc., 14th.* **2,** 487. The influence of chemical reactions in the DC arc on spectroscopic results.
Rykalin, N. N. (1968). *Vestn. Akad. Nauk SSR,* No. 3, 33. Plasma processes for obtaining and treating materials.
Thring, M. W. (1966). *Pure Appl. Chem.* **13,** 329. Plasma engineering.
Vukanovic, V., Pavlovic, B., and Ikonomov, N. (1967). *Proc. Int. Colloq. Spectrosc., 14th.* **2,** 515. Composition of the atmosphere of an arc discharge from the spectrochemical viewpoint. (a) Influence of nitrogen- and H_2O-atmospheres.

VII. Vapor-Phase Metallurgy

A. Introduction

1. *Background*

The main question to be considered in the following discussion concerns the role of high temperature vapors in metallurgical processing. There is an increasing interest in the development of new pyrometallurgical and vapometallurgical procedures where high temperature vapors are involved, as opposed to more conventional condensed phase processes.† Much of the current impetus to seek such alternative processing methods arises from the existing forces-for-change of energy and mineral shortages, together with the desire for an increased environmental control during metallurgical operations.

A survey of relatively recent patent literature, relating to the commercial preparation of inorganic materials, indicates that at least 10% of the processes involve high temperature vapor-phase interactions as an essential part of the process (see Sittig, 1968). There are also many additional processes that involve gas–solid reactions at high temperatures, but without appreciable volatilization of the metals.

† See, for example, the comparison of the early symposium—*Disc. Faraday Soc.* **4** (1948), and the more recent metallurgical discussions edited by Jones (1972) and by Kubaschewski (1972), in addition to the reviews of Sale (1971) and Fletcher (1969).

The general principles of extractive metallurgy, and also an extensive bibliography, may be found in the book of Habashi (1969). Some of the physicochemical aspects of process metallurgy are also discussed in the articles edited by St. Pierre (1959, 1961). Discussions of vapor-phase metallurgy, in particular, may be found in the book of Powell *et al.* (1966).

As recently as 1962, the concern was expressed that existing metallurgical processes are based almost entirely on the findings of empirical research efforts (Bakish *et al.*, 1962). That is, there is very little fundamental understanding of the processes at the molecular level. The still inadequate degree of understanding that currently exists will be revealed by the ensuing discussion.

Basically, for high temperature vapometallurgical processes, the controlling factors are the thermodynamic, kinetic, and mass transport system characteristics (see Chapter 1, Section V). Discussions of these basic aspects have been given elsewhere in connection with the related topics of Chemical and Physical Vapor Deposition Processes (Section II), Chemical Transport Along a Temperature Gradient (Section III), Enhanced Volatility Processes and Complex Vapors (Section IV), Plasma Chemistry (Section VI), Ore Genesis (Section VIII), and particularly the discussion of Gas–Solid Reactions (Chapter 2).

2. *Scope of Application for High Temperature Vapors*

Taking a broad view, the following areas may be identified where high temperature vapors play a metallurgical role.

a. *Reduction of Ore to Metal by Gas–Solid Reactions*

The metallurgical operations usually involved in the conversion of ore to pure metal are: ore dressing, opening the ore, recovery, metal preparation, and metal refining to the pure state. The extractive procedures used may be either wet, i.e., aqueous, or dry processes and in the latter case high temperature gas–solid reactions are usually involved. Dry ore-reduction processing commonly involves a reaction of solid metal oxides with reducer gas ($H_2 + CO + CO_2 + H_2O$)— as derived from the in situ partial oxidation of coal. In this process the free energy change ΔF for the production of CO by oxidation of C(s) becomes more favorable, i.e., more negative, with increasing temperature, T. This is mainly a consequence of the positive reaction entropy and its increasing contribution to ΔF as T increases. On the other hand, the oxidation of C(s) to CO_2 has an insignificant entropy change, and hence little temperature dependence. Now the standard free energy change for the production of a solid metal oxide from metal and oxygen, becomes less negative with increasing T, owing to an entropy decrease. Hence, the chemical stability of the metal oxides decreases relative to CO with increasing temperature. The reduction of metal oxides to their metallic condition using CO, as derived from C(s), is therefore favored by high temperatures. However, the general usefulness of this process type is limited by the formation of stable carbide phases and also volatile suboxides, such as Al_2O. In this instance, the formation of high temperature vapors is undesirable to the metallurgical process.

b. Volatilization and Transport of Metal-Containing Species, Such As Halides and Sulfides

Gas–solid reactions can also be utilized for the desirable production of a volatile metal-containing compound, and this is particularly the case for halogenation and carbonylation processes. In effect, to borrow a phase attributed to Lord Kelvin, high temperature vaporization processes can

provide wings to metals.

Individual metals can then be separated by techniques such as fractional distillation and disproportionation. A list of metallurgical processes where high temperature vapor species are known to be significant is given in Table 3.9. Both commercial and patented—potentially commercial—processes are indicated.

c. The Use of Vapor Transport and Vapor Pressure Techniques for Determining the Thermodynamic Activities of Condensed Metallurgical Phases

Komarek (1972) has recently reviewed this aspect. An interesting example involves use of the equilibrium reaction

$$2\text{Al (in alloy)} + \text{AlF}_3(\text{s}) = 3\text{AlF},$$

where the vapor pressure of AlF is used to determine the activity of aluminum in the alloy (Gross *et al.*, 1959).

d. Laboratory Experiments for Simulating and Testing Ore-Forming Theories

In addition to the discussion of orogenesis given in Section VIII, an interesting demonstration of this point can be made in connection with the following moon-rock-dating problem. Silver (1972) argues that the observation of apparent inconsistencies in the dating of moon rocks by the various isotopic ratio methods—including the use of Pb isotopes—may be due to high temperature vapor transport of volatile metals such as lead. Most of the variation in the dating results can be explained in terms of a gain or loss of lead at some time in the rock sample's history. Silver (1972) demonstrated experimentally that an in vacuo volatilization of lead from

TABLE 3.9

Examples of Vapor Phase Related High Temperature Metallurgical Processes[a]

Process	Reference[b]
Fire refining copper	Jeffes and Jacob (1972)
Oxidation and vaporization of impurities, e.g., S, Pb, As, Sb, using oxidizing flame heating at 1200°C	
Copper TORCO segregation process[c,d]	Brittan (1970)
$CuO(s) + Cu_2O(s) + NaCl(s) + C(s) \rightarrow Cu_3Cl_3 \rightarrow Cu(s)$, at 825°C	
Ferrous chloride PRMS process for iron powder production[c]	Rigg (1972)
$H_2 + FeCl_2(c) \rightarrow Fe(c) + HCl$, ~800°C; $FeCl_2$ and $MnCl_2$ vapor transport occurs	
Pyrite-cinder processing	Remirez (1968), Fletcher (1969)
$FeS_2(ore) \xrightarrow{\text{roast/air}}$ pyrite-cinder (iron oxide + impurities); pyrite-cinder + $CaCl_2(c) \xrightarrow{\sim 1250°C}$ chloride vapors of Cu, Zn, Pb, Au, Ag, Bi, As; the remaining ferrous residue is then suitable as a blast furnace feed	
Zirconium extraction from zircon sand[c]	O'Reilly *et al.* (1972)
$ZrO_2(s) + 2C(s) + 2Cl_2 \rightarrow ZrCl_4 + 2CO$	
Iodide processes[c]	
e.g., crude $Zr(s) + 2I_2 \xrightarrow{\text{hot}} ZrI_4 \xrightarrow{\text{very hot}}$ pure $Zr(s) + 2I_2$	
Float glass process[c]	Hayes (1972), Pilkington (1969)
Formation of SnO and SnS vapor species undesirable to glass production	
Zinc blast furnace Imperial Smelting Process[e]	Lumsden (1972)
$ZnO(s) + CO = Zn + CO_2$	
Pig iron blast furnace	Vidal and Poos (1973), Ward (1962)
Problems of silica reduction and vapor transport of Mg and alkali metals, e.g., $SiO_2(c) + C(c) \rightarrow SiO + CO$; $MgO(c) + C(c) \rightarrow Mg + CO$; $Na_2O(c) + C(c) \rightarrow 2Na + CO$	

TABLE 3.9 (*Continued*)

Process	Reference[b]
Selenium extraction from copper tailings	Dutton *et al.* (1971)
Vaporization of Se from crude, followed by oxidation, i.e., $Se + O_2 \rightarrow SeO_2$, 500–600°C	
Additional patented processes[f]	
Mo metal from H_2 reduction of $MoCl_5$ vapor	
Ta metal from H_2 reduction of $TaCl_5$ vapor	
Cu-plating from argon transpiration of CuCl vapor (Cu_3Cl_3) at 240–290°C followed by H_2 reduction	
$FeCl_3$ from ore using $C + Cl_2$ reactants[c]	
$CoCl_2$ from ore using $FeCl_3$ vapor as a chlorinating agent at 850°C (see Section IV)	
$CrCl_3$ from ore using $C + Cl_2$ at 850°C and a distillation separation of $CrCl_3$ and $FeCl_3$[c]	
V_2O_5 from ore using HCl or Cl_2 at 1000°C followed by distillation of $VOCl_3$ and steam oxidation to give V_2O_5	
AlAs(s) from $Al(s) + As(c) + I_2 \rightarrow AlAs(s)$	
Al(s) from AlN and Al_2O_3, i.e., $2AlN(s) = 2Al + N_2$; $Al_2O_3(s) + 4Al \underset{\text{cool}}{\overset{\text{heat}}{\rightleftarrows}} 3Al_2O$ (i.e., basically a disproportionation of Al_2O)	
$MoO_3(s)$ from $MoS_2 + O_2 \rightarrow MoO_3 \text{ vapor} \xrightarrow{\text{cool}} MoO_3(s)$	
Various titanate crystals from flames	

[a] At least the first eight listed processes are known to be in commercial use.

[b] The references listed are not necessarily the primary source but serve rather as recent examples of basic studies relating to the processes indicated.

[c] See more detailed discussion in the main text.

[d] Nickel metal can also be separated from the normally intractable iron laterite ore, using an analogous procedure to that for Cu, at temperatures of 800–1000°C (Iwasaki *et al.*, 1966).

[e] The kinetics of the reverse reaction is of particular interest in connection with the unwanted oxidation of Zn in cooler regions of the furnace (Stott and Fray, 1972).

[f] See the compilation of Sittig (1968); the commercial status of these processes is not known.

lunar samples occurs over the temperature range of 550–980 °C. Since these temperatures are well below those expected to have been present in the early moon history, the vapor transport of lead could have occurred, as suggested.

3. *Future Prospects*

As has been suggested by Kellogg (1966), the recently developed body of information concerning high temperature vapors provides a basis for extending their application to metallurgical problems. However, we should recognize that this basic foundation is far from being complete.

For practical metallurgical utility, vapor transport systems should have vapor pressures of the order of 0.1 atm. This requires an extrapolation of most existing vaporization data, as obtained under Knudsen effusion conditions ($\lesssim 10^{-4}$ atm), over a pressure interval of about three orders of magnitude. For this purpose, accurate heat-of-formation data is required. Also, for many metallurgical systems a knowledge of multicomponent high temperature vapor interactions will be necessary. For example, the prospect of separating SnS from an ore containing PbS, by vaporization, may be complicated by the possible formation of complex species such as $SnPbS_2$. In the absence of experimental data for most complex systems, it is useful to develop correlative predictions using the known data for systems with fewer components. For instance, the conditions leading to the formation of $SnPbS_2$ can be determined from the reaction

$$(SnS)_2 + (PbS)_2 = 2(SnPbS_2)$$

using analogous thermodynamic arguments to those given in Section IV for complex halide systems. From the rules given in Section IV we can show that for the above metal sulfide exchange process

$$\Delta H \sim 0 \text{ kcal mol}^{-1}$$

and

$$\Delta S \simeq 1.37 \text{ cal deg}^{-1} \text{ mol}^{-1},$$

where the entropy change is attributed to the reduction in symmetry which results from the formation of a presumably cyclic $SnPbS_2$ species. Hence the equilibrium constant has a predicted magnitude of 1.9 at 1100 K. This compares very favorably with the known experimental value of 1.6 (Colin and Drowart, 1962).

Systems that produce vapor species in a reduced valence state such as

$$Al_2O_3(c) + Al(c) \rightarrow 3Al_2O,$$

$$AlCl_3 + 2Al(c) \rightarrow 3AlCl,$$

or

$$SnO_2(c) + Sn(c) \rightarrow 2SnO,$$

are potentially useful for metal preparation by a low temperature disproportionation of the reduced valence species—that is, by a reversal of the above reactions at reduced temperature. Such oxide systems also exhibit a sharp maximum in vapor pressure at a critical oxygen pressure. This observation has important implications in the possible utility of metal-fuming processes, as the control of oxygen pressure may be used to selectively extract the metal of interest.† For example, it can be thermodynamically shown that indium could be produced, as In_2O, by a fuming process at temperatures of 1500–1600 K and at about 10^{-9} atm oxygen pressure (Kellogg, 1966). Another aspect to this critical oxygen dependence is demonstrated by the vacuum distillation of minor elements from liquid ferrous alloys where the removal of Si and Al is particularly sensitive to the oxygen content of the melt, because of the ready formation of the volatile suboxides SiO and Al_2O (Olette, 1961). The Pb–O_2 system represents a somewhat anomalous case in that the rate of release of Pb-containing species to the vapor phase is almost independent of the oxygen pressure, and hence the partial pressure of O_2 cannot be used to control the lead vaporization. This results from atomic-Pb being a predominant vapor species at low O_2 pressures, and at high O_2 pressures PbO, $(PbO)_2$, $(PbO)_4$, and $(PbO)_3$ are major vapor species.

The thermodynamic aspects of a potential process for extracting tin, in the form of the sulfide, from low-grade oxide ore have been evaluated by Kellogg (1966). The basic process would involve the interaction of S_2 vapor with $SnO_2(s)$, i.e.,

$$SnO_2(s) + S_2 = SnS + SO_2.$$

An optimum set of reaction conditions is considered to be a temperature of 1300 K, an S_2 pressure of 10^{-2}–10^{-1} atm, and an SO_2 pressure of about 10^{-1} atm. An upper partial pressure limit for sulfide species is set by the need to prevent formation of a liquid SnS phase which could be troublesome in a fluidized bed process. In actual practice, the oxygen activity would be reduced by the addition of carbon (coal) to the tin oxide feed ore.

Another aspect of tin species volatility occurs with the commercial float glass process. The volatility of SnS and SnO can lead to undesirable faults in the glass, and stringent control over the sulfur and oxygen impurity levels is therefore required in the tin–glass liquid system. The species

† An example of a commercial fuming process is given by the recovery of rhenium as Re_2O_7 vapor from the flue gas stream of molybdenite roasting plants, e.g., see Coudurier *et al.* (1970).

volatilities are controlled by reactions such as

$$Sn(l) + O[Sn(l)] = SnO$$

and a feasible method of reducing SnO formation would be to lower the activity of Sn(*l*) by alloying with a metal such as Au, as was indicated by Hayes (1972). In the absence of alloying, typical partial pressures of SnO and SnS over the molten system are determined by thermodynamic calculation to be 2.47×10^{-6} and 1.4×10^{-4} atm, respectively.

B. Chlorination Processes

1. Scope of Application

As chemical reagents, chlorine and hydrogen chloride are relatively inexpensive, particularly in comparison with the other halogens where Br_2, I_2, and F_2 are factors of about 6, 30, and 60 times more expensive than Cl_2, respectively. The two main chemical advantages to be gained by chlorination are the high solubility of metal chlorides for hydrometallurgical processing, and the high metal chloride volatility for vapo- or pyrometallurgical processing. Also, the prospect of regenerating the chlorinating reagent should allow for closed-loop operations.

One of the limitations associated with chlorination is the difficulty in recovering chlorine from some of the halogenated gangue minerals and also the undesirable formation of large amounts of iron chlorides. Dechlorination of ferric chloride, for example, is therefore an important part of chlorination metallurgy in general. Henderson *et al.* (1972) have recently shown that ferric chloride dechlorination by the process

$$2FeCl_3(c) + \tfrac{3}{2}O_2 \rightarrow Fe_2O_3(s) + 3Cl_2$$

can be 95% effective at 650 °C, or alternatively at 500 °C in the presence of a NaCl catalyst. The chlorination of sulfide ores can also lead to an inefficient use of the halogen owing to the formation of sulfur chlorides, e.g., see Ingraham and Parsons (1969).

Corrosion of the reactor by the halogen gases also represents a practical limitation. Nickel or its alloys, frequently used as container materials, have an upper limit operating temperature of about 500 °C. Paulson and Oden (1971) have found that MgF_2 has a greater resistance to corrosion in wet or dry Cl_2 atmospheres, and that a temperature increase of about 100 °C over that allowed by Ni is possible. On the other hand, CaF_2 and SrF_2 were found to be less halogen resistant than Ni.

Several early examples of ore chloridization processes, such as the use of NaCl for the production of $PbCl_2$ as a vapor from sulfide ore concentrates, are discussed in the reviews of Ralston (1924) and Oldright (1924). Existing commercial chlorination processes include extraction from the ore of rare metals—e.g., Ti, Zr, and Nd—and also the more common metals of Cu, Fe, Pb, and Zn, and the detinning of tin plate, e.g., see Remirez (1968). Processes that have been demonstrated to have commercial feasibility include the separation from their ores of Ni (Vucuroic and Miodrag, 1969; Iwasaki *et al.*, 1966), W, Nb, Ta (Henderson, 1964), Au, Al, Cr, the rare earth, and the transuranium elements, in addition to the processing of nuclear fuel—as discussed in Chapter 6, Section III.E.

The prospect of metal purification by gas chromatographic separation of metal chlorides has also been demonstrated recently, e.g., see the review of Guiochon and Pommier (1973). In particular, the use of $AlCl_3$ complexing (see Section IV) has allowed the volatilization of rare earth and transuranium elements for chromatographic separations at the relatively low temperatures of 200–250 °C (Zvarova and Zvara, 1969, 1970a, b).

While chlorination processes have thus far been of a relatively modest scale, metallurgists predict future use of this general approach on a much larger scale. In view of such a prospect, a supplemental bibliography of the more pertinent recent research literature (patents excluded) may be of value and is given as an appendix to this discussion of vapor-phase metallurgy.

2. *Rare Metal Extraction*

Ores containing minerals of the rare metals Ti, Zr, Hf, U, or Be may be mixed with carbon and reacted with chlorine gas at high temperature (700–1000 °C) to produce gaseous metal chloride products, e.g.,

$$MO_2(s) + 2C(s) + 2Cl_2 \rightarrow MCl_4 + 2CO.$$

Flow charts describing these processes have been given by Jamrack (1963). In the absence of carbon these reactions are thermodynamically unfavorable. Thus for the example of ZrO_2, $\Delta F_{1100} = +36.9$ kcal mol^{-1}, whereas in the presence of carbon $\Delta F_{1100} = -58.8$ kcal mol^{-1}.

The minerals may be in the form of the oxides or as complex mixed oxides, such as zircon ($ZrSiO_4$), ilmenite ($FeTiO_3$), monazite (rare earth and thorium phosphates), and beryl ($3BeO \cdot Al_2O_3 \cdot 6SiO_2$). However, in the case of ilmenite, an inefficient use of chlorine results from the undesirable production of $FeCl_3$, e.g., see Patel and Jere (1960). Hence, the simple oxide rutile (TiO_2) is preferred for the chlorination process. In this instance, the vapor-phase steps include, chlorination of rutile ore (TiO_2 + impurities), condensation of impurity $FeCl_3$ at 200°C, and fractional distillation to remove VCl_3, $SiCl_4$, CCl_4, $COCl_2$, and HCl from the $TiCl_4$

product. For the similar case of Zr extraction, the relatively involatile $ZrCl_4$ is purified by sublimation.

A method for the separation of zirconium and hafnium chlorides, which have very similar volatilities, relies on the reaction

$$ZrCl_2(s) + ZrCl_4 \xrightarrow{400\ ^\circ C} 2ZrCl_3(s),$$

which does not have a hafnium analog, and hence $HfCl_4$ remains in the gas phase. At a higher temperature, $ZrCl_3$(s) can release $ZrCl_4$ by the reverse disproportionation reaction. An account of the extractive metallurgy of Zr and Hf, and the state of the industry, has been given by Schlechten (1968). The chlorination kinetics for the oxides of these rare metals has recently been determined by Landsberg *et al.* (1972).

The chlorination-type process has been suggested as a technique for extracting $ThCl_4$ from monazite at about 1000 °C. By-products would be the volatile chlorides of P, Al, Fe, Pb, Mn, Cu, and the condensed rare earth chlorides which are relatively nonvolatile at this temperature.

3. *Reduction to Metal*

Less-common metals such as Ti, Zr, Nb, U, V, and Be are frequently prepared by high temperature reduction processes. These processes usually involve a reduction of the oxides, the fluorides, or the chlorides by common metals such as Mg, Ca, or Na, e.g.,

$$4Na + MCl_4 \rightarrow 4NaCl(c) + M(c).$$

The slag, e.g., NaCl and soluble impurities, can be removed either by vaporization or, more commonly, by leaching. These reductions take advantage of the relatively high free energies of formation for the metal halides of Groups IA and IIA. However, the stability of subhalide and oxyhalide intermediates, particularly for V and Zr, necessitates careful control of the reactor conditions.

The Imperial Chemical Industries batch process for the production of Ti utilizes a reaction of Na-vapor with $TiCl_4$ vapor, at 850–900 °C, in a mild steel reactor. In an effort to develop a continuous process, rather than the relatively expensive batch-type procedure, Fuhs (1959) investigated the diffusion flame interaction of $TiCl_4$ vapor with liquid Na. The production of some metallic Ti was noted in this experiment. A detailed account of Ti metallurgy, including halide reduction and flame processes, has been given by Barksdale (1966).

The commercial Kroll process involves the use of liquid Mg to reduce $TiCl_4$, and the spongy Ti product contains Mg and $MgCl_2$ impurities which may subsequently be removed by vacuum evaporation processes. Takeuchi *et al.* (1961) found that if Mg and $TiCl_4$ were both allowed to react in the vapor phase, and the Ti product condensed on Ti ribbons, a massive form of relatively pure Ti could be obtained. It should be noted that this gas-phase reaction is not particularly favorable thermodynamically, but that the continuous condensation of Ti allows the reaction to proceed more readily.

Owing to the high thermodynamic stability of HCl, one can also use H_2 for the reduction of metallic chlorides to the metal. For the case of $MoCl_6$ and WCl_6, the gas-phase reaction with H_2 generates a flame, and, by controlling the metal halide partial pressures, it is possible to produce very fine metallic particles, i.e., 0.01–0.1 μm, of metallurgical utility (Lamprey and Ripley, 1962). Oxley and Campbell (1959) have demonstrated the feasibility of producing metals on a continuous basis by H_2 reduction of metal chloride vapors in a fluidized bed reactor, using metal granules as the deposition surface. From kinetic and thermodynamic studies, Rigg (1973) has shown that the production of Co and Ni by hydrogen reduction of the chlorides should be even more favorable than the analogous PRMS ferrous chloride reduction process (see Section B.6). Boron has also frequently been produced by reduction of the gaseous chloride or bromide using H_2 or Zn metal as a reducing agent—see Niselson *et al.* (1971).

4. *The Halomet Process*

A recently suggested modification of the more conventional type of chlorination process, known as the Halomet process, takes advantage of the ability of a metal chloride to chlorinate the oxide of another metal which has a higher chloride affinity (Anon., 1971a). The chloride affinity is essentially based on the free energy differences between the metal chlorides and the oxides. According to Othmar and Nowak (cited in Anon., 1971a), the series follows the order given in Table 3.10. This series does not necessarily follow the same order as that for the carbothermic chlorination process, e.g.,

$$TiO_2(s) + 2C(s) + 2Cl_2 = TiCl_4 + 2CO,$$

and is derived from a consideration of the free energies for reactions such as

$$PbCl_2 + Na_2O(s) = 2NaCl + PbO(s), \qquad \Delta F_{1000} = -65.3 \text{ kcal mol}^{-1}.$$

Thus, in this example, since ΔF is negative, Na would fall higher in the series than Pb. It follows that one would first have to chlorinate the Na_2O(s) component of an ore before the other oxides (i.e., those below Pb) in the series could be converted to volatile metal chlorides.

As an example of the Halomet process, consider the following possible modification of the chlorination process for ilmenite ore ($FeTiO_3$). Follow-

TABLE 3.10

Oxide Chlorination Series (→)

1300 °C	Cu	Pb	Zn	Co	Ni	Sn	Fe	Ti	Ge	Al	Si	V	P	S
800 °C	Pb	Cu	Zn	Co	Sn	Ni	Fe	Ti	Ge	Al	Si	S	V	P

ing the initial normal carbothermic chlorination, the resulting $FeCl_3$–$TiCl_4$ mixture is allowed to react with a second batch of ore, where $TiCl_4$ reacts with the iron oxide component to form ferric chloride and titania (TiO_2). The ferric chloride is then allowed to pass out of the reactor and the residue consists of a titania-enriched gangue, which may be chlorinated to produce a relatively pure $TiCl_4$. This cycle may be repeated for an increase in purity.

It seems probable that the success of the pyrite-cinder processing treatment (see Table 3.9) is due to the lower chloride affinity of the oxides of Fe, as compared with those of Zn, Cu, and Pb, for example.

5. *The Copper Segregation Process*

The TORCO copper segregation process is based on a chemical interaction between Cu-ore, coal or coke, and sodium chloride at temperatures of 750–850 °C, which leads to the deposition of metallic copper on the carbonaceous material, e.g., see Brittan (1970) and cited literature. This process is advantageous in its ability to extract Cu from ores (e.g., vermiculite) that are refractory to more conventional separation techniques.

Basically, the process is carried out in two stages. First the copper oxide- or sulfide-containing ore is heated to a temperature of about 825 °C. Then a mixture of NaCl and coke or coal is added to the heated ore and the copper segregation proceeds. From Brittan's (1970) simulated process studies the chemical mechanism is believed to involve the reactions

$$NaCl(c) + \tfrac{1}{2}H_2O + \tfrac{1}{2}SiO_2(c) = HCl + \tfrac{1}{2}Na_2SiO_3(c),$$

$$3HCl + \tfrac{3}{2}Cu_2O(c) = Cu_3Cl_3 + \tfrac{3}{2}H_2O,$$

$$Cu_3Cl_3 + \tfrac{3}{2}H_2 = 3Cu(s) + 3HCl.$$

Similar reactions involving CuO(s) are also likely. The second of these reactions is believed to be slow and therefore rate controlling. That gas transport is not a rate-limiting factor is attributed to the self-agitation developed in the roaster bed by the liberated gas which retards boundary layer formation. The source of H_2 is either carbon reduction of H_2O or the coal itself.

An interesting aspect of this process is the small amount of NaCl used which is much less than the stoichiometric requirement. Evidently the HCl generated is involved in a cyclic process.

A more recent study by Gross and Stuart (1972) supports the general features of this mechanism with the exception of the nature of the Cu-containing vapor species. Apparently, most of the copper is transported to the vapor phase in the form of the mixed-halide complexes $CuNaCl_2$ and Cu_2NaCl_3, rather than as Cu_3Cl_3. For instance, at 830 °C, about 68% of the Cu is transported in a complex halide form. This observation is in accord with the general predictions of complex formation made in Section IV. Similarly, the analogous segregation of Ag may be the result of mixed metal chloride formation.

In conclusion, one should take note of the possible future extension of this process to the segregation of other metals from their ores, particularly nickel (Dor, 1972).

6. *Iron Powder Production*

A commercial process, known as the Peace River Mining and Smelting (PRMS) ferrous-chloride-reduction process, utilizes *hydrometallurgical* HCl leaching of managaniferous iron ore, or scrap, followed by H_2 reduction of the ferrous chloride crystals, e.g., see Rigg (1972). The reduction process is carried out at a sufficiently high temperature, e.g., ~800 °C, for vapor transport of $FeCl_2$ and the $MnCl_2$ impurity to occur and interfere with the separation of Mn and Fe. The possible formation of complex vapor species incorporating both $FeCl_2$ and $MnCl_2$ (see Section IV) may be a relevant, though as yet untested, factor in this regard. A key factor in the separation process is the unfavorable reaction free energy for the H_2 reduction of $MnCl_2$ in the condensed phase as compared with condensed $FeCl_2$. However, the reduction of $MnCl_2$ vapor, or $MnCl_2$–$FeCl_2$ vapor complexes, is thermodynamically favorable, and the generation of Mn-containing vapor species is therefore undesirable in this iron separation process.

7. *Aluminum Production*

The prospect of extracting and refining aluminum by chlorination processes has been recognized for several decades, e.g., see the review of Hermann (1970). However, a scale-up from laboratory or pilot plant operation to a commercial process has thus far not been successful.

One process given strong consideration by commercial interests is based on the disproportionation of AlCl, i.e.,

$$AlCl_3 + Al(1) \underset{1350\ K}{\overset{1450\ K}{\rightleftarrows}} 3AlCl.$$

In this process, AlCl is formed from a crude aluminum source and then transported to a relatively cool region where disproportionation results in the production of a purified form of the metal. The corrosive nature of AlCl vapor appears to have been one of the drawbacks in the commercialization of this process. An unexpected halogen transport of other impurity metals was also a limiting feature. Recent work by Dewing (1970) has revealed a possible explanation for this separation difficulty. That is, the demonstrated ability of $AlCl_3$ to complex with and transport relatively nonvolatile halides, such as $MnCl_2$ and $FeCl_2$, may have been a contributing factor (see Section IV).

Under flow-reactor conditions, the rate of conversion of $AlCl_3$ to AlCl by the above reaction is given by the relation

$$V = 1.28P^{4/3}(AlCl_3)\exp(-11{,}800/RT) \quad \text{gm Al cm}^{-2}\ \text{min}^{-1},$$

(Kikuchi *et al.*, 1969). Hence, for the conditions of 1 atm $AlCl_3$ and 1450 K, the rate of Al extraction and transport from the crude Al(1) would be about 0.02 gm cm^{-2} min^{-1}.

8. *Tin Chlorination*

The chlorination of molten tin is of interest for the preparation of stannic chloride, as the commercial process of chlorinating granulated tin is relatively inefficient. At 260 °C the rate of chlorination of molten tin is about twice as high as for the chlorination of granulated tin at 70°C. The degree of chlorine conversion is about unity at 350 °C and the reaction activation energy is about 10.5 kcal $mole^{-1}$ (Zvezdin *et al.*, 1970).

9. *Gold Chlorination*

There is currently a considerable interest in the prospect of extracting gold from its ores by chloride volatilization processes, e.g., see Hager and Hill (1970) and cited literature. Experimental feasibility has been demonstrated and some of the basic thermodynamics determined. The optimum conditions of chlorine concentration and temperature are rather stringent due to the presence of competing processes, as indicated in Fig. 3.10. Note that the equilibrium production of volatile Au_2Cl_6 from 1 atm Cl_2 is at a maximum at about 250°C.

For an elemental mixture of Au and Fe, the chloride transport of Au is enhanced by a factor of about 25 (at 332 °C and $pCl_2 = 0.66$ atm) and the

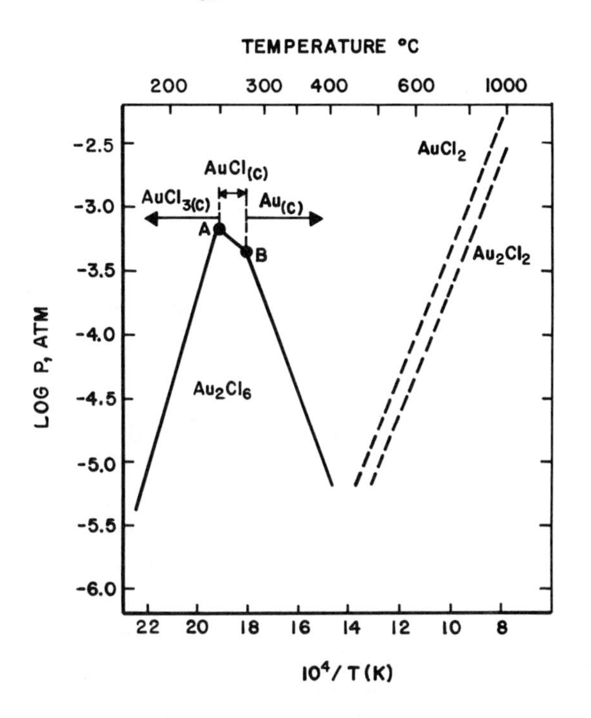

Fig. 3.10 Vapor pressure of gold–chloride species as a function of temperature at 1 atm Cl_2 pressure (Hager and Hill, 1970).

mass spectrometric identification of $AuFeCl_6$ in this system suggests the overall reaction

$$2Au(s) + Fe_2Cl_6 + 3Cl_2 = 2AuFeCl_6,$$

to be responsible for the enhanced volatility (Hager and Hill, 1970).

10. Sponge Chromium

A classic example of the application of fundamental thermodynamic principles, and carefully evaluated basic data, to the design of a metallurgical process is given by the US Bureau of Mines investigation of the production of sponge chromium by a chlorination-reduction process (Maier, 1942). The process, based on the modynamic data and demonstrated experimentally on a laboratory scale, is similar to an existing commercial chlorination process for Ti.

The Cr_2O_3 oxide-based ore is chlorinated in the presence of carbon to reduce the oxygen activity. In this process the other ore components, which are primarily the oxides of Al, Fe, and Mg, are also chlorinated. The Al_2Cl_6 and Fe_2Cl_6 components are much more volatile than $CrCl_3$ and can, in principle, be removed by distillation at a temperature of several hundred degrees centrigrade. However, some loss of $CrCl_3$ may result at this stage due to the possible formation of volatile Al_2Cl_6–$CrCl_3$ and Fe_2Cl_6–$CrCl_3$ complexes (see Sections III.D and IV). The other significant component, $MgCl_2$, is at least an order of magnitude less volatile than $CrCl_3$, and hence $CrCl_3$ can be extracted by distillation using a temperature of about 900 °C.

An important restriction on the process conditions is dictated by the need to prevent formation of the very volatile $CrCl_4$ and the relatively non-volatile $CrCl_2$ chlorides. As the formation of condensed $CrCl_2$ represents a more severe problem, in practice, the operating conditions utilize an excess of the stoichiometric quantity of chlorine. In the temperature region of 900 °C, it can be shown from the free energies of formation of $CrCl_2$(c) and $CrCl_3$ that about 10% excess chlorine is required to prevent the formation of $CrCl_2$(c).

In practice, the rate of chlorination was found to vary with the particle size of the oxide and also with the reactor design. One of the rate-limiting features of the process is probably the relatively low diffusivity of the high molecular weight vapor products $CrCl_3$, Fe_2Cl_6, and Al_2Cl_6. A practical solution to obtaining a suitable reaction rate was provided by a carrier particle method. The technique involves coating carrier particles, such as silica, with a mixture of ore and carbon.

The reduction of $CrCl_3$ to Cr-sponge using H_2 was thermodynamically predicted and experimentally demonstrated. Continuous removal of the HCl product effectively prevented a back reaction. It was also demonstrated that, to prevent hydrolysis of the chloride, considerable care should be taken to eliminate H_2O as an impurity at this stage of the process.

C. Iodide Processes

1. Background

The development of iodide processes resulted from a need for high purity metals in the modern technology areas of atomic energy, missiles and aircraft, and electronic devices. Small amounts of impurity can markedly affect the neutron absorption properties of metals, such as Zr, used in nuclear reactor technology. Similarly, the conductance of semiconductors can be highly sensitive to impurity levels of greater than a few parts per billion. With iodide processing, the formation of single crystals is also possible and this can provide materials of superior strength and oxidation resistance in particular crystal orientations.

For many years iodide processing was the sole source of high purity Zr. Iodide processes are also known to be suitable for the purification of the following elements:

Ti, Hf, Th, V, Nb, Ta, Cr, Cu, Ag,

Fe, B, Ge, Y, Ni, U, W, Pa, and Si.

However, the processes involve relatively expensive batch procedures and are only used where a high purity product is required. The iodine cost is usually negligible as in a closed system only a small amount of iodine is required to transport large quantities of metal.

An iodide process generally involves a selective reaction of I_2 vapor with metal, in crude form, to produce a volatile metal iodide species which is transported by diffusion to a hot wire and then thermally dissociated to the metal but in a purer state. The resulting iodine atoms recombine and eventually diffuse back to the crude metal source to complete the cycle, e.g.,

$$M(s) + 2I_2 \xrightarrow{T_1} MI_4 \xrightarrow{T_2} M(s) + 4I,$$

where $T_2 \gg T_1$, and

$$4I \rightarrow 2I_2.$$

The thermal dissociation step is strongly endothermic but at high temperatures this tends to be offset by the favorable $T\,\Delta S$ contribution to the free energy, particularly for the more polyatomic vapor species such as MI_4.

The following conditions must be met by the hot wire technique:

- the metal has a volatile iodide,
- the metal melting point is greater than the filament temperature,
- the metal has a low vapor pressure at the filament temperature, and
- the metal iodide must be readily decomposable at the filament temperature—hence the use of iodide rather than the other more stable halides.

In many instances, the deposition of metal is diffusion limited. For instance, from the appropriate transport equation, e.g., see Section III.C.3, the observed iodide transport of Fe, via the FeI_2 transport species, at 1 atm and 1173 K, can be quantitatively accounted for by assuming a diffusion controlled transport and an equilibrium condition at both ends of the temperature gradient. We should note that a significant dissociation of I_2 to atomic I, with subsequent recombination, can be expected under these temperature and pressure conditions. This presence of a dissociation-recombination chemical reaction along the temperature gradient can lead to enhanced thermal and mass transport effects (see Chapter 1, Section V.B and Chapter 6, Section II.C).

For the low pressure free molecular flow regime, i.e., $<10^{-3}$ atm, the observed deposition rates for Zr are similar to those predicted from the Knudsen equation, with the assumption that every iodide molecule striking the hot filament is converted to the metal.

An extensive account of iodide processes has been given by Rolsten (1961). Also, the chemical transport aspects of these processes have been considered in Section III and the following remarks are therefore brief.

2. *Limitations*

The main fundamental limiting aspects of iodide processes are the inability of iodine to react with many metals in an oxide form and the instability or high dissociation pressure of some of the metal iodides—such as those of Cr, V, Fe, Co, and Ni—at their vaporization temperatures. Nevertheless, it is still possible to produce metals such as Cr in an exceptionally pure form by iodide transport. Practical limits also exist for the filament temperature and the vapor pressure of the metal iodide. The filament temperature range is set by the heterogeneous decomposition temperature of the metal iodide species and by the vapor pressure of the metal itself. A lower pressure limit for the metal iodide is determined by the need to compensate for vaporization of metal from the filament. An upper pressure limit sometimes exists due to the formation of stable subiodide species, which result in "chewing" or loss of metal from the filament, e.g.,

$$TiI_4 + Ti(s) \rightarrow 2TiI_2.$$

Also, high rates of transport and low filament temperatures are undesirable as they can result in the production of amorphous metal deposits with lower purity.

Most iodide processes involve the tetrahalide as the transport species, as dissociation of the lower valence species is thermodynamically less favorable. For instance, in the case of the monoiodides there is only a small entropy change associated with the dissociation reaction

$$MI \rightarrow M(s) + I,$$

and thermal decomposition is therefore difficult. The monoiodides of aluminum and the rare earths are sufficiently stable that a hot wire decomposition to the metal is expected to be difficult. However, a cooling of these reduced valence species should lead to disproportionation to form the metal together with a higher valence iodide species. If the monoiodide is polymeric, e.g., $(CuI)_3$, then the decomposition, as given by the reaction

$$(CuI)_3 \rightarrow 3Cu(s) + 3I,$$

has sufficient entropy production to favor deposition of the metal. Similarly dimers, such as $(FeI_2)_2$, are favorable species for decomposition, since

$$(FeI_2)_2 \rightarrow 2Fe(s) + 4I$$

is a high entropy process.

3. *Examples*

For the case of Ti, an optimum rate of production is obtained with a filament temperature of about 1500 °C, a TiI_4 pressure of about 20 Torr, and a wall—or crude Ti—temperature in the region of 100–200 °C. The rate of deposition is strongly dependent on the vaporization temperature. That is, from 0 to ~150 °C, a monotonic increase in Ti production with temperature is found (via TiI_4 species), but from 150 to 350 °C, this production decreases to a negligible level. This apparent loss of volatility is thought to be due to the formation of a nonvolatile solid TiI_2. At 470 °C this solid vaporizes, probably as molecular TiI_2 and TiI_4, and a monotonic increase in Ti deposition again occurs. In fact this higher temperature reaction allows a higher rate of metal production than the lower temperature stage (Runnalls and Pidgeon, 1952).

High purity Zr is produced by a similar process, e.g., see Shelton (1968b). Titanium–zirconium alloys can also be prepared by the iodide process, provided the normal vapor pressure of the tetraiodides is modified by the presence of reduced activity compounds such as $CuTiI_4(s)$ and $Cu_{0.25}Zr_{0.75}I_3(s)$ (Shelton and Holt, 1972).

For the production of pure Nb metal, the reaction of iodine with crude Nb to produce volatile niobium iodide species is carried out at temperatures of 300–500 °C (Rolsten, 1959). The iodide vapor is then dissociated on a Nb filament at about 1000 °C to yield Nb metal, and iodine is returned to the cycle. A nonmonotonic dependence of deposition rate on the temperature of the crude Nb surface is also found, as shown in Fig. 3.11. The increasing rate of deposition over the temperature interval 300–380 °C can be explained in terms of an increase in the rate of vaporization of a condensed iodide, probably NbI_3. The nature of the vapor species is not known but the likely vaporization reactions are

$$NbI_3(c) \rightarrow NbI_3$$

and

$$2NbI_3(c) \rightarrow NbI_4 + NbI_2(c).$$

It appears that the latter reaction predominates as the maximum rate of transport is approached at 380 °C. The loss of volatility above this temperature can be attributed to the stability of $NbI_2(c)$. At the higher temper-

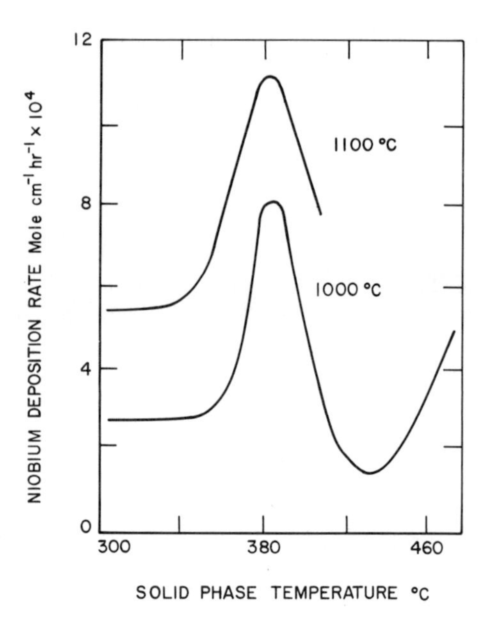

Fig. 3.11 Plots of Nb deposition rate as a function of condensed niobium iodide temperature for filament temperatures of 1000 and 1100 °C (Rolsten, 1959).

ature regime of 440–470 °C, vaporization or decomposition of this solid occurs, i.e.,

$$NbI_2(c) \rightarrow NbI_2$$

and

$$2NbI_2(c) \rightarrow NbI_4 + Nb(c)$$

or

$$2NbI_2(c) \rightarrow NbI_3 + NbI(c).$$

It is important to note that the phenomenon of a maximum source temperature effect on the filament deposition rate, as typified by the example in Fig. 3.11, may have an alternative explanation. For the Zr–iodide process the rate of deposition maximizes at about 260 °C sample temperature. This was originally attributed to a loss of volatile ZrI_4 at higher temperatures, due to the formation of a stable ZrI_3 condensed phase, i.e.,

$$3ZrI_4 + Zr(c) = 4ZrI_3(c).$$

However, Shelton (1968a) has shown that the observed dependence of deposition rate on source temperature can be explained by a physical mass transport model. The rate-limiting step in the overall deposition process is then considered to be the rate of transport of iodine returning from the filament to the crude source. As the source temperature is increased up to 260 °C, the pressure of ZrI_4 increases and the iodine near the filament is subjected to an increasingly denser medium. Eventually, the reduction in iodine diffusivity offsets the advantage to be gained from a higher concentration of ZrI_4 vapor species and a decreased rate of deposition results.

D. Fluoridization Processes

The breakdown of the ore beryl represents an example of a fluoride breakdown process. At temperatures of 750–850 °C the following reaction occurs:

$$3BeO \cdot Al_2O_3 \cdot 6SiO_2(s) + 6Na_2SiF_6(s) \rightarrow 3Na_2BeF_4(s) + 2Na_3AlF_6(s) + 9SiO_2(s) + 3SiF_4.$$

However, the formation of SiF_4 gas is undesirable as it results in a removal of fluorine from the system. This may be offset by the presence of NaF, where

$$SiF_4 + 2NaF(s) \rightarrow Na_2SiF_6(s).$$

In practice, then, one uses the following reaction:

$$3BeO \cdot Al_2O_3 \cdot 6SiO_2(s) + 2Na_2SiF_6(s) + Na_2CO_3(s) \rightarrow 3Na_2BeF_4(s) + 8SiO_2(s) + Al_2O_3(s) + CO_2.$$

A process for aluminum extraction using MgF_2 as a fluorinating reagent has been recently demonstrated on a laboratory scale by Layne *et al.* (1972). Basically, the process involves a fluoridization-vaporization reaction

$$MgF_2(l) + 2Al \text{ (crude alloy containing Fe, Si, Ti, C)} \rightarrow 2AlF + Mg; \quad 1500\text{–}1650\ °C,$$

followed by vapor transport of AlF and Mg to a cool region (≤ 1350 °C) where a reversal of the above reaction occurs, to yield a relatively pure Al liquid which is immiscible with the molten MgF_2.

E. Vacuum Metallurgy

A brief review of the development and future trends of vacuum metallurgy has been given by Moore (1958). Important vaporization processes in this area include the distillation and reduction procedures for rare metal preparation, vaporative separation of zinc from lead, and the production of Mn by an arc-carbon reduction of rhodonite ore with selective condensation of the Mn. Now that high temperature and vacuum chamber systems are commonplace, there is an increasing interest in applying distillation techniques to metallurgical problems, e.g., see St. Clair (1958) and Trombe and Male (1968). For a discussion of "normal" temperature distillation principles see, for example, Leslie (1966).

In practice, the low pressure distillation rate for high temperature materials depends on the rate of thermal energy transfer to the surface. The selectivity of distillation depends primarily on the relative volatility and molar ratios of the metal components. Partial pressures of about 10^{-3} atm provide a reasonable rate of distillation. If the mixture is a liquid, the activity and diffusion properties of the components also affect the observed distillation rate.

When the distillation is carried out in a walled container, the rate of distillation may also depend on the rate of diffusion along the pressure gradient and is then given by the expression

$$J = DM(P_1 - P_2)/L,$$

where

D is the effective diffusion constant,
M the molecular weight,
P_1 the partial pressure of vapor at evaporating surface,
P_2 the partial pressure of vapor at condensing surface, and
L the effective distance between evaporating surface and condenser.

For the case where diffusion occurs through a residual gas background, the distillation rate is

$$J = (D'M/L) \log[(P_3 - P_2)/(P_3 - P_1)],$$

where

P_3 is the pressure of residual gas, and
D' the apparent diffusion constant of the vapor through the residual gas.

Some additional remarks concerning distillation may be found in the appendix of Section IV.

As an example of high temperature vacuum distillation, Murphy *et al.* (1970) have calculated and measured the molecular distillation rates and separation factors for Sm-containing metal systems.

F. Industrial Fumes

The emission of gaseous and particulate material from industrial processes, such as smelting, is of practical concern and a few examples are considered here.

1. *Cryolite Emissions*

In the electrolytic production of aluminum, the cryolyte bath produces emissions of HF, $NaAlF_4$ vapor and particulates of Na_3AlF_6 (cryolite) and $Na_5Al_3F_{14}$ (chiolite). It is thought that the chiolite may result from a disproportionation of $NaAlF_4$ vapor (Cochran *et al.*, 1970).

2. *Iron Oxide Emissions*

The process of oxygen steelmaking is accompanied by the release of orange-brown fumes of hematite (Fe_2O_3). This is currently a problem in terms of pollution control in the steel industry (Anon., 1971b). The amount of fume released is far in excess of that expected by a simple diffusion controlled vaporization of Fe followed by oxidation. This fume is corrosive to the furnace refractories and is also difficult to separate from the exhaust gases. Hence, there is considerable interest in the mechanism for the enhanced release of material to the gas phase (Ellis and Glover, 1971).

The reactive-diffusion model (see Section IV.F) accounts for only part of this enhanced apparent vaporization. Photographic and fume-collection laboratory scale experiments indicate that the majority of fume is produced by a mechanical action, rather than by vaporization. The mechanical action is due to the rapid oxidation of carbon in the steel with the production of CO bubbles. These bubbles burst and throw Fe droplets into the oxidizing atmosphere. Many of these droplets are observed to "explode" in the gas phase and produce a fine submicron-size oxide fume. The fume production is therefore proportional to the oxygen partial pressure and the use of a less oxidizing atmosphere should reduce the fuming. This is observed in practice.

3. *Lead Fumes*

An example of the application of thermodynamics to the evaluation of lead sulfide roasting has been given by Kellogg and Basu (1960). Under typical roaster gas compositions, i.e., SO_2 pressures of 10^{-1}–10^{-2} atm and O_2 pressures of 10^{-1}–10^{-2} atm, at 1100 K, $PbSO_4$ is by far the most stable condensed phase and most of the fume-lead is likely to be present in this form.

The melting of lead glasses involves a change in the glass properties (on recooling), because of a loss of PbO by volatilization of the melt (Matousek and Hlavac, 1971). In addition to the composition and temperature, the factors that affect the loss of PbO are the rate of diffusion of PbO through the melt to the surface and the rate of surface vaporization.

The rate-limiting step appears to be the diffusion of PbO through the melt and diffusion coefficients of the order of 10^{-6} cm^2 sec^{-1} are observed (Matousek, 1972).

Lead oxide volatilization and entrainment into the atmosphere can also occur in the fire refining of copper, e.g., see the thermodynamic evaluation by Jeffes and Jacob (1972). Lead emissions are also generated by melting scrap.

4. *Glass-Melt Emissions*

In batch glass preparation, the initial mixture may contain the following salts and oxides—SiO_2, Al_2O_3, CaO, and alkali metal carbonates, sulfates, nitrates, fluorides, and chlorides. The heating and melting of such mixtures involves the release of common gases such as CO_2, SO_3, SO_2, H_2O and the less common species Na_2O, K_2O, B_2O_3, PbO, SiF_4, BF_3, HF, HCl, etc. In special glasses, transition metal oxides may also be present, as color additives. Frequently these batch mixtures are heated by direct contact (and by radiative heat transfer) with a natural gas-air flame, and the melts may achieve temperatures of around 1600 °C. Hence, very high concentration levels of the post combustion gases CO_2, CO, H_2O, and N_2 may be in contact with the molten glass. Thus there exists a wide range of fume-producing chemistry (but virtually uncharacterized) involving the interaction of vapors and gases with complex glass melts.

Vapor fractionation of silicate melts has been observed by Walter and Giutronich (1967). For a melt with an 82% SiO_2 content in contact with the atmosphere, at 2800 °C, the general order of volatility of the melt components is found to be

$$K_2O > SiO_2 > Na_2O > FeO > MgO \gtrsim Al_2O_3 \gtrsim CaO.$$

5. *Toxic Emissions*

Other troublesome emissions include the toxic products of zinc from heated galvanized iron, KCN from blast furnace operations, and Hg, together with other toxic metals released as an impurity from coal combustion.

Appendix. Some Additional Recent Literature on Chlorination Metallurgical Processes

Alfredson, P. G., and Carter, F. R. (1968). Aust. At. Energ. Comm., AAEC/TM (Rep.) AAEC/TM 431. Review of methods for zirconium production with suggested process improvements.

Athavale, A. S., and Altekar, V. A. (1971). *Ind. Eng. Chem., Process Res. Develop.* **10,** 523. Kinetics of selective chlorination of ilmenite using hydrogen chloride in a fluidized bed.

Athavale, A. S., and Altekar, V. A. (1970). *Proc., Int. Miner. Process. Congr., 9th.* p. 278. Chemical beneficiation of minerals by selective chlorination in fluidized beds.

Babenko, A. R. (1969). *Nauch. Tr., Perm. Politekh. Inst.*, No. 52, 156. Chlorination kinetics and the thermodynamics of copper and zinc ferrites.

Fomin, V. K., Zvezdin, A. G., and Ketov, A. N. (1971). *Zh. Prikl. Khim. (Leningrad)* **44,** 416. High temperature chlorination of tin.

Frents, G. S., Tratsevitskaya, B. Y., and Novikova, E. I. (1969). *Issled. Protssessov Met. Tsvet. Radk. Metal.* 131. Salt chlorination of a tin-tungsten concentrate.

Hussein, M. K., and El-Barawi, K. (1971). *Inst. Mining Met., Trans.*, Sect. C80, March 7–11. Chlorination and beneficiation of Egyptian chromite ores.

Ivanov, V. A., Kishnev, V. V., and Lisovskii, D. I. (1969). *Sb. Mosk. Inst. Stali Splavov*, No. 56, 32. Chlorination of titanium slags in a melt and setting up the mathematical model.

Ivantsova, V. I., and Ivashentsev, Y. I. (1971). *Zh. Fiz. Khim.* **45,** 1343. Temperatures for the start of the reaction of germanium, tin and lead oxides with chlorine.

Ivantsova, V. I., and Ivashentsev, Y. I. (1969). *Obogashch. Met. Tsvet. Met.*, No. 3, 112. Chlorination mechanism of germanium, tin, and lead oxides.

Ivashentsev, Y. I., Gerasimova, V. B., Gerasimov, P. A., Dubnyak, M. I., and Rechkunova, E. M. (1970). *Izv. Vyssh. Ucheb. Zaved., Tsvet. Met.* **13,** 79. Chorination of titanium, zirconium, and hafnium dioxides with chlorine.

Jorgensen, F. R., and Dorin, R. (1971). *Ind. Eng. Chem. Prod. Res. Develop.* **10,** 339. Production of aluminum chloride from aluminum-zinc alloy.

Kopinashvili, N. E., and Buchukuri, Ya. G. (1971). *Soobshch. Akad. Nauk Gruz. SSR* **63,** 629. Production of manganese chloride by chlorination of iron-manganese ores.

Lythe, R. G., and Prosser, A. P. (1969). *Inst. Mining Met., Trans.*, Sect. C78, Dec., p. C206. Catalytic effects in the chlorination of cassiterite.

Monk, H., and Fray, D. J. (1973). *Trans. Inst. Min. Met. C* **82,** 161. Purification of zinc chloride produced by chlorination of zinc sulphide ores: 1—removal of iron.

Morozov, I. S., and Fefelova, G. F. (1971). *Zh. Prikl. Khim. (Leningrad)* **44,** 1161. Equilibrium of the chlorination of chromium and iron oxides.

Motornaya, G. A., Gnatyshenko, G. I., and Abdeev, M. A. (1967). *Sb. Nauch. Tr., Vses. Nauch.-Issled. Gornomet. Inst. Tsvet. Metal.*, No. 10, p. 8b. Effect of the composition of pyrite cinders on extraction of non-ferrous metals during chloride sublimation.

Otsuka, T., and Murata, A. (1969). *Denki Kagaku Oyobi Kogyo Butsuri Kagaku* **37,** 729. Production of hafnium metal from hafnium oxide by chlorine process.

Pimenov, L. I., Mikhailov, V. I., and Khudyakov, I. F. (1971). *Tsvet. Metal.* **44,** 15. Conditions for the chlorination of a nickel matte.

Privol'nev, A. T. (1969). *Nov. Isslad. Tsvet. Met. Obogashch.*, p. 37. Interaction of iron silicates and hydrosilicates with gaseous hydrogen chloride.

Shelton, R. A. J. (1971). *Metall. Rev.*, No. 151, part II, 84. (D) Chloride-volatilisation processes in extractive metallurgy.

Sidorenko, A. P. (1970). *Sb. Tr., Vses. Nauch.-Issled. Proekt. Inst. Titana* **6,** 41. Chlorination of lump titanium slag.

Tripsa, I., Iatan, N., and Dumitree, S. L. (1969). *Stud. Cercet. Met.* **14,** Rom. 51994. 3 pp. Addn. to Rom. 51993. Treatment of pyrite ashes containing high concentrations of zinc and copper.

Vucurovic, D., and Spasic, M. (1969). *Glas. Hem. Drus., Beograd.* **34,** 541. Chlorination

of Yugoslavian nickel-iron ores for their complex processing. I. Chlorination of natural ores.
Zak, M. S., Leizerovich, G. Y., and Maiskii, O. V. (1970). *Tsvet. Metal.* **43,** 13. Gas phase in the fluidized-bed chloridizing roasting of pyrite cinders.
Zyryanov, M. N., Khlebnikova, G. A., and Mikhailov, B. N. (1971). *Izv. Vyssh. Ucheb. Zaved., Tsvet. Met.* **14,** 69. Degree and rate of chlorination of gold by hydrogen chloride.
Zyryanov, M. N., Nivin, A. P., and Poletaev, S. V. (1971). *Tsvet. Metal.* **44,** 34. Effective form of iron during chloride sublimation of unyielding gold-containing concentrates.
Zyryanov, M. N., Khlebnikova, G. A., and Mikhailov, B. N. (1970). *Izv. Vyssh. Ucheb. Zaved., Tsvet. Met.* **13,** 60. Degree and rate of chlorination of lamellar gold.

VIII. Ore Genesis

A. Geological Background

According to Krauskopf (1967a), "The origin of ore deposits is one of the great unsolved problems in geology."

It is generally agreed that ores are derived from cooling magmas, which are basically silicate melts containing dissolved H_2O and metal oxides or, to a lesser extent, sulfides.† The typical ppm concentrations of metals known to be present in igneous and sedimentary rocks is more than sufficient to account for the formation of ore deposits, provided a suitable means exists for the separation and concentration of the ore metals (Krauskopf, 1967b). The means by which the ore metals are separated from the silicate melt and transported and deposited as an ore metal concentrate has been the subject of much argument and speculation, e.g., see Krauskopf (1967a) and Sullivan (1957). Basically, two modes of metal transport are considered in current theories of ore genesis. These involve gas-phase volatility processes on the one hand and liquid, i.e., aqueous, phase solubility on the other.

During the final stages of magmatic (basaltic or granitic) crystallization, the fluid contains primarily water plus the alkali feldspar components of sodium, potassium, aluminum, and silica. Also present are the volatiles, water-solvated compounds, and elements whose ions do not fit readily into the structures of growing silicate minerals. The actual state of the aqueous fluid is in doubt, as the estimated conditions of pressure and temperature for magmatic crystallization are in the region of the critical

† The oxygen fugacity in magmatic fluids is too low for the stabilization of sulfates; this is supported by the laboratory studies of Fincham and Richardson (1954), for example.

point for H_2O, i.e., 217 atm and 374 °C. Also, the effect of solute on this critical point is not known. Hence, arguments have been proposed for both a hydrothermal (i.e., subcritical) and a pneumatolytic (supercritical) medium for metal transport and ore deposition.

Under the latter conditions of a gaslike state, thermodynamic calculations indicate the feasibility of chemical interaction between the gas-phase components and the condensed metal oxide or sulfide phase. This interaction leads to the production of volatile metal-containing species in sufficient quantity for significant ore-body formation.

A major source of conceptual difficulty arises when the hydrothermal mechanism is considered, owing to the insolubility of metals in sulfur containing solutions, e.g., see Barnes (1967). The familiar ore metals of Pb, Zn, Cu, Ag, Mo, and Hg occur naturally in sulfide form. However, such sulfides are extremely insoluble in hydrothermal solutions, e.g., $\sim 10^{-7}$ gm liter^{-1}, and it can be shown that solubilities of greater than about 10^{-5} gm liter^{-1} would be required for any significant deposition of ore to occur. Krauskopf (1967a), among others, contends that the simple hydrothermal hypothesis fails to give a reasonable interpretation of geological facts. However, the bulk of evidence supports the contention that the *final* stages of ore deposition are from liquid-medium hydrothermal solutions, and there is general concurrence on this point despite the impasse regarding metal sulfide insolubility.

The role of volatility, then, is considered to be of greatest importance in the initial stages of metal separation from magmas or zones of metamorphism.

Insufficient quantitative information about the state of the magma, the thermodynamics or kinetics of volatility processes, and the absence of solubility data for geochemical conditions, requires, as White (1963) has stated, that

> all possibilities for origin of an ore transporting fluid should be considered, and reliable criteria for distinguishing between different origins must be developed.

B. Physicochemical Conditions for Ore Genesis

The observed regularity in the deposition sequence of minerals may be taken as evidence of the influence of well-defined physicochemical factors, rather than random events, in determining ore deposition. Given the long time scale of most geological events, and the elevated temperature and pressure conditions, it is reasonable to assume that thermodynamic stabilities will determine the chemistry and location of ore deposits—as derived from magmatic fluids. The identities of the various mineral phases, and their known free energies of formation, will then serve as an indication of the ore-forming conditions of temperature, pressure, and chemical composition.

A field investigation of mineral assemblages in ore bodies is instructive as it provides information about the actual thermodynamic conditions of ore formation. The basic igneous residue of the magmatic separation process is comprised mainly of the high stability oxides CaO, MgO, BaO, Al_2O_3, ZrO_2, and TiO_2, together with SiO_2. The separated ore forming fluid contains elements whose oxides or silicates are of lower stability and these tend to deposit at lower temperatures, i.e., at the periphery of the main igneous body. A temperature gradation is also evident in these oxyphile ore deposits, and zones of high, medium, and low formation temperature can be classified from geological evidence.

High temperature mineral zones are characterized by the presence of oxide and silicate minerals of W, Mo, Sn, and Bi. Medium temperature zones tend to contain sulfide minerals of Cu, Pb, Zn, Ag, As, Sb, Fe, and elemental Au, in addition to quartz and the carbonates of Mg and Fe. Sulfide minerals of Ag, Sb, and Hg tend to be prevalent in the lower temperature regions together with elemental Au and Ag and various gangue minerals.

The temperatures at which ore deposition occurs may be estimated very approximately from the known presence or absence of certain minerals; for instance, pyrite is unstable at temperatures of greater than 740 °C. On this basis, we find that most vein and replacement ore deposits were formed over the temperature interval of 50–550 °C.

Geological evidence indicates that most ore deposits form at depths of less than 20 km and this sets the possible conditions of hydrostatic pressure as being in the range of ~200–2000 atm. If the system was a closed one, the chemistry would be modified by the high lithostatic pressure according to the thermodynamic relation between free energy, pressure (P), and volume (V), i.e.,

$$\frac{d\,\Delta F}{dP} = \Delta V.$$

C. Composition of the Geothermal Fluids

> We have only vague and general ideas about the nature and amounts of the volatiles which constitute our problem (Krauskopf, 1967a).

The evidence for the composition of geothermal fluids is necessarily indirect, as an actual sample cannot be obtained from the subterranean levels at which ore formation occurs. The nature of the volatiles has therefore been inferred from the composition of volcanic gases, and of fluid inclusions found in minerals of igeneous rocks and in rocks located around intrusive bodies such as granite batholiths. In-so-far as they are derived from the silicate magma, volcanic gases provide valuable information regarding the initial conditions for ore deposition. These gases are present in the lava to the extent of several percent prior to their escape, when the pressure is reduced to a near ambiant level. Chemical analyses indicate H_2O as the predominant component, followed usually by SO_2, N_2, and CO_2 with lesser amounts of CO, H_2, Ar, H_2S, S_2, SO_3, HCl, HF, CH_4, NH_3, and boron [probably as $B(OH)_3$], in addition to trace quantities of other metals such as Cu and Na (White, 1963).

The relative amounts of these component gases vary considerably with sampling location, and this is thought to be indicative of temperature variations at the source. For instance, SO_2 and H_2 are favored over H_2S and CO_2 by high gas temperatures and such observations can be rationalized by equilibrium thermodynamics (Ellis, 1957). In some instances the gas composition probably represents that of a frozen equilibrium. That is, the reactions were quenched at a higher temperature than that observed at the point of sampling.

In addition to the gas composition as observed at ambiant pressures, the source gas also contains vapor material which condenses in the form of so-called sublimates at the mouth of the fumaroles where the gases are commonly sampled. These sublimates contain many of the common ore metals and are present in the form of chlorides and sulfates of Na, K, Ca, Mg, Al, and Fe, and the oxides of Si and Fe. Minor amounts of compounds containing B, P, As, Zr, Cu, Pb, Mn, and Sn and trace amounts of Li, Be, Ag, Ni, Co, V, Mo, Ga, Ge, Ti, Zr, Cr, Cd, Sb, Bi, Sr, Ba, Se, Te together with F, Br, and native sulfur are also found. From these analyses, the metal concentration of the fluid can range from a few ppm to about 10^3 ppm.

Other analytical evidence for the release of metal compounds as vapor, or aerosols, from volcanic basalts has been summarized by Naboko (1959). In the initial stages of surface lava flow, where $T \sim 1000$–800 °C, the metals of Na, K, Fe, and Cu apparently are released from the lava as chlorides of unspecified molecular composition. The condensed sublimates can be enriched relative to the basalt source by as much as 6000 times for the case of Cu. Lesser enrichments are also found for the elements:

Li, Be, Pb, Sn, Ag, Zn, Co, Ni, Zr,

Mo, Bi, Ga, Te, Cr, V, Ba, and Sr

present in chloride condensates. Intense releases of Si and Fe, in association with fluorine, are also observed. An enrichment of V is found in association with sulfurous gases. Evidence for a transport effect in the presence of superheated steam was also noted—particularly for silicon; in deposits characteristic of this mode of silicon transport, compounds of the metals As, Pb, Cu, Ag, Zn, Co, Ni, Mo, Bi, Tl, V, Li, and Sc were also found. At least some of the observed transport is in the form of aerosols and their transport over distances of 2–3 km has been observed.

From these observations no quantitative indication of relative importance of vapor phase metal transport can be given although, as was stated by Naboko (1959),

> Numerous observations of fumarolic activity of volcanoes provide evidence of the transportation of metals having been done in a gaseous phase.

From emission spectra of the volcanic flame of several volcanic eruptions, CuCl has been observed as a vapor species, and it is thought that most of the vapor-phase copper exists in this form (Murata, 1960). The temperatures of these magmatic vapors or flames is in the range of 1000 to 1200 °C.† Some inconclusive spectroscopic evidence for the presence of S_2 was also obtained by Murata (1960). However, the major emitting species in volcanic flames is generally observed to be Na.

Given this indirect terranean information about the chemical content of fluids derived from magmatic separation processes, the problem is to extend the data to the subterranean conditions of high pressure and elevated temperature. The various arguments underlying the selection of a representative subterranean gas composition have been detailed by Krauskopf (1959). He considers a typical composition for a magmatic gas to be

Gas:	H_2O	CO_2	H_2S	HCl	N_2	HF
Pressure (atm):	1000	50	30	10	10	0.3

This composition will vary according to the oxygen pressure and temperature conditions, and carbon may also be present as CH_4, sulfur as SO_2 or S_2, and nitrogen as NH_3.

The presence of oxide-forming metals in the magma maintains a very low O_2 pressure which may be estimated from the absence of haematite (Fe_2O_3) but presence of magnetite (Fe_3O_4) in high temperature deposits. Hence, the equilibrium

$$\tfrac{1}{2}O_2 + 2Fe_3O_4(s) = 3Fe_2O_3(s)$$

† Spectroscopic evidence suggests that volcanic flames are supported by hydrogen burning in air (Cruikshank *et al.*, 1973).

TABLE 3.11

Calculated Equilibrium Composition of Magmatic Gas[a]

	627°C		827°C	
O_2	−14[b]	−23	−8	−17
H_2O	3.0	2.7	3.0	2.9
H_2	− 1.5	2.7	−1.9	2.5
HCl	1.0	1.0	1.0	1.0
HF	− 0.5	− 0.5	−0.5	− 0.5
H_2S	− 1.0	1.0	−4.6	1.0
S_2	− 4.4	− 8.8	−8.8	− 6.5
SO_2	1.0	−10.2	1.0	− 6.9

[a] Data of Krauskopf (1967a); the initial composition is taken as H_2O = 1000 atm (i.e., 4–10 km depth), HCl = 10 atm, $H_2S + S_2 + SO_2$ = 10 atm and HF = 0.3 atm.
[b] Represents log pressure in atm.

sets the upper O_2 concentration limit. A lower O_2 concentration limit may be set by the absence of metallic Pb; i.e., the reaction

$$\tfrac{1}{2}O_2 + Pb(s) = PbO(s)$$

sets the O_2 concentration. From the free energies of these and similar processes, the O_2 field may be set at

$$10^{-14}\text{–}10^{-23} \text{ atm at } 627\ ^\circ\text{C}$$

and

$$10^{-8}\text{–}10^{-17} \text{ atm at } 827\ ^\circ\text{C},$$

e.g., see Krauskopf (1967a). The primary significance of the oxygen field is that it determines the mineral form of the metal in its condensed state; that is, whether Pb, for instance, is present as PbS, PbO, $PbSO_4$, or $PbSiO_3$.

For the conditions of this oxygen field, one can calculate the equilibrium composition of the magmatic gas. A typical set of calculated composition data is given in Table 3.11. The temperatures of 627 and 827°C represent magma conditions of near complete crystallization and mainly molten state, respectively.

D. Chloride Volatility Theory of Ore Formation

1. *Background*

Several volatility theories, sometimes referred to as a collector mechanism, have been proposed to describe the separation and transport of ore-forming metals from magmas (Edwards, 1956). These include volatilization of the metals in their elemental form or as the sulfide or chloride. Given the present knowledge about the thermodynamic stability of these metal-containing vapors under "typical" gas compositions, such as those of Table 3.11, we can exclude from consideration the presence of elemental or metal sulfide species, with a few exceptions—namely, Hg, SnS, CuS, SbS, AsS, and PbS.

In recent years Krauskopf (1957, 1963, 1964, 1965, 1967a, b) has been an ardent proponent of the chloride volatility theory and a summary of his main arguments is given here.

For a granitic batholith—that is a dome-shaped igneous intrusion without an apparent base—having typical dimensions of $100 \times 10 \times 10$ km and containing 4% H_2O at 1000 atm pressure, a significant ore formation of 1000 ton, or greater, would require a total metal halide pressure of about 10^{-7} atm. This pressure then represents the lower limit, below which chloride volatility would not contribute significantly to ore formation.

2. *Thermodynamic Predictions*

An examination of volatility data for the pure metal chloride salts gives, at best, only a crude indication of the possible grouping of elements in geothermal vapor transport. A more realistic indication of which metals are likely to be transported as chloride vapors is given by thermodynamic calculations including reactions of HCl with actual geological phases such as metal oxides, sulfides, or silicates. For instance, the conversion of ZnS(c) to volatile species is determined by the free energies of the set of reactions

$$\mathrm{ZnS(c)} + 2\mathrm{HCl} = \mathrm{ZnCl_2} + \mathrm{H_2S},$$

$$\mathrm{Zn} + 2\mathrm{HCl} = \mathrm{ZnCl_2} + \mathrm{H_2},$$

$$\mathrm{ZnS(c)} = \mathrm{ZnS},$$

$$\mathrm{ZnS(c)} = \mathrm{Zn} + \tfrac{1}{2}\mathrm{S_2},$$

$$\mathrm{H_2} + \tfrac{1}{2}\mathrm{S_2} = \mathrm{H_2S}.$$

As was indicated in the previous section, the range of oxygen and sulfur fugacities expected in the gas phase can be set from the known composition of volcanic gases and the form of the minerals found in nature. Also, the HCl content of the gases can be reasonably well estimated. The interrelation between the various gas components and the condensed metal, in its most stable form, can then be defined from the known free energies of formation for the various phases. These phase relationships are shown by the example of Fig. 3.12. The sulfur and O_2 fugacities determine the stability regimes of the condensed phases which, in this example, are SnO_2, SnS_2,

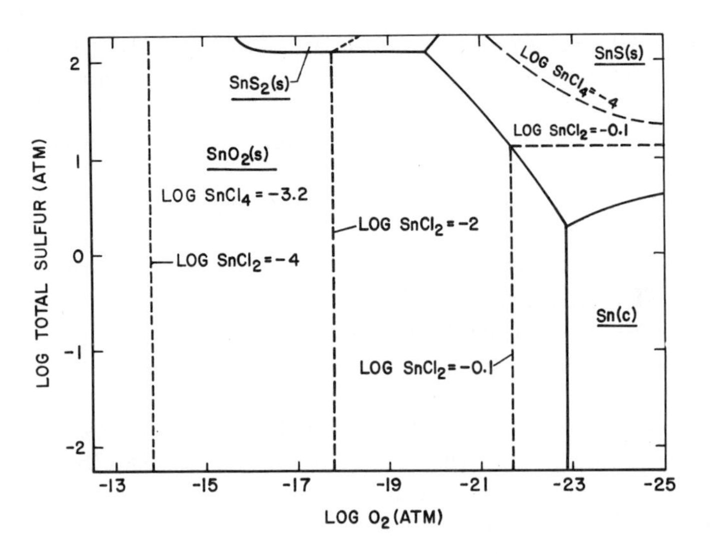

Fig. 3.12 Stability diagram for Sn compounds at 627 °C and in the presence of 1000 atm H_2O and 10 atm HCl; the solid lines separate the various condensed phases and the broken lines indicate partial pressure isobars of tin halide species (Krauskopf, 1964).

SnS, and Sn. The calculated interaction of HCl (10 atm) with each of these phases to yield $SnCl_2$ or $SnCl_4$ as vapor species can vary appreciably with oxygen and sulfur concentration, as shown in Fig. 3.12. This is the result of competing thermodynamic processes such as

$$SnO_2(s) + 4HCl = SnCl_4 + 2H_2O,$$

$$SnCl_4 + H_2 = SnCl_2 + 2HCl,$$

$$H_2 + \tfrac{1}{2}O_2 = H_2O,$$

$$H_2 + \tfrac{1}{2}S_2 = H_2S,$$

$$H_2S + O_2 = H_2 + SO_2.$$

Thus, for low O_2 fugacity and high sulfur content, $SnCl_2$ is present at a relatively high partial pressure. A reversal of these conditions favors the presence of $SnCl_4$ but with a somewhat lower partial pressure.

Given the criterion of metal halide pressure greater than 10^{-7} atm to be of ore-forming significance, it is apparent from Fig. 3.12 that Sn can be transported in the vapor phase for any expected O_2 and sulfur fugacity at a temperature of 627 °C, corresponding to the final stage of magmatic crystallization.

The amounts of various volatile metal chlorides formed under conditions of high O_2 (10^{-14} atm) and low sulfur (10^{-1} atm) content and as a function of HCl pressure are shown in Fig. 3.13. Note that Cu, Zn, Cd, Pb, and Sb form greater amounts of chloride vapor than Sn under similar conditions. Also Bi, As, and Hg form even greater amounts of volatile chlorides than the elements represented in Fig. 3.13. For low HCl concentrations, the nonchloride species of Sb_4O_6, MoO_3, and Cd are the most significant vapor components. At high HCl concentrations, condensed chloride salts such as $PbCl_2$ and AgCl are stable. The chloride vapor content is then determined by the vapor pressures of the corresponding condensed salts and is therefore invariant with HCl pressure.

For conditions of fixed HCl pressure and variable O_2 pressure, the metal halide volatilities are relatively invariant with O_2 pressure as shown by Fig. 3.14.

Recognizing that the order of metal halide volatilities can sometimes change, depending on the varying conditions of HCl, O_2, ($S_2 + H_2S + SO_2$) pressures and the temperature, it is possible to select groups of metals on the basis of a similar degree of chloride volatility as follows:

(a) U, Be, Ti, V, Cr, Zr;

this group does not participate in significant chloride vapor formation.

(b) Mn, Fe, Co, Ni, and possibly V;

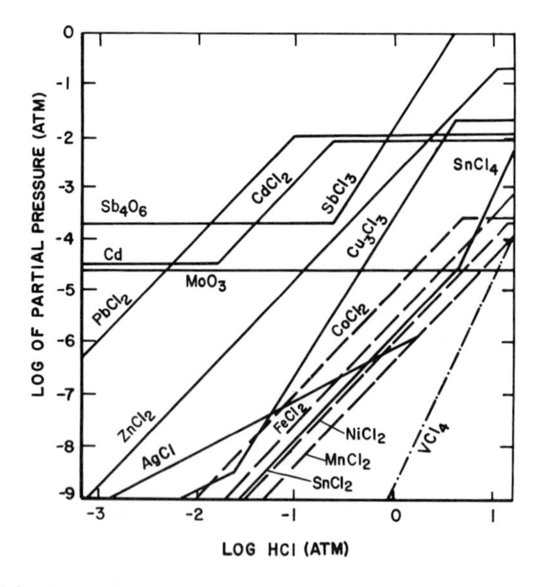

Fig. 3.13 Volatilities of metal containing species as a function of HCl pressure, where pressure $O_2 = 10^{-14}$ atm, total sulfur pressure $= 10^{-1}$ atm and $T = 627\ °C$ (Krauskopf, 1964).

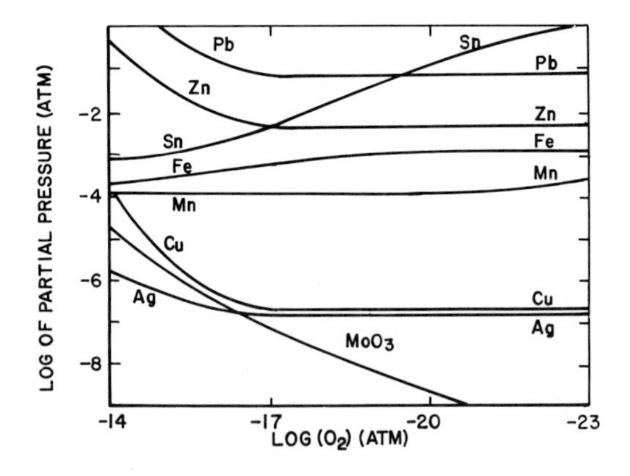

Fig. 3.14 Partial pressures of metal chloride compounds (except for MoO_3) as a function of O_2 pressure, for HCl = 10 atm and T = 627 °C (Krauskopf, 1963).

this group would not be transported as chloride vapors except for conditions of very high HCl concentration.

(c) Zn, Cd, Hg, Sb, As, Sn, Pb, Bi, Cu, and possibly Ag;

the elements of this group form a sufficient quantity of chloride vapor to be likely candidates for vapor transport.

(d) Mo, W;

the oxides or hydrates (hydroxides) are more significant volatile species than the halides and vapor transport is possible—even in the absence of HCl.

As was emphasized by Krauskopf (1964), the inadequate condition of the basic thermodynamic data, from which Figs. 3.13 and 3.14 are derived, does not warrant a more quantitative discussion of the chloride volatility concept of ore genesis. Also, the likely importance of liquid phase chemistry in the final stages of ore deposition may very well obscure some of the effects of vapor transport. The final fate of the volatile chlorides is probably determined by hydrolysis reactions such as

$$2FeCl_3 + 3H_2O \rightarrow Fe_2O_3(s) + 6HCl$$

or

$$2AlCl_3 + 3H_2O + 3SO_3 \rightarrow Al_2(SO_4)_3(s) + 6HCl.$$

3. *Geological Test of Theory*

The following pieces of geological evidence may be cited as support for the chloride vapor transport theory of heavy metal ore genesis.

- The geologically observed separation of the chalcophile metals (Sn, Pb, Zn, Cd, Hg, Cu, Ag, As, Sb) from other metals is consistent with the higher volatility of the chalcophile chlorides, as shown, for example, in Fig. 3.13. The alkali- and alkaline earth-metals Li, Na, Ba, Ca, and Mg, which tend to remain as silicate minerals in the magma residue, each form relatively nonvolatile chlorides.
- Heavy metal deposits [i.e., elements of group (c)] in volcanic sublimates are usually found in higher concentration than the nonvolatile metals, i.e., groups (a) and (b).
- The occurrence of Ni but very little Co in high temperature magmatic segregation deposits is consistent with the somewhat higher volatility of $CoCl_2$ than $NiCl_2$ (for similar HCl concentrations) which would presumably result in its nondeposition at high temperatures.
- The sporadic occurrence of V in volcanic sublimates can be rationalized by the formation of VCl_4 under occasional conditions of high HCl concentration, e.g., see Fig. 3.13.
- The almost complete separation of Hg from other metals in nature can be accounted for by the much greater volatility of elemental Hg, and its chlorides.
- The common association of Sb and As in ore deposits is consistent with their similar chloride volatilities.
- The formation of "haloes" of Hg, Sb, and As in the rock and soil environment around ore deposits is consistent with vapor transport away from the hot center and deposition at a concentric isothermal low temperature zone.
- The common association of Pb, Zn, Cd, and Fe in sulfide ores is consistent with their similar chloride volatilities.
- The thermodynamic stability of the metal chlorides, even in the presence of H_2S, supports the halide vapor transport theory, since, if a solution phase were involved, the stability and low solubility of metal sulfides would detract from the transport of metals from a magmatic source to an ore deposition zone.
- It is well known that ore deposits of Sn, W, and Mo tend to be present in close spatial association with granite intrusions, e.g., see Stemprok (1963). One of the suggested modes of formation for these deposits involves vapor transport through fissures in the granite body. The chloride volatility of Sn, the oxide volatility of Mo, and the oxy-hydroxide volatility of W could account for the formation of such deposits. Also, the known

stability of $SnWO_4$, $SnWO_3$, and the Mo-analogs, as vapor species at temperatures as low as 900 °C (Verhaegen *et al.*, 1965) could lead to an additional cotransport of these metals at high temperatures.

Apparent anomalies in the chloride vapor transport mechanism include the following observations.

- The location of Cu and Ag, relative to other more volatile metals, in low temperature ore deposits cannot be readily accounted for by chloride volatility as the vapor pressures are too low, even at 600 °C.
- Likewise, the low chloride volatility of Mn is not consistent with its observed position above Zn and Pb in zonal sequences.
- The observation of Ca and Mg in fumarole sublimates is difficult to reconcile with their very low chloride volatilities.

Some of these apparent discrepancies may be the result of the neglect of other possible vapor solvating interactions such as hydration (see Section F to follow). Also, the general lack of data regarding fluoride volatility has necessitated, for the most part, the neglect of this possible additional source of vapor transport. Another source of uncertainty concerns the possible interaction between the metal halides to form complex species, each containing several different metals, e.g., see Section IV. Further discussion of this point is given below (see Section E). It is also possible that some metal components may be carried by the gases in the form of aerosols, e.g., see Krauskopf (1967a).

E. Possible Role of Complex Formation

It has been suggested that at the high pressures (~1000 atm) and moderate temperatures (~600 °C) associated with post magmatic endogenous mineral-forming processes, minerals may be transported in the form of complex compounds (Ganeev, 1962, 1963; Beus, 1958, 1963). The stability of high temperature complex metal halide species (see Section IV) has been considered as supporting evidence for such a hypothesis (Beus, 1963).

Critical phenomena are most likely present in high temperature post magmatic solutions and, under such conditions, complexes are likely to be thermodynamically stable and present as neutral molecular species rather than as the separated anionic and cationic components. The stability of most high temperature complexes decreases with decreasing temperature and hence, on cooling, the fluid should reach a condition where the dissociation of complexes results in an increased acid activity of the solution. Thus, below criticality the aqueous phase probably becomes acidic and this would correspond to the beginning of the hydrothermal post magmatic process.† Similarly, a sudden pressure decrease would also favor the dissociation of complexes and an increased acidity of the solution.

In order to demonstrate the possible significance of complex formation to the metal chloride transport mode of ore genesis, consider the following two examples of apparent anomalies between halide volatility and geological observation.

Example 1. The occurrence of Cu together with Pb and Zn is difficult to reconcile with its low volatility, except for conditions of unusually high HCl concentration. Clearly, if volatility is the operating transport mechanism, an unknown volatile Cu-containing species must be formed. In view of the many mixed metal halide vapor complexes that are now known to form (see Section IV), it is reasonable to suggest the possible transport of Cu in the form of hypothetical species such as $CuPbCl_3$ or $CuZnCl_3$. There is, in fact, some indirect transpiration evidence to suggest that a vapor-phase interaction—and volatility enhancement—occurs in the binary systems of $ZnCl_2$–$CuCl_2$‡ and $CuCl_2$–$PbCl_2$ (Tarasenkov and Klyachko-Gurvich, 1936).

According to the criteria for complex formation given elsewhere (Section IV), the apparent vapor pressure of CuCl (or Cu_3Cl_3) would be enhanced by association with the more volatile species $ZnCl_2$ and $PbCl_2$. As an indication of the possible extent of this enhancement, consider the hypothetical interaction of CuCl and $PbCl_2$ to form $CuPbCl_3$. From Fig. 3.13 typical concentrations of CuCl and $PbCl_2$ are 10^{-6} and 10^{-2} atm, respectively. According to the thermodynamic generalizations for complex formation given in Section IV, it can be shown that the partial pressure of $CuPbCl_3$ will be of the order of 10^{-3}–10^{-4} atm. Thus, the vapor-phase Cu content is increased to a concentration level well above the minimum requirement for ore-forming significance. The concurrent formation of Cu, Zn, and Pb in ore deposits may then be explained in terms of such complex formation.

Example 2. The high concentrations of NaCl and KCl found in fumarole sublimates could also be due to their transport as complexes of the more volatile metal chlorides. Binary salt mixtures of NaCl–$PbCl_2$, NaCl–$SnCl_2$, NaCl–$ZnCl_2$, NaCl–$CdCl_2$, and the potassium analogs, are each known to form suitably volatile complex species of the type $NaMCl_3$, e.g., see Section IV. The interested reader may wish to follow this suggestion to a quantitative conclusion about the amount of alkali halide

† For a discussion of complexing in the hydrothermal process see Helgeson (1964).

‡ $CuCl_2$ is relatively unstable with respect to CuCl and vaporizes as the CuCl or Cu_3Cl_3 species.

transported by such complexes and under the hypothetical conditions for ore genesis.

F. Gas-Phase Hydration

The chloride volatility theory, with its neglect of complex halide interactions and possible hydration or hydroxylation, provides a lower limit indication of the possible extent of vapor-phase transport in ore formation. From the discussion of Chapter 2, Section VIII.E, it is apparent that many of the metal oxide, sulfide, or halide, magma components are soluble in steam at 1000 atm. This gas solubility often exceeds the lower partial pressure limit of 10^{-7} atm for a significant ore body formation. The high steam solubility of the normally involatile salts KCl and NaCl could well account for their occurrence in fumarole sublimates. Unfortunately, steam solubility data are available only for a few inorganic solids. However, from the thermodynamic rationalization of these data in terms of hydrate and hydroxide formation, as given in Chapter 2, Section VIII.E, it is apparent that many ore components could be transported by these solubility effects. Also, the likelihood of a highly complex vaporlike chemical state under these conditions can be appreciated from the already demonstrated complex molecular character of steam itself (see Chapter 2, Section VIII.E.2).

It is clear that the combined possibilities for additional metal transport by complex halide formation and hydration provide an interesting basis for an extension of the chloride volatility theory of ore genesis.

One can appropriately conclude this brief discussion with the comment of Ganeev (1963)

> To solve the problem of hydrothermal origin of minerals on the basis of complex compounds, the main difficulty lies in our limited knowledge of their thermal stability under high-temperature and high-pressure conditions.

This remark applies both to complex halides and hydrates and to sub- and supercritical conditions.

IX. Metal Halide Lamps

Metal halide lamps are essentially chemical reactors in which molecular transport and vapor deposition processes play a key role in their sustained operation. There are two types of metal halide lamps. The first type is essentially a modified tungsten incandescent lamp, the function of the halogen component being to control filament degradation by regenerative cycles. The second type represents a relatively recent development where the light emission is believed to result from the formation of excited metal halide molecular species in a molecular arc. These halide lamps are used commercially in photography, outdoor lighting, large space interior lighting, and car headlights.

A. Tungsten–Halogen Incandescent Lamps

The efficiency of converting electricity to light with an incandescent filament is a rapidly increasing function of temperature. If, for example, the filament temperature is increased from 2800 to 3300 K, the efficiency doubles. However, the lamp lifetime decreases markedly with increasing temperature due to the presence of "hot spots" along the filament. Transport of tungsten can occur along the temperature gradient (see Section III) to the cooler parts of the filament—leading eventually to a break in the filament. It has been found that halogen-containing additives can reverse this transport-effect and thereby increase the filament lifetime; e.g., see Zubler and Mosby (1959), and the review of Rabenau (1967).

The presence of trace amounts of H_2O, present as an impurity, can lead to an irreversible chemical transport of W from the hot filament to cooler regions of the lamp, such as the walls—e.g., see Almer and Wiedijk (1971). This has been attributed to the formation at the filament of a volatile oxide or hydroxide species which diffuses to cooler regions and disproportionates back to the metal. In view of the difficulty of completely eliminating H_2O from such systems, the solution to this problem requires the presence of an additional transport agent which can reverse the process. That is, the transport additive should react with the tungsten wall deposit to form a vapor species which diffuses to the hot filament, is decomposed to W metal, and is then returned to the cycle. Iodine has been found to be a suitable transport agent. Originally its function was believed to involve the reactions (Zubler and Mosby, 1959):

at wall,

$$W(s) + 2I \rightarrow WI_2,$$

and at filament,

$$WI_2 \rightarrow W(s) + 2I.$$

As the dissociation of the metal iodide is endothermic, the metal is deposited at the region of highest temperature and serves to repair "hot spots" at the filament. More recently it has been found that the additional presence of trace quantities of O_2 is necessary for successful operation of the tungsten halogen cycle and that WO_2I_2, rather than WI_2, is the transport species (Dettingmeyer *et al.*, 1969).

Tungsten–bromine incandescent lamps are also used commercially, the bromine being introduced initially in the form of an alkylbromide such as CH_2Br_2. Thermochemical calculations, where local thermodynamic equilibrium conditions are assumed, indicate that $WOBr_4$ and WO_2Br_2 are the likely transport species in such lamps (Yannopoulos and Pebler, 1971, 1972).

The other halogens, e.g., chlorine and fluorine, when present initially in the form of salts such as NaCl, K_3TlCl_6, cryolite, MgF_2, and CoF_2, have also proven to be effective catalysts for W-filament regeneration.

Thermodynamic calculations appear to be useful in predicting the gas solubility and direction of chemical transport for tungsten in these lamp systems. For the W–O–Br–C–H equilibrium system, the considerable number of stable high temperature molecular species and competing reactions that are likely to be present results in a rather complex variation of total tungsten vapor pressure with temperature, as shown in Fig. 3.15. The vapor species considered here were

$$W,\quad WO,\quad WO_2,\quad WO_3,\quad W_2O_6,\quad W_3O_9,\quad W_3O_8,$$
$$W_4O_{12},\quad WO_2(OH)_2,\quad WOBr_4,\quad WO_2Br_2,\quad WBr_5, \text{ and } WBr_6$$

(Yannopoulos and Pebler, 1972). Other possible species, such as WBr_4 and WBr_2, were excluded from consideration owing to a lack of basic thermodynamic data. According to Rabenau (1967), reactions involving the low valence-state species, e.g., W and WO, are predominant in the immediate vicinity of the filament, while the high-valent species, e.g., WO_2Br_2, are important in the cooler regions such as the wall.

From the results of Fig. 3.15, it is evident that for the temperature range likely to exist between the walls ($\sim$800 K) and the filament ($\sim$2500 K), several inversions occur in the dependence of total tungsten vapor pressure on temperature. It follows from these results, and Schafer's chemical transport rules (see Section III) that for the temperature regions of 500–1100, 1100–2000, and 2000–3500 K, the predicted direction of tungsten transport will be down the temperature gradient, up-gradient, and down-gradient, respectively. In practice, filament temperatures are usually in the range ot 2400–3400 K, except at the relatively cool electrodes. For the hypothetical concentration conditions defined by Fig. 3.15, tungsten should be transported toward the region of minimum total pressure for tungsten species, i.e., to the filament position corresponding to a temperature of around 2000 K. Clearly, it would be unwise to operate the filament at temperatures of greater than 2000 K *for this case.* This temperature of minimum total pressure can, in principle, be controlled by adjusting the

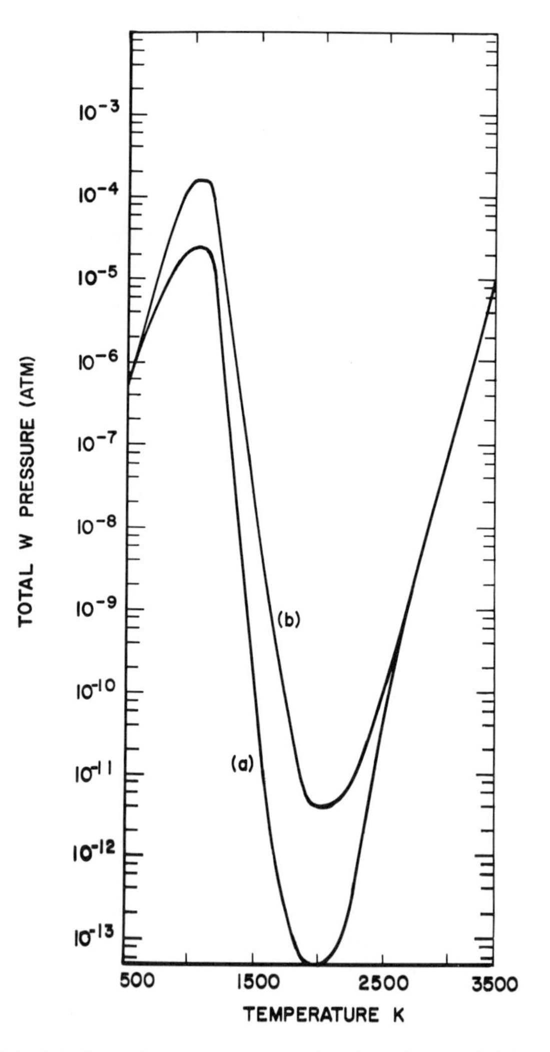

Fig. 3.15 Calculated total vapor pressure for tungsten-containing species in the W–O–Br–C–H system where O/Br = 0.01(a) or 0.10(b), O/C = 0.02(a) or 0.20(b), H/Br = 1.0; and for initial equivalent partial pressures of C = 1×10^{-3}, Br = 2×10^{-3}, O = 2×10^{-5}(a) or 2×10^{-4}(b), and H = 2×10^{-3} atm (Yannopoulos and Pebler, 1972).

chemical composition of the total system, e.g., see Yannopoulos and Pebler (1972).

Additional recent examples of multicomponent equilibria calculations for W–halogen systems have been given by Hangos *et al.* (1972) and Neumann (1973a, b).

In conclusion, it should be recognized that an alternative means of reducing tungsten transport, and thereby increasing the lamp lifetime, is to increase the pressure of buffer gas. This reduces the rate of vaporization (Harvey, 1972) and the extent of axial diffusion transport (Horster *et al.*, 1971). Typically, the pressure of inert gas used in halogen lamps is in the range of 5–10 atm. However, W-transport also occurs in these high pressure lamps due to temperature gradients and convection, hence filament regenerative cycles are still required.

B. Molecular Arc Lamps

Molecular arc lamps containing metal halides rely on light emission from excited molecular species rather than incandescent metal filaments. The variety of metal halides that can be used in such lamps has an advantage in that different color properties can be selected, depending on the electronic structure of the emitting species. Other advantages over conventional lamps include a relatively high power output (i.e., 10^2–10^3 watts) and good efficiency (i.e., $\sim$30%) in the use of electrical power for the production of a very good quality visible light output.

The commercially available arc-lamps utilize halide—usually iodide—combinations of indium, thallium, sodium, and lithium; or indium, thallium, and dysprosium; or scandium and sodium (Waymouth 1971a, b). The halides of tin and aluminum have also been used in light sources—e.g., see Gallo (1971), Grabner and Pilz (1967), Schirmer and Seehawer (1967), Springer and Taylor (1971), Zollweg and Frost (1969), Speros and Caldwell (1972), and Shaffner (1971). An extensive coverage of the practice and theory of metal halide arc- and other electric discharge-lamps has been given by Waymouth (1971a).

These metal halide lamps operate at high temperatures, in the range of 4000–6000 K at the arc center, and with fused quartz walls at temperatures of about 1000 K. Steady-state pressures are maintained in the range of 0.2–10 atm. The high molecular collision frequencies that pertain to this pressure and temperature regime, and the constant volume operation, increase the likelihood that local thermodynamic equilibrium conditions exist in these lamps. However, at the lower pressure region and with low arc currents, this equilibrium condition is not achieved, e.g., see Shaffner (1971) and cited references.

A particular problem associated with the use of halides in arcs results from the high electron affinity of halogen species. If volatile halogens are present during ignition of the lamp, the current carrying electrons may be scavenged from the system by electron capture. As the hydrogen halides are the only volatile species present at start-up temperatures, molecular halogen being precluded by the use of excess metal, it is only necessary, initially, to prevent hydrolysis of the metal halides by the elimination of H_2O from the system. However, once the system reaches a temperature of several hundred degrees centigrade, a mercury–iodine system, for example, will form HgI_2 in the vapor phase. Dissociative electron capture by this species will result in electron losses and lead to higher ignition voltages (see Waymouth, 1971a, b).

C. Tin–Halide Arc-Lamps

As an example of the present status of the theoretical evaluation of metal halide arc-lamps, consider systems containing tin halides—such as those described by Shaffner (1971) and Speros and Caldwell (1972).

The calculation of radiation output, thermal and electrical conductivity requires as basic input a knowledge of species concentration and temperature versus distance profiles between the arc and the walls of the lamp. It is possible to measure temperature and emission intensity profiles for some of the excited species by use of emission spectroscopy, but generally not species concentration profiles (Springer and Taylor, 1971). However, with the assumption of a local thermodynamic equilibrium, it is possible to derive hypothetical partial pressure or concentration profiles for the molecular species that are expected to be present.

The basic input data for the calculation of species concentrations are the thermodynamic functions of the type available in the JANAF Thermochemical Tables (JANAF, 1971). However, for many of the important halide-arc species, the data are either unknown or rely on a number of estimation procedures. For example, one of the transport cycles considered to be important in the lamps described by Speros and Caldwell (1972) involves a disproportionation of the species WCl_2, which is formed at the W-electrode to form WCl_4. No experimental thermochemical data are available for WCl_2, and using the JANAF estimates (JANAF, 1971), one can readily show that the concentration of WCl_2, relative to WCl_4, is uncertain by at least two orders of magnitude. Thus, as was pointed out by Speros and Caldwell (1972), it is impossible to provide any quantitative basis for the suggested material cycle involving the WCl_2 species.

A typical example of the calculated composition profiles for a multiple metal halide lamp, containing Hg, Ar, SnI_2, and $SnBr_2$, is given in Fig. 3.16. Note that the diatomic halide species SnI and SnBr are predicted to be present, at significant concentrations, over a wide temperature interval. Similar calculations on arcs containing tin chloride indicate the species

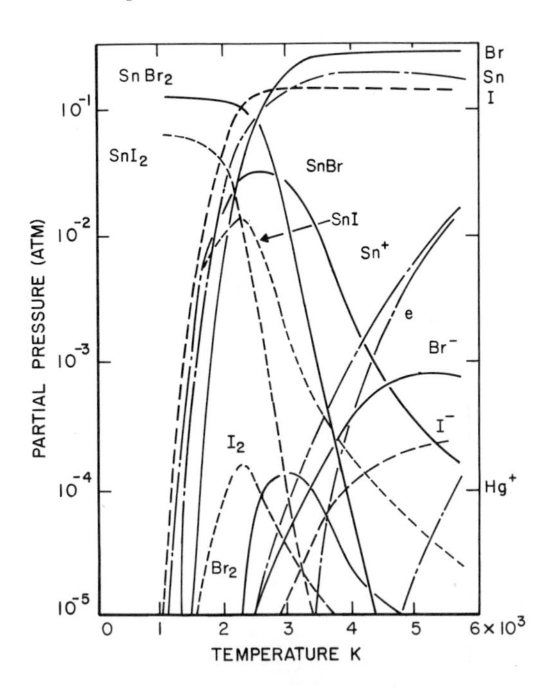

Fig. 3.16 Calculated species partial pressure vs temperature profiles for the Hg–Ar–$SnBr_2$–SnI_2 arc lamp system; data from Shaffner (1971).

SnCl to be a major molecular species over the temperature interval of 3000–6000 K (Speros and Caldwell, 1972). The calculations also indicate that an increase in the pressure of buffer gas, for example Ar, should increase the concentration of SnCl in the arc.

While such thermodynamic calculations are instructive for predicting the distribution of molecular species in a state of local thermal equilibrium, they do not reveal the source of the light emission. Spectroscopic studies by Springer and Taylor (1971) on the continuum emission, made at a number of wavelengths in the visible-uv region, suggest that SnCl is a principal source of radiation. A profile of the SnCl emission at 650 nm (6500 Å) is given in Fig. 3.17. From intensity measurements of Cu line spectra, these workers also derived a temperature profile as shown in Fig. 3.17.

The source of the electronic excitation of SnCl has not been established. However, from the data of Fig. 3.16, it is clear that the high concentration of electrons, atomic-halogen, and atomic-tin could result either in electron impact excitation of SnCl or excitation by atomic recombination processes which should most certainly be prevalent under the arc conditions. The

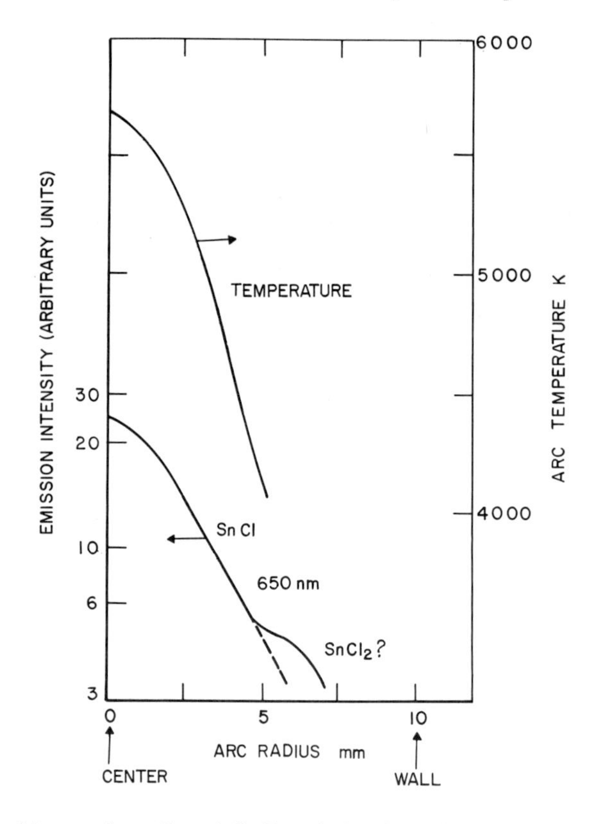

Fig. 3.17 Measured profiles of SnCl emission intensity and arc temperature vs distance from arc center to the wall for a lamp containing 14 mg Hg, 25 mg $HgCl_2$, 20 mg HgI_2, 35 mg Sn, 25 Torr Ar, and 2 mg CuI_2 (as a spectroscopic temperature indicator), and with an arc gap of 45 mm operating at 400 W alternating current and 5.5 amp (root mean square); data of Springer and Taylor (1971).

emission continuum is thought to be due to high pressure broadening of band spectra. In contrast to this model, for the group III metal iodide lamp systems Waymouth (1971a) has suggested that the spectral continuum arises from transitions occuring between an upper, somewhat bound, triplet state and a lower repulsive triplet state of the diatomic metal iodide. It is also conceivable that the spectral continua may arise from negative ion species, e.g., Al^- in the case of aluminum chloride discharge-lamps (Schirmer and Seehawer, 1967).

The successful steady-state operation of a lamp requires not only the generation of species with special thermochemical and spectral properties, but certain material balances must also be maintained by chemical cycling processes. As was the case for the tungsten and carbon incandescent lamps, the material cycling processes serve the purpose of preventing wall deposits and electrode losses. A detailed discussion of such cycles and transport processes in metal halide lamps has been given by Speros and Caldwell (1972).

In the presence of an oxygen impurity, a substantial transport of tungsten occurs from the lower to the upper electrode, which is a hotter region. This has been explained by a cycle involving the formation at the lower electrode of WO_2Cl_2, which decomposes in the arc to the atomic species. It is postulated that atomic-W is then preferentially transported from the arc onto the upper electrode, and the more mobile lighter elements O and Cl diffuse radially from the arc, rather than vertically, and tend to return to the walls and eventually to the lower electrode. A number of other cycles also operate simultaneously and a subtle balance of diffusion and thermochemical properties apparently determines the successful maintenance of the lamp. A priori calculations of these balances are not presently possible and this is due primarily to the lack of basic data.

Mention should also be made of the formation of pools of molten halides in the cool regions of these lamps. The calculation of species partial pressures therefore requires a knowledge of the component activities in these molten mixtures. In the absence of experimental data these activities are assumed to be those of an ideal solution (Shaffner, 1971). It is known that binary halide mixtures, where the cations have the same charge, e g., Sn^{2+} and Hg^{2+}, form solutions that are not too far from ideality. However, mixed valence systems can have very nonideal properties, e.g., see Lumsden (1966).

One can conclude that many of the limitations to the calculation of properties, and hence the optimization of design and operating parameters, for metal halide lamps result from deficiencies in the basic thermochemical, kinetic, and spectroscopic data. Even the identity of the emitters is not firmly established.

4

Vapor–Phase Aspects of Corrosion at High Temperature

I. Introduction

The general problem of high temperature corrosion most frequently involves gas–solid interactions with the formation of undesirable condensed phase products. In the following discussion, emphasis will be given to those corrosion problems that appear to be vapor-phase related; a more general account of corrosion may be found in the review of Fontana (1971).

Only in special circumstances is the volatility of product species a predominant factor in the corrosion process. Examples in this category include the vaporization losses of W, Re, Mo, and Cr that may occur in an oxidizing atmosphere, of SiO_2 in a reducing atmosphere, and of BeO and SiO_2 in water vapor atmospheres. In addition to the surface recession of material by vaporization of corrosion products, the diffusion of vapor species through fissures in an oxide film may lead to fracture of the film and the loss of a normally protective coating.

For some practical systems, corrosion appears as a multiphase problem. This is particularly the case for the "fire side" corrosion of power plant boilers, or for gas turbine blades, where the corrosive agents are initially in the solid (for coal) or liquid (for fuel oil) state, but are vaporized during the combustion process, transported to the cooler metal surfaces of the boiler, or turbine blade, and then condensed to form a reactive slag.

The need for an improved oxidation protection for high temperature components, such as turbines, is particularly evident in the aerospace industry. Also, future space programs, such as the space-shuttle orbiting laboratory endeavor, require materials that can sustain repeated use particularly with respect to atmospheric reentry heating. Hence, the common use of ablative nose cones in disposable space vehicles may give way to a radiative heat loss system in multiple reentry vehicles. Leading edge temperatures of about 1760 to 1900°C, and use of zirconium diboride or silicon carbide ceramics, are anticipated (Korb and Crockett, 1970). The use of superalloys, such as Inconel, for rocket construction is also likely and the corrosive effects of NaCl, as derived from sea air, on these alloys is currently of concern (Paton *et al.*, 1973). Materials problems associated with various rocket engines have been outlined by Hessing (1964), and Holzman (1969) has discussed corrosion problems for rockets in general.

Owing to the multidisciplinary nature of the corrosion problem, some of the discussion given elsewhere in this book, particularly the chapters on materials transport (Chapter 3) and energy systems (Chapter 6), also relates to corrosion. Corrosion at high temperatures is frequently a gas–solid phenomenon and a basic discussion of gas–solid reactions may be found in Chapter 2.

For purposes of discussion, it is convenient to classify high temperature corrosion in terms of corrosion with or without vaporization of the products, but in practice both of these modes of corrosion may be simultaneously present.

II. Corrosion without Vaporization

A. Metal Oxidation

A common form of corrosion involves a gas–solid oxidation of metals and alloys with the formation of an oxide, or, less frequently, a sulfide or sulfate scale. This scale may serve as a protective coating for the metal against further oxidation, depending on its melting point, vapor pressure, and impedence to gas diffusion to the metal surface.

The fundamentals of metal oxidation without vaporization have been discussed in detail by Kofstad (1964, 1966, 1970), Wood (1970), and

Logani and Smeltzer (1971). Also, Allen (1970b) has reviewed the oxidation behavior of the refractory metals. The basic controlling factors in this type of corrosion are the free energy of metal oxidation and the diffusion properties of the oxide film. When lattice diffusion through the oxide film is rate limiting, the oxidation rate is inversely proportional to the oxide thickness and the oxidation has a parabolic time dependence. According to the important Wagner theory, both the chemical and electrical potentials of the oxide are contributing factors in such a diffusion process. The importance of an electrical contribution has been verified experimentally from the influence of externally applied currents on the oxidation process.

B. Steam Corrosion

Investigations of the corrosion of base-metal alloys, and the metals Ni, Fe, Nb, Zr, Cu, and Mg, indicate that steam is usually more corrosive than air, but at the temperatures and pressures of interest, that is $<500°C$ and ~ 200 atm, respectively, no indication of a vapor-phase effect is found (Berry, 1970). This is not unexpected for such relatively low temperature conditions. However, see the discussion of Chapter 2, Section VIII.E for an indication of the likely importance of gas–solid solubility effects, particularly at higher steam pressures.

A rather specialized example of a rapid form of steam corrosion can occur in nuclear reactor excursions, i.e., accidents, where an unintended interaction between the water coolant and the hot fuel—or alternatively liquid Na—is possible. The chemical aspects of this interaction have been reviewed by McLain (1969) and, though important, are likely to be outweighed by other factors. Of greatest concern is the steam oxidation of the Zircaloy fuel pin cladding, leading to the production of H_2 and a possible H_2–air explosion. The radiolytic decomposition of water in the presence of a neutron source can also lead to the generation of H_2. It is noteworthy that the use of explosion inhibitors, also known as hydrogen–oxygen recombiners, is either practiced or being considered for the reduction of this hazard.†

Considerable computer modeling of steam–metal and steam–UO_2 interactions under simulated nuclear excursion conditions has been carried out. The basic phenomenological, i.e., nonmolecular, kinetic input data are apparently satisfactorily known for this application, and the production of new vapor species does not appear to be an important factor here. However, other reactor materials, such as boron carbide control rods, can react with

† For a general account of chemical inhibition of combustion processes see Chapter 5, Section V.

steam at temperatures of greater than about 450°C to form volatile boric acid, for example. The relatively minor role of these chemical reactions in the reactor accident is related to their low energy output as compared with that provided by the nuclear fission process. Additional discussion of steam-related reactions in nuclear fusion systems may be found in Chapter 6, Section III.D.

C. Corrosion by Fuel Impurities

1. The General Problem

The high temperature corrosion of metals that are in contact with fossil fuel–air combustion gases, i.e., CO_2, H_2O, O_2, CO, and N_2, is particularly enhanced by the presence of fuel impurities and most notably by S, V, Cl, Na, and K; e.g., see Diamant (1971), Whittingham (1971), Johnson and Littler (1963), Krause (1959), and Slunder (1959). Vanadium is an important impurity in fuel oil and can be present, in the form of a soluble organic complex, at concentration levels of the order of 500 ppm. Alkali metals are usually more important as corrosive reagents in coal-fueled combustion systems. They are present as the chloride salts or as minerals such as feldspars at concentrations of the order of 1000 ppm. Sulfur can be an important corrosive agent in either coal or oil combustion systems. Information regarding the mineral content of fossil fuels may be found in the reviews of Reid (1971b), Nelson (1959), and Ward (1973).

In passing, it is of interest to note that Ba is also present in coals at concentrations of several thousand ppm and, as has been discussed elsewhere (Chapter 5, Section VI), this concentration level is sufficient to catalyze combustion processes such as smoke oxidation.

The "fire-side" corrosion of steel–steam boilers in fossil-fueled power plants is, in part, a vapor-phase problem. In practice, the burnt gas, which has a pressure of $\sim$1 atm, or greater, and temperatures of 1200–1600°C, transports fuel impurities as refractory oxide particles and also as high temperature species to the boiler surfaces. From the papers and discussions contained in Johnson and Littler (1963), the principal vapor species are believed to be NaOH, Na, KOH, K, SO_2, SO_3, HCl, V_2O_5, V_2O_4, and in some instances SiS, SiS_2, and SiO. Upon condensation of these particles and vapor species at the stainless-steel boiler surfaces, alkali metal sulfates or pyrosulfates (e.g., see Nelson and Cain, 1960) and vanadates can form. The deposition of these complex alkali sulfates is thought to be controlled by the rate of mass transfer between the gaseous and condensed phases (Linville and Spencer, 1973).

These complex salt deposits are usually molten and are highly corrosive under the practical operating conditions of metal temperatures up to about 600°C. They also serve as a binder for the post combustion ash which consists primarily of oxides of silicon and iron. The resulting slag deposit, in addition to being corrosive, reduces the heat transfer between the combustion gas and the steam contained within the boiler tubing, thereby lowering the thermal efficiency of the power generating system. The ability of the slag to dissolve the protective oxide scales from metal surfaces limits the lifetime and high temperature capabilities of boilers. A practical upper temperature limit for steam boilers occurs at about 565°C. In gas turbines metal surface temperatures are considerably higher, i.e., up to about 900°C. Below temperatures of about 700°C, the corrosion rate is controlled by liquid-phase processes but at higher temperatures, where the molten sulfates decompose, gas-phase corrosion becomes predominant.

2. *Mechanistic Understanding*

Very little is known about the release of fuel impurities to the vapor phase during combustion, and their subsequent condensation to form sulfate and oxide slag. However, now that many of the component mineral fuel impurities have been identified, it is likely that thermodynamic vaporization models will be developed for an improved understanding of the initial stages of coal combustion.

One suggested reaction scheme, leading to the formation of corrosive metal sulfates, is as follows (Dunderdale *et al.*, 1963). The Na content of coal is considered to be present in the form of NaCl, and this salt has a sufficiently high vapor pressure to enter the vapor phase during the combustion process. However, there is some question that in the presence of silicates a high proportion of the sodium may react, prior to vaporization, and be bound up in the form of a condensed phase silicate (Boow, 1972). For a pulverized coal burning system, the time scale allowed for particle combustion and material transport to the boiler walls is typically in the range of 1–5 sec. It is well established that NaCl particles can be decomposed to elemental Na in premixed flames within a period of about 1 msec. Also, Halstead and Raask (1969) have found that the NaCl component of pulverized coal is virtually completely vaporized during the combustion process. The pyrite (FeS_2) component releases a substantial fraction of its sulfur content and, under dew point conditions, the sodium and sulfur flame components condense primarily as Na_2SO_4 or $K_3Fe(SO_4)_3$ with very little condensed NaCl resulting. These observations are basically in agreement with the predictions of equilibrium thermodynamics.

At the molecular level, following vaporization to produce NaCl species, it is considered likely that radical reactions such as

$$NaCl + H \rightarrow Na + HCl,$$

occur during combustion. Then, in the cooler post-flame region, NaOH is formed according to the reaction

$$Na + H_2O = NaOH + H.$$

This is probably followed by the processes

$$SO_2 + O\ (+M) \rightarrow SO_3\ (+M),$$

$$SO_3 + NaOH \rightarrow (NaHSO_4)_{g\ or\ l},$$

$$NaHSO_4 + NaOH \rightarrow (Na_2SO_4)_{g\ or\ l} + H_2O,$$

where M is either a gaseous third body, such as CO_2, or alternatively a surface.

That the production of SO_3 depends primarily on the presence of O-atoms, as suggested in the above reaction scheme, is supported by the results of flame species concentration profile determinations in a H_2S–O_2–N_2 flame using a microprobe sampling chemical analysis technique (Levy and Merryman, 1965). Also, known O-atom inhibitors, such as NH_3, reduce the formation of SO_3, and commercial use is made of this effect. In practice, it appears that part of the production of SO_3 occurs in the post-reaction-zone combustion gases and the remainder at a surface. It is known that solids containing Fe_2O_3 have a strong catalytic effect on the conversion of SO_2 to SO_3.

In stoker-fired coal burning installations, where the fuel environment is less oxidizing, there is some evidence that silica is transported to the furnace walls and the following overall reactions are believed to be involved:

$$SiO_2(s) + FeS(s) + 2C(s) \rightarrow SiS + Fe(c) + 2CO,$$

$$2SiO_2(s) + FeS_2(s) + 2C(s) \rightarrow 2SiS + Fe(c) + 2CO_2.$$

The volatile SiS species then condenses on the cooler furnace walls where it apparently reconverts to SiO_2 and SO_2 in the presence of oxygen, i.e., see the review of Krause (1959).

The enhanced corrosion rates found for stainless steels containing Cr and Ni in the joint presence of chloride- and sulfur-containing species are believed to result from the production of volatile metal chlorides, e.g.,

$$24KCl(c) + 10Cr_2O_3(s) + 9O_2 = 12K_2CrO_4(s) + 8CrCl_3$$

(Alexander, 1963). The loss of the protective chromic oxide or chromium-iron spinel surface then allows sulfate oxidation of the iron to take place, i.e.,

$$3Fe + SO_4^{2-} \rightarrow Fe_3O_4 + S^{2-}.$$

In the absence of halogen, the chromium alloy component can usually retard the effects of sulfate corrosion by the production of a stable chromium sulfide layer. Protection up to temperatures of at least 750°C is found. However, in the presence of chlorides, corrosion proceeds at temperatures as low as 400°C. This type of corrosion problem also applies to marine gas turbines where sea water constituents, such as sodium chloride and sodium sulfate, tend to be associated with the air intake to the combustion system and this further aggravates the corrosive effects of fuel impurities, e.g., see Goward (1970).† Sulfur-related corrosion of alloys can also occur in the absence of alkalis or halogens but at higher temperatures, i.e., about 870 °C for Ni–Cr alloys (Viswanathan and Spengler, 1970).

3. *Corrosion Inhibitors*

The addition of chemical additives to the fuel can have beneficial effects in retarding fire-side corrosion, in addition to reducing the amount of sulfur oxides released to the atmosphere.

The corrosive effects derived from sulfur impurities in fuel oil can be markedly reduced by keeping excess air to a minimum since this reduces the concentration of O-atoms available for SO_3 production. Alternatively, the addition of magnesium metal (e.g., see Laxton, 1963) or its hydroxide (e.g., see Holland and Rosborough, 1971) may be used to reduce the sulfate corrosion problem. The refractory nature of the magnesium oxide and magnesium vanadate deposits formed appears to be the beneficial factor in the use of magnesium additives. Calcium oxide additives are believed to prevent the formation of the highly corrosive alkali metal iron trisulfates by the formation of relatively harmless condensed double sulfates such as $K_2SO_4 \cdot 2CaSO_4$. With the formation of protective oxide wall coatings an increase of about 20 °C in boiler operating temperature is possible (Holland and Rosborough, 1971). These types of additives are finding some commercial application.

Metal additives also appear to have an inhibiting effect on the formation of SO_3 *within* the flame (Ivanov *et al.*, 1973). The presence of naphthenates of Mn, Pb, Na, K, Mg, Zn, Al, Ba, and Co, in high-sulfur liquid fuels, can reduce the SO_3 flue gas concentration by a factor of 2 for additive concentrations of the order of 20 ppm by weight. A catalytic interference with O-atoms is believed to be responsible for this effect (see Chapter 5, Section II.L).

† See also the evaluation of this problem by the National Materials Advisory Board (NMAB-260, May 1970, AD870745).

D. Corrosion of Nuclear Reactor Materials

The corrosion of reactor materials, such as austentitic stainless steel and various base-metal systems, in the presence of liquid metals and fused salts has been reviewed by Simons (1970). Also, the recent use of silicide coatings to protect the refractory metals of interest as reactor materials has been reviewed by Allen (1970a, b). Corrosion where vaporization of products occurs does not appear to be an important factor under normal reactor conditions. Even the formation of an oxide scale, or other condensed phase corrosion processes, is of secondary consequence to the effects of radiation damage, e.g., see Chapter 6, Section III.C.

III. Corrosion with Vaporization

A. Refractory Metals and Silicon

The refractory metals are considered as W, Re, Os, Ta, Mo, Ir, Nb, Ru, Hf, Rh, V, and Cr, with melting points ranging from 3410 to 1875 °C. Silicon is also considered in the present discussion as its use, in combination with refractory metals, can lead to refractory alloys or coatings, but it may also suffer from corrosion by vaporization of products. Many refractory metal high temperature corrosion processes involve gas–solid interactions and a more basic discussion of these reactions may be found in Chapter 2. The thermodynamic aspects of refractory metal corrosion have been reviewed recently by Gulbransen (1970). The nature of the oxidative corrosion of metals is largely dependent upon the properties of the oxide scale such as its melting point, vaporization or dissociation pressure, and diffusion rates. For example, Ta and Nb exhibit a poor resistance to oxidation, owing to the relatively low melting point of their oxides. On the other hand, Os, Re, Mo, and W form solid oxides, at temperatures of less than 1000 °C, which are extremely volatile, and this is a predominant factor in the corrosion of these metals.

The high temperature chemical compatibility of refractory metals in contact with ceramics, is of interest in connection with rocket nozzle design (Sprague, 1968). For instance, thermodynamic free energy calculations for such systems indicate that Al_2O_3 is more suitable than BeO in contact with the same metal and that the least interaction of these ceramics is found with W. Other systems of compatibility interest include Al_2O_3–HfN and Al_2O_3–TaC.

1. The Tungsten Example

For the case of tungsten, it is possible to describe the corrosive process in terms of the vapor pressure of the volatile oxide species, as shown in Fig. 4.1. At oxide pressures of greater than about 10^{-3} atm, gas diffusion has an important limiting effect on the rate of vaporization, and hence corrosion.† This rate-limiting aspect is of particular interest in the modeling of W-recession rates for space vehicle reentry applications. For example, in connection with the development of such models, Bartlett (1968) has determined the rate of W-recession in oxygen atmospheres at pressures of 10^{-6}–10 atm and at temperatures of between about 1320 and 3170 °C. The results are represented by the Fig. 4.2. Above about 2000 °C the tungsten oxide species have a sufficiently high vapor pressure to form a boundary layer which lowers the concentration of oxygen at the tungsten surface and

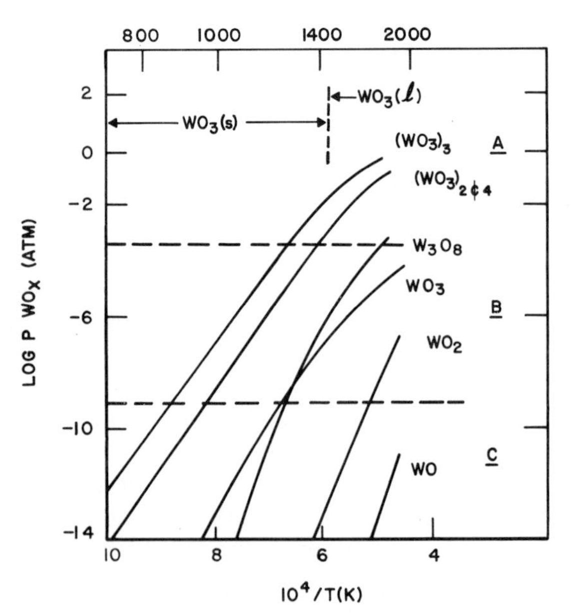

Fig. 4.1 Oxidative corrosion of W in 0.05 atm O_2, expressed in terms of calculated W-containing species partial pressures as a function of temperature; the broken horizontal lines indicate approximate boundaries defining different corrosive processes; the boundary at $\sim 10^{-3.3}$ atm separates a regime of chemical and diffusion-limited processes from one of solid oxide scale formation, i.e., at $<10^{-3.3}$ atm; below $\sim 10^{-9}$ atm the oxide scale has an insignificant vaporization loss (adapted from Gulbransen, 1970).

† For a discussion of W-oxidation at lower pressures see Chapter 2, Section IV.C.

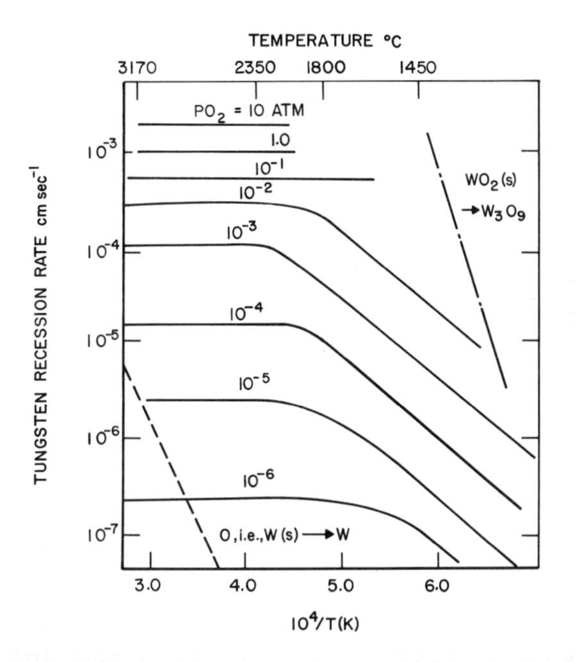

Fig. 4.2 Measured W recession rates as a function of oxygen pressure and temperature; the horizontal lines at high temperatures are in agreement with those calculated from a gas diffusion rate-limiting model (adapted from Bartlett, 1968).

the recession rate becomes approximately temperature independent. This effect has been modeled by Bartlett (1968) using a free-convection theory. It is noteworthy that the use of small surface areas, i.e., <1 cm^2, tends to eliminate these rate-limiting diffusion effects (Gulbransen *et al.*, 1964).

2. *The Silicon Example*

A recent survey of the reactions of metals and alloys with gases at elevated temperatures, leading to scale formation, has been given by Rapp (1970). Regions of stability, with respect to vaporization, for the metals and their oxide scales may be defined in terms of "Kellogg-type" diagrams which basically contain isothermal free energy information. As these diagrams serve to demonstrate the importance of a detailed knowledge of the vaporization processes, in connection with the oxidative corrosion of metals, it is useful to consider an example in detail, as follows.

In systems containing silicon, a partially reducing atmosphere tends to be corrosive because of the high volatility of the lower valence species, SiO, relative to SiO_2. On the other hand, for systems containing refractory metals such as W, Mo, or Cr, the reverse is true. For example, in oxidizing

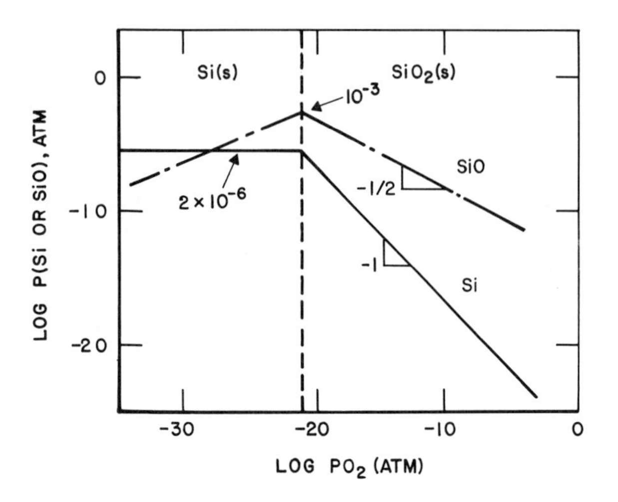

Fig. 4.3 Vaporization in the Si–O_2 system at 1500 K (adapted from Rapp, 1970).

atmospheres, Cr forms the species CrO_3 which is appreciably more volatile than either CrO_2 or CrO. The main thermodynamic reactions leading to the loss of Si by vaporization in an atmosphere which is not strongly oxidizing are

$$Si(s) + \tfrac{1}{2}O_2 = SiO, \tag{1}$$

$$Si(s) = Si, \tag{2}$$

$$SiO_2(s) = SiO + \tfrac{1}{2}O_2, \tag{3}$$

and

$$SiO_2(s) = Si + O_2. \tag{4}$$

A knowledge of these reactions and their equilibrium constants may be used to define the isothermal dependence of vaporization pressures on the oxygen activity, as shown by Fig. 4.3. Similar plots may also be constructed for the refractory metals, e.g., see Rapp (1970). The vertical line in the Fig. 4.3 represents the stability dividing line, i.e., O_2 activity, between the solids Si and SiO_2, and is obtained from the free energy of formation for each condensed phase; note that the indicated slopes of $-\frac{1}{2}$ and -1 are determined by the stoichiometric oxygen dependences of reactions (3) and (4), respectively.

B. Noble Metal Coatings

In contrast with the refractory materials of graphite, tungsten, and tantalum—which form stable oxide vapor species such as CO, CO_2, WO_3,

TaO, and TaO_2—the noble metals, for instance Pt and Ir, tend to form oxide vapor species of relatively low stability.

At about 600 °C, Ir forms a stable oxide film in air which vaporizes at temperatures of about 1000 °C. The vapor species are believed to be IrO_2, IrO_3, and Ir_2O_2, depending on the conditions of temperature and pressure, e.g., see Wimber (1971). However, these species tend to be unstable at higher temperatures and the oxygen dissociation pressure of the iridium oxides attains a value of 1 atm at 1105 °C. Above this temperature the loss of Ir by oxide vaporization is sufficiently low to allow Ir to be used in oxidizing systems. For instance, Ir containers have been found to be suitable for vaporization of the oxides ZrO_2, ThO_2, and CeO_2 at temperatures approaching 2500 °C (Kaufman *et al.*, 1967).

The use of Ir coatings on more readily oxidizable, but less expensive, materials such as C, W, and Ta is of potential importance for high temperature components in space reentry vehicles, gas-cooled (e.g., CO_2) nuclear reactors, and rocket motors. Hence, there is considerable interest in the thermodynamics, kinetics, and mechanism of Ir-oxidation, e.g., see Wimber (1971). Under stagnant-gas conditions it appears that the rate-limiting step in the oxidation process is the diffusion of IrO_2 and IrO_3 species through the gaseous boundary layer. A review of refractory metal coating techniques and properties has been given by Leeds (1968).

C. Ceramics

Ceramics such as alumina and silica are relatively inert to corrosive atmospheres, and they form the basis for many of the coatings used to protect metal components such as gas turbine blades. The upper temperature limit of their application is usually determined by the rate of oxidant diffusion through the coating to the metal substrate and by vaporization of the oxide. It is likely that improved ceramic coatings could result from the use of binary or ternary oxide systems where a reduction in vaporization rate can be expected on thermodynamic grounds.

A few examples where application limitations can occur due to vaporization are indicated as follows.

1. Sulfur Corrosion

Corrosion of oxide ceramics by sulfur-containing gases can occur at high temperatures. In practical situations, such as the combustion of a sulfur-containing fossil fuel, the sulfur is usually present together with oxygen, and the relative stabilities of the metal oxides and sulfides can determine the sulfur contribution to corrosion in a non-rate-limited system (Alcock and

Zador, 1971). As was found to be the case for steel corrosion (Section II.C), the S, V, and Na fuel impurity combination also has detrimental corrosive and mechanical effects on furnace refractories such as magnesite or aluminosilicates, e.g., see Pressley (1970).

The nature of sulfur corrosion can vary significantly with temperature. For example, sulfation of alumina can occur at temperatures of less than 1000 K owing to the stability of the solid $Al_2(SO_4)_3$ under oxidizing conditions. At higher temperatures, where this compound is unstable, some corrosion due to the formation of vapor species, such as $(AlS)_2$ and AlS, could occur (Ficalora *et al.*, 1968a). However, owing to the highly negative heat of formation for the solid oxide Al_2O_3, sulfur pressures of the order of 1 atm would be required to achieve a significant degree of corrosion by vaporization of the sulfides (Alcock and Zador, 1971).

Silica is much more subject to sulfur corrosion than alumina owing to the relatively high stability of the SiS species, i.e., D_0 SiS $=$ 149 whereas D_0 AlS $=$ 88 kcal mol^{-1}—see the review of Hauge and Margrave (1972). On the other hand, as there are no stable sulfates of silica, this oxide is relatively more resistant to sulfur attack in the presence of oxygen.

Alcock and Zador (1971) have noted that there is a general correlation between the stability of sulfide phases, relative to the oxides, and the radius of the metal cation. As the cation radius increases, the relative stability of sulfide to oxide increases. The same relationship can be found in the dissociation energies of the diatomic metal chalcogenide vapor species (Hauge and Margrave, 1972; Ficalora *et al.*, 1968a).

2. *Silicon Carbide*

Silicon carbide and its composites have oxidation-resistant applications in nuclear reactors and in the aerospace field. Two kinds of oxidation behavior can be defined, one where a relatively passive film of solid SiO_2 is formed and the other where gaseous SiO is produced. This latter behavior leads to a rapid corrosion and the conditions under which it occurs are therefore of interest. At high oxygen pressure the production of SiO is reduced due to the process

$$SiO + \tfrac{1}{2}O_2 \rightarrow SiO_2(s),$$

i.e., see Fig. 4.3. Antill and Warburton (1970) found that water vapor was even more corrosive than oxygen at temperatures of 1000–1300 °C, and it was suggested that perhaps hydroxyl ions were generated at the gas–solid interface. The possible formation of a neutral hydroxide species, such as the hypothetical species $Si(OH)_2$, could also be a contributing factor (see Chapter 2, Section VIII).

3. *Zirconium Boride*

Another ceramic material with potential applications similar to silicon carbide is zirconium diboride and its composites—e.g., with SiC or C. Corrosive breakdown of this material results from the production of ZrO_2 solid and B_2O_3, or other boron oxide vapors, e.g., see Graham (1970). For a pressure of about 0.3 atm and at temperatures of less than 1100 °C, very little vaporization of B_2O_3 is found. Also, for conditions where the production of nonvolatile oxides predominate, the usual parabolic dependence of degree-of-oxidation on time is observed and diffusion through a protective oxide film is apparently the rate-determining process. However, for conditions where vaporization is possible a more rapid, almost linear, time dependence is found.

4. *Beryllia*

Beryllia (BeO) has been used as a ceramic neutron moderator for propulsion reactors using an oxidizing coolant, e.g., see Rothman (1968). While BeO is resistant to degradation by oxygen or hydrogen and also to self-vaporization, it is strongly affected by the presence of water vapor, and this is thought to be due to the reaction

$$BeO(s) + H_2O = Be(OH)_2.$$

Thermodynamic data for this process are available (JANAF, 1971) and allow a maximum rate of corrosion to be calculated. However in practice, the overall rate of corrosion appears to be limited by the diffusion rates of H_2O to the surface and of $Be(OH)_2$ from the surface boundary layer. The following diffusion-vaporization model fits experimental data at moderate flow rates of air and water vapor along a BeO surface:

$$N = (DM/RTx)\,(Kp_w - p_B),$$

where

N is the loss of BeO in gm cm^{-2} sec^{-1},
D the diffusion coefficient of $Be(OH)_2$ in air (estimated),
R the gas constant,
T the temperature,
x the boundary layer thickness,
K the equilibrium constant for the above reaction,
p_w the partial pressure H_2O,
p_B the partial pressure of $Be(OH)_2$, and
M the molecular weight of $Be(OH)_2$.

5

Combustion

I. Introduction

Combustion science and technology is an extremely diverse subject and the following literature sources are recommended. An elaboration of the basic principles may be obtained from sources such as Bradley (1969), Fristrom and Westenberg (1965), Williams (1965), Strehlow (1968), Minkoff and Tipper (1962), Lewis and Von Elbe (1961), and Lawton and Weinberg (1969). A selected bibliography on combustion up to 1964 may also be found in Fristrom and Westenberg (1965), where the cited references to the specialized AGARD reviews and the International Symposia Proceedings of the Combustion Institute (1953) are of particular interest. Hougen *et al.* (1959) have discussed the general principles of dynamic combustion systems, such as internal-combustion engines, gas turbines, ram jets, and chemical rockets. The application of flames and plasmas to high temperature chemical processing has been reviewed by Brown *et al.* (1968)—see also Chapter 3, Section VI. Industrial flames and their characterization have recently been discussed by Chedaille and Braud (1972).

The empirical combustion science of incineration has been described in the book edited by Corey (1969).

It is informative to follow the development of the combustion field as represented, for example, by the International Symposia of the Combustion Institute held over the past 20 years. Of particular note is the increased application of spectroscopic techniques in recent years, and this has resulted in a more microscopic and fundamental understanding of combustion phenomena. For example, the ability to monitor and characterize reaction intermediates has allowed a better, although by no means yet definitive, understanding of the chemical mechanisms for combustion problems such as nonintended explosions and fires, knocking in internal combustion engines, and the production of soot or gaseous pollutants.

A particularly noteworthy contribution of the basic science of high temperature vapors to an applied combustion problem, has been in the calculation of rocket performance parameters from basic thermochemical and kinetic data for the reactants, intermediates, and products. Even solid rocket-fuel systems, such as Al or Mg plus perchlorates, apparently involve vapor-phase reactions in the combustion process, e.g., see Brzustowski and Glassman (1964).

The control of combustion from the point of view of efficient utilization of energy sources, minimization of air pollution, and fire prevention, represent the most urgent immediate problems. In this connection, the use of inorganic additives, such as Ba and Fe compounds to suppress smoke formation, K and Ba to increase electrical conductivity, Sb, P, and halogen compounds for flame speed reduction or flame extinction, and Pb additives for knock prevention, exemplifies the variety of inorganic systems that are currently being utilized to modify combustion processes.

Another area of concern, which involves inorganic components in combustion systems, is the fate of mineral impurities in coal combustion. The most undesirable impurities, from the point of view of toxic products entering the atmosphere as ash and dust, include:

Pb, Cd, Hg, As, Se, Sb, and Be,

present at ppm concentration levels, and

Ni, Zn, Mn, V, S, Cl,

present at one or more orders of magnitude greater concentration. Only for the case of sulfur is there any reasonable body of literature and understanding of the combustion chemistry of such inorganic components, i.e., see the extensive review of Cullis and Mulcahy (1972). However, even in this case there are many unsettled questions, such as the existence of either an isomeric or dimeric form of SO_2.

In the area of heterogeneous combustion, dust explosions and fires are common industrial hazards, particularly in flour milling, woodworking, coal mining, and metal or chemical production. However, the chemical properties of dust flames have received little attention and no propagation mechanism has been established (Palmer, 1973). Similarly, the potential hazard of a nuclear fission fast reactor loss-of-coolant accident, where molten Na and water may interact explosively, and the established hazard of molten metal "explosions" in the aluminum industry, are areas of current practical concern. Water also interacts explosively with some molten salts, e.g., NaCl and AgCl (Rosenburg, 1970). The mechanisms for these phenomena are also unknown, but apparently involve heterogeneous combustion.

The development of lasers over the past decade has further broadened the scope of combustion science (Schwar and Weinberg, 1969). In particular, the recent development of flame systems as a chemical source for laser energy, has given new impetus to the spectroscopic and kinetic characterization of flames. Lasers have also been used to initiate or support combustion, particularly for less conventional systems such as light metals in an oxidizing environment. The combustion of certain light metals, such as Be, Al, and B is of interest as they can be used to enhance the performance of solid rocket propellants. One might also mention the use of lasers to initiate nuclear fusion which is an extreme form of combustion, e.g., see Chapter 6, Section IV.

The presence of charged species in combustion systems, even in the absence of ionizable additives, has long been known from the observation of a physical deflection of flame gases in the presence of electric fields and, in recent years, practical interest has developed in these charged species. For instance, the problem of a marked attenuation in the electromagnetic radiation signals used for the communication and guidance of rocket and atmospheric reentry vehicles, results from the presence of free electrons in rocket exhausts and in the region of ablating nose cones. Hence, in order to devise means for controlling the electron and ion concentrations in these systems, many fundamental studies dealing with the identity of the charge carriers and their kinetics and thermodynamics have been carried out, e.g., see Jones *et al.* (1963) and Shuler and Fenn (1963). It has been found that both electron–ion recombination and electron–molecule attachment processes can significantly reduce the free electron concentration in rocket exhausts (Balwanz, 1965). The high electron affinities and thermal stabilities found for a number of metal inorganic species in flames is particularly pertinent to the question of electron scavenging. A heterogeneous attachment of electrons to alumina particles in rocket exhausts has also been reported (Balwanz, 1965). However, this may not necessarily result in free-electron depletion because alumina droplets are believed to have low work functions, i.e., ~5–6 eV, and they may therefore act as a source of electrons in high temperature combustion systems (Gatz *et al.*, 1961).

It has also been noted that soot in flames has a positive charge and that there is a correlation between the total concentration of ions resulting from the addition of ionizable metals, particularly Ba and Cs, and the degree of metal-catalyzed smoke reduction.

Several other useful aspects of the charged nature of flames include: use of the charged species for energy transport in the electrical augmentation of flames, e.g., see Lawton and Weinberg (1969); the use of flame ionization in gas chromatography detectors, which relies intrinsically on the generation of chemi-ions in hydrocarbon-containing flames, e.g., see Bocek and Janak (1971); and the use of rockets containing additives such as Ba which are released as ions (probably molecular) into the upper atmosphere for mapping magnetic lines of force (Rosenberg, 1966) in addition to other space applications (Haerendel *et al.*, 1967). The spectroscopic character of these upper atmosphere releases has therefore been of special interest, e.g., see Batalli-Cosmovici and Michel (1971) and Chapter 7, Section II. Another charge-related application of some potential utility involves the magnetohydrodynamic generation of electric power by using the conventional combustion of fossil fuels—particularly coal—as a high temperature bath for the thermal ionization of a seed material, such as potassium. A detailed discussion of the high temperature chemistry aspects of magnetohydrodynamic energy systems is given in Chapter 6, Section VII.

In the discussion to follow, Sections IV–IX deal with combustion systems where high temperature inorganic species are of key practical significance. Sections II and III provide fundamental background to the later sections of this chapter and also to the related discussions of Chapter 3, Section IV; Chapter 6, Sections VI and VII; and Chapter 7, Section II.C.

II. Molecular Aspects of Flames—Flame Chemistry

A. Flames in General

1. Introduction

The combustion of either organic or inorganic systems under certain initial conditions of composition, temperature, and pressure, and operating conditions of material and energy transport can result in the stabilization of flames. A "normal" flame is a thermal wave which propagates at subsonic velocities and is driven by exothermic chemical reactions. Suitable flame-generating reactions may be created either by the decomposition of metastable monopropellants, e.g., N_2H_4, O_3, H_2O_2, CH_3NO_2, or by the combination of fuels such as H_2, CH_4, C_2H_2, CO, C_2N_2, B_2H_6, Al, Mg, Ti, $TiCl_4$, and $Al(CH_3)_3$ with oxidants such as O_2, N_2O, Br_2, Cl_2, F_2, $HClO_4$, or ClO_3F. However, the fact that a chemical mixture is thermodynamically metastable with respect to certain hypothetical products of an exothermic reaction is not a sufficient criterion for flame production. For practical purposes, the potentially combustible mixture must have a composition that falls within empirically determined flammability limits (Lovachev *et al.*, 1973). Flammability-limit data for most common combustible mixtures are well known and may be found in the data compilations of Zabetakis (1965) and Coward and Jones (1952).

Most flames are sustained by rapid chemical reaction, on a millisecond time scale, under constant pressure and variable volume conditions. In contrast, propellant systems are essentially constant-pressure constant-volume processes with relatively long residence times and a closer approach to chemical equilibrium. Explosions, on the other hand, are constant-volume variable-pressure processes and are two or three orders of magnitude faster than other combustion phenomena.

The macroscopic observables of a flame, such as temperature, burning velocity, and flammability composition limits, can, in principle, be rigorously related to the flame microstructure, or molecular chemistry, which is determined by the basic thermodynamic, chemical-kinetic, and transport properties of the various flame species. The connection between the micro- and macroscopic properties is provided by the equations of continuity, motion, energy balance, and the diffusion equation, e.g., see Hirschfelder *et al.* (1954, p. 768), Frank-Kamenetskii (1969), and Williams (1965).

2. *Radiation and Temperature*

Most of the radiation from a flame is chemiluminescent, rather than thermal, and is emitted primarily in the infrared. However, this energy loss by radiation is relatively small and most of the chemical energy is converted into translational energy. Hence, observed flame temperatures are usually not very different from the adiabatic temperature, i.e., not more than 5% lower.

The adiabatic flame temperature is strictly a thermodynamic quantity and is given by

$$\Delta H_0 + \int_0^{T_u} (C_p)_u \, dT = \int_0^{T_b} (C_p)_b \, dT,$$

where

ΔH_0 is the heat of reaction,
$(C_p)_u$ the heat capacity of reactants,
$(C_p)_b$ the heat capacity of products,
T_u the temperature of reactants, i.e., initial temperature, and
T_b the adiabatic final flame temperature.

In practice, the calculation of flame temperature is an iterative procedure, because some of the equilibrium products are in a dissociated form at the flame temperature. Discussions of the calculation and measurement of flame temperatures may be found in the text of Gaydon and Wolfhard (1970), and in the review of Snelleman (1969).

3. *Flame Speed or Burning Velocity*

The terms flame speed and burning velocity tend to be used synonomously. The chemical and physical structure of a flame is determined to a large extent by the burning velocity, which, in turn, is determined by the chemical-kinetic and transport properties of the flame. It is thus a fundamental flame property having both practical and theoretical significance.

The macroscopic significance of burning velocity is readily apparent if we consider a stationary flame system. Flames having spatial stability can be produced using a flameholder, such as a burner, which serves as a heat sink and allows the flame reaction zone to stabilize downstream from the burner. The flame shape is then determined by the balancing of preflame gas flow patterns and the burning velocity which results from gas acceleration out of the reaction zone. In a stabilized premixed laminar flow flame, fed by a burner tube, the flame reaction zone takes on a shape such that the burning velocity, i.e., gas velocity normal to the plane of the reaction zone, matches the gas flow velocity. This allows a stationary reaction zone to persist and the burning velocity S_u is related to the volume flow $\dot{V}$ of unburnt gas and the flame front area A_f by

$$\dot{V} = S_u \times A_f.$$

Nonstationary, or turbulent, flames usually result when the input gas flow rate is in appreciable excess of the burning velocity. Blow-off occurs at the extreme of this condition.

Burning velocity is related to the reaction rates and heats which are temperature and pressure dependent. Usually an increase in temperature or a decrease in pressure results in a higher burning velocity. For instance, the CH_4–air flame has a maximum burning velocity of $S_u = 45 \pm 2$ cm

$\sec^{-1}$, at an equivalence ratio of 1.07,† and this velocity doubles for about a 100° increment in the initial gas temperature. The decrease in velocity with increasing pressure P for an equivalence ratio of 1.0 is given by the relation

$$S_u = 43P^{-0.5}$$

(Andrews and Bradley, 1972).

Flames with relatively high burning velocities tend to be less affected by a pressure change. This is a result of the predominance of second order, or bimolecular, processes in fast flames (Gilbert, 1956; Smith and Agnew, 1956). High burning velocities are usually associated with high temperature flames. However, there are a few cases where very fast flame reactions may lead to excessively high burning velocities at moderate temperatures. For example, the stoichiometric diborane–oxygen flame (pressure ~0.01 atm) has a velocity of about 2500 cm $\sec^{-1}$ whereas H_2- or C_2H_2-fueled flames of similar temperature have burning velocities that are more than an order of magnitude lower (Wolfhard *et al.*, 1964). Furthermore, burning velocities of nonchain branching systems such as the $CO–N_2O$ flames are relatively low, i.e., about 40 cm $\sec^{-1}$, even though the flame temperatures are high, i.e., ~2800 K (Kalff and Alkemade, 1972).

4. *Chemical Structure*

It is remarkable that the basic chemical structure of a flame was demonstrated more than a century ago (Faraday, 1957). In his famous study of the chemical history of a candle, Faraday demonstrated the formation of spatially distinct chemical zones in a flame, i.e., preflame, reaction-zone, and burnt-gas regions. The reaction zone was found to be a region where "an intense chemical action takes place." The chemical zones, as identified by Faraday, correspond respectively to an area of limited chemical reaction, a region of excessive rapid reaction where about 90% of the energy is released, and a region where slow recombination of fragments and relaxation of internal excitation occurs. Frequently, the rapid, or primary, reaction zone contains an excess concentration of radicals, ions, and excited molecular species, with respect to the equilibrium levels.

The current status of flame chemistry and structure has been the subject of numerous reviews and several monographs, i.e., see Cotton (1970), Jenkins and Sugden (1969), Bradley (1969), Schofield and Broida (1968),

† The term equivalence ratio refers to the ratio of the amount of fuel present in the initial combustion mixture to that required for a stoichiometric reaction, i.e., in this instance, for the process

$$CH_4 + 2O_2 \rightarrow CO_2 + 2H_2O.$$

Fristrom (1966), Fristrom and Westenberg (1965), Wehner (1964), and Fenimore (1964). As will become evident in the more detailed discussion to follow, flame chemistry, in many instances, is synonymous with the high temperature chemistry of radical species, particularly H, O, and OH.

Most of the chemistry in flames is the result of bimolecular processes. Ternary processes as factors are only 10^{-3}–10^{-4} as frequent at atmospheric pressure and predominate only in the postflame deexcitation region. The presence of an excess concentration of radicals in the reaction zone is then a consequence of their production, by bimolecular processes, being more rapid than their loss, by ternary recombination reactions. The excess ionic and excited species concentrations found are the result of chemiexcitation reactions involving the flame radicals. Processes leading to the production of excited species in flames have been reviewed by Sugden (1962).

The spatial characteristics of atmospheric premixed laminar flow flames, such as the visually distinct luminous reaction zone of hydrocarbon flames, are a direct result of the steep concentration, i.e., $\sim 10^2$ mole fraction cm^{-1}, and temperature, i.e., $\sim 10^5$ °C cm^{-1}, gradients present in the region of the reaction zone. These large gradients enhance the contribution of the transport processes of mass diffusion and thermal conduction to the flame chemistry. However, an approximately constant pressure is maintained throughout the flame.

The contribution of species diffusion to flame structure and chemistry is important primarily for low molecular weight species such as H and H_2. The importance of H-atom diffusion into the preflame gas region was recognized even before the microscopic structure of flames had been established (Tanford and Pease, 1947; Tanford, 1947). It was suggested that this diffusion, rather than heat conduction, was the important factor in initiating combustion, and this is in accord with the present-day understanding of flame structure. Radiation effects on flame chemistry are usually negligible except for systems which contain particulate matter such as smoke. As a result of the effects of molecular diffusion, the species flux, i.e., species flow per unit area per unit time, is determined by the concentration, i.e., species per unit volume, and also by the diffusion velocity, i.e., species flow per unit time. In a flow system, such as a flame, the net reaction rate is proportional to the spatial derivative of the flux.

Typical time scales for processes in 1-atm flames are indicated in Table 5.1. It is evident that species in the reaction zone may be vibrationally excited. Rotational excitation occurs to a lesser extent, though OH is a known case where rotational temperatures, i.e., populations, are above the equilibrium values, particularly in the region of the reaction zone, e.g., see Broida and Heath (1957).

Flame chemical structure also depends on the total pressure. For ex-

TABLE 5.1

Typical Time Dependence for Processes in 1-atm Flames[a]

Gas velocity	$= 10^2$–10^3 cm sec^{-1}
Diffusion velocity	
H atoms	= 100 cm sec^{-1}
O atoms	= 10 cm sec^{-1}
Reaction zone thickness	$= 10^{-2}$ cm
∴ Residence time in reaction zone	$= 10^{-4}$–10^{-5} sec
Chemical reaction (bimolecular)	$\leq 10^{-5}$ sec (for 20–30 kcal mol^{-1} activation energy)
Relaxation of excited species	
translational	$= 10^{-8}$ sec
rotational	$= 10^{-8}$–10^{-7} sec
vibrational	$= 10^{-6}$–10^{-3} (10^{-4} typical) sec
electronic (chemiluminescence decay)	$\geq 10^{-3}$ sec

[a] $T \sim 2000$ K.

ample, the thickness of the reaction zone varies inversely with pressure. Higher pressure flames, particularly those at greater than 1 atm, tend to be hotter and closer to chemical equilibrium because of the collision nature of relaxation processes. The phenomenological, i.e., nonmolecular, aspects of high pressure flames (>1 atm) have been reviewed by Cummings (1957).

5. *Flame Applications*

Aside from the intrinsic importance of the flame itself, flames are useful as crucibles both for high temperature chemistry studies of additives and for the synthesis of inorganic materials. In particular, flames obtained by the combustion of H_2–O_2–N_2 gas mixtures, at 1 atm, are well characterized, and they are generally useful as a source of high temperature (1500–2500 K), of radicals, i.e., H, OH, and O, and of superheated steam. The dis-equilibrium in the burnt-gas region near the reaction zone also allows kinetic studies to be made. As the gas velocity is practically constant in the burnt-gas region, then the distance coordinate is a simple function of time and kinetic factors are therefore readily determined. The time scale of chemical reactions in flames is in the range of 10^{-1}–10^{-6} sec and this bridges that available from gas-flow reactor studies and shock wave or flash photolysis studies. Thus flames constitute a useful kinetic medium.

In addition, the equilibrium reached well downstream of the reaction zone enables thermodynamic studies to be made.

The problems of container interactions, usually encountered in the study or preparation of materials at high temperatures, are also eliminated by the use of a flame medium. Thus, as was suggested by Padley (1969), flames can be convenient laboratories for studies of high temperature chemistry. For instance, the equilibrium postflame gas region has been used as a heat bath for the thermodynamic study of many metal hydroxides, oxides, and halides, as discussed later in this section.

The ability to calculate flame temperatures and also energy losses due to radiation, conduction, and incomplete equilibration of the reaction products, has led to the suggested application of flames as primary (Snelleman, 1968a) or secondary (Snelleman, 1968b) temperature standards.

The high temperature and high radical concentration characteristics of flames have been of great significance for quantitative chemical analysis using flame spectrophotometry. In practice, a small amount of the compound to be analyzed is introduced to the flame, where the combined effects of high temperature and high radical concentration serve, in many instances, to reduce the additive to its elemental constituents for atomic absorption analysis. Also, the excessive electronic excitation that the elements may have in the vicinity of the flame reaction zone allows one to use the emission spectra for analytical purposes, as discussed in Section IV.

At sufficiently high metal concentrations, condensation of the more refractory inorganic additives occurs in flames, and since both the temperature and concentration can be controlled, it is possible to grow single crystals from flames, or alternatively, to generate a ceramic spray, such as SiO_2, ZrO_2, or Al_2O_3, for coating applications, e.g., see Ballard (1963), Sittig (1968), Frolov *et al.* (1971), Svirskiy and Pirogov (1971), and Hecht (1972). Several commercial chemical reactor applications include the production of TiO_2 particles from $TiCl_4$ in H_2 flames and the conversion of UF_6 gas to solid UO_2 in H_2 flames. Similarly, the commercial production of pyrogenic or fumed silicas is achieved by the combustion of $SiCl_4$ in a premixed H_2–air flame with quenching of the products in an air stream. Factors relating to the particle sizes obtained by such processes have been considered by Ulrich (1971). The titanates of Ba, Ca, Ni, and Zn may also be prepared in single-crystal form using flame processes, e.g., see Sittig (1968).

A number of areas of practical importance, where flame chemistry plays a significant role, are indicated in the chart of Table 5.2. Specific examples from these areas will be considered in the ensuing discussion. The problem of "fire side" corrosion in fossil-fueled combustors has already been considered in Chapter 4.

TABLE 5.2

Practical Applications of Flame Chemistry

Materials preparation	Corrosion	Combustion	Inhibition	Analysis	Ionization
Crystal growth	Energy systems	Air pollution	Flammability	Spectro-photometry	Ionization detectors
Ceramic spraying		Energy systems	Fires		Radar interference
					MHD

B. Premixed Flames

Premixed laminar-flow flames are, from both an experimental and theoretical point of view, more convenient systems for the study of combustion chemistry than, for example, are diffusion flames or turbulent flames. They are therefore much better understood than any other combustion system and form the main basis for discussion in this chapter.

The premixing of fuel with oxidant, prior to entering a combustion regime, allows the course of the combustion process to be controlled primarily by rapid chemical reactions rather than by a slow diffusive mixing of fuel and oxidant in the vicinity of the flame itself. This rapid-reaction characteristic can often result in the production of excessively large concentrations of radical species relative to the thermodynamic equilibrium level. This, in turn, gives rise to many interesting and important phenomena, such as chemiluminescence, chemi-ionization, excess dissociation of molecular species, a reduced tendency for smoke production in hydrocarbon systems, and relatively high rates of flame spread through a combustible medium. An indication of the scientific and technological consequences of such phenomena will be developed in later sections of this chapter, particularly Sections IV–VIII.

A fundamental characteristic of premixed flames, and indeed any combustion system, is the spatial distribution of each chemical species in the flame. This distribution will take the general form indicated in Fig. 5.1. At atmospheric pressure conditions the concentration of species in the postflame, i.e., burnt-gas, region usually is determined by the equilibrium thermodynamic properties of the system. The chemical structure of the reaction zone is determined by the reaction rates of many competing reactions and the preflame-region chemistry is influenced both by reaction

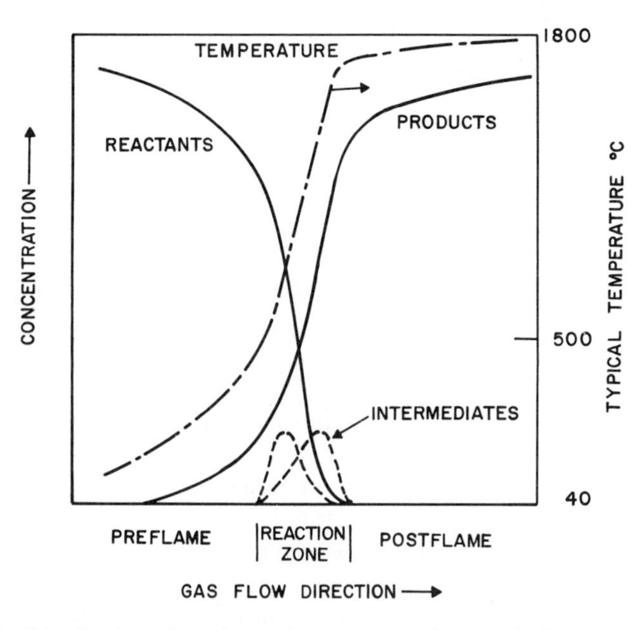

Fig. 5.1 Distribution of species and temperature in a typical premixed flame at atmospheric pressure.

rates and also transport of energy and species out of the reaction zone into this region.

An example of the distribution of radical intermediates in low pressure flames is given in Fig. 5.2. In this particular case the intermediate species

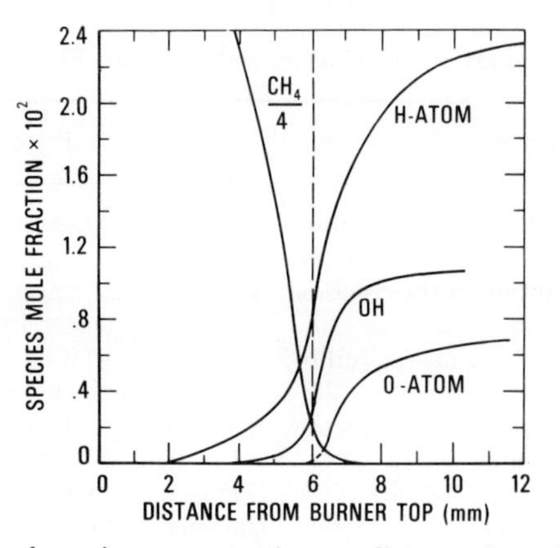

Fig. 5.2 Profiles of species concentration vs distance from burner in a reduced pressure premixed flame of composition 9.6% CH_4–21.3% O_2–69.1% Ar (adapted from Lazzara *et al.*, 1973).

concentrations remain at a high level in the postflame region owing to the lower radical recombination reaction rates which result from the low pressure condition.

C. Diffusion Flames

1. Diffusion Flames in Comparison with Premixed Flames

For fuel-oxidant flame systems where the components are physically separated in the preflame condition, the rate-limiting step in the combustion process is determined by the rate of interdiffusion of fuel and oxidant to the flame zone. Flames sustained under these conditions are known as diffusion flames. Barr (1955) has summarized the macroscopic conditions of oxidant and fuel flow, whereby different diffusion flame types appear, i.e., convective, turbulent, smoky, etc. The mathematical treatment of diffusion flame structure where reaction kinetics and diffusion coefficients are explicitly taken into account, is given, for example, by the work of Clarke and Moss (1970).

A comparison of the general features of diffusion and premixed flames is made in Table 5.3. The equilibrium condition at the reaction zone of diffusion flames results from the low rate of diffusive mixing of fuel and oxidant which allows time for the relaxation processes to occur. Also, for hydrocarbon–air diffusion flames, it is known that the fuel is pyrolyzed prior to coming into contact with the oxygen. Such a system tends to be of

TABLE 5.3

Comparison of Diffusion and Premixed Flames

Diffusion	Premixed
Chemical diffusion is the rate-determining process	Chemical reaction is the rate-determining process
The rate of O_2 consumption in the reaction zone is typically 10^{-5} mol cm^{-3} sec^{-1}	The rate of O_2 consumption in the reaction zone is typically 4 mol cm^{-3} sec^{-1}
Species are probably at, or near, equilibrium in the reaction zone	Excess dissociation and excitation are present in the reaction zone
Reaction zone temperature is close to the maximum adiabatic value	Reaction zone temperature is somewhat less than the maximum value
One cannot define a burning velocity	One can define a burning velocity
Chemical inhibition pushes the reaction zone into the air, or oxidant, region	Chemical inhibition pushes the reaction zone downstream from the source of flowing reactants

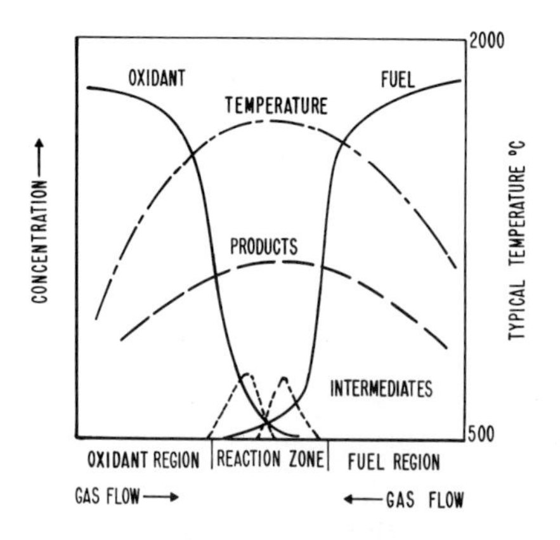

Fig. 5.3 Distribution of species and temperature in a typical diffusion flame at atmospheric pressure.

low chemical potential, which contrasts with the high chemical potential and rapid breakdown in the reaction zone for premixed flames.

The general chemical structure of a diffusion flame is represented in Fig. 5.3. As the distribution of intermediates is largely controlled by their thermodynamic equilibria with fuel and oxidant, then the fuel-related intermediates, e.g., CH_3, will be located nearer the fuel-flame boundary and vice versa for oxidant-related species, e.g., OH or O.

Composition profiles, obtained by probe sampling with subsequent mass spectrometric and gas chromatographic analyses, have been obtained for methane diffusion flames in air at 1 atm (Smith and Gordon, 1956). The results indicate the following:

- The methane is consumed very early and is essentially at zero concentration at the flame edge
- N_2 is found to have a very high concentration even at the center of the fuel column.
- Similarly, O_2 is found to be present everywhere in the flame and, at the center of the fuel column, concentrations of the order of 1% are found.
- When the products H_2O, CO_2, and CO are taken into account, the total oxygen, i.e., free and bound, content of the fuel-zone region is much greater than that allowed by the O_2 to N_2 air ratio. Evidently the diffusion coefficient of O_2 in the combustion product–fuel mixture is greater than that in air.
- The observed degree of combustion in the fuel column is greater than that predicted by the equilibrium water–gas reaction. Also, the early consumption of CH_4 is much greater than that allowed by a purely thermal

pyrolysis process. It must be concluded that the decomposition is promoted by the presence of oxygen.

In these analyses no oxygenated hydrocarbon species, such as formaldehyde, were found. To summarize, results such as these indicate that the diffusion flame resembles a fuel-rich premixed flame. It is likely that this is also the case for "real fires," as supported by pyrolyzing substrates, and hence fuel-rich premixed flames may be used to model such "real fires."

2. *Metal-Fueled Diffusion Flames*

The combustion of metals usually falls in the category of diffusion flames. An interest in the fundamental structure of metal diffusion flames derives from their practical utility in areas such as rocket propulsion, flash lamps, and exploding wire ignition devices, e.g., see Sections VIII and IX. These flames have also found application in the production of chemi-excited high temperature species for fundamental spectroscopic studies.

a. The Mg–O_2 Diffusion Flame Example

A number of fundamental flame structure studies have used the Polanyi low pressure diffusion flame technique, e.g., see Polanyi (1932), and Rapp and Johnston (1960). As an example let us consider the Mg + O_2 + Ar (excess) diffusion flame, where the observed macroscopic flame structure has been interpreted in terms of either a unimolecular or bimolecular rate process (Markstein, 1963). In this study, it was assumed that the major flame products were MgO species and MgO particles. The experimental conditions used were: total pressures of 2–10 Torr, oxygen pressures of about 1 Torr, and Mg vaporizer temperatures in the region of 650 °C—leading to a metal vapor pressure of several Torr.

The basic assumptions of the technique are

- the local radiation intensity is proportional to the rate of reaction;
- the flame has spherical symmetry, i.e., the diffusion and chemical reaction rates are fast compared with the flow of metal from the nozzle at large radii from the nozzle; and
- the concentration of gas in the atmosphere surrounding the fuel, i.e., Mg-metal in this instance, is effectively constant.

In practice, concentration profiles in space are determined from the intensity of light produced at various regions of the flame. A significant limitation in relating the macroscopic kinetic observations to the microscopic molecular processes is the uncertainty regarding the nature of the species present, as evidenced by the discussion following the paper of Markstein (1963).

Macroscopic kinetic data may be derived from the following relations:

$$C_N = (A/r)\exp(-\omega r) \qquad \text{and} \qquad \omega^2 = kC_A{}^n/D_N,$$

where

C_N and C_A are the concentrations of nozzle (i.e., fuel) and atmosphere (i.e., oxidant) reactants respectively,
A is a constant for radii r large compared with the orifice radius,
k the overall rate constant,
n the reaction order, and
D_N the fuel diffusion coefficient.

For large radii, the integration of the above C_N expression along a light emission path which is perpendicular to the nozzle axis is given approximately by

$$I = \text{const}\ r^{-1/2}\exp(-\omega r).$$

This integral is equivalent to the light emission intensity along a similar path as detected by exposure on a photoplate. The unknown ω can then be determined from the slope of the line obtained when $\log(r^{1/2}/\tau)$ vs r is plotted for various film exposure times τ. Rate constants may then be determined from the measured ω provided C_A and D_N are known.

For the Mg + O_2 + Ar (excess) example, the flame temperature is estimated to be about 1000 K and the binary diffusion coefficient of Mg in Ar was taken as

$$D_{Mg,Ar} = 1108/p\ \text{cm}^2\ \text{sec}^{-1},$$

where p is the Mg pressure in Torr. In principle, the reaction order n can be obtained from a logarithmic plot of ω^2/p vs pO_2 (since $D \propto p^{-1}$). However, for the present case, the experimental uncertainty was such that it was not possible to distinguish between $n = 0$ and $n = 1$, and the following rate constants were calculated, at about 1000 K, for both unimolecular and bimolecular processes, i.e., 2.5×10^3 sec^{-1} and 4×10^{11} cm^3 mol^{-1} sec^{-1}, respectively. For the unimolecular case it is believed that the reaction involves an oxidation reaction at the surface of MgO particles. It should be noted that if the system were at equilibrium then the concentration of MgO product species in the vapor phase would be insignificantly small.

More recent studies of the Mg–O_2 diffusion flame have been made by Markstein (1969) and by Deckker and Rao (1970), which support a homogeneous, rather than heterogeneous, reaction mechanism. The work of Deckker and Rao (1970) also established experimental temperature profiles. For a vaporizer, i.e., Mg vapor, temperature of 610 °C the flame has a temperature of about 560 °C at the apparent center. As the results fit the Polanyi dilute diffusion flame model it is believed that the flame processes are homogeneous and occur initially in the vapor phase.

b. Chemiluminescence Studies

Low pressure diffusion flames generated by the reactions of active species, such as H-atoms or N-atoms, with normally stable molecular species are useful experimental systems for spectroscopic studies of intermediate species. Typical examples of interest to the high temperature chemist are the production of chemiexcited B_3 by the reactions of H-atoms with BCl_3 and of NSe by the reactions of N-atoms with selenium chlorides (Vidal *et al.*, 1969).

Diffusion flames of alkali metal atoms and metal halide or oxyhalide vapors have been exploited by Palmer and co-workers for chemiluminescence studies of "high temperature species" such as SnCl, $SnCl_2$, $GeCl_2$, BO, and BO_2 (Naegeli and Palmer, 1968; Tewarson *et al.*, 1969; Tewarson and Palmer, 1967). In addition to providing spectroscopic information, difficult to obtain by other means, such as the location of the low-lying triplet state of $GeCl_2$ (i.e., see Hastie *et al.* (1969b)), some insight into reaction mechanisms of diffusion flames is also provided by these studies.

D. Unusual Flames

1. Introduction

One usually associates flames with the combustion of hydrocarbons, or H_2, in air or oxygen. However, there exist many other possible fuel-oxidant combinations and examples of some of the more exotic homogeneous flame systems are given in Table 5.4. Some of these lead to very high temperature flames, as indicated in Table 5.5. These temperatures may be compared with that of a stoichiometric H_2–O_2 flame, i.e., 3077 K, or of a CH_4–air flame with a temperature of 2232 K.

As with hydrocarbon flames, some of these unusual flame systems undergo a multistage combustion process. For example, premixed flames of

TABLE 5.4

Examples of Unusual Flames

Oxidant	Fuel	Oxidant	Fuel	Oxidant	Fuel
O_2	C_4N_2	ClO_3F	H_2	O_2	$Al(CH_3)_3$
O_2	C_2N_2	ClF_3	H_2	F_2	CCl_2F_2
F_2	H_2	ClF_3	CO	F_2	NH_3

TABLE 5.5

Very High Temperature Flames

Reaction	Pressure (atm)	Observed temperature[a] (K)	Calculated temperature (K)	Reference
$H_2 + F_2 \rightarrow 2HF$	1	4300 ± 150	4300	Wilson *et al.* (1951)
$(CN)_2 + O_2 \rightarrow 2CO + N_2$	1	4640 ± 150	4810 ± 50	Conway *et al.* (1953)
$C_4N_2 + 2O_2 \rightarrow 4CO + N_2$	1		5261	Kirshenbaum and Grosse (1956)
	10		5573	
$C_4N_2 + \frac{4}{3}O_3 \rightarrow 4CO + N_2$	1		5516	
	40.82		6100	

[a] The sun was used as a comparison emission source for temperature measurement.

trimethylaluminum–oxygen involve spatially distinct oxidation of the Al and hydrocarbon components of the fuel (Vanpee *et al.*, 1970).

2. *The CCl_2F_2–F_2 Flame Example*

Combustion systems containing fluorine are of special interest on account of their high exothermicity, i.e., flame temperature, which usually results from the production of the stable species HF as a combustion product. Also, the recently developed HF chemical lasers rely on fluorine combustion reactions.

One of the few fluorine-supported flames that has been experimentally subjected to a species analysis is the dichlorodifluoromethane (CCl_2F_2)–F_2 partially premixed system (Homann and MacLean, 1971a). From the species profiles shown in Fig. 5.4, it appears that the flame consists of two chemically distinct combustion zones. In the first stage of combustion the intermediate products ClF and $CClF_3$ appear, and in the second stage these intermediates are consumed with the formation of Cl_2 and CF_4 as the final combustion products. From the observed species profiles and reaction-energetics arguments, the following reaction mechanism has been suggested:

$$F + CCl_2F_2 \rightarrow CClF_3 + Cl, \tag{1}$$

$$Cl + F_2 \rightarrow ClF + F, \tag{2}$$

$$F + CClF_3 \rightarrow CF_4 + Cl, \tag{3}$$

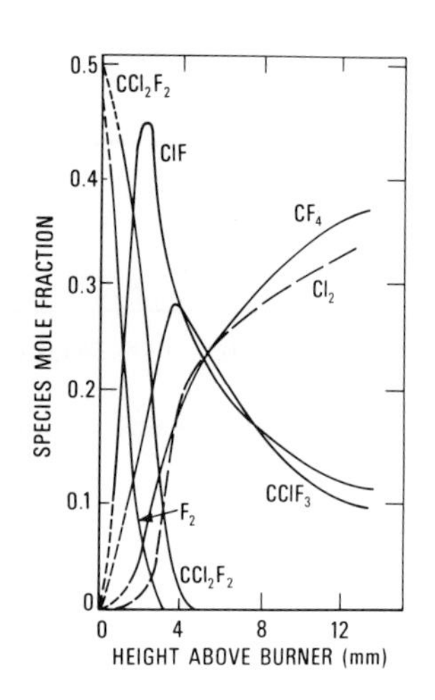

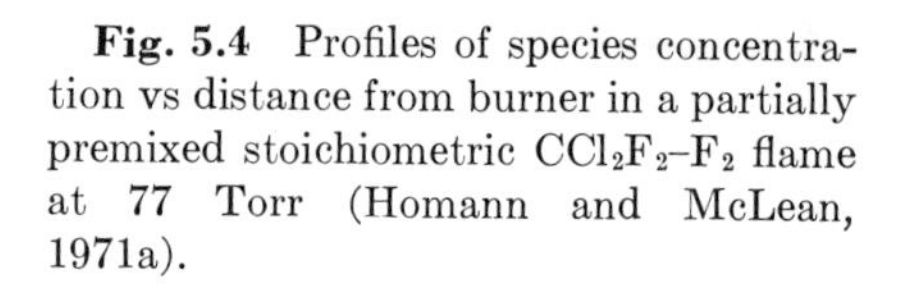

Fig. 5.4 Profiles of species concentration vs distance from burner in a partially premixed stoichiometric CCl_2F_2–F_2 flame at 77 Torr (Homann and McLean, 1971a).

and

$$Cl + ClF \rightleftarrows Cl_2 + F. \tag{4}$$

Other possible reactions were excluded by their high endothermicity. The four reactions listed above are either close to thermal neutral or they are exothermic.

Of the species participating in the reaction scheme, only the F atom was experimentally undetected. This inability to detect F atoms is probably due to the fast nature of reaction (4) in the back direction. It is believed that this reaction is balanced, i.e., is in local equilibrium, and hence

$$[F]/[Cl] = K_4[ClF]/[Cl_2].$$

From known values of the equilibrium constant K_4, it can be shown that $[F] > [Cl]$ in the first combustion stage, and vice versa for the final combustion stage.

These authors have also reported on similar studies using the flames: H_2–F_2, H_2–F_2–NH_3, NH_3–F_2, C_2H_2–F_2, and C_2H_4–F_2 (Homann and MacLean, 1971b).

E. Flame Theory

1. Introduction

A basic objective of flame theory is to provide a rigorous connection between macroscopic flame phenomena such as temperature, heat release, burning velocity, thermal and electrical conductivity, and smoke production, and the microscopic character as defined by species concentrations and their basic kinetic, thermodynamic, transport, and spectroscopic properties. With the availability of high-speed computing techniques it now becomes feasible to optimize the design of practical combustion systems without recourse to an expensive trial and error construction procedure. In view of this prospect there is considerable interest in the development and testing of theoretical flame models. To remain consistent with the overall objectives of this monograph the following discussion will outline flame theory only to the extent required for an appreciation of the parameters and basic data involved in its application.

Essentially, the problem of theoretically accounting for observable flame properties involves a hybridization of reaction kinetics and transport phenomena. The importance of reaction kinetics to flame propagation is demonstrated, for example, by the pronounced effect of trace quantities of chemically active additives on macroscopic flame properties such as flame speed, e.g., see Section V.

There are several levels of sophistication in utilizing flame theory, depending on whether one needs a completely ab initio calculation of flame properties or whether a combination of theory and experiment will suffice. Thus, for example, if the temperature and the concentration of the major reactants are known it is possible, via a steady-state approximation, to deduce either radical concentration or rate-constant data.

A common usage of flame theory involves its reverse empirical application. That is, given sufficient experimental data on species concentrations, temperature profiles, and diffusion coefficients, it is possible to determine time derivatives and hence reaction rates and a kinetic flame model.

The calculation of flame properties strictly from basic data represents the highest level of sophistication in flame theory—requiring the solution of complex differential equations representing conservation of species and energy. As has been discussed by Williams (1965), the utilization of basic flame theory involves some mathematical approximations; for example, the assumption of a steady state in a laminar-flow flame system. The approximations appear to be reasonable for the few cases where sufficient data are available for a comparison of theory and measurement.

An introductory discussion of flame structure, both at the macro- and microscopic levels, has been given by Cotton (1970). More basic discussions of flame theory may be found in Wehner (1964), Fristrom and Westenberg (1965), Williams (1965), Strehlow (1968), Fristrom (1966), and Hirschfelder *et al.* (1954). Some specific examples of application of the theory may be found in the above-mentioned works and the reports of Dixon-Lewis and co-workers (1965, 1967, 1968, 1970, 1970a, b), Fenimore and Jones (1959), Wilson *et al.* (1969), Browne *et al.* (1969), Spalding and Stephenson (1971), and earlier cited work.

2. *The Steady-State Approximation for Reaction Intermediates*

The so-called steady-state approximation for flame intermediates represents a first-order approximation to the exact solution of the differential flame equations. The basic assumption is that the intermediate species concentration is independent of time, i.e., the individual species net reaction rate is zero. This assumption is equivalent to the condition of a long species residence time relative to its chemical time, i.e., see Millan and DaRiva (1960). The attainment of a steady-state condition relies, then, on fast reaction rates relative to both the time of passage of the species through the flame and the rate of diffusion to regions of lower concentration. It also depends on the radical concentrations being small relative to the major flame components.

In view of the complexity associated with a complete solution of the flame differential equations, it is desirable to determine under what conditions the far simpler steady-state solution of species concentrations pertain. Such steady-state criteria have been developed, and it has been shown that they hold for the hydrogen–oxygen flame type but not for the hydrogen–bromine flame (Millan and DaRiva, 1960). In this latter case the Br-atom concentrations apparently cannot be determined with accuracy from a steady-state treatment. However, as shown by the complete flame analysis of Spalding and Stephenson (1971), the departure of the steady-state concentrations from the rigorous values is not excessively large—being about 30% or less

As an example of the application of the steady-state approximation, consider the H_2–Br_2 flame. It is generally agreed that the following set of reactions should properly define the kinetics of this flame system:

$$Br_2 \underset{k_5}{\overset{k_1}{\rightleftarrows}} 2\ Br,$$

$$Br + H_2 \xrightarrow{k_2} HBr + H,$$

$$H + Br_2 \xrightarrow{k_3} HBr + Br,$$

and

$$H + HBr \xrightarrow{k_4} H_2 + Br.$$

Under steady-state conditions the rates of formation and loss for a particular species, e.g., H, will be equivalent and the following concentration relation can be determined:

$$[H] = \frac{k_2[Br][H_2]}{k_3[Br_2] + k_4[HBr]}.$$

From this, and similar relationships for the other species, together with the known rate constants, one can predict the concentrations of the intermediate species H and Br at any stage of the reaction, e.g., see Anderson (1957). Unfortunately, for this relatively simple system no experimental data on the intermediate species concentrations have ever been obtained, and hence a test of the calculated data cannot be made. As this is one of the more simple flames available to test various aspects of flame theory, it presents a challenge for the experimentalist to provide the necessary comparative data.

The experimental difficulties involved in the measurement of flame radicals has generally hampered the progress of flame theory. However, a recent determination of the ubiquitous HO_2 species in flames allows a semiquantitative test of the steady-state approximation (Hastie, 1974). In this example the flame conditions are such that the formation and loss of HO_2 are dominated by the reactions

$$H + O_2 + M = HO_2 + M, \tag{1}$$

$$HO_2 + H = 2OH, \tag{2}$$

where M represents the major nonreactive species such as N_2 and H_2O. It follows from the steady-state approximation that the concentration of

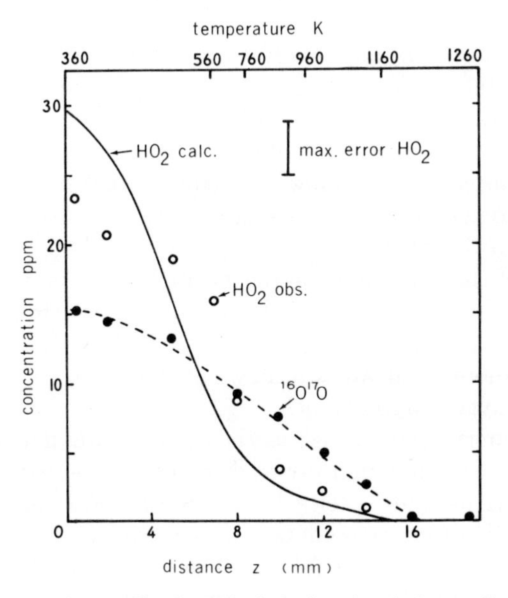

Fig. 5.5 Concentration profiles for HO_2 (calculated and observed) and O_2 (observed) in a relatively cool fuel-rich H_2–O_2–N_2 flame at 1 atm (Hastie, 1974).

HO_2, neglecting diffusion effects, may be calculated from

$$[HO_2] = (k_1/k_2) [O_2] [M].$$

A comparison of calculated and measured HO_2 concentrations in a H_2–O_2–N_2 flame is given in Fig. 5.5. The uncertainties in k_1 and k_2 lead to an uncertainty of about a factor of 2 in the calculated HO_2 concentrations which agree with the experimental data within this error limit.

3. *Determination of Kinetic Models from Experiment*

For the simple model flame system of a premixed laminar-flow flat flame, it is theoretically possible to transform basic reaction rate and diffusion data to observables such as flame speed and species concentration profiles, providing the correct reaction mechanism, i.e., set of elementary kinetic reactions, is considered. This ab initio calculation of flame properties involves approximate integrations of the partial differential equations for species and energy conservation. A critical discussion of general methods may be found in the text of Williams (1965). The papers of Dixon-Lewis (1970a, b) and Spalding and Stephenson (1971) deal with the specific cases of H_2–O_2–N_2 and H_2–Br_2 flames, respectively.

To test the validity of the choice of a reaction mechanism, a comparison of the calculated properties with those obtained by direct measurement is required. As the basic reaction-rate data are often inadequate, the application of flame theory usually involves the derivation of reaction-rate data from experimental species concentration versus distance profiles through the flame. This reverse application of flame theory is mathematically more tractable—involving an experimental determination of differential terms rather than an integration of differential equations. A further simplification is usually provided by invoking the steady-state hypothesis but extended to include diffusion effects.

Examples of this kind of flame-theory application may be found in the low pressure CH_4–O_2 flame studies described by Fristrom and Westenberg (1965), the low pressure H_2–N_2O flame studies of Fenimore and Jones (1959), or the more recent work of Pownall and Simmons (1971) on 1-atm lean propane–oxygen–argon flames.

For a stationary system, such as the experimental flat flame, it is possible to measure concentration versus distance profiles normal to the flame front for the various flame species, i.e., see Section F. In order to determine kinetic information from these data it is necessary to derive the mass flow fractions G_i for each species as follows. The total mass flow of gas through the flame is constant along the flame and is experimentally known. It is also equivalent to ρv, where ρ, the density, may be calculated at each flame position from the known temperature, pressure, and composition. Hence, the mass average gas velocity v may be deduced. In passing, we should note that the mass flow is a fundamental parameter in flame theory, being the eigenvalue solution to the conservation of mass and energy equations. Diffusion also contributes to the species mass flow. The diffusion velocity for species i through the medium is given by

$$V_i = -\frac{dX_i}{dz}\left(\frac{D_{ij}}{X_i}\right),$$

where

X_i is the mole fraction;
dX_i/dz is the change in mole fraction with distance z normal to the flame front; and
D_{ij} is approximated, in practice, as the binary diffusion coefficient between species i and the major flame component j which is usually N_2.

For pressure and temperature scaling of diffusion data the relation

$$D \propto (P)^{-1}$$

and the empiricism

$$D \propto T^{-1.8}$$

are used, respectively.

The fraction of mass flow, for species i, is then related to experimental quantities by the equation

$$G_i = (X_i)\left(\frac{M_i}{\bar{M}}\right)\left(\frac{v + V_i}{v}\right),$$

where M_i is the molecular weight for species i and $\bar{M}$ is an average molecular weight for the gas. The net rate of species production and disappearance is then given by

$$K_i = \frac{dX_i}{dt} = \frac{\rho v}{M_i}\frac{dG_i}{dz},$$

where dG_i/dz is positive or negative for species production or disappearance, respectively. These net reaction rates should not be identified directly with the reaction rates of individual molecular processes. However, given the concentration profiles for all species present, it is possible to identify the individual molecular processes and their specific rate constants. For instance, if a reaction involving species i leads to the formation of a product species j; that is,

$$i + m \rightarrow j + n,$$

then

$$-\frac{dX_i}{dt} = \frac{dX_j}{dt}.$$

This equality may be used to test which product species corresponds to a particular reactant and hence the kinetic reaction may be identified. Also,

$$\frac{dX_j}{dt} = k[X_i][X_m],$$

from which the elementary reaction-rate constant k may be determined.

Alternatively, given the elementary reactions and the rate constants, together with the appropriate diffusion data, it is possible to compute the species concentration profiles. As an example, one may deduce concentrations for the radical species OH if it is assumed that the following reaction dominates the conversion of CO to CO_2, i.e.:

$$OH + CO \rightarrow CO_2 + H$$

then

$$[OH] = K_{CO_2}/k[CO],$$

where K_{CO_2} is the net rate of reaction of CO_2 as derived from measured concentration profiles for CO_2, [CO] is the measured concentration of CO and k is the literature rate constant for the above reaction, e.g., see Pownall and Simmons (1971). Usually this type of calculation is more complicated than this example, as the steady-state production of a species may result from many competing elementary reaction steps.

A check on the experimental data and the diffusion calculations is given by the mass balance requirement of

$$\sum_i \frac{n_i G_i}{M_i} = \text{const}$$

(i.e., at all positions of z), where n_i is the number of atoms of a particular element in a molecular species i. Difficulty in achieving a satisfactory mass balance, particularly for H_2 and H_2O, is evident from the results of Fristrom and Westenberg (1957).

A summary of the basic flame relationships used in developing a kinetic interpretation of species profile data is given in Table 5.6. Of particular note is the requirement of second derivative spatially dependent data, i.e., dX_i^2/dz^2. Such data are difficult to obtain for 1-atm flames where the reaction zone is very narrow, i.e., ~0.1 mm. Hence, for quantitative flame structure analysis, use is made of low pressure flames where the reaction

TABLE 5.6

Some Basic Flame Relationships[a]

Diffusion velocity (normal to flame front)	$V_i = -\frac{dX_i}{dz}\left(\frac{D_i}{X_i}\right)$
Fractional mass flow	$G_i = X_i \frac{M_i}{\bar{M}}\left(\frac{v + V_i}{v}\right)$
Net reaction rate	$\frac{dX_i}{dt} = \frac{dG_i}{dz}\left(\frac{\rho v}{M_i}\right)$
Reaction mechanism, e.g., for $i + m \rightarrow j + n$	$-\frac{dX_i}{dt} = \frac{dX_j}{dt} = k[X_i][X_m]$
Mass balance	$\sum_i \frac{n_i G_i}{M_i} = \text{const}$

[a] Notation: X_i = mole fraction,
V_i = diffusion velocity,
v = gas velocity,
ρ = density,
z = distance,
D = diffusion coefficient,
M = molecular weight,
n_i = number of atoms of an element in species i, and
k = reaction rate constant.

zone may be thicker by several orders of magnitude, as might be expected from the reduction of collision frequency and hence reaction rates. Alternatively, low burning velocity flames, which also tend to yield a relatively thick reaction zone, may be used where higher pressures are desired. In reduced pressure systems, the diffusion velocities are comparable with the gas velocity, and the determination of mass flow—i.e., flux—profiles must include the effect of mass diffusion.

The importance of diffusion in atmospheric flames is mainly limited to the very low mass species, such as the H-atom. As shown by the ab initio flame theory data in Fig. 5.6, the H-atom fluxes through a H_2–O_2–N_2 flame are determined primarily by ordinary concentration diffusion. For the other heavier flame species, such as N_2, O_2, and H_2O, the convective flux tends to determine the transport contribution to the total flux profiles (Dixon-Lewis, 1970a)

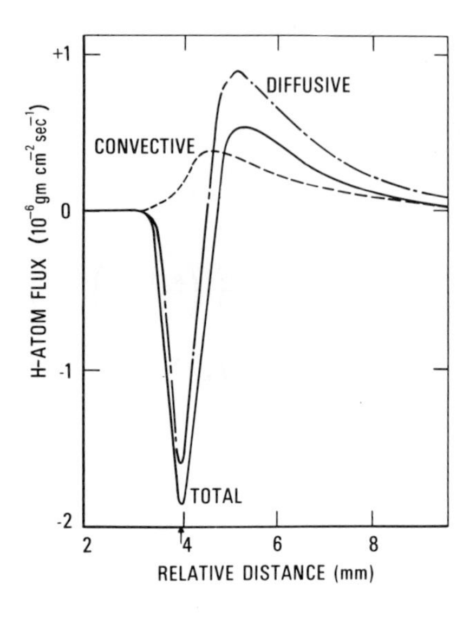

Fig. 5.6 Calculated fluxes of H-atoms, as a function of distance from burner, in the flame $H_2(0.188) + O_2(0.046) + N_2(0.766)$; the arrow ($\sim$4 mm) indicates the approximate position of the reaction zone (Dixon-Lewis, 1970a).

4. *Examples of Reaction Mechanisms*

The reaction mechanisms which provide the best agreement between flame theory, at various levels of sophistication, and experimental data are indicated in Table 5.7. The mechanism for relatively cool H_2–O_2–N_2 fuel-rich flames at 1 atm is supported by the good agreement found between calculated and measured concentration and temperature profiles (Dixon-Lewis, 1970a). For hydrocarbon flames, only the predominant reactions are indicated in Table 5.7 since complete reaction schemes have not yet been determined. Of course, many of the reactions included in the H_2–O_2–N_2 mechanism will also be important in hydrocarbon flames.

5. *Pressure Scaling*

As various flame studies utilize flames at different pressures, it is necessary to recognize the effect of pressure on flame structure. From the form of the flame equations and gas kinetic theory, the effect of pressure on flame propagation can be determined, e.g., see Hirschfelder *et al.* (1954) and Fristrom and Westenberg (1965). Specifically, the various factors determining flame propagation have the following relationships to pressure, P:

- for binary diffusion $D_{ij} \propto P^{-1}$,
- thermal conductivity is approximately pressure independent,
- chemical reaction rate $\propto P^{\beta}$, where β is the reaction order and is usually 2 for 1-atm flames; for a mass rate-of-flow of m_1 at 1 atm the mass rate m at pressure P is given by

$$m = P^{\beta/2} m_1,$$

- the flame velocity is given by

$$v_0 \propto m/P,$$

i.e., for second-order reactions v_0 is pressure independent and for third-order reactions

$$v_0 \propto P^{1/2},$$

- the reaction zone thickness varies approximately as the inverse of the pressure.

The validity of such pressure scaling relationships has been verified for 0.05- and 0.1-atm CH_4–O_2 flames, e.g., see Fristrom and Westenberg (1965).

TABLE 5.7

Flame Reaction Mechanisms

Cool-rich H_2–O_2–N_2:	$OH + H_2$	$= H_2O + H$
	$H + O_2$	$= OH + O$
	$O + H_2$	$= OH + H$
	$H + O_2 + M$	$= HO_2 + M$
	$H + HO_2$	$= OH + OH$
	$H + HO_2$	$= H_2 + O_2$
	$H + H + M$	$= H_2 + M$
Hot H_2–O_2:	$H + O_2$	$= OH + O$
	$H + H_2O$	$= H_2 + OH$
	$O + H_2$	$= OH + H$
	$H + O + M$	$= OH + M$
	$H + OH + M$	$= H_2O + M$
	$O + H_2O$	$= 2OH$
	$H + H + M$	$= H_2 + M$
	$O + O + M$	$= O_2 + M$
Hot-rich CH_4–O_2:	$CH_4 + H$	$= H_2 + CH_3$
Hot-lean CH_4–O_2:	$CH_4 + OH$	$= H_2O + CH_3$
Cool CH_4–O_2:	$CH_4 + O_2$	$= HO_2 + CH_3$

F. Characterization of Species in Flames by Direct Methods

1. Introduction

The principal methods of detection and concentration measurement for reactive species in flames utilize either optical, i.e., visible, uv and to a lesser extent far-uv and ir, or mass spectroscopy. The present brief discussion deals primarily with those aspects of spectroscopy which are peculiar to flame studies.

For the study of flame species, the selection of either an optical or a mass spectroscopic technique is influenced by the comparison of general capabilities as shown in Table 5.8. The two major limitations are, that the mass spectroscopic approach suffers from possible perturbations of the flame by the insertion of a probe and that the optical spectroscopic technique is limited mainly to the detection of atomic or simple molecular species. In practice, the combined results of both techniques are often of greatest value.

2. Optical Spectroscopic Methods

The spectroscopic techniques used with flames are similar to those described elsewhere for other high temperature vapor-phase systems, e.g., see Margrave (1967a). Detailed discussions of practical flame spectroscopy may be found in Mavrodineanu and Boiteux (1965) and also Gaydon and Wolfhard (1970). Both emission and absorption of radiation are utilized to identify and quantify flame species, and some discussion of these techniques has been given elsewhere in connection with analytical flame spectroscopy (Section IV).

TABLE 5.8

Flame Species Characterization Techniques—General Capabilities

Optical spectroscopy	Mass spectroscopy
Detection of most atoms and diatomic species	Detection of most species including polyatomics
Radical–detection limited to OH by direct means	All radicals detected, e.g., H, OH, O, CH_3, HO_2, CHO
Detects excited states	—
Fair spatial resolution	Good spatial resolution
No perturbation of system	Some perturbation by probe
Not affected by condensed particles	Cannot sample in presence of condensed particles

For purposes of the following discussion on flames containing metal additives, it is sufficient to note that the concentrations of atomic flame additive species are commonly determined from emission spectral intensities, and known f numbers, using an internal nonreactive standard such as Na.

An example of the combined use of optical absorption and emission techniques is provided by the study of species concentration profiles for C_2, CH, and OH in low pressure ($\sim$2 Torr) acetylene–oxygen flames, which made use of light modulation phase sensitive detection and multiple reflection techniques (Bulewicz *et al.*, 1970). In addition to basic flame structure studies, optical spectroscopic techniques have also proven to be useful for monitoring the OH radical species in systems of practical interest such as the internal combustion engine, e.g., see Smith and Starkman (1971).

3. *Oscillator Strengths for Molecular Flame Species*

Electronic excitation of molecular species is common in flames and other high temperature systems, such as the reentry wakes of space vehicles and stellar atmospheres. The oscillator strength, or f number, of such emissions is therefore of interest for the determination of species concentrations. These data are well established for atomic systems, but very few oscillator strength determinations have been made for high temperature molecular species.

A recent example of the determination of oscillator strengths for an excited species in flames is provided by the study of the A $^2\Sigma$–X $^2\Sigma$ band system for AlO (Vanpee *et al.*, 1970). The use of hot ($\sim$4200 K) cyanogen–oxygen flames (30 Torr), in this case, simplified the flame chemistry. That is, the loss of Al-atoms, and hence AlO molecules, owing to the formation of species such as $Al(OH)_2$, or $Al_2O_3(c)$, was prevented. The strength of the emission was determined by comparison with a standard incandescent lamp using a sector disk for matching. In order to determine the concentration of AlO, it was necessary to assume that Al and AlO were the only Al-containing species present, and concentrations of:

$$[\mathrm{Al}] = 0.58 \times 10^{13} \text{ atoms cm}^{-3},$$

and

$$[\mathrm{AlO}] = 2.26 \times 10^{13} \text{ molecules cm}^{-3},$$

were indicated. A value of 0.0037 was then determined for f electronic. When this value is compared with those of other experimental or theoretical determinations, which range from 0.0035 to 0.13, it is apparent that the use of oscillator strengths for determining molecular species concentrations is necessarily of low reliability, e.g., see Michels (1972).

4. *Modern Spectroscopic Techniques*

The recently developed technique of laser magnetic resonance spectroscopy has been applied to the identification of the CH radical in a low pressure oxyacetylene flame (Evenson *et al.*, 1971). The 118.6-μm-wavelength radiation output of a water vapor laser was absorbed by a pure rotational transition of CH. A magnetic field sweep allowed the measurement of Zeeman splittings, which provided an unambiguous identification for the radical. A number-density concentration of about 10^{12} cm^{-3} was found and the signal-to-noise level was 260 to 1. In this experiment the flame is located within the laser cavity and some difficulty can therefore be expected in obtaining the spatial resolution necessary for species profile determinations.

The use of a dye laser intercavity absorption technique for the analysis of flame species has also been recently demonstrated, as discussed elsewhere (see Section IV.F).

Conventional laser-Raman spectroscopy has found very limited application in the monitoring of flame species. However, the use of stimulated-Raman techniques should allow for greater sensitivity and versatility and such techniques are beginning to find application in flame studies (Regnier and Taran, 1973).

5. *Mass Spectroscopic Methods*

The mass spectroscopic technique has been applied for several decades, with varying degrees of success, to the analysis of flames. Examples of some of the more important developments in applying the technique may be found in the chronology given in Table 5.9. Matthews (1968) and Homann (1967b) have reviewed the applications of mass spectrometry to flame analysis, and Knuth (1973) has considered similar applications in the related area of internal combustion engines.

Generally, the link between flame and mass spectrometer is achieved by use of a small-orifice probe. A small amount, i.e., $\sim 2\%$, of the flame gas is withdrawn through the probe. The gas is then quenched to room temperature, or lower, either by collisions within the cooled probe or by a rapid adiabatic expansion of the gas into a vacuum space. In the former instance, which is known as microprobe sampling, the reactive flame intermediates are destroyed and the analysis deals only with the stable molecular constituents of the flame. With the latter condition, the rapidity of the quenching process ($\sim 10^{-7}$ sec and ~ 10 collisions) prevents any major loss of reactive species, and the expanding gas jet may be skimmed to yield a molecular beam which can then be analyzed by a mass spectrometer. This effectively real-time technique, known as molecular beam mass

TABLE 5.9

Chronology of Flame Mass Spectrometry for Uncharged Species

Author	Year	Comments
Eltenton	1947	a,c,e,f
Foner and Hudson	1953	a,c,e,f
Westenberg and Fristrom	1960	a,d
Fenimore and Jones	1964	a,d
Dixon Lewis and Williams	1963	b,d
Homann *et al.*	1963	a,c,e,f
Milne and Greene	1966	b,c,e,f
Hastie	1973	b,c,e f
Williams and Wilkins	1973	a,c,e,f
Peeters and Mahnen	1973	a,c,e,f
Lazzara *et al.* / Biordi *et al.*	1973	a,c,e,f
Knuth	1973	b c,e

[a] Low pressure flames.
[b] Atmospheric pressure—or greater—flames.
[c] Molecular beam sampling.
[d] Microprobe sampling.
[e] Signal modulation techniques used.
[f] Direct evidence given for radicals or other flame intermediates.

spectrometric sampling, is uniquely valuable for measuring reactive species in flames and, as indicated in Table 5.9, has found increased application in recent years.

The two major probe methods then, are the microprobe and molecular beam techniques. A third, and less frequently used variation of these techniques, is the scavenger probe method which, by indirect means, analyzes radical species, e.g., see Fristrom (1963a). The requirements basic to this method are that the probe sampling, mixing, and scavenger reaction rates must be rapid compared with any other reactions that the radical may undergo. By this technique, concentration profiles for flame radicals have been determined using the following scavenger reactions:

$$H + C_nCl_{2n+2} \rightarrow HCl + C_nCl_{2n+1},$$

$$O + NO_2 \rightarrow O_2 + NO,$$

and

$$CH_3 + I_2 \rightarrow CH_3I + I.$$

The products are monitored continuously by mass spectrometric detection. A concentration calibration may be achieved by comparison with the results of a more conventional visual titration of atoms generated by a discharge. The technique appears to be satisfactory for O-atom analysis but fewer H-atoms than expected were detected.

a. Probe Effects

A critical feature of the mass spectrometric approach is the possible effect of the sampling probe on the flame chemistry, and there has been much discussion on this point. The conditions under which a perturbation of the sampled gas composition can be expected have been mathematically modeled recently by Yanagi and Mimura (1972). Earlier theoretical treatments of probe perturbations have been reviewed by Fristrom and Westenberg (1965). Some experimental aspects of the early probe quenching techniques have been discussed by Halpern and Ruegg (1958). More recently, Cuthbert (1966) has reviewed the various factors that influence the sampling of high pressure gases into a mass spectrometer. The theoretical criteria for extracting a representative molecular beam from a high pressure high temperature combustion system have also been considered in detail recently by Knuth (1973).

The probe-related perturbations that we must generally be concerned with are

- the effect of probe insertion on flame structure,
- a change in the apparent flame chemistry at the probe entrance—e.g., by boundary layer formation, and
- a change in the gas composition between the mass spectrometer side of the probe orifice and the mass analyzer itself.

Perturbations of the latter type can be predicted to some extent from well-known gas-dynamic relations. However, in order to deal with the sum total of perturbation effects, it is usually necessary to systematically vary the experimental sampling conditions, such as orifice diameter, to establish the presence or absence of significant perturbation effects. In general, one can expect probe perturbation effects to be more critical in the region of the reaction zone of flames, where steep species-concentration and temperature gradients are present. A few examples of cases where perturbation effects have been noted are as follows.

Spurious ions, such as $H_5O_2^+$, have been observed to form in the flame-probe boundary layer, e.g., see Hayhurst and Telford (1971) and the more detailed discussion in Section III.C. In low pressure flame studies, a marked cooling effect, i.e., $\sim$200 °C temperature reduction, of the probe on the flame gas has been observed by Homann (1967b). Also, the position of the flame reaction zone, relative to the burner top, has been noted to change in the presence of a relatively cool probe. These heat-transfer related effects appear to be much less significant for higher pressure flames, i.e., $\sim$1 atm, and for low thermal conductivity hot ceramic probes (Hastie, 1975).

Fortunately, it can be shown from gas-dynamic theory (Knuth, 1973) and experimental observation (Hastie, 1974) that molecular aggregation effects, of the type frequently found in room-temperature atmospheric pressure gas sampling experiments constitute a negligible perturbation in flame sampling.

Evidence for a minimal perturbation of atmospheric flame chemistry by the molecular beam sampling technique is provided by an indirect observation of postflame H-atom concentration profiles in comparison with those from the nonperturbing CuH emission technique (Hayhurst and Kittelson, 1972c). Also, the concentration ratio $[H]/[H_2O]$ was found to be undisturbed by probe sampling. As the chemical reactivity, boundary layer, and free jet expansion properties of H and H_2O are quite different, it is evident that very little chemical perturbation resulted during the sampling process. The concentration ratio $[H]/[H_2O]$ was determined indirectly by use of the balanced reaction

$$Ca^+ + H_2O = CaOH^+ + H,$$

and the ratio $[CaOH^+]/[Ca^+]$, as measured mass spectrometrically, from which

$$[H_2O]/[H] = [CaOH^+]/[Ca^+]K,$$

where K is the known reaction equilibrium constant. Incidentally, it follows that $[H]$ may also be determined by this technique.

Additional evidence for a minimal probe effect, in molecular beam sampling of 1-atm flames for mass spectrometric analysis, has been given in connection with Br-atom determinations in halogen-doped flames (Hastie, 1973a), HO_2 determinations in clean H_2–O_2–N_2 flames (Hastie, 1974) and H-atom determinations in H_2–O_2–N_2 flames (Hastie, 1975). Likewise, the quartz microprobe sampling of atmospheric propane–air flames containing sulfur additives shows reasonable agreement between calculated and observed species concentrations in the equilibrium flame region (Johnson *et al.*, 1970).

G. Characterization of Species in Flames by Indirect Methods: The Family of Flames Approach

The indirect identification of compounds and reactions in the burnt-gas region of H_2–O_2–N_2 flames is greatly facilitated by the use of groups of flames. For instance, use of a group of isothermal flames, where the temperatures are essentially the same but the concentrations of the major flame components differ, allows one to test mass action relationships and hence, establish the molecularity of reactions and the probable identity of product species. Flames of constant fuel-to-oxidant ratio but different temperature, such as may be produced by the cooling effect of added N_2, may also be used to characterize reactions since the Van't Hoff equation should hold under equilibrium conditions for the correct choice of species, e.g., see Sugden (1956) and Reid and Sugden (1962). To obtain data on these various flame groups a number of flame families are utilized. In each family the ratio $[H_2]/[O_2]$ is varied and $[N_2]/[O_2]$ is fixed. A new family is given by a different $[N_2]/[O_2]$ ratio.

The type of information that results from the use of flame families is indicated by the following few examples. More examples will be revealed in later discussions.

- In the investigation of the mechanism of electronic excitation, it is possible to distinguish between collisional or thermal activation and various modes of chemiluminescent excitation, e.g., see Reid and Sugden (1962).
- It is possible to determine the thermal nature of species ionization, and hence derive the heats of formation for flame ions, e.g., see Jensen and Padley (1966a).
- It is possible to characterize reactions involving H-atoms and alkali metals from which chemical methods for determining the concentration of H-atoms have been developed, e.g.:

$$Li + H_2O \rightleftarrows LiOH + H \qquad \text{and} \qquad Na + HCl \rightleftarrows NaCl + H$$

(Bulewicz *et al.*, 1956). Similar studies have provided information on the mechanisms of formation, and the thermodynamic properties, of metal oxide and hydroxide species in flames, e.g., see Padley and Sugden (1959b) and also Section II.K.

The "family-of-flames" approach is usually used in conjunction with an atomic emission spectroscopic monitoring of free-atom concentration changes for the various flame conditions, e.g., see Jenkins and Sugden (1969) and Schofield and Sugden (1971). The emission intensity I is given by

$$I = Ah\nu[M](g'/g_0)\exp(-h\nu/kT),$$

where

A is the transition probability,
g' and g_0 are statistical weights of upper and ground states, respectively, {alternatively the oscillator strength or f value may replace $A(g'/g_0)$},

h is the Planck constant,
k is the Boltzmann constant,
ν is frequency of the emission line, and
[M] is the species concentration.

This expression is also sometimes given in terms of spectral wavelength rather than frequency (e.g., see Section K.2.).

The loss of free atoms, owing to compound formation, is more readily determined from spectral emission intensities when compared with the emission of another atom which does not undergo chemical reaction, such as Na. In this case, for two elements of concentration $[M_1]$ and $[M_2]$, we have

$$\frac{[M_1]}{[M_2]} = \frac{I_1}{I_2}\frac{A_2}{A_1}\frac{\nu_2}{\nu_1}\left(\frac{g_0}{g'}\right)_1\left(\frac{g'}{g_0}\right)_2 \exp\left(\frac{-h(\nu_2 - \nu_1)}{kT}\right).$$

Note that with some experimental systems, the relative intensity data I_1/I_2 may require normalization to include the effect of different spectral sensitivities for the detector at the various wavelengths. Another precaution to be noted is the need to use low concentrations of the additives M_1 and M_2, i.e., $\sim 10^{-6}$ atm, such that self-absorption effects are absent. Suppose only M_1 forms a compound M_1X and the ratio $[M_1X]/[M_1]$ is given by ϕ, then

$$\frac{[M_1X] + [M_1]}{[M_2]} = \frac{[M_1](1 + \phi)}{[M_2]} = \frac{[M_1]_0}{[M_2]_0}.$$

The zero subscripts are for the total amount of each element initially added to the flame (note $[M_2] = [M_2]_0$). As these initial additions are known quantities, the value of ϕ, i.e., the relative concentration of the compound M_1X to M_1, can be determined from spectral measurements of I_1 and I_2.

From the general relation between I and [M], it follows that for constant [M], a plot of $\log I$ vs T^{-1} should be linear. Such plots can be used to infer whether compounds of M are forming. The slope of the line also indicates the magnitude of the thermal excitation energy ($h\nu$). An indication of compound formation, between the added element and the natural flame constituents, can usually be determined from similar plots, using families of flames. The examples given in Fig. 5.7 indicate that Na is present as the free atom whereas Li is apparently involved in compound formation with the flame constituents. This compound can be shown to be LiOH. The family-of-flames effect has similarly been used to demonstrate the

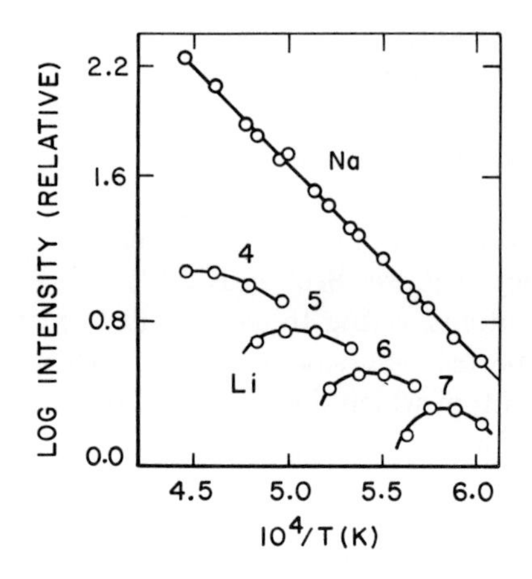

Fig. 5.7 Variation of emission intensity for the resonance lines of Na and Li as a function of flame temperature for various H_2–O_2–N_2 flames at 1 atm and at 1.25 msec from the reaction zone; groups of flames with N_2/O_2 ratios of 4, 5, 6, and 7 are shown; each group is made up of flames with H_2/O_2 ratios of 2.5, 3, 3.5, and 4, reading towards lower temperatures (Schofield and Broida, 1968).

formation of GaOH and InOH, and the absence of TlOH, in H_2–O_2–N_2 (~2000 K) flames (Bulewicz and Sugden, 1958a).

The method is satisfactory only when at least 20% of the added metal undergoes compound formation, and the assignment of species identity becomes less reliable if more than one compound is present. An example of a complex system is given by the case of Ba in H_2–O_2–N_2 flames where BaO, BaOH, and $Ba(OH)_2$ are believed to form, e.g., see Cotton and Jenkins (1968) and also Section K.3.

The reactions by which local equilibrium is achieved for a compound can also be inferred from plots of log K vs T^{-1}, where K is the reaction equilibrium constant. If an incorrect, or nonbalanced, reaction has been considered, a nonlinear plot will result. This is particularly evident if the balancing reaction is termolecular and a bimolecular process was considered in the plot. The slope of such curves should also be consistent with the reaction enthalpy derived from a third-law treatment of the equilibrium constant data.

Another source of information regarding possible compound formation is the "curve of growth" for the element in question; that is, the variation of thermal emission intensity with the concentration of element added to the flame. Examples of such curves, and their analysis in terms of new species production, have been given by Bulewicz and Padley (1973).

H. Hydrogen-Atom Detection in Flames

1. Techniques

The detection of H-atoms in flames exemplifies the application of both direct and indirect measurement techniques, and a summary of the more commonly used methods and the key reactions is given in Table 5.10. Somewhat similar techniques to those discussed here have also been used for the radical species OH and O, e.g., see the reviews of Schofield and Broida (1968) and Pungor (1967). However, the measurement techniques and accumulated data appear to be better established for the H-atom. In fact, some determinations of OH first involve a measurement of the H-atom concentration; then an assumption of local equilibrium for a reaction of known equilibrium constant, such as

$$H + H_2O = H_2 + OH,$$

where the H_2O and H_2 concentrations are known quantities, allows OH to be determined. Therefore, the present discussion of specific radical detection methods and results emphasizes the H-atom.

The electron spin resonance method of H-atom detection in flames has been utilized only on a few occasions and primarily at low pressures. Measurements in 1-atm flames have been limited by poor spatial resolution (Bennett *et al.*, 1970). The use of a quartz microprobe to extract H-atoms and also O-atoms from low pressure (0.1-atm) flames has provided better

TABLE 5.10

H-Atom Detection Methods in Flames

Direct or physical methods	
Electron spin resonance	
Mass spectrometry—molecular beam sampling	
Chemical methods with spectroscopic detection	
Li method:	$Li + H_2O \rightleftarrows LiOH + H$
Na chlorine method:	$H_2 + Cl \rightleftarrows HCl + H$
	$Na + HCl \rightleftarrows NaCl + H$
CuH Emission:	$Cu + H + X \rightleftarrows CuH^* + X$
Chemical methods with mass spectroscopic detection	
Reactive additive approach:	$H + D_2O \rightleftarrows HD + DO$
	$H + D_2 \rightleftarrows HD + D$
	$H + N_2O \rightleftarrows N_2 + OH$
Scavenger probe technique:	$H + CCl_4 \rightleftarrows HCl + CCl_3$

spatial resolution but some loss of radicals occurred during sampling (Westenberg and Fristrom, 1965).

The application of molecular beam sampling of flames, with mass spectrometric detection, for the determination of H-atom concentrations was relatively unsuccessful until recently (see Table 5.9). This technique readily provides relative H-atom concentration data, but requires use of an alternative calibration method for absolute measurements. For hot 1-atm flames, i.e., $T \gtrsim 2000$ K, where thermodynamic equilibrium is readily established in the burnt-gas region, it is convenient to use calculated (see Chapter 1, Section V.A) equilibrium H-atom concentrations, made at the *experimental* temperature condition, for an absolute calibration of the mass spectrometric data.

Another mass spectrometric molecular beam sampling approach, recently demonstrated by Hayhurst and Kittelson (1972c) utilizes the following balanced reaction for Ca-containing flames:

$$Ca^+ + H_2O = CaOH^+ + H,$$

from which the H-atom concentration is

$$[H] = K[H_2O][Ca^+]/[CaOH^+].$$

The ion concentration ratio $[Ca^+]/[CaOH^+]$ is a measurable mass spectrometric quantity, and as the equilibrium constant K and the steam concentration $[H_2O]$ are known, then $[H]$ may be determined.

The CuH emission technique allows a determination of $[H]$ closer to the reaction zone than most methods, since chemiluminescence does not appear to be significant, at least for 1-atm flames. This method is a relative one and must be calibrated against other absolute methods (Bulewicz and Sugden, 1956).

The Li, or Li–Na, method utilizes a comparison of the resonance emission intensities of Li and Na (Bulewicz *et al.*, 1956). This represents an example of use of the indirect technique for species identification, i.e., LiOH, as described in the previous Section II.G. The comparison of intensities yields a value for the ratio

$$\phi = [LiOH]/[Li].$$

Then, according to the reaction indicated in Table 5.10, the H-atom concentration will be given by

$$[H] = K[H_2O]/\phi.$$

Values of the equilibrium constant K may be obtained from sources such as the JANAF Thermochemical Tables (JANAF, 1971). The concentration term [H_2O] can be calculated approximately, i.e., usually better to than 10% uncertainty, by assuming a negligible dissociation of H_2O at flame temperatures. Alternatively, one can consider explicitly the effect of dissociation by carrying out a multicomponent-equilibrium free-energy minimization calculation of species concentration.

This frequently used Li method is believed to be accurate to ±20% on an absolute basis and ±5% on a relative scale (Phillips and Sugden, 1960). Much of this error could be eliminated if the dissociation energy of Li–OH were known to better than ±1 kcal mol^{-1} uncertainty, and also if the spectroscopic parameters of LiOH were known rather than estimated. These are the basic data from which values of K have been calculated (i.e., via $-RT \ln K = \Delta H - T \Delta S$). At low temperatures, i.e., <1600 K, the chemiluminescence of these alkali metal atoms is comparable with thermal excitation, and the method is then invalid. McEwan and Phillips (1965) suggest that a direct measurement of the Li concentration by absorption may be used to circumvent this problem. However, other workers have experienced some difficulty with this suggested approach (Halstead and Jenkins, 1967).

The alternative, and somewhat similar, NaCl technique (see Table 5.10) can suffer in accuracy when the H-atom concentration is low, due to chemical perturbation of the flame by the halogen. Relatively high halogen concentrations (e.g., 1%) are required with this method (see Bulewicz *et al.*, 1956).

For lean flame conditions the NaCl and CuH methods are not suitable but the LiOH and an analogous NaOH method appear to be reliable (McEwan and Phillips, 1967). However, for very lean conditions the formation of new species, e.g., LiO_2, could interfere with the determination of ϕ.

The chemical mass spectrometric methods involve the addition of a small quantity of a reactant, e.g. D_2 or D_2O, to the flame, which reacts with H according to the reactions indicated in Table 5.10. Microprobe sampling is then used to determine the stable-component concentrations, i.e., HD, and from the known reaction rate constants the H-atom concentration may be derived (Fenimore and Jones, 1958).

The scavenger probe technique, listed in Table 5.10, has been discussed in Section II.F.5.

2. H-Atom Concentration Data

Using these various techniques, H-atom concentrations have now been obtained for a considerable variety of flames. As this information is basic

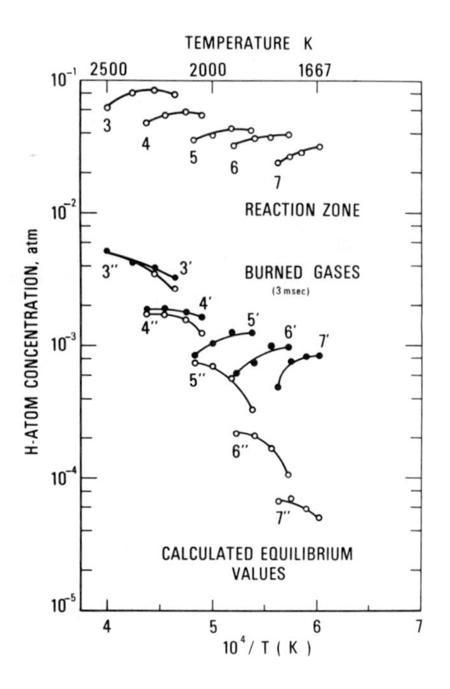

Fig. 5.8 Hydrogen-atom concentrations in H_2–O_2–N_2 flames; each group or family of flames is characterized by a fixed N_2/O_2 unburned gas ratio of magnitude 3–7 as indicated in the figure; within each family the H_2/O_2 ratio has values of 2.5, 3.0, 3.5, and 4.0, reading in the direction of decreasing temperature (Padley, 1959; Jenkins and Sugden, 1969).

to the discussions to follow, on flame chemistry and its applications, then it is useful to consider available data on H-atom concentrations.

The fuel-rich H_2–O_2–N_2 flames are relatively well characterized in terms of their temperature and radical properties, and the variation of H-atom concentration with flame composition and location, i.e., time downstream of the reaction zone, is given in Fig. 5.8. Concentrations at the reaction zone are at least two orders of magnitude greater than the equilibrium values, which are attained at burnt-gas regions corresponding to a time delay of between about 3 msec and 8 msec. Note that the 3 msec burnt-gas region tends to reach equilibrium only for the hotter flames, i.e., in the temperature region of 2000 K and greater. The rate of approach of the H-atom concentrations to the equilibrium level is determined primarily by the rates for the elementary recombination processes:

$$H + OH + M \rightarrow H_2O + M^* \quad \text{and} \quad H + H + M \rightarrow H_2 + M^*,$$

where the third body, M, is a major gas component such as H_2O. As these rates are known then it is possible to define the H-atom concentration as a function of time, or distance, into the burnt-gas region for flames—such as those represented in Fig. 5.8. Further discussion of these kinetic aspects may be found in Section II.L.

For stoichiometric and lean H_2–O_2–N_2 flames the H-atom concentrations are appreciably less than for the above fuel-rich case, and OH tends to be the predominant radical species, as shown in Table 5.11. The concentration of OH in these flames is in excess of the equilibrium values, typically by two to three orders of magnitude—see also the similar data of Kaskan (1958).

Hydrocarbon flames give H-atom concentrations that are one or two orders of magnitude less than for corresponding H_2 flames, as shown in Table 5.12. Also, the addition of only about 1% CH_4 to a fuel-rich H_2 flame results in a 50% reduction in H-atom concentration. It is particularly noteworthy (see Table 5.12) that the H-atom concentration may even be less than the equilibrium value—even for the soot-free conditions used by

TABLE 5.11

Concentrations of H and OH in Stoichiometric and Lean Hydrogen Flames at 1 atm[a]

H_2 : O_2 : N_2			Temperature (K)	H (atm)	OH (atm)
2.0	1	3	2250	5.5(3)[b]	1.2(2)
2.0	1	4	2070	2.9(3)	8.6(3)
2.0	1	5	1895	2.7(3)	7.1(3)
2.0	1	6	1740	1.9(3)	5.1(3)
2.0	1	7	1690	2.8(3)	4.7(3)
1.5	1	3	2040	6.7(4)	9.7(3)
1.5	1	4	1810	5.8(4)	5.0(3)
1.5	1	5	1530	1.5(4)	1.6(3)
1.5	1	6	1495	1.3(4)	1.3(3)
1.5	1	7	1470	9.1(5)	9.8(4)
1.0	1	3	1550	8.5(5)	2.0(3)
1.0	1	4	1380	2.4(5)	1.5(3)
1.0	1	5	1300	2.6(5)	5.2(4)
1.0	1	6	1280	3.2(5)	4.6(4)
0.5	1	3	1100	3.1(6)	1.3(4)

[a] Data from McEwan and Phillips (1967); measurements made at 0.37 msec downstream from reaction zone.

[b] 5.5(3) stands for 5.5×10^{-3} atm.

TABLE 5.12

Hydrogen-Atom Concentrations in the Burnt-Gas Region of Rich Hydrocarbon Flames at 1 atm[a]

Temperature (K)	Fuel	Equivalence ratio	[H] (atm $\times 10^4$)	[H]/[H]e
1700	C_2H_2	1.59	2.2	1.0
1775	C_2H_2	1.59	4.7	1.0
1850	C_2H_2	1.59	7.0	0.7
1600	C_3H_8	1.24	1.3	1.4
1700	C_3H_8	1.24	2.2	1.0
1700	C_3H_8	1.57	2.3	0.8
1650	CH_4	1.18	0.6	0.6
1700	CH_4	1.18	1.3	0.7
	(mainly hydrogen)			
1455	7.9% $C_2H_2 + H_2$	1.26	34	150
1455	19.0% $C_2H_2 + H_2$	1.21	11	65
1345	1.2% $C_2H_2 + H_2$	1.31	49	970
1345	1.6% $C_2H_2 + H_2$	1.40	37	590
1345	1.1% $CH_4 + H_2$	1.36	45	720
1345	Pure H_2	1.27	93	1600

[a] Data from Fenimore and Jones (1958) and for flames of fuel + O_2 + N_2; the last data column indicates the ratios of [H] observed to [H]e for the calculated thermodynamic equilibrium condition.

Fenimore and Jones (1958). This point receives further discussion in Section VI.

I. Local Equilibrium and Balanced Reactions

The characterization of reactions in flames is greatly simplified if a local equilibrium can be established. A local equilibrium condition requires that the reaction of interest be sufficiently rapid, both in the forward and reverse directions, for the reaction to be balanced. The significance of a balanced reaction condition to the flame chemist is that the *relative* concentrations of the various species on either side of the reaction can be defined from an equilibrium constant for the reaction. However, the *absolute* concentrations may well be very different from the general equilibrium levels.

As the collision frequency, in 1-atm flames, is about 3×10^9 sec^{-1}, this is sufficiently high to allow the relaxation and equilibration of many bimolecular processes. Arguments concerning the predominance of particular reaction types and the conditions leading to balanced reactions have been given by Sugden (1956). These arguments have more recently been extended by Jensen (1972), who finds that the application of such reaction-selection criteria for the case of rocket exhaust calculations is more stringent than for the 1-atm laboratory flames from which the criteria were developed. Uncertainties in the reaction rate data make it difficult to determine whether or not one can assume balancing of reactions.

In 1-atm flames there exists a competition between binary and ternary reactions, e.g.,

$$\mathrm{M + HZ \rightleftarrows MZ + H} \quad \text{and} \quad \mathrm{M + Z + X \rightleftarrows MZ + X.}$$

From kinetic reasoning, Sugden (1956) suggests that for a dissociation energy limit of $D_{\mathrm{MZ}} < 72$ kcal mol^{-1}, e.g., if MZ is CuH, the ternary reaction predominates. Both of these reaction types can usually achieve a balanced condition in less than 10^{-3} sec, which is within the time scale available in the burnt-gas region of 1-atm flames.

Reactions that are close to being thermoneutral, particularly where the activation energy is less than kT (k is the Boltzmann constant) can achieve a balanced condition on a much shorter time scale as shown by the following example. Consider the reaction

$$\mathrm{H_2O + H \rightleftarrows OH + H_2,}$$

which occurs in 1-atm H_2–O_2–N_2 flames. The components H_2 and H_2O are present in excess concentration and can therefore be taken as constant. Let $[\mathrm{OH}]_t$ and $[\mathrm{OH}]_\tau$ be the concentrations of OH at time t and at the relaxation time τ, respectively. The rate of appearance of OH is then

$$\frac{d[\mathrm{OH}]}{dt} = k_{\mathrm{f}}[\mathrm{H}]$$

and the integrated form is

$$([\mathrm{OH}]_\tau - [\mathrm{OH}]_t)/[\mathrm{OH}]_\tau = \exp(-t/\tau)$$

and

$$\tau = (k_{\mathrm{f}}[\mathrm{H_2}] + k_{\mathrm{r}}[\mathrm{H_2O}])^{-1},$$

where k_{f} and k_{r} are the rate constants for the forward and reverse reactions, respectively. From the known rates of the forward and reverse reactions it can be shown that $\tau \sim 10^{-6}$ sec (Jenkins and Sugden, 1969).

Several typical examples of known balanced reactions in the reaction zone or burnt-gas flame regions, are as follows:

$$\mathrm{H_2 + OH \rightleftarrows H_2O + H,}$$

$$\mathrm{H_2 + O \rightleftarrows OH + H,}$$

and

$$OH + O \rightleftarrows O_2 + H.$$

This set of reactions allows the establishment of a local equilibrium on a 1-μsec time scale between the important flame propagating radicals H, OH, and O. Therefore, steady-state relative concentrations can be derived from the general thermodynamic equilibrium values which are readily obtained by calculation, i.e.:

$$[OH]/[H] = [OH]_e/[H]_e \quad \text{and} \quad [O]/[H] = [O]_e/[H]_e,$$

where the subscript refers to the equilibrium concentration values. It follows that it is only necessary to measure the absolute concentrations of one of these radicals, usually H, in order to determine the concentrations of the others.

It is encouraging that the species H and OH are known to be linked in both premixed laminar flow, and turbulent H_2-fueled, flames; i.e., the reaction

$$H_2 + OH \rightleftarrows H_2O + H$$

is balanced in both flame types (Halls and Pungor, 1969). This means that, to a certain extent, one can use premixed laminar-flow flames to model the flame chemistry of more practical turbulent flame systems.

The concept of balanced reactions is also particularly useful for predicting the steady-state concentration levels of flame additives which may undergo rapid forward and reverse reactions with nonequilibrium concentration levels of flame radicals. For example, the composition of S-containing flames is most probably controlled by the low heat-change reactions

$$SH_2 + H \rightleftarrows SH + H_2,$$

$$SH + H \rightleftarrows S + H_2,$$

$$S + H_2O \rightleftarrows SO + H_2,$$

$$SO + H_2O \rightleftarrows SO_2 + H_2,$$

$$SO + SH \rightleftarrows S_2 + OH,$$

(Demerdache and Sugden, 1963). The concentration of the various sulfur species can then be determined from thermodynamic equilibrium calculations, but including the observed nonequilibrium concentrations of H-atoms and OH-radicals in the computation.

A further example of local equilibrium is given by the constant ratio of [MO]/[M] species where M is a metal in the burnt-gas region of

H_2–O_2–N_2 flames. Since the following reactions are likely to be balanced

$$M + H_2O \rightleftarrows MO + H_2,$$

$$M + OH \rightleftarrows MO + H,$$

$$M + H_2O \rightleftarrows MOH + H,$$

and as $[H_2O]/[H_2]$ is effectively constant, then the ratio $[MO]/[M]$ should also appear constant, e.g., see Reid and Sugden (1962).

J. Inorganic Molecular Species Identified in Flames

The explosion of evidence regarding the presence of new molecular species in high temperature vapors over the past several decades, e.g., see Margrave (1967a) and Hastie *et al.* (1970), has been paralleled by the "identification" of numerous similar species in flames. A listing of these flame species is given in Table 5.13. With the exception of the several natural flame molecular species given, i.e., HO and HO_2, the listed species are for atmospheric pressure flames containing, for the most part, trace quantities, i.e., $\sim 10^{-6}$ atm, of the elemental additives. Although not indicated in the table, a high fraction of atomic species may also be present, together with the molecular components. In many instances, particularly in the burnt-gas postflame region, the relative concentrations of the various molecular and atomic species can be approximately predicted from the appropriate reaction equilibrium constants. A detailed discussion of free-atom fractions may be found in the section on analytical flame spectroscopy (Section IV), where the reduction of molecular to atomic species is of prime importance in this application.

The concentrations of the various molecular species formed by additives in flames can vary considerably between the various elements. For example, Li can be present as LiOH to the extent of about 90% of the total Li concentration, whereas the corresponding hydroxides of Na, K, Rb, and Cs tend to be negligible, relative to the atoms. This particular result is reflected in the data of Fig. 5.7, where the free-atom concentration of Li relative to Na varies from about 0.25 to 0.1—depending on the flame temperature and composition. A high proportion of an alkaline earth metal in flames may also be present in the form of hydroxide and oxide species, e.g., see Section II.K. Also, the alkaline earth chlorides and fluorides are well-known stable flame species. The alkaline earth phosphates are apparently difficult to decompose in flames though no determination of the molecular species has yet been made. However, other salts containing polynuclear anions, such as nitrates and sulfates, are readily decomposed in flames.

The hydrides CuH, PtH, AuH, and TlH form less than 1% of the total metal present, but are well known on account of their readily identifiable atomic-like spectra.

With regard to the species listed in Table 5.13, one should not infer the presence of ground-state species from emission spectra, unless the emission is known to be thermal. As a case in point, the emission intensities for many excited MOH species, where M is a metal, result from nonthermal reactions of the type

$$M + OH + N_2 \rightarrow MOH^* + N_2,$$

and

$$MO + H + H_2O \rightarrow MOH^* + H_2O,$$

e.g., see Reid and Sugden (1962). We should also note that the various species indicated in Table 5.13 were not always determined in the same general flame location, i.e., preflame, reaction zone, or postflame, nor under similar chemical conditions, i.e., fuel rich or fuel lean. Hence, some caution should be exercised in extrapolating these observations to other conditions. In addition to this, an element of uncertainty exists for those species where only indirect or inferred evidence is given for their presence.

It is reasonable to assume that this list of flame species will grow considerably with the future development of improved characterization techniques, and particularly in the areas of mass spectroscopy and laser spectroscopy.

K. Dissociation Energies from Flame Studies

1. *Introduction*

Much of the original bond dissociation energy data for diatomic high temperature species was determined from flame spectrophotometric studies. A summary of this early work has been given by Gaydon (1968). The present discussion will therefore be brief and will emphasize two factors affecting the determination of dissociation energies for flame species. These are, first, the importance of polyatomic species—a factor only recognized in recent times and, second, the high degree of interaction of many metals with the natural flame components—giving rise primarily to the formation of metal oxide and hydroxide species.

Both the second- and third-law thermodynamic methods may be applied to the determination of reaction heats and dissociation energies for flame species; preference is usually given to the latter method. The measurement problem reduces to one of identifying the species and determining their concentrations under various conditions of flame composition and

TABLE 5.13

Inorganic Molecular Species Identified in Flames

Species[c]	Detection method	Reference
AgH	[a]	[a]
AlO	[a]; Emission spectra in H_2–O_2–N_2 flames	Newman and Page (1970, 1971)
AlO	Emission in C_2N_2–O_2 flames	Vanpee *et al.* (1970)
$Al(OH)_2$	[d]	Jensen and Jones (1972)
AlO	Mass spectrometric analysis, lean H_2–O_2 flame	Farber *et al.* (1973)
AlO_2		
AlOH		
AlO(OH)		
AuH	[a]	[a]
BO_2	[a]; Emission in C_2N_2–O_2 flames containing BCl_3	Tischer and Scheller (1968)
HBO_2	Emission and absorption spectra in postflame region	Kaskan and Millikan (1962)
BO_2		
BaBr	[d]	[f]
BaCl	[d]	Gurvich and Ryabova (1964a); [f]
$BaBr_2$	[d]	[f]
BaBrOH	[d]	[f]
$BaCl_2$	[d]	Schofield and Sugden (1971); [f]
BaClOH	[d]	[f]
BaF	[d]	Gurvich and Ryabova (1964b)
BaO	[a];[d]	Kalff and Alkemade (1970)
BaOH	[a];[d]; emission in postflame gases of H_2–O_2–N_2 flames	Cotton and Jenkins (1968); Reid and Sugden (1962); [f]
$Ba(OH)_2$	[d]	Cotton and Jenkins (1968)
BeO	[a]	
BrO	[b]; emission in mainly fuel-rich flames	Phillips and Sugden (1960)
CF	Addition of CF_4 to flame gave CF emission but CF_2 not seen	Broida (1954)
CF	Emission in low pressure O_2-ethylene or-hydrogen flames	Gordon (1967)
CCl		
CaBr	[d]	[f]
CaCl	[d]	[f]
$CaBr_2$	[d]	Schofield and Sugden (1971); [f]
$CaCl_2$	[d]	
CaBrOH	[d]	[f]
CaClOH	[d]	[f]
CaO	[a];[b]	
CaOH	[a], emission in postflame gases	Reid and Sugden (1962); [f,g]
CaOH	[d]	Cotton and Jenkins (1968)
$Ca(OH)_2$	[d]	

TABLE 5.13 (*Continued*)

Species[c]	Detection method	Reference
CbO (i.e., NbO)	[a]	
CeO	[a]	
ClO	[b]	
CoO	[a]	
CrBr	[d]	[e]
CrCl	[d]	
CrI	[d]	
CrO	Absorption spectra	Bulewicz and Padley (1971b)
CrO_2, $HCrO_3$	Inferred from composition changes observed for Cr and CrO	Bulewicz and Padley (1971b)
CrO, CrO_2, CrO_3, H_2CrO_4, $KHCrO_4$, K_2CrO_4	Mass spectrometric analysis	Farber and Srivastava (1973b)
CsBr	[d]	[e]
CsCl	[d]	
CsF	[d]	
CsI	[d]	
CsOH	[d]	Jensen and Padley (1966b)
CuCl	[b]	
CuH	[b]; emission	Bulewicz and Sugden (1956b)
CuOH	[b]; emission	Bulewicz and Sugden (1956a)
EuOH (?)	[a]	
FeO, FeOH	[a]; [b]; [d]; low pressure lean flames	Linevsky (1971); Jensen and Jones (1973)
$Fe(OH)_2$	Mass spectrometric analysis	Farber *et al.* (1974)
GaBr	[d]	[e]
GaCl	[d]	
GaI	[d]	
GaO	[d]	Gurvich and Ryabova (1964c); Bulewicz and Sugden (1958); Kelly and Padley (1971b)
GaOH	[d]	
HO		See summary of Schofield and Broida (1968)
HO_2	Mass spectrometric analysis	Hastie (1974)
IO	[b]; emission in mainly fuel-rich flames	Phillips and Sugden (1960)
InBr	[d]	[e]
InCl	[d]	
InF	[d]	
InI	[d]	
InO	[d]	Gurvich and Ryabova (1964c)

TABLE 5.13 (*Continued*)

Species[c]	Detection method	Reference
InOH	[d]	Bulewicz and Sugden (1958); Kelly and Padley (1971b)
KBr	[d]	[e]
KCl	[d]	
KF	[d]	
KI	[d]	
KOH	[d]	Jensen and Padley (1966b); Kelly and Padley (1971b)
KO_2	[d]; lean H_2–O_2–N_2 flames	Kaskan (1965)
LaO	[a]	
LiBr	[d]	[e]
LiCl	[d]	
LiF	[d]	
LiI	[d]	
LiO	Inferred from spectroscopic observation of depletion of Li atoms in dry CO–O_2 flames (Li + CO_2 $\rightleftarrows$ LiO + CO)	Dougherty *et al.* (1971)
LiOH	[d]	Jensen and Padley (1966b); also review of Zeegers and Alkemade (1970); Kelly and Padley (1971b)
LuO	[a]	
MgO	[a]; band spectra, also used to calculate flame temperature	Nazimova (1967)
MgO, MgOH	[a]; emission spectra	Bulewicz and Sugden (1959)
MnBr	[d]	[e]
MnCl	[d]	
MnI	[d]	
MnO	[a]; [b]	Padley and Sugden (1959b); also Reid and Sugden (1962)
MnOH	Emission in postflame gases	
$KHMoO_4$, H_2MoO_4	Indirect evidence from ions and electrons studies in H_2–O_2–N_2 flames	Jensen and Miller (1969b, 1971)
$KHMoO_4$, K_2MoO_4, H_2MoO_4, MoO_3	Mass spectrometric analysis	Farber and Srivastava (1973a)
NaBr	[d]	[e]
NaCl	[d]	
NaF	[d]	
NaI	[d]	
NaO_2	[d]; lean H_2–O_2–N_2 flames	Kaskan (1965)

TABLE 5.13 (*Continued*)

Species[c]	Detection method	Reference
NaOH	[d]	Jensen and Padley (1966b)
NiBr	[d]	[e]
NiCl	[d]	[e]
NiH	[b]	
NiI	[d]	[e]
NiO	[a]	
PO	Mass spectrometric observation	Hastie (1972, 1973a)
PO_2	Mass spectrometric observation	Hastie (1972, 1973a)
HPO_2	Mass spectrometric observation	Hastie (1972, 1973a)
P_2	Mass spectrometric observation	Hastie (1972, 1973a)
PN	Mass spectrometric observation	Hastie (1972, 1973a)
PO	Emission spectra in low pressure O_2-ethylene/hydrogen flames	Gordon (1967)
HPO	Emission in postflame gases of low pressure fuel-rich H_2 flames	Fenimore and Jones (1964)
PbO	[b]; [d]	Friswell and Jenkins (1972)
ReO	Mass spectrometric observation	Farber *et al.* (1974)
ReO_2	Mass spectrometric observation	Farber *et al.* (1974)
ReO_3	Mass spectrometric observation	Farber *et al.* (1974)
$ReO_3(OH)$	Mass spectrometric observation	Farber *et al.* (1974)
RbBr	[d]	Jensen and Padley (1966b); Kelly and Padley (1971b);[e]
RbCl	[d]	Jensen and Padley (1966b); Kelly and Padley (1971b);[e]
RbF	[d]	Jensen and Padley (1966b); Kelly and Padley (1971b);[e]
RbI	[d]	Jensen and Padley (1966b); Kelly and Padley (1971b);[e]
RbOH	[d]	
SH	[b]	
SO	[b]	
SO_2	[b]	
HSO_2	Indirect arguments from Na D–line chemiluminescence observations	Kallend (1967)
S_2	[b]	
$SbBr_3$	Mass spectrometric detection in 1-atm CH_4–O_2 flames	Hastie (1973c)
$SbCl_3$	Mass spectrometric detection in 1-atm CH_4–O_2 flames	Hastie (1973c)
SbO	Mass spectrometric detection in 1-atm CH_4–O_2 flames	Hastie (1973c)
ScO	[a]	
SiO	[b]; equilibrium calculation using thermochemical data	Chester *et al.* (1970)
SnO	Emission from oxycoal gas flame containing $SnCl_2$	Joshi and Yamdagni (1967)
SnO	[a]; emission spectra in H_2–O_2–N_2 flames	Bulewicz and Padley (1971a)
SnOH	[a]; emission spectra in H_2–O_2–N_2 flames	Bulewicz and Padley (1971a)
SrBr	[d]	[f]

TABLE 5.13 (*Continued*)

Species[c]	Detection method	Reference
$SrBr_2$	[d]	[f]
SrCl	[d]	Gurvich *et al.* (1971); [f]
SrBrOH	[d]	[f]
SrClOH	[d]	Ryabova *et al.* (1971); [f]
$SrCl_2$	[d]	Schofield and Sugden (1971)
SrF	[d]	Ryabova and Gurvich (1964)
SrF_2	[d]	
SrO	[a]	
SrOH	[a]; [d]; also emission	Gurvich *et al.* (1971); Reid and Sugden (1962); [f]; [g]
SrOH	[d]	Cotton and Jenkins (1968)
$Sr(OH)_2$	[d]	
TlBr	[d]	
TlCl	[d]	[e]
TlF	[d]	
TlH	Emission bands	Bulewicz and Sugden (1958)
TlI	[d]	[e]
UO_4H_2	Indirect from metal atom concentration changes	Kelly and Padley (1971a)
VO	[a]	
VO		
H_2VO_3	Mass spectrometric detection	Farber and Srivastava (1973b)
K_2VO_3		
$KHVO_3$		
$KHWO_4$	Indirect evidence from ions and electrons studies in H_2–O_2–N_2 flames	Jensen and Miller (1970)
H_2WO_4		
WO_3		
$KHWO_4$		
K_2WO_4	Mass spectrometric analysis	Farber and Srivastava (1973a)
H_2WO_4		
WO_3		
YO	[a]	
YbO	[a]	
YbOH (?)	[a]	
ZrO	[a]	

[a] Cited in Mavrodineanu and Boiteux (1965); detection method of emission band spectra in region 3179–9195 Å using acetylene–air, or O_2, flames.

[b] Cited in Gaydon (1948), emission band spectra.

[c] Listed in alphabetical order of the metal, or the least electronegative atomic component.

[d] Inferred from optical spectroscopic study of free-atom depletion.

[e] Bulewicz *et al.* (1961), using H_2–O_2–N_2 flames.

[f] Gurvich *et al.* (1973), using H_2–O_2–N_2 flames.

[g] Van der Hurk (1973).

temperature. Frequently, the only species determined by direct observation is a metal atom, and the presence of molecular species containing the metallic element is inferred from the atom concentration changes, e.g., see Gurvich and Ryabova (1964a, b) and Gurvich *et al.* (1973).

The concentration of an atomic species can be derived from the intensity of its discrete emission spectrum, as indicated previously (see Section II.F). However, the spectra of molecular species can rarely be used to determine their concentration, as the oscillator strengths are usually unknown. The concentration of a molecular species is therefore obtained indirectly from the difference between the total and the observed atomic concentrations (see Section II.G). Absolute calibrations of the atomic concentration may be made by comparison of spectral intensities with a standard lamp.

For the simple case where MX is a species of the type MO, MOH or MH, its concentration may be determined by use of the reaction

$$\mathrm{M} + \mathrm{X} \rightleftarrows \mathrm{MX},$$

and the concentration of the flame radical X, as may be obtained via a determination of H using the Li method and the balanced-reaction relations that exist between the flame radicals, as discussed previously (Section II.I).

By monitoring atom concentrations, Gurvich and Ryabova (1964b) determined the dissociation energy, $D_0(\mathrm{BaF})$, from the reaction

$$\mathrm{Ba} + \mathrm{HF} \rightleftarrows \mathrm{BaF} + \mathrm{H},$$

where the HF concentration is obtained from the reaction

$$\mathrm{H_2} + \mathrm{F} \rightleftarrows \mathrm{H} + \mathrm{HF},$$

for which the equilibrium constant and the H_2, H, and total F concentrations are known. From the third-law treatment of data, they found $D_0(\mathrm{BaF}) = 147 \pm 7$ kcal mol^{-1}. Similar experiments indicated that $D_0(\mathrm{BaCl}) = 118 \pm 5$ kcal mol^{-1} (Gurvich and Ryabova, 1964a). These dissociation energies are significantly greater, i.e., by 10 kcal mol^{-1} or more, than those obtained by other proven methods such as Knudsen cell-mass spectrometry, e.g., see Hastie and Margrave (1968b). The explanation for this type of discrepancy is in the interpretation of which molecular species are involved in the observed loss of free metal atoms from the system. As will be shown in the examples to follow, the molecular composition for metal additives can be surprisingly varied. When the proper accounting of chemical species is made the "measured" concentration of BaCl, relative to Ba-atoms, for example, is substantially decreased, and a lower bond energy is determined, i.e., 106.8 kcal mol^{-1} (Gurvich *et al.*, 1973) as compared with the earlier incorrect value of 118 kcal mol^{-1}.

2. Polyatomic Species—the $BaCl_2$ Example

Most studies of metal additives in flames have assumed monatomic and diatomic species to be the major constituents. Usually this assumption was partly supported by the observation, in emission, of atomic line spectra or diatomic band spectra. However, the observation of emission band spectra generally provides only semiquantitative evidence for the importance of the molecular species and does not preclude the possible presence of more complex species.

As indicated above, the development of reliable alternative techniques for determining bond dissociation energies, such as the Knudsen effusion mass spectrometric method, has made apparent a number of discrepancies in data obtained by flame techniques. The most likely source of serious error in determining dissociation energies by flame methods is an incorrect accounting of all of the species present. That is, the nature and distribution of species in flames may be more complex than assumed.

The recent reinvestigation of alkaline earth metal–halogen interactions in 1-atm H_2–O_2–N_2 flames by Schofield and Sugden (1971) provides a very good example of the application of flame techniques to systems where more than one metal-containing molecular species is present. This study also resolves earlier discrepancies between dissociation energies obtained by the flame method and the mass spectrometric Knudsen effusion method. Contrary to previous interpretations, the diatomic MX species, where M is Ca, Sr, or Ba and X is Cl or Br, is of minor importance as compared with the MX_2 species in the burnt-gas region.

More recently, Gurvich *et al.* (1973) have also reinvestigated these metal–halogen systems with similar results except for the inclusion of mixed hydroxyhalide species in the analysis, as discussed in a later section (K.4).

The study of Schofield and Sugden (1971) is representative of the many similar spectroscopic investigations of additives in the burnt-gas region of 1-atm flames carried out by Sugden and co-workers using the "family-of-flames" approach. Essentially, the technique used is similar to that developed earlier for the determination of dissociation energies of diatomic metal–halide species (Bulewicz *et al.*, 1961). The basic experimental requirements are

- the use of a shielded practically flat flame where the distance–time conversion is readily obtained and the temperature is approximately constant in the burnt-gas region;
- the use of a number of families of well-characterized H_2–O_2–N_2 fuel-rich flames with differing stoichiometries and final temperatures;

- the use of atomic emission intensity ratios for the reacting metal and a nonreacting standard, such as Na, for the determination of relative concentrations of reactant and product species;
- the use of very low additive concentrations, e.g., $\sim 10^{-8}$ mole fraction of metal and $\sim 10^{-2}$ mole fraction of halogen, such that the bulk flame properties and radical concentrations are essentially unperturbed and a negligible self-absorption of emission occurs; and
- that measurements be made sufficiently downstream of the reaction zone for the emission to be thermal in origin and for bimolecular reactions of relatively low exo- or endothermicity to be balanced.

The basic relationship from which concentration terms are derived for elements denoted as a and b, is given by

$$\frac{I_a}{I_b} = \left(\frac{N_a}{N_b}\right)\left(\frac{\alpha_a}{\alpha_b}\right)\left(\frac{f_a}{f_b}\right)\left(\frac{\lambda_b}{\lambda_a}\right)^3 \exp\left[-\frac{hc}{kT}\left(\frac{1}{\lambda_a} - \frac{1}{\lambda_b}\right)\right] \tag{1}$$

where

I is the emission intensity of resonance line,
N the number of ground state atoms cm^{-3},
α the spectral sensitivity of detector system at wavelength λ,
f the f value of the optical transition,
h the Planck constant,
k the Boltzmann constant,
c the velocity of light, and
T the temperature.

Consider the case where M is Ba and X is Cl. In the absence of the halogen, a number of new species result from the presence of Ba alone, and these are known to be $Ba(OH)_2$, BaOH, BaO, $(BaOH)^+$, and Ba^+. As the concentration of positive ions is several orders of magnitude lower than the other species they can be neglected. The total amount of added Ba can then be expressed as

$$[Ba]_0 = [Ba] + [BaO] + [BaOH] + [Ba(OH)_2].$$

The standard additive used is Na, where no molecular species are formed, and the total amount added can be expressed as

$$[Na]_0 = [Na] + [Na^+].$$

As the flames in this study have temperatures in the range of 2134 K–2534 K, then $[Na^+]$ is negligible. Also, NaCl is an unstable species in these particular flames.

Equation (1) then becomes

$$(1 + \phi) = \left(\frac{I_{\mathrm{Na}}}{I_{\mathrm{Ba}}}\right)\left(\frac{\alpha_{\mathrm{Ba}}}{\alpha_{\mathrm{Na}}}\right)\frac{[\mathrm{Ba}]_0}{[\mathrm{Na}]_0}\left(\frac{f_{\mathrm{Ba}}}{f_{\mathrm{Na}}}\right)\left(\frac{\lambda_{\mathrm{Na}}}{\lambda_{\mathrm{Ba}}}\right)^3 \times \exp\left\{-\frac{hc}{kT}\left(\frac{1}{\lambda_{\mathrm{Ba}}} - \frac{1}{\lambda_{\mathrm{Na}}}\right)\right\}, \quad (2)$$

where

$$\phi = \frac{[\mathrm{BaO}]}{[\mathrm{Ba}]} + \frac{[\mathrm{BaOH}]}{[\mathrm{Ba}]} + \frac{[\mathrm{Ba(OH)_2}]}{[\mathrm{Ba}]}.$$

From this Eq. (2) and the emission intensity measurements for Na and Ba, one obtains values for $(1 + \phi)$, and the barium free-atom concentration [Ba] is then given by the relation

$$[\mathrm{Ba}]_0 = (1 + \phi)\,[\mathrm{Ba}].$$

The concentration ratios contained in ϕ are considered to be linked via the balanced reactions

$$\mathrm{Ba + H_2O \rightleftarrows BaO + H_2},$$

$$\mathrm{Ba + H_2O \rightleftarrows BaOH + H},$$

and

$$\mathrm{BaOH + H_2O \rightleftarrows Ba(OH)_2 + H}.$$

The addition of small amounts of halogen apparently does not affect the balancing of these reactions, and therefore the ratios, [Ba]/[BaO], etc., are unaffected by such additions, even though a considerable reduction in the absolute Ba-atom concentration is observed. This reduction in Ba level is assumed, for the moment, to be associated with the formation of a molecular metal–halide species BaY, of composition to be determined, where Y contains at least one halogen atom. Thus, in the presence of halogen, the distribution of Ba is given by

$$[\mathrm{Ba}]_0 = (1 + \phi)\,[\mathrm{Ba}]' + [\mathrm{BaY}],$$

where $[\mathrm{Ba}]'$ is the concentration of free Ba in the presence of halogen. Also, from Eq. (1),

$$\frac{[\mathrm{Ba}]}{[\mathrm{Ba}]'} = \frac{I}{I'} = 1 + \frac{[\mathrm{BaY}]}{[\mathrm{Ba}]'\,(1 + \phi)}, \quad (3)$$

where the primed terms are for the flames containing halogen species.

The fate of the halogen additive in these flames can be represented by

$$[\mathrm{Cl}]_0 = [\mathrm{Cl}] + [\mathrm{HCl}] + [\mathrm{BaY}].$$

It can be shown that the concentration of molecular chlorine is negligible at the given flame temperatures and additive concentrations. Also, as the additive concentrations are chosen such that $[\mathrm{Ba}]_0 \ll [\mathrm{Cl}]_0$, one can neglect the effect of BaY on halogen abstraction and therefore the halogen content is given by

$$[\mathrm{Cl}]_0 = [\mathrm{Cl}] + [\mathrm{HCl}].$$

The species Cl and HCl are linked in these H_2-rich flames by the balanced reaction

$$\mathrm{Cl} + \mathrm{H_2} \rightleftarrows \mathrm{HCl} + \mathrm{H} \tag{4}$$

and hence

$$[\mathrm{Cl}]_0 = [\mathrm{HCl}]\left(1 + \frac{[\mathrm{H}]}{[\mathrm{H_2}]K_4}\right), \tag{5}$$

where K_4 represents the equilibrium constant.

As the interaction of Ba and Cl is linked to the chemistry of the flame itself, the identity of BaY can be determined from its relationship to changes in flame composition and temperature. The possible BaY species are BaCl, BaClOH, BaOClOH, BaOCl, and $BaCl_2$. Each of these species should exhibit a different dependence on the flame conditions.

Consider, for example, the hypothetical case where BaCl is a major species. Its relation to the flame chemistry can be derived from balanced reactions such as

$$\mathrm{Ba} + \mathrm{HCl} \rightleftarrows \mathrm{BaCl} + \mathrm{H}. \tag{6}$$

It follows from the relations (3) and (5) and the reactions (4) and (6) that

$$\frac{I}{I'} = 1 + \frac{K_6[\mathrm{Cl}]_0}{[\mathrm{H}]\,(1+\phi)\,(1+[\mathrm{H}]/[\mathrm{H_2}]K_4)}. \tag{7}$$

Hence, for a particular flame, a plot of the experimental ratios I/I', versus the amount of Cl added, i.e., $[\mathrm{Cl}]_0$, should be linear. It can be shown that this should also be the case for the other molecular species which contain no more than one Cl atom.

For the case where $BaCl_2$ is the important species, a balanced condition can be achieved via reactions such as

$$\mathrm{BaCl} + \mathrm{HCl} \rightleftarrows \mathrm{BaCl_2} + \mathrm{H},$$

and as Ba and BaCl have already been suggested as attaining a balanced condition, one can consider the local-equilibrium reaction

$$\mathrm{Ba} + 2\mathrm{HCl} \rightleftarrows \mathrm{BaCl_2} + 2\mathrm{H}. \tag{8}$$

It follows that

$$\frac{I}{I'} = 1 + \frac{K_8[\mathrm{Cl}]_0^2}{[\mathrm{H}]^2(1+\phi)\ (1 + [\mathrm{H}]/[\mathrm{H_2}]K_4)^2} \tag{9}$$

from which a parabolic dependence of I/I' on $[\mathrm{Cl}]_0$ would follow. As a parabolic dependence is observed in practice, the presence of some $BaCl_2$ is indicated. The importance of $BaCl_2$ relative to species containing a single halogen atom is given by

$$I/I' = 1 + \mathrm{A}[\mathrm{Cl}]_0 + \mathrm{B}[\mathrm{Cl}]_0^2,$$

where A describes the contribution of monohalide compound formation and B the dihalide. The parameters A and B can be obtained from a computer fitting of the experimental I/I' vs $[\mathrm{Cl}]_0$ parabolic curves.

A more sensitive test of the relative importance of these species is obtained by studying the effect of different flame temperatures on the emission intensities. Consider the dependence of equilibrium constants on temperature for the various possible reactions. For the case of BaCl predominance, it follows from Eq. (7) that the equilibrium constant K_6 can be represented by

$$K_6 = \mathrm{A}[\mathrm{H}](1+\phi)(1 + [\mathrm{H}]/[\mathrm{H_2}]K_4).$$

Similarly, for the case of $BaCl_2$ formation it follows from Eq. (9) that K_8 can be represented by

$$K_8 = \mathrm{B}[\mathrm{H}]^2(1+\phi)(1 + [\mathrm{H}]/[\mathrm{H_2}]K_4)^2.$$

The values of ϕ, H, and H_2 are obtained from earlier work on Ba-doped flames without a halogen component, and A and B are obtained from the observed dependence of I/I' on $[\mathrm{Cl}]_0$. If either of the species BaCl or $BaCl_2$ predominates, then plots of log K_6 or log K_8 vs T^{-1} should be linear. Plots for the BaCl and $BaCl_2$ test cases are shown in Fig. 5.9, where it is apparent that, within the experimental scatter, the dihalide case has the necessary linear dependence and the monohalide case does not (however see the discussion of Section K.4). Hence, $BaCl_2$ is considered to be the predominant metal halide species. Similar arguments were used by Schofield and Sugden (1971) to rule out other barium chloride species.

The measured equilibrium constant, K_8, when combined with the appropriate free energy functions and heats of formation, leads to a value for the atomization energy of $BaCl_2$ at 0 K of 217.9 ± 5 kcal mol^{-1}, which compares very well with a recent mass spectrometric Knudsen effusion

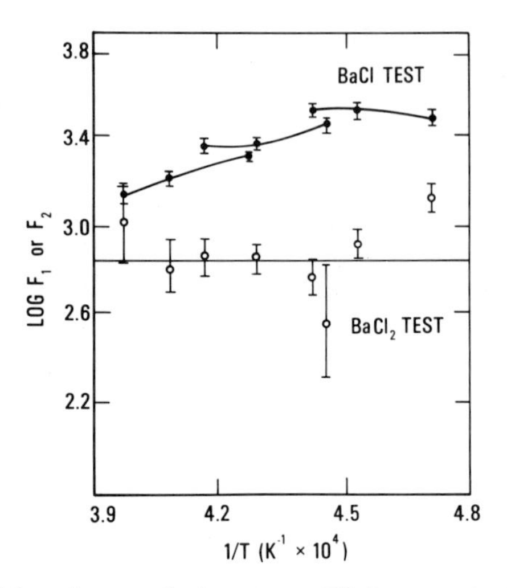

Fig. 5.9 Plots of functions equivalent to equilibrium constants (see text) for the hypothetical formation of $BaCl_2$ or BaCl as a function of temperature; data of Schofield and Sugden (1971).

value of 217.2 kcal mol^{-1}—see Schofield and Sugden (1971) for further detail. The corresponding value of 223 ± 6 kcal mol^{-1}, obtained by Gurvich *et al.* (1973) and using similar—though not identical—spectroscopic techniques is also in general agreement. Similar agreement between results, as obtained by mass spectrometric Knudsen effusion and optical spectroscopic flame techniques, is found for the other alkaline earth dihalide systems. A summary of dissociation energy data for alkaline earth metal mono- and dihalides is given in a following Section K.4.

3. *Metal Hydroxides in Flames*

The combustion of H_2–O_2–N_2 mixtures to yield superheated steam, and over a considerable concentration and temperature range, provides one with a unique opportunity for the study of the hydroxylation of metal oxide species. Metal hydroxide species are of importance in materials transport processes, such as those used for single crystal preparation, e.g., see Chapter 3, or geochemical processes—such as may develop in subterranean volcanic activity or nuclear explosions. The solubility of refractory materials in high temperature steam is also significant (see Chapter 2, Section VIII).

a. Hydroxides of Ca, Sr, and Ba

The methods used for hydroxide species identification and thermodynamic characterization are similar to those outlined above for the $BaCl_2$

example, where a combination of isothermal studies on different flame compositions and fixed composition studies at various temperatures suffices—in principle—to characterize the *major* flame species and reactions.

From the work of Cotton and Jenkins (1968), the principal balanced flame reactions to be considered are

$$M + H_2O \rightleftarrows MOH + H, \tag{1}$$

$$M + 2H_2O \rightleftarrows M(OH)_2 + 2H, \tag{2}$$

$$M + OH \rightleftarrows MO + H, \tag{3}$$

and

$$H_2 + OH \rightleftarrows H_2O + H, \tag{4}$$

where M is Ca, Sr, or Ba. For the steady-state condition, the parameter ϕ, defined as

$$\phi = [M]_0/[M] - 1,$$

is then given by

$$\phi = \frac{K_1[H_2O]}{[H]} + \frac{K_2[H_2O]^2}{[H]^2} + \frac{K_3}{K_4}\frac{[H_2O]}{[H_2]}.$$

The equilibrium constant K_4 is known. Experimentally, ϕ is measured as a function of [H] and values of K_1, K_2, and K_3 are derived from this data.

For these flame studies, the principal M-containing species were determined to be $M(OH)_2$ and MOH with very little MO or M being present. Earlier flame studies on these systems usually assumed MO or MOH to be the important species. As the $Ba(OH)_2$ thermodynamic data are found to be in reasonable agreement with the results of mass spectrometric studies, i.e., see Stafford and Berkowitz (1964), the more recent interpretation appears most reasonable. These data are also supported by the results of Schofield and Sugden (1965). Additional support for this interpretation arises from the very good agreement found between the flame and mass spectrometric data for the MX_2 halide species, where the $M(OH)_2$ data are necessarily included in the flame analysis, as was indicated in the previous section.

A summary of dissociation energy data for these and similar dihydroxide species is given in Table 5.14.

b. Hydroxides of Al

The composition of the vapor phase in aluminum–oxygen systems is known to be complex—involving the presence of the species AlO, AlO_2, Al_2O, Al_2O_2, and possibly others, e.g., see JANAF (1971). Hence, it is not surprising that the chemistry of aluminum in H_2–O_2–N_2 flames is also ap-

TABLE 5.14

Dissociation Energies of Dihydroxides in Flames[a]

Species	D_0 M–(OH)$_2$ kcal mol^{-1}	Comments and references
$Be(OH)_2$	251 ± 6	Not studied in flames, JANAF (1971)
$Mg(OH)_2$	190 ± 3	
$Ca(OH)_2$	203.8	Cotton and Jenkins (1968)
$Sr(OH)_2$	202.2	
$Ba(OH)_2$	212.5	
$Al(OH)_2$	256	Jensen and Jones (1972)
$Fe(OH)_2$	194 ± 5	Farber *et al.* (1974)

[a] Dissociation energy is for the process

$$M(OH)_2 = M + 2OH,$$

and referred to 0 K.

parently complex. From several recent flame studies the species AlO, AlO_2, Al(OH), AlO(OH), and $Al(OH)_2$ have been suggested as being important flame species. However, there is some disagreement concerning the relative importance of the various hydroxide species and also AlO. This uncertainty is reflected in the calculation of widely disparate values for the bond dissociation energy of the AlO species (this may be seen in Table 5.18).

Jensen and Jones (1972), using similar flame techniques to those described above for the alkaline earth metal halide and hydroxide examples, have indirectly determined $Al(OH)_2$ to be the principal flame species, in the presence of lesser amounts of Al and AlO, for fuel-rich H_2–O_2–N_2 flames at temperatures in the region of 2500 K. From their thermochemical analysis, the $Al(OH)_2$ species shows a somewhat surprising degree of stability, i.e., the average Al–OH bond dissociation energy is $\bar{D}$(Al–OH) ~ 128, as compared with $\bar{D}$(Al–F) ~ 131 and $\bar{D}$(Al–Cl) ~ 94 kcal mol^{-1}. This result seems anomalous when compared with the relative hydroxide to chloride dissociation energies for the alkali metal and alkaline earth metal species where

$$\bar{D}(\text{M–OH}) \lesssim \bar{D}(\text{M–Cl}) \ll \bar{D}(\text{M–F}).$$

Likewise, for BX_3, where X is OH, F, or Cl, the average B–X bond energies are 132, 153, and 104 kcal mol^{-1}, respectively; i.e., the hydroxide to fluoride ratio is 0.86 and the hydroxide to chloride ratio is 1.27 (data from JANAF, 1971). From these comparisons one would expect the average Al–OH bond energy to be at least 15 kcal mol^{-1} less than observed. Further discussion along these lines is deferred to a later section (K.5).

The study of Jensen and Jones (1972) also indicated an anomalously high dissociation energy for AlO of 142 ± 5 kcal mol^{-1}. The usual explanation for an anomalously high apparent stability for a molecular flame species, as determined by indirect flame spectrophotometric methods, is the misassignment of species. Such was the case for the already discussed alkaline earth systems. Indeed, a recent direct mass spectrometric analysis of a lean H_2–O_2 flame, containing aluminum, suggests that a misassignment of species may have occurred (Farber *et al.*, 1973).

The Al-containing species "identified" in the mass spectrometric study were Al, AlO, AlO_2, Al(OH), and AlO(OH), with Al(OH) being the predominant hydroxide component. No evidence for $Al(OH)_2$ was obtained although the signal-to-noise levels were such that a concentration of about 20% $Al(OH)_2$, relative to the total of Al-containing species, would be undetectable. The observed concentrations of AlO and Al were also found to be in accord with those expected from the accepted literature value for the dissociation energy of AlO (i.e., the low value of Table 5.18). Thus it would *appear* that the indirect spectrophotometric studies did not consider the proper complement of species in their analysis, as was suggested by Farber *et al.* (1973). However, we need to consider the possibility of marked changes in relative abundance for the various Al-containing flame species under conditions of excess H_2—as in the spectrophotometric experiments. For instance, the species AlO(OH), observed in the lean flames of Farber *et al.* (1973), could be converted to $Al(OH)_2$ under fuel-rich conditions by a low free energy change process such as

$$AlO(OH) + OH = Al(OH)_2 + O.$$

This follows from the expected several-fold increase in OH concentration relative to O—and hence $Al(OH)_2$ relative to AlO(OH)—with increasing H_2 concentration. Similarly, $Al(OH)_2$ would be favored relative to AlO_2 by an increased H_2 concentration in the flame.

It is perhaps relevant to note that the Al–OH bond energy for AlOH, as determined from the mass spectrometric study, is equal to the average Al–OH bond energy in $Al(OH)_2$, as determined by Jensen and Jones (1972), namely, 128 kcal mol^{-1}. Based on an analogy with the alkaline earth metal hydroxides, this correspondence in bond energies for mono- and dihydroxide species is to be expected (i.e., see Table 5.16). Thus, in this respect, the spectrophotometric and mass spectrometric results are consistent, and the "anomalous" stability for $Al(OH)_2$ may well be realistic.

The practical implications of such a high stability for $Al(OH)_2$, and perhaps other systems such as $Ln(OH)_2$ where Ln is a rare earth element, warrant an independent verification, preferably by a direct mass-spectroscopic identification of the $Al(OH)_2$ species under fuel-rich flame conditions. It would also appear desirable to eliminate possible ion-source electron impact ambiguities, arising from processes such as

$$e + Al(OH)_2 \rightarrow AlO(OH)^+ + H$$

by using the lowest possible ionizing electron energy (i.e., <20 eV).

c. Monohydroxides

The techniques used to determine dissociation energies for metal monohydroxide flame species are similar to those already considered. A summary of the dissociation energy data is given in Table 5.15 (see also Table 5.16, Section K.4, for the alkaline earth metal hydroxide dissociation energies).

4. *Mixed Hydroxyhalide Species*

It follows from the established formation of metal dihydroxides and dihalides in flames, and from the similarities in their bond energies, that mixed hydroxyhalide species, i.e., MX(OH) where X is Cl or Br, can be expected to form under suitable flame conditions. Analogous mixed halide species are well known in high temperature vapor systems (Hastie, 1971). As has been discussed elsewhere (Chapter 3, Section IV.D), their formation is favored by the reduced symmetry—and hence higher rotational entropy—associated with their production from the symmetrical dihalide

TABLE 5.15

Dissociation Energies of Metal Monohydroxide Flame Species

Species	D_0 M–OH kcal mol^{-1}	Comments and references
LiOH	104.5 ± 1	Selected value of Zeegers and Alkemade (1970)
LiOH	103 ± 2	Kelly and Padley (1971b); H_2–O_2–CO_2 flames; 2nd law
NaOH	79 ± 2	
KOH	84 ± 2.5	
RbOH	86 ± 3	
GaOH	104 ± 2	
InOH	90 ± 2	
TlOH	<72	
GaOH	102 ± 5	Bulewicz and Sugden (1958); H_2–O_2–N_2 flames; 2nd and 3rd law values
InOH	86 ± 7	
FeOH	100 ± 3	Linevsky (1971); low pressure, H_2–O_2–N_2 flames
SnOH	≲85 ± 9	Bulewicz and Padley (1971a)

mixture. If we apply a similar argument to the formation of MX(OH) species by the equilibrium process

$$MX_2 + M(OH)_2 = 2MX(OH),$$

it follows that a positive reaction entropy of about 1.4 cal deg^{-1} mol^{-1} can be expected. As a first degree of approximation we can assume the reaction enthalpy to be negligible. This follows from observations on mixed halide systems and is also borne out by the hydroxyhalide results discussed shortly. Hence the reaction free energy is given mainly by the entropy effect and an equilibrium constant of about 2 is then derived. That is, the partial pressure of MX(OH) is about 1.4 times the geometric mean of the partial pressures for $M(OH)_2$ and MX_2. Thus, from this simplified model, we can expect a priori that MX(OH) would be a major species relative to $M(OH)_2$ and MX_2.

Indirect flame spectrophotometric experimental evidence has recently been obtained, suggesting the presence of MX(OH) species in atmospheric pressure fuel-rich H_2–O_2–N_2 flames, where M is Ca, Sr, or Ba and X is Cl or Br (Ryabova *et al.*, 1971; Gurvich *et al.*, 1973). These studies provide equilibrium constant and reaction enthalpy data defining the formation of MX(OH), and it is apparent that this species can predominate over MX or MX_2 in such flames. Also, as expected, the formation of MX(OH) from MX_2 and $M(OH)_2$ is found to be an approximately thermal neutral process. This follows from the data of Ryabova *et al.* (1971) for SrCl(OH), and the data of Cotton and Jenkins (1968) and Schofield and Sugden (1971) for the corresponding dihydroxides and dihalides, respectively. Hence, the expectation that the formation of hydroxyhalide species is analogous to the mixed halide case appears to be experimentally verified.

A comparison of bond dissociation energies for the hydroxyhalide flame species with those for the corresponding halides and hydroxides is given in Table 5.16.

It is pertinent to examine the degree to which the several different flame studies on the alkaline earth metal dichlorides and hydroxides agree. From the comparison of data given in Table 5.16, it is apparent that there is satisfactory agreement between the various studies on the bond energies of MOH, MCl, and MCl_2. The most notable disagreement is the absence of $M(OH)_2$ and MCl(OH) in the data interpretations of Gurvich *et al.* (1973) and of Schofield and Sugden (1971), respectively.

Consider one particular flame studied by each research group, namely, the atmospheric pressure flame defined by mole ratios of $H_2/O_2 = 3.5$ and $N_2/O_2 = 2$ and a final temperature of about 2450 K. For the addition of Ba in amount equivalent to a total metal-containing species pressure of 3.3×10^{-7} atm, and of chloride in amount equivalent to an HCl pressure

TABLE 5.16

Dissociation Energies, in kcal mol^{-1}, of Alkaline Earth Metal Hydroxides, Halides, and Hydroxyhalides in Flames[a]

Species \ Metal, M	Ca	Sr	Ba
$M(OH)_2$[c]	203.8	202.2	212.5
MOH[b]	96	95	111
MOH[c]	103.7	103.0	114.3
MOH[f]	103	102	113
MCl[b,e]	94.1	95.6	106.8
MBr[b]	75.7	78.4	87.5
MCl_2[b]	215	210	223
MCl_2[d]	211	209	218
MBr_2[b]	187	186	196
MBr_2[d]	190	—	—
MCl(OH)[b]	208	200	214
MBr(OH)[b]	189	186	204

[a] Dissociation energies are for the processes

$$MX = M + X \quad \text{or} \quad MX_2 = M + 2X\,,$$

where X = OH, Cl, or Br, and are referred to 0 K; uncertainties are typically in the range ± 2–6 kcal mol^{-1}.

[b] From Gurvich *et al.* (1973).

[c] From Cotton and Jenkins (1968); these data also agree, within experimental error, with the similar results of Schofield and Sugden (1965).

[d] From Schofield and Sugden (1970).

[e] These data agree well with the nonflame results of Hildenbrand, as cited in Schofield and Sugden (1966).

[f] Kalff and Alkemade (1973), uncertainties $\lesssim \pm 2.5$ kcal mol^{-1}.

of 1.3×10^{-3} atm, the equilibrium constant data of Gurvich *et al.* (1973) indicate the following approximate metal-containing species concentrations:

Species:	Ba(OH)	BaCl(OH)	$BaCl_2$	BaCl	Ba
Concentration (%):	67.5	15.4	15.2	1.8	0.1

Under basically the same conditions, the equilibrium constant data of Schofield and Sugden (1971) indicate only 0.9% $BaCl_2$, and a much lower amount of BaCl. Also, from the nonhalogen flame data of Cotton and Jenkins (1968), the concentrations of $Ba(OH)_2$ and BaOH should be similar in the flame under consideration. However, the formation of BaCl(OH) would tend to reduce the significance of $Ba(OH)_2$ as well as $BaCl_2$ and by a similar degree in each case. The observation of $BaCl_2$ and BaCl(OH) in similar amount is in accord with the a priori symmetry arguments given earlier. However, the complete absence of $Ba(OH)_2$ is difficult to rationalize.

From the available evidence one must conclude that the major metal-containing species in these flames are

$$MX(OH), \quad MX_2, \quad M(OH)_2, \quad \text{and} \quad M(OH),$$

but that their relative concentrations are subject to question. At this point an earlier comment concerning the difficulty in assigning species and their concentrations by indirect flame methods should be reiterated. It is probably also pertinent to note that the data contained in the $BaCl_2$ test curves of Fig. 5.9, given earlier, show a departure from the expected model for the cooler flames.

5. *Correlation of Hydroxide and Halide Dissociation Energies*

For purposes of evaluating and estimating metal hydroxide stabilities it is possible to use the corresponding halides for comparison, as indicated in Table 5.17. From these comparisons, the following empiricism is sug-

TABLE 5.17

Comparison of Metal Hydroxide and Metal Halide Dissociation Energies[a,b]

Dissociation energy ratio		
MOH/MCl	MOH/MF	Comments
0.86 ± 0.03		For M = Li, K, Cs, Ga, and In
	0.72 ± 0.03	For M = Li, K, Cs, Ga, and In
0.98 ± 0.04		For $\bar{D}_{M-OH}$ {in $M(OH)_2$}, where M = Mg, Ca, Sr, Ba
1.05 ± 0.03[c]		For M = Fe, Co, and Ni
1.14 ± 0.04[c].		For M = Be
	0.83 ± 0.04[c]	For M = Be

[a] Data from Bulewicz and Sugden (1958), Cotton and Jenkins (1968), Gurvich *et al.* (1973), and the JANAF Thermochemical Tables (JANAF, 1971).

[b] It appears that nonmetal hydroxides have slightly higher relative stabilities, e.g., for the species $TeO(OH)_2$ and $TeCl_4$, the ratio M–OH/M–Cl = 1.28.

[c] Nonflame data of Belton and Jordan (1967) for the dihydroxides.

gested for the estimation of M–OH bond energies, i.e.:

$$D_{\mathrm{M-OH}} = 0.5D_{\mathrm{M-Cl}} + 0.4D_{\mathrm{M-F}},$$

with an uncertainty in $D_{\mathrm{M-OH}}$ of ± 5 kcal mol^{-1}. The only known possible exception to this empirical relation is for the Al—OH bond energy, as discussed previously.

6. *Metal Oxides in Flames*

a. MO_2 Species in Flames

Only a few examples of polyatomic metal oxide species in flames are known—see Table 5.13. BO_2 is believed to be present, in addition to HBO_2, in the B-containing flames studied by Kaskan and Millikan (1962). A second-law treatment of the reaction

$$\mathrm{OH + HBO_2 \rightleftarrows H_2O + BO_2}$$

indicated a standard heat of formation of

$$\Delta H_f{}^0\ \mathrm{BO_2} = -84\ \mathrm{kcal\ mol^{-1}}.$$

Kaskan (1965) has also determined, indirectly, the presence of KO_2 in lean flames with post-combustion gas compositions of,

$$\mathrm{O_2(1\text{–}6\%) + N_2(50\text{–}80\%) + H_2O}.$$

The species formation reaction is considered to be

$$\mathrm{K + O_2 + M \rightarrow KO_2 + M}.$$

It should be noted that the observed concentration of KO_2 is much greater than for KOH and this is contrary to that for an equilibrium situation as predicted from the data given by the JANAF Thermochemical Tables (JANAF, 1971).

b. MO Species in Flames

The presence of MO species in flames is well known from their band spectra, and their dissociation energies have been reviewed by Brewer and Rosenblatt (1969). Some more recent determinations are given in the Table 5.18, where the discrepancy between the dissociation energy of AlO as determined from flame studies, and that determined by other methods is of particular note. The values obtained from the flame studies appear to be too high.

TABLE 5.18

Metal Monoxide Dissociation Energies from Flame Studies

Species	D_0 (kcal mol^{-1})	Comments and references
FeO	100 ± 2	D_{298}, low pressure H_2–O_2–N_2 flames; Linevsky (1971)
CaO	87 ± 4.5	Kalff and Alkemade (1973)
SrO	94 ± 2.5	
BaO	120 ± 1	Kalff and Alkemade (1970, 1973); dry CO–N_2O flames containing $BaCl_2$; 2nd and 3rd law
MgO	87 ± 2.5	Zeegers *et al.* (1969); 3rd law
CrO	113 ± 2.5	
FeO	103 ± 2.5	
MnO	97 ± 2.5	
LiO	81 ± 3	Dry CO + O_2 + N_2 flames ($T \sim 2295$ K); Li + $CO_2 \rightleftarrows$ LiO + CO; Dougherty *et al.* (1971).
AlO	144 ± 8	Newman and Page (1970, 1971); 2nd law, indirect method
AlO	142 ± 5	Jensen and Jones (1972); indirect method
AlO	119.5 ± 5	Farber *et al.* (1973); mass spectrometric analysis of lean H_2–O_2–N_2 flames
AlO	118.5 ± 1	Gole and Zare (1972); cross beam chemiluminescence, Al + $O_3 \rightarrow$ AlO + O_2; result compares favorably with mass spectrometric and shock tube data
MgO	96 ± 3	Cotton and Jenkins (1969); third law analysis of Mg + OH $\rightleftarrows$ MgO + H; Brewer and Rosenblatt (1969) give 78 ± 7 kcal mol^{-1}; see also Zeegers *et al.* (1969) above
PbO	91.3 ± 2	Friswell and Jenkins (1972); lean H_2–O_2–N_2 flames; 2nd law; indirect evidence for PbO

L. Kinetic Studies in Flames

1. Introduction

Flames provide a useful kinetic medium for the production, excitation, and containment of high temperature species, e.g., see the review of Schofield and Broida (1968). The temperatures of flames available for kinetic studies cover a range of 1000–6000 K, but are more typically in the region of 1500–2500 K. This is a useful region as it connects the high temperature shock tube data with the results of low temperature photolysis and discharge studies. The useful range of available flame pressures is about 10^{-4} to 50 atm. Where the primary function of the flame is to serve as a high temperature bath, it is desirable to simplify the chemistry of the flame itself and hence many kinetic studies have utilized the relatively simple H_2–O_2–N_2 flames, burning at 1 atm pressure.

The time scale in flames may be obtained by the intentional introduction of refractory particles, e.g. 5 μm MgO, into flames in the presence of a light source. This allows a visual observation and photographic recording of the gas flow patterns and hence a determination of the flame speed. The time base for particle velocity measurements is provided by modulation of the light source, e.g. with a rotating sector of known speed. A recent evaluation of this particle track technique has been given by Kumar and Pandya (1970); see also Fristrom and Westenberg (1965). Alternatively, for a simple flame geometry, the flame gas velocity may be determined from the unburnt-gas flow rate which is multiplied by a volume expansion factor derived from the known changes in temperature and number of moles on combustion. Kelly and Padley (1969a) find agreement to within 4% for these two techniques of converting flame distances to a time scale. In practice, the aerodynamic features of the laboratory flame are usually simplified by the use of a one-dimensional shielded flat flame. A detailed discussion of laboratory burners may be found in the text of Gaydon and Wolfhard (1970).

For studies where the excessive excitation and dissociation provided by the flame reaction zone is of interest, it is convenient to extend the physical dimensions of the reaction zone, which is usually only a few tenths of a mm thick for 1-atm flames. This is normally achieved by using low pressure flames, e.g., see Fristrom and Westenberg (1965) and Kaskan (1958). The degree of chemical or thermal nonequilibrium that may be found under flame conditions is determined by the time scale associated with various reaction and relaxation processes. Typical time scales for atmospheric flame conditions may be found in Table 5.1. In low pressure flames, radiative nonequilibrium may also be important.

2. *Radical Recombination Kinetics*

The excess concentration of radicals present in the region of the reaction zone permits determination of radical recombination kinetics by monitoring directly the concentration decay as a function of distance, and hence time, downstream of the reaction zone. These three-body radical recombination processes may also be indirectly monitored by observing the chemiluminescence of additive species which results from reactions such as

$$H + H + Tl \rightarrow H_2 + Tl^*,$$

and

$$H + OH + Pb \rightarrow H_2O + Pb^*,$$

or

$$\text{M} + \text{OH} \rightarrow \text{MOH}^* \quad (\text{M} = \text{alkali metal}).$$

It is sometimes possible to follow the radical decay process to completion, that is to an equilibrium level of concentration, as with 1-atm H_2–O_2–N_2 flames at $T > 2000$ K. This has an advantage in that it is possible to calculate the equilibrium concentrations from basic thermodynamic data and hence calibrate or verify the nonequilibrium measurement scale.

The techniques used for species characterization are essentially the same as those discussed for the thermodynamic studies of dissociation energies (Section II.K). For typical 1-atm-flame studies, each centimeter of distance along the principal gas flow axis corresponds to a time change of about 10^{-3} sec.

With a termolecular process, such as

$$\text{H} + \text{H} + \text{M} \rightarrow \text{H}_2 + \text{M},$$

the rate of H-atom depletion is given by

$$-\frac{d[\text{H}]}{dt} = 2k_0[\text{H}]^2, \qquad \text{i.e.,} \qquad \frac{d[\text{H}]^{-1}}{dt} = 2k_0,$$

where k_0 is the reaction rate constant. Therefore, the slope of the $[\text{H}]^{-1}$ vs t curve yields a value for the rate constant k_0. Examples of typical results from which k_0, or similar rate constants, may be derived are given in Fig. 5.10. For these examples, the recombination process

$$\text{H} + \text{OH} + \text{M} \rightarrow \text{H}_2\text{O} + \text{M}$$

is also contributing to the loss of H-atoms and the slopes of these plots may be interpreted to yield the rate constants for both of the above recombination reactions (Bulewicz and Sugden, 1958b).

Using Na as a third body and monitoring the intensity of chemiluminescence for the Na D-lines, Padley and Sugden (1958) determined radical recombination rates of the order of 10^{-32} mol^{-2} cm^6 sec^{-1} in H_2–O_2–N_2 flames at about 2000 K. Similar results were found using the LiOH, NaCl, and CuH techniques for the monitoring of radicals (Bulewicz and Sugden, 1958b). It is of interest to note that for the case where one of the H-atoms is replaced by OH in the recombination process, the rate constant is increased by a factor of 24. These techniques rely on the absence of a catalytic effect by trace quantities of Na, Li, or Cu on the recombination processes. Cases where catalysis by a third body does occur have been found in recent years and a number of examples are considered in the following sections.

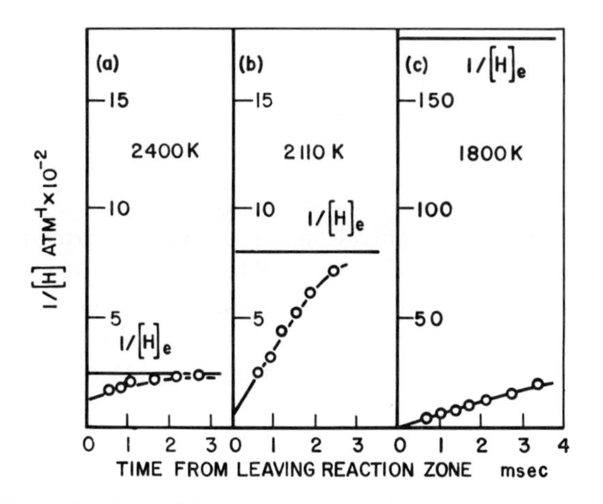

Fig. 5.10 Plots of reciprocal H-atom concentration as a function of time from the reaction zone in the burnt-gas region of fuel-rich H_2–O_2–N_2 flames; the equilibrium reciprocal concentration level is indicated by the horizontal lines (Bulewicz and Sugden, 1958b).

3. *Catalysis of Radical Recombination by SO_2*

SO_2 catalyzes the recombination of both H and OH and the mechanism in fuel-rich flames is believed to be

$$H + SO_2 + M \rightarrow HSO_2 + M, \tag{1}$$

$$H + HSO_2 \rightarrow H_2 + SO_2, \qquad \text{and} \qquad OH + HSO_2 \rightarrow H_2O + SO_2.$$

e.g., see Kallend (1967, 1972). The effect of this catalysis on radical concentrations is indirectly demonstrated by the Na D-line chemiluminescence intensity versus distance profiles shown in Fig. 5.11. We should note that the reaction

$$OH + SO_2 + M \rightarrow HOSO_2 + M$$

is also possible, particularly in non-fuel-rich flames, and is indistinguishable from the above reaction scheme, owing to the balanced nature of the reaction

$$OH + H_2 \rightleftarrows H_2O + H$$

(Wheeler, 1968). The overall catalysed recombination rate constant is $k_{2000\ K} \simeq 10^{-31}$ mol^{-2} cm^6 sec^{-1}. From the form of Na *D*-line chemiluminescence intensity profiles, such as those given in Fig. 5.11, it is believed that reaction (1) is the rate-determining step of the catalytic process.

The addition of 1% SO_2 to the flame is sufficient to reduce the steady-state concentration level of H-atoms, in the region of the reaction zone, by

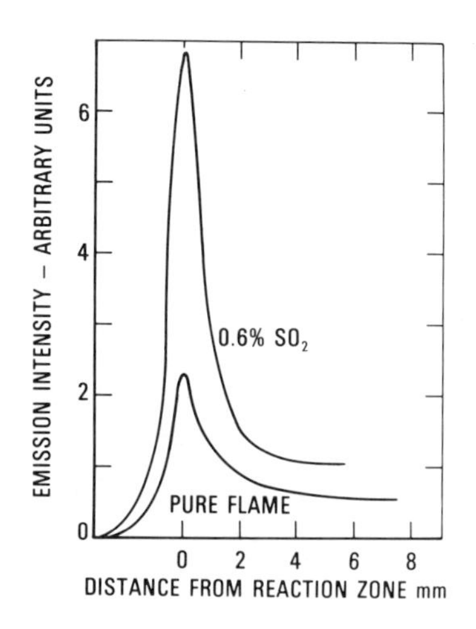

Fig. 5.11 Na D-line chemiluminescence intensity versus distance profiles showing, indirectly, the effect of 0.6% SO_2 on radical recombination in the flame $H_2/N_2/O_2$ = 4/6/1 (Kallend, 1967); note Na serves as the M of reaction (1) in this instance.

about an order of magnitude. Hence the flame propagation is inhibited by the presence of SO_2. However, in practice, SO_2 is a far less effective inhibitor than conventional extinguishants, such as brominated fluorocarbons (see Section V).

It should be noted that other S-containing species such as S_2, SO, SH, and H_2S may also be present in these systems, i.e., see Kallend (1972).

4. *Catalysis of Radical Recombination by Metal Additives*

a. Scope of the Effect

In the presence of certain metal additives a catalytic effect on H-atom recombination is found which is fundamentally similar to that for the SO_2 example, but greater in extent by several orders of magnitude (Bulewicz and Padley, 1970a; Bulewicz *et al.*, 1971; Cotton and Jenkins, 1971). In fact, the metals can show such an effect even at concentrations as low as 0.1 ppm, but below this concentration level no effect is noted.

The catalytic effect is conveniently monitored by using the Li technique to follow the H-atom concentration (see Section II.H). It has been established that the presence of a small, i.e. $<10^{-6}$ atm, amount of Li does not affect the rate of H-atom recombination (Bulewicz *et al.*, 1956). Hence, by monitoring the thermal Li-emission intensity as a function of time above the flame reaction zone, one obtains, from a plot of $[Li]^{-1}$ vs time, a measure of the uncatalyzed normal rate of H-atom recombination. If the ad-

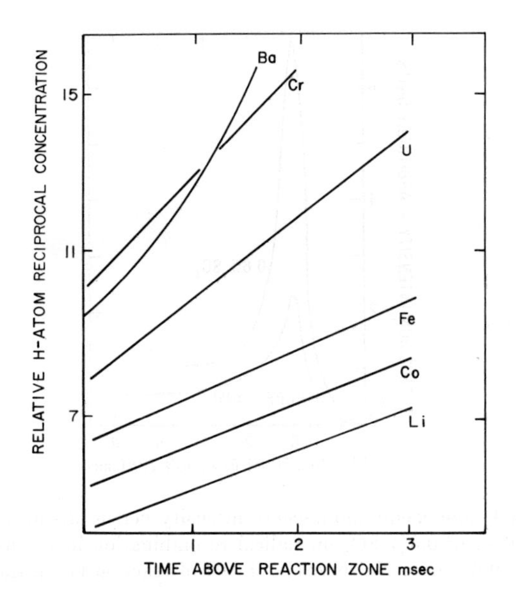

Fig. 5.12 Curves of relative H-atom reciprocal concentration, as given by the reciprocal Li resonance emission intensity (arbitrary units), versus time above the reaction zone, in the presence of the various metals indicated; the plots are displaced vertically for clarity; the flame used is given by, $H_2/O_2/N_2 = 3/1/6$ and $T = 1860$ K.

dition of another metal to the flame increases this normal recombination rate, the Li-emission intensity, which is a measure of the free Li-atom concentration, will show a corresponding decrease due to the reduced concentration of H-atoms. This follows from the basic reaction of the Li technique, i.e.,

$$H + LiOH \rightleftarrows Li + H_2O,$$

from which it is evident that a reduction in the quantity of H-atoms leads to a loss of Li-atoms due to LiOH formation. Thus, in the presence of a catalytic increase in the rate of H-atom recombination, the curves of $[Li]^{-1}$ vs time will show a greater than normal slope. Such an effect is shown by the curves given in Fig. 5.12. The slopes of these curves may be taken as a measure of the effectiveness of the various catalytic agents. Hence, from the curves of Fig. 5.12, it appears that the order of catalytic effectiveness is

$$Cr > U \gg Fe > Co.$$

We may convert this raw data to absolute H-atom concentrations, i.e. see Section II.H, by the relation

$$[H] = K[H_2O]\{([Li]_{tot}/[Li]) - 1\}^{-1},$$

where K is the known equilibrium constant for the reaction

$$Li + H_2O \rightleftarrows LiOH + H.$$

This leads to data of the type shown in Fig. 5.13. where the significant catalytic effect of Cr is more readily apparent.

The efficiency of the catalytic effect may be defined as

$$E = k_{obs}/k_{uncat},$$

where k is the H-atom recombination rate constant, as derived from the slopes of curves such as those given in Fig. 5.12. Values of E have been determined for an appreciable number of metals, in fuel-rich H_2–O_2–N_2 flames, by Bulewicz and Padley (1970a). The following metals showed little or no catalytic effect, i.e., $E \simeq 1.0$, for concentrations at least *up to* 10^{-6} *mole fraction*:

V, Ni, Cu, Zn, Ga, In, Tl,

La, Ce, Li, Na, K, Rb, and Cs.

A measurable catalytic effect was found for the following elements, with the

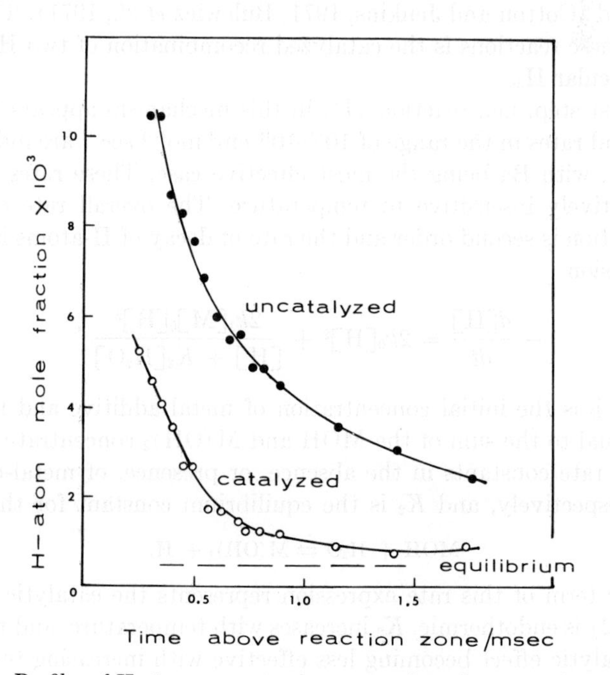

Fig. 5.13 Profiles of H-atom concentration versus time above reaction zone showing the catalytic effect of Cr ($\sim 10^{13}$ cm^{-3}) on the excess H-atom concentration for the 1-atm flame $H_2/O_2/N_2 = 3/1/5$ and $T = 2020$ K (Bulewicz and Padley, 1971b).

E parameter given in parenthesis:

Cr (2.8), U (1.82), Ba (1.75), Sn (1.6),

Sr (1.35), Mn (1.30), Ca (1.25), Mg (1.25),

Fe (1.20), Mo (1.16), Co (1.10), and Pb (1.10).

These results were obtained for the flame H_2–O_2–N_2, with mole ratios of, 3:1:6, T = 1860 K, and with total metal concentrations of about 5×10^{12} cm^{-3}, i.e., approximately 10^{-6} mole fraction.

b. Mechanisms

For the case where M is Ca, Sr, or Ba, the radical recombination results can be interpreted in terms of the reaction scheme:

$$MOH + H \rightarrow MO + H_2, \quad (1)$$

$$MO + H_2O(+X) \rightarrow M(OH)_2(+X),$$

$$M(OH)_2 + H \rightleftarrows MOH + H_2O,$$

where X is a third-body major gas component which may, or may not, be involved (Cotton and Jenkins, 1971; Bulewicz *et al.*, 1971). The overall result of these reactions is the catalyzed recombination of two H-atoms to yield molecular H_2.

The first step, i.e., reaction (1), in this mechanism appears to be rate limiting and rates in the range of 10^{12}–10^{13} $cm^3\ mol^{-1}\ sec^{-1}$ are indicated for various M, with Ba being the most effective case. These rates are found to be relatively insensitive to temperature. The overall rate of H-atom recombination is second order and the rate of decay of H-atoms is given by the expression

$$-\frac{d[H]}{dt} = 2k_0[H]^2 + \frac{2k_1[M]_0[H]^2}{[H] + K_2[H_2O]},$$

where $[M]_0$ is the initial concentration of metal additive and is approximately equal to the sum of the MOH and $M(OH)_2$ concentrations, k_0 and k_1 are the rate constants in the absence, or presence, of metal-containing species, respectively, and K_2 is the equilibrium constant for the reaction

$$MOH + H_2O \rightleftarrows M(OH)_2 + H. \quad (2)$$

The latter term of this rate expression represents the catalytic effect. As reaction (2) is endothermic, K_2 increases with temperature, and this results in the catalytic effect becoming less effective with increasing temperature and practically nonexistent at 2000 K. The observed catalytic effects were for flames having temperatures of 1570 and 1800 K.

The catalytic effect for Sn appears to be similar to that for the alkaline earth metals. For additions of 10^{-5}–10^{-6} mole fraction of Sn, SnO is the major flame species and SnOH, for which band spectra were tentatively assigned, is also present but to a lesser extent (Bulewicz and Padley, 1971a). It is thought that SnOH*, in an excited electronic state, is the key catalytic intermediate according to the reaction scheme:

$$\mathrm{SnO}(^1\Sigma) + \mathrm{H}(^2\mathrm{S}) + (\mathrm{X}) = \mathrm{SnOH}^*(\mathrm{A}) + (\mathrm{X}),$$

$$\mathrm{SnOH}^*(\mathrm{A}) \text{ crosses to } \mathrm{SnOH}^*(\mathrm{B}),$$

and

$$\mathrm{SnOH}^*(\mathrm{B}) + \mathrm{H}(^2\mathrm{S}) \rightarrow \mathrm{SnO}(^1\Sigma) + \mathrm{H}_2(^1\Sigma).$$

The notations (A) and (B) represent different excited electronic states of SnOH.

The strong catalytic effect of Cr is mechanistically more complex and evidence of both heterogeneous and homogeneous processes has been found, with the latter being predominant (Bulewicz and Padley, 1971b). Using the "family-of-flames" procedure for species identification, evidence was obtained for the presence of Cr, CrO, CrO_2, and $HCrO_3$ as the main Cr-containing species. It is thought that the homogeneous catalytic process may be similar to that indicated for Sn.

In conclusion, it is of note that these catalytic effects provide an explanation for some of the interference effects found in analytical flame photometry (Bulewicz and Padley, 1973).

III. Ion Chemistry in Flames

A. Introduction

Practical problems associated with the electrical character of combustion gases, such as the attenuation of radar signals by electrons in rocket exhausts and reentry wakes, and gas conductivity limitations in magnetohydrodynamic devices, have resulted in an increased interest in the ion chemistry of flames over the past decade. This is evidenced by the significant number of fundamental reviews that have appeared during this period; i.e., Mukherjee *et al.* (1962), Calcote (1963), Van Tiggelen (1963), Sugden (1963, 1965), Miller (1968), Gaydon and Wolfhard (1970), Feugier (1970), Calcote and Miller (1971), Bocek and Janak (1971), and Miller (1973); in addition to the comprehensive text of Lawton and Weinberg (1969).

A significant fraction of the subject of flame ion chemistry involves both positively and negatively charged high temperature inorganic additive flame species; this results from their low ionization potentials and high electron affinities, respectively. Even given the meager amount of thermodynamic data available for these species, it is apparent that they are likely to be important modifiers of the charged character of practical combustion systems, such as those mentioned above.

The presence of charged species in clean flames has long been known from the ability of flames to conduct an electric charge. In recent years, with the advent of molecular-level analytical techniques, the identity and concentrations of these charged species have been established. Free electron—and total positive ion—concentrations of the order of 10^{11} species cm^{-3}, i.e., $\sim 10^{-7}$ atm partial pressure, are typical. This concentration level is many orders of magnitude greater than that allowed by equilibrated collisional ionization processes and is the result of chemionization reactions. Flames fueled by H_2 or CO have very low ion concentrations, i.e., less than 10^6 ions cm^{-3}, whereas hydrocarbon flames may have positive ion concentrations of up to 10^{13} ions cm^{-3}. In fact, the ionization resulting from hydrocarbon additions to H_2 flames forms the basis of flame ionization detectors used for gas chromatography (Bocek and Janak, 1971).

The concentration regime for ions is many orders of magnitude below that of the flame propagating radical intermediates, i.e., 10^{15}–10^{16} radicals cm^{-3}. Therefore, even allowing for the relatively fast nature of ion–molecule reactions, the presence of ions should have a negligible effect on the *overall* flame propagation chemistry. The physical character of flames can, however, be affected by perturbations of the charged components, such as may be produced by an external electric field. Such fields can influence the combustion energy, mass transfer, and extinction limits of flames, by virtue of induced ionic wind effects, e.g., see Miller (1973), Parker and Heinsohn (1968), Lawton and Weinberg (1969), and cited references. An influence of electric fields on the smoke level and its transport in flames has also recently been demonstrated. However, such effects are found to be of secondary importance as compared with the primary radical controlled combustion processes. Indeed, no significant effect of direct electric fields on the normal burning velocity was found in recent studies of hydrocarbon–air flames (Bowser and Weinberg, 1972).

In passing, it is of interest to note that high temperature vapors themselves can also exhibit an appreciable electrical conductivity, even in the absence of a flame environment. This apparently can result from the thermodynamic stability of complex ions, such as $AlCl_4^-$ and $NaAlCl_3^+$ in the vapor phase (Dewing, 1971).

B. General Techniques†

Total ion effects in flames are conveniently monitored *in situ* by oppositely biased electrical probes. Electron concentrations have been measured using Langmuir probes and microwave or rf conductivity methods (see Section F.1 to follow). The spatial resolution of such methods is relatively low, being in the range of several millimeters to several centimeters. For a discussion of the Langmuir probe technique, and its application to ions in flames, see, for example, King (1963), Calcote (1962), and Gaydon and Wolfhard (1970). Electrostatic probe techniques have been developed for monitoring total positive ion concentrations over the range of 10^5–10^{14} ions cm^{-3}.

A recently developed rotating probe technique for determining total positive ion concentrations of additives in flames has been described by Kelly and Padley (1969a, b, 1971a). The semiempirical expression relating ion concentration to probe current is given by

$$N_i = A'I^{1.3}V_p^{-0.65},$$

where N_i is the total positive ion concentration; A' a constant for a given probe, ion, flame velocity, and temperature and I the current that flows when a potential difference V_p—typically several hundred volts—is applied between two electrodes. The electrodes consist of a platinum spherical probe, of about 1-mm radius, and a mesh electrode assembly which rotate through the flame plasma, thereby reducing space-charge buildup. The probe has been demonstrated over a charged species concentration range of 5×10^7–5×10^{12} ions cm^{-3}. This may be compared with the microwave cavity resonance technique, where the sensitivity range is 2×10^9–2×10^{11} electrons cm^{-3}. The probe may be calibrated with known ionizable additives, such as Cs or Rb, but taking into consideration the losses due to the formation of CsOH and RbOH.

The optimum technique for determining individual ion concentrations is the mass spectrometric molecular beam probe-sampling method. The principal investigations of ions in clean (i.e., unseeded) flames by mass spectrometry have been carried out by the groups of Calcote (1963), Van Tiggelen (1963), and Sugden (1963) and both 1-atm and low pressure flame types have been studied. Similar studies of seeded flames have been particularly fruitful—yielding new information on the ionization potentials,

†A review of flame plasma diagnostic techniques, with an extensive bibliography, has appeared very recently (Jensen and Travers, 1974).

electron affinities, and kinetics of charged high temperature species, e.g., see the review of Miller (1973).

The mass spectrometric ion-sampling system of Hayhurst and Sugden (1966) has an estimated efficiency of 2% ion detection; that is, two out of a hundred ions entering the sampling orifice are detected. In practice, ion concentrations as low as 10^5 ions cm^{-3} can be detected. This may be compared with the molecular beam sampling of neutrals with mass spectrometric analysis, where the combined effects of beam losses and a low electron impact ionization probability yield an efficiency of about 10^{-5}%; that is, concentrations of only about 10^{10} neutrals cm^{-3} may be detected.

C. Probe Effects on Ion Sampling

For most of the early mass spectrometric work, the use of pinhole leaks in diaphragms, rather than conical probes, led to some spurious results due to perturbing boundary layer and free jet expansion chemical phenomena. Systems developed more recently, such as that of Hayhurst *et al.* (1971), attempt to reduce the importance of sampling anomalies, which are particularly noticeable for 1-atm sampling.

The earlier discussion of probe effects on neutral species sampling for mass spectrometric analysis, i.e., see this chapter, Section II.F, is also pertinent to the special case of ion sampling. This general problem of probe perturbations is, however, more severe for ion chemistry on account of the relatively high reaction rates for ion–molecule reactions, as will become evident in later discussion.

The "observation" of the series of species ($H_3O^+ \cdot nH_2O$), where $n = 1, 2, \ldots$, in flames is now known to be a probe-related sampling artifact. H_3O^+ is an established flame ion but the adduct species $H_3O^+ \cdot nH_2O$ are known to be of low stability and should not exist under flame conditions. Similarly, NH_4^+, observed in the sampling of hydrocarbon flames, is not a flame ion but most likely results from the catalytic production of NH_3 at the probe surface, followed by the reaction

$$H_3O^+ + NH_3 \rightarrow H_2O + NH_4^+,$$

e.g., see Bascombe *et al.* (1962).

Boundary layer effects at the tip of the sampling probes, which are cool relative to the flame gases, are believed to be responsible in part for the appearance of secondary species such as $H_3O^+ \cdot nH_2O$. These effects are found to be very sensitive to the orifice diameter which controls the flow of gas and hence the boundary layer thickness, in addition to the time required to attain a frozen expansion condition. For orifice diameters of less than 0.002 in., and flames at 1 atm, very significant boundary layer effects can occur, e.g., see Bascombe *et al.* (1962) and Hayhurst and Sugden (1966). However, these effects are relatively unimportant for the larger orifice diameter probes, i.e., ~0.002 in., and greater. A detailed theoretical analysis and discussion of these effects has been given by Hayhurst and Telford (1971).

It is encouraging to note that the presence of probe-related aggregation effects did not apparently perturb the relative intensities of the flame-additive species Ca^+:$CaOH^+$ and Sr^+:$SrOH^+$ formed in seeded H_2–O_2–N_2 flames (Hayhurst and Kittelson, 1972a).

D. Positive Ions in Unseeded Flames

1. Scope of Application

Flames derived from H_2–O_2–N_2 gas mixtures are, for practical purposes, free of charged species, although the ion H_3O^+ is usually observed owing to the presence of trace hydrocarbon impurities. Subsequent discussion will indicate the relationship of H_3O^+ to hydrocarbon flame chemistry. The hotter flames, such as C_2N_2–O_2, which have been studied relatively little, should contain appreciable concentrations of NO^+, even at equilibrium. Hydrocarbon flames contain the most significant concentration of positive ions, electrons, and, to a lesser extent, negative ions. As with the neutral species chemistry, the ion chemistry of these flames is prolific and complex.

Total positive ion concentrations are known from Langmuir probe experiments. Using this technique, Calcote (1962) has found that low pressure (33 Torr) flat flames of premixed air and acetylene, methane, propane, or ethylene, contain about 10^{-7} mole fraction of positive ions, and the maximum concentration occurs for the near-stoichiometric fuel–air compositions. Profiles of ion concentration versus distance along the principal gas flow axis show a steep increase at the reaction zone, followed by a gradual decrease in the burnt-gas region. This is typical of biomlecular formation and termolecular—or slow—recombination processes. It is also found that the rate of total positive ion formation by chemionization is about 2×10^{15} ions cm^{-3} sec^{-1} at 1 atm. This rate is equivalent to an interaction of 10^{-2} mole fraction of one component with about 10^{-11} mole fraction concentration of another and with zero activation energy.

The analysis of individual hydrocarbon ion species, by mass spectrometry, is simplified by the use of H_2–O_2–N_2 flames containing small quantities, i.e., ~1%, of a hydrocarbon, such as acetylene, and typical positive ion profiles for such a flame are shown in Fig. 5.14. The 37- and 55-amu species, i.e., $H_3O^+ \cdot H_2O$ and $H_3O^+ \cdot 2H_2O$, respectively, are believed to form in the

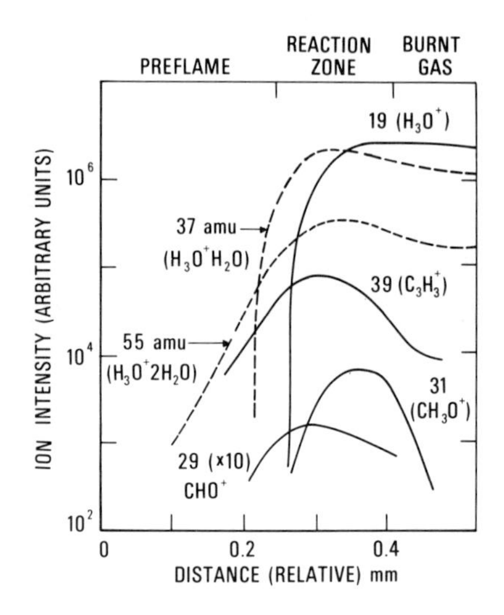

Fig. 5.14 Profiles of positive ion intensities versus distance along flame for the 1-atm flame $H_2/O_2/N_2 = 1/0.3/1$, $T = 2300$ K, and containing 1% C_2H_2; a sampling hole diameter of 0.002 in. was used for the mass spectrometric analysis (Bascombe *et al.*, 1962).

flame-probe boundary layer and are not representative of the flame chemistry. The predominant negatively charged species in this flame is the free electron. Similar studies by Calcote and Reuter (1963), but using low pressure flames and sampling conditions which would appear to be relatively free of spurious free jet expansion and boundary layer effects, also revealed the presence of the ubiquitous ion $H_3O^+ \cdot H_2O$.

As is indicated in Fig. 5.14, the natural ion chemistry of hydrocarbon-containing flames is dominated by the presence of H_3O^+, which is the major ion in the reaction zone and burnt-gas regions. Rocket flown mass spectrometers in the *D* region have also shown H_3O^+ to be important in the ionosphere, e.g., see Burke and Miller (1970).

2. H_3O^+ Formation and Decay

The key species in hydrocarbon ion chemistry is believed to be the relatively low concentration CHO^+ intermediate, e.g., see Fig. 5.14 and also Calcote and Miller (1971) and Peeters *et al.* (1969). This ion is generated by the chemionization process,

$$CH + O \rightarrow CHO^+ + e^-,$$

which is almost thermoneutral. H_3O^+ is then generated by the very fast reaction

$$CHO^+ + H_2O \rightarrow H_3O^+ + CO.$$

The rapidity of this CHO^+ depletion process, i.e., $\sim 10^{-11}$ cm^3 sec^{-1}, explains the low steady-state concentration level of the CHO^+ species.

The slow decay of H_3O^+ in the burnt-gas region is believed to result from the recombination reaction

$$H_3O^+ + e^- \rightarrow H_2O + H.$$

Another process leading to H_3O^+ depletion is considered to be

$$H_3O^+ + CH_2O \rightleftarrows CH_3O^+ + H_2O.$$

The formation of $C_3H_3^+$, as a significant species, could then result from the reaction

$$CH_3O^+ + C_2H_2 \rightleftarrows C_3H_3^+ + H_2O.$$

A discussion of similar secondary reactions involving other $C_xH_yO_z^+$ species may be found in Calcote and Miller (1971).

Thus CH and O are the key radical precursor species to the formation of the prolific variety of chemi-ions in hydrocarbon flames. Also, as the CH radical exists primarily at the reaction zone, then the production of CHO^+, H_3O^+, and the other secondary ion species, effectively ceases downstream of the reaction zone.

E. Positive Ion Chemistry of Flame Additives

1. Scope of Application

As was shown to be the case for neutral additives, flames provide a useful medium for the kinetic and thermodynamic study of ionizable additives. Chemionization processes involving additives may be utilized to determine kinetic data for molecular ions. Also, the equilibrium character of the burnt-gas region of atmospheric pressure flames allows thermodynamic studies involving molecular ions to be made, from which ionization potential and electron affinity data may be derived.

For 1-atm flames at 2500 K, and containing 1% of an ionizable component with an ionization potential of 8–10 eV, which is a typical range for many high temperature species, the equilibrium concentration of positive ions is in the range of 1.2×10^{10}–1.2×10^8 ions cm^{-3}, i.e., about 10^{-8} mole fraction and less. This concentration range is well within the detection capabilities of ion-probe and mass spectrometric techniques, and hence the potential utility of flames for the study of foreign ions is evident.

A convenient set of tables, indicating the calculated equilibrium concentrations of metal ions in flames, has been given by Woodward (1971).

2. *Positive Ion Formation Processes*

Typical profiles of total ion concentration versus time above the reaction zone in H_2–O_2–N_2 flames, containing metal additives, are given in Fig. 5.15. The maximum concentration of positive ions, i.e., $\sim 10^9$ ions cm^{-3}, is about a factor of 10^4 lower than the total concentration of the additive species.

For Na and Sr (not represented in the Fig. 5.15) the ion-forming mechanisms are considered to be

$$Na + X \rightarrow Na^+ + e^- + X$$

and

$$SrO + H + X \rightarrow SrOH^+ + e^- + X,$$

respectively (Kelly and Padley, 1969b). More recent mass spectrometric studies have verified the formation of molecular ions such as $SrOH^+$ and $CaOH^+$, and the rates have been determined for the reactions

$$M + OH \rightleftarrows MOH^+ + e^- \qquad \text{and} \qquad MO + H \rightleftarrows MOH^+ + e^-,$$

where M is Sr or Ca (Hayhurst and Kittelson, 1972b).

For the metals indicated in Fig. 5.15 a different, but not yet established, mechanism is suggested by the form of the observed profiles. It is believed

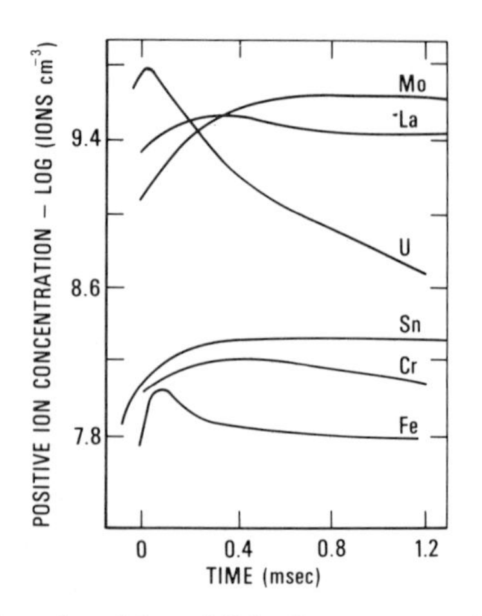

Fig. 5.15 Profiles of total positive additive-ion concentration versus time from the reaction zone, for the flame $H_2/O_2/N_2 = 4/1/3$, $T = 2170$ K; neutral metal additive concentrations of $\sim 2.6 \times 10^{13}$ cm^{-3} were used (Kelly and Padley, 1969b).

that, in these cases, the presence of condensed metal oxide particles has an influence on the positive-ion characteristics of these flames. A mechanism favored by Kelly and Padley (1969b) involves the vaporization of solid oxide followed by production of MOH^+ or MO^+ species.

A more recent analysis of some of these systems has been made using the "family-of-flames" approach, and strong evidence is given for the following gas-phase mechanisms (Kelly and Padley, 1970, 1971a). For Ca, Sr, and Ba the following mechanism is most consistent with the total ion data:

$$M(OH)_2 + H \rightleftarrows MOH + H_2O,$$

$$MOH + H \rightleftarrows MO + H_2,$$

and

$$MO + H \rightleftarrows MOH^+ + e^-.$$

For U as an additive, the probable ion production mechanism is

$$H_2UO_4 + 3H \rightleftarrows HUO_3^+ + e^- + H_2 + H_2O.$$

Using the second law thermodynamic treatment of data, reaction enthalpies and thence ionization potentials were derived from these and similar studies, as shown in Table 5.19.

Similar studies on 1-atm H_2–O_2–N_2 flames, at temperatures in the range of 2200–2800 K, and containing either the alkali metals or Ga, In, and Tl, indicate the ionization process to be

$$M + X \rightarrow M^+ + e^- + X,$$

with an activation energy virtually identical to the ionization potential of the metal M (Kelly and Padley, 1969a). The process has an unusually high cross section, i.e., ~10,000 Å^2, which is about a factor of 10^3 greater than simple kinetic theory would predict. Rate constants for ionization and recombination are given in Table 5.20. Note the monotonic inverse dependence of ionization rate constant on ionization potential. From more recent rate constant measurements for the alkali metals, Ashton and Hayhurst (1973) have shown that this relationship between ionization rate constant and potential is given by

$$k_i = (9.9 \pm 2.7) \times 10^{-9} T^{1/2} \exp(-IP/RT).$$

This process of ionization is slow, with half-lives of about 0.1 to 1 msec, and hence very few ions are formed in the reaction zone. Most of the ions then are formed in the burnt-gas region.

The high cross section for the reaction

$$Na + X \rightarrow Na^+ + e^- + X,$$

TABLE 5.19

Ionization Potentials from Flame Studies[a]

Species	IP (eV)
CaOH	5.9 ± 0.1[b]
SrOH	5.55 ± 0.1[b]
BaOH	5.25 ± 0.1[b]
HUO_3	<3.9
SnOH	7.0 ± 0.6[c]

[a] From Kelly and Padley (1970, 1971a).

[b] Similar results were obtained, using a slightly less accurate mass spectrometric technique, by Hayhurst and Kittelson (1972a) and also by Jensen (1968).

[c] Calculated from $\Delta H_0 = 100 \pm 6$ kcal mol^{-1} for the reaction: SnO + H = $SnOH^+$ + e, as given by Jensen (1969a), together with an estimate, by the present author, of D_0 Sn – OH = 85 ± 10 kcal mol^{-1}, which is supported by an upper limit value from Bulewicz and Padley (1971a), and using D_0 SnO = 126 kcal mol^{-1} (Brewer and Rosenblatt, 1969). Note that the IP of the electronically similar species SnF is also known to be 7.0 eV, i.e., see the review of Hastie and Margrave (1968a).

TABLE 5.20

Rate Constants for Metal-Atom Ionization (k_i) and Metal-Ion Electron Recombination (k_r) in Flames[a]

M	IP[b] (kcal mol^{-1})	k_i (sec^{-1})	$10^9 k_r$[c] (cm^3 molecule^{-1} sec^{-1})
Tl	141	0.29	4.7
Ga	138	0.36	8.6
In	133	0.87	1.8
Na	118.5	55	7.3
K	100	1600	5.5
Rb	96.5	2500	4.0
Cs	89.5	12,000	4.8

[a] Data from Kelly and Padley (1969a) for the flame H_2 (3.5) + O_2 (1) + N_2 (2).

[b] Literature IP data.

[c] k_r calculated from k_i and the equilibrium Saha constant; satisfactory agreement with values measured experimentally for Ga, In, and Tl was found.

was originally thought to be due to the reaction scheme:

$$Na + H_2O \rightarrow Na^+\cdot H_2O + e^- \qquad \text{(slow)}$$

and

$$Na^+\cdot H_2O + X \rightarrow Na^+ + H_2O + X \qquad \text{(fast)}$$

where X is a third-body major flame species (Sugden, 1965). However, more recent investigations (Kelly and Padley, 1969a) suggest that the ionization process probably results from an excited metal atom species. Also, from the observation of similar rates of ionization in hydrogen-free CO flames, one can exclude consideration of hydrates or similar intermediates from the mechanism of alkali metal ion formation (Hollander, *et al.*, 1963).

The formation of alkali metal positive ions is catalyzed by the presence of halogens and also by alkaline earth metals. The halogen atom, e.g., Br, catalyzes the process

$$M + 2H \rightleftarrows M^+ + e^- + H_2,$$

via the reactions

$$M + Br \rightleftarrows M^+ + Br^-,$$

$$Br^- + H \rightleftarrows HBr + e^-,$$

and

$$HBr + H \rightleftarrows H_2 + Br$$

(Hayhurst and Sugden, 1967). The catalytic effect of alkaline earth metals, e.g., Sr, is considered to result from the reaction

$$SrOH^+ + Na \rightarrow SrOH + Na^+$$

(Schofield and Sugden, 1965). Synergistic effects, in increasing the rate of approach of ionization to the equilibrium level, have also been observed for combinations of additives, such as Na and Cr or Na and U (Kelly and Padley, 1971a).

While some of these ionization processes apparently involve flame radicals, i.e., H-atoms, the number of species involved is relatively very low and no significant effect on the overall radical chemistry is likely. In fact, flame inhibition tests, where both an alkali metal and a halogen are present, indicate an antagonistic effect between these additives, rather than the synergistic effect suggested by the above observations (see Section V for more detail).

3. H_3O^+–Metal Atom Charge-Exchange in Flames

Metal atoms can catalyze the reduction of an excess H_3O^+ concentration to the equilibrium level by the charge-exchange process:

$$M + H_3O^+ \rightarrow M^+ + H_2O + H,$$

where M is an alkali metal or Tl. For Cr, Pb, Mn, Cu, and Fe, the exchange occurs by the reactions

$$MO + H_3O^+ \rightarrow MOH^+ + H_2O,$$

$$M + H_3O^+ \rightarrow MOH^+ + H_2,$$

and

$$MOH^+ + H \rightleftarrows M^+ + H_2O.$$

Direct mass spectrometric identification was made for the various positive ions. The charge-exchange rate constants are in the range 0.5–50 $\times 10^9$ ions cm^{-3} sec^{-1} (Hayhurst and Telford, 1970). In these studies the concentration of additive used in the flame is not specifically mentioned, but presumably it was below the 1 ppm level, above which condensed species formation and sampling orifice plugging problems can occur.

4. *Positive Ion–Electron Recombination Processes*

The excess positive ion concentrations produced near the reaction zone by chemionization processes, such as

$$M + H_3O^+ \rightarrow M^+ + H_2O + H,$$

allow one to utilize the decay of these charged species in the burnt-gas region for the determination of ion–electron recombination kinetics. Jensen and Padley (1967) find that the principal source of decay is by the ion–electron recombination process

$$M^+ + e^- + X \rightleftarrows M^* + X,$$

where M* is an excited form of M. For such a second-order decay process, the reciprocal electron concentration varies linearly with distance—i.e., time—along the flame, and the slope of the curve yields a value for the recombination rate constant k_r. Where M is Mn, Pb, Cs, Rb, K, Na, and Li, k_r has values in the range 7.5–9.0 $\times 10^{-9}$ cm^3 $molecule^{-1}$ sec^{-1}. In this study metal concentrations of about 10^{13} atoms cm^{-3}, and atmospheric H_2–O_2–N_2 flames at temperatures of about 2200–2500 K, were used. These rate of recombination results are similar to those given in Table 5.20. They are slightly temperature dependent due to a small negative activation energy, which is typically several kcal mol^{-1} in magnitude.

F. Negatively Charged Species in Flames

1. *Electrons*

Practically all of the negative charge component of clean, i.e., unseeded, flames is in the form of free electrons. However, in seeded flames, these electrons may be effectively scavenged by additives which lead to the formation of species having high electron affinities, i.e., greater than about 2 eV.

The light mass of the electron allows it to absorb energy effectively from electromagnetic radiation. The excess translational energy imparted to the electron is then dissipated by collisions with neutrals in the combustion gas. Thus the absorption of high frequency (e.g., microwave) radiation, which in practice can lead to a troublesome attenuation of radar signals, may be used as a measure of the electron concentration. For example, in the microwave cavity resonance and related techniques, the electron concentration $[e^-]$ may be determined from the electrical conductivity σ of the cavity, i.e.,

$$\sigma = ([e^-]q^2/m)/\{W/(\omega^2 + W^2)\},$$

where

q is the electronic charge,
m the electron mass,
W the collision frequency of electrons with gas molecules, and
$\omega = 2\pi f$, where f is the frequency of microwave radiation,

e.g., see Padley and Sugden (1962). Alternatively, the technique may be simplified by using an internal calibrant, such as Cs^+, and the electron concentration is then given by

$$[e^-] = (1/C)[(I_0/I)^{0.5} - 1],$$

where C is the "cavity constant"—obtained usually by seeding with known quantities of cesium, I_0 is the rectified current from the output loop of the cavity in the absence of ionizing additive, and I is the current in the presence of additive, e.g., see Jensen and Padley (1967).

A more recent technique for determining electron concentrations in flames, monitors the cyclotron resonance line intensity, e.g., see Bulewicz and Padley (1969). The transmission of microwave power—48 kMHz at 2-kHz modulation—through the flame is determined in the presence of a mutually perpendicular magnetic field, which is swept over the cyclotron resonance line, i.e., at about 17,000 gauss, then

$$[e^-] \propto \beta,$$

where β is the decibel attenuation of the center of the cyclotron resonance line.

As a practical example, microwave attenuation measurements have been made on rocket exhausts and profiles of electron concentration determined (Balwanz, 1965). Also, the ionization potential determinations of CaOH, SrOH, and BaOH, already mentioned, were derived from electron concentration measurements for the process

$$M + OH \rightarrow MOH^+ + e^-,$$

where M = Ca, Sr, and Ba (Jensen, 1968). In this study, [OH] was determined via the measurement of [H] by the CuH technique.

2. *Negative Ions*

Negative ions are relatively unimportant in unseeded flames, their concentration being at least two orders of magnitude less than that of the electrons and also the principal positive ion H_3O^+, e.g., see Calcote *et al.* (1965). In pure H_2–O_2–N_2 flames, at 1 atm, no negative ions have been observed at the reaction zone, but downstream in the burnt-gas region negative ion concentrations of about 10^{10} ions cm^{-3} have been observed (Knewstubb and Sugden, 1962). These ions include OH^-, O^-, O_2^-, and a number of unidentified ions which may have formed during the sampling process.

More recently, Knewstubb (1965) has reported on negative ion results for a 1-atm fuel-rich acetylene flame, using relatively high sampling rates, i.e., $\gtrsim 0.002$ in. sampling orifice diameter. It is thought that for small sampling rates the boundary layer problem causes an appreciable loss of negative ions. Negative ion concentrations were found to be comparable with those for positive ions in the preflame and reaction zone regions but were negligible in the burnt-gas region.

Feugier and Van Tiggelen (1965) observed negative ions in stoichiometric neo-C_5H_{12}–O_2–51% N_2 flames. The species C_2H^-, O_2^-, $C_2H_2^-$, C_2^-, and CO_2H^- were tentatively identified. These ions disappeared in the presence of about 1% of a chlorine-containing additive which effectively scavenges free electrons due to the higher stability of Cl^-.

3. *Thermodynamic Studies of Additive Negative Ions in Flames*

The addition of readily ionizable metals such as potassium to 1-atm fuel-rich H_2–O_2–N_2 flames, at temperatures of ~1800–2500 K, provides a source of free electrons in the burnt-gas region, i.e., up to 1.2×10^{12} electrons cm^{-3} for 10^{-6} atm of potassium. When additional elements, such as B, W, or Mo, are added to the flame, this free electron concentration is substantially depleted and the formation of negatively charged metal-containing ions is inferred (Jensen, 1969b). Indirect evidence for the formation of negative ions has also been obtained from the observation of electron losses in low pressure acetylene–oxygen flames when the additives Cl_2, Br_2, NH_3, H_2S, and C_2N_2 are present (Bulewicz and Padley, 1969).

The formation of the ions Cl^-, Br^-, HS^-, and CN^- was inferred from these results.

For the case where the additive M is W or Mo, the negative ions have been identified by mass spectrometry to be HMO_4^- and MO_3^- (Jensen and Miller, 1969b, 1970). These ions are believed to form by the reactions

$$e^- + H_2MO_4 \rightleftarrows HMO_4^- + H$$

and

$$HMO_4^- + H \rightleftarrows MO_3^- + H_2O.$$

A reduction in the concentration of free K-atoms, observed in these studies, is attributed to the reaction

$$K + H_2MO_4 \rightleftarrows KHMO_4 + H.$$

The evidence for the participation of H_2MO_4 and $KHMO_4$ in these reaction schemes is as follows:

- H_2MO_4 is calculated to be the most thermodynamically favored species under the experimental flame conditions;
- equilibrium constants from these, or dependent, reactions expressed in logarithmic form vary linearly with the reciprocal of flame temperature and plausible second and third law heats of reaction result from this data;
- mass action-type experiments show that the loss of K is associated with a species containing only one K atom.

As an example of the mass action type of species verification, consider the loss of K to be the result of the formation of a compound KY. Then, the total K content $[K]_c$ in the flame is given by

$$[K]_c = [K] + [K^+] + [KOH] + [KY].$$

Let

$$\phi = [KOH]/[K] = K_R[H_2O]/[H],$$

where K_R is the known equilibrium constant for the reaction

$$K + H_2O = KOH + H.$$

As $[H_2O]$ and $[H]$ are known quantities, then ϕ can also be determined. The concentration ratio of KY to free K can then be given as

$$[KY]/[K] = \{[K]_c - (1 + \phi)[K] - [K^+]\}/[K],$$

where $[K]$ is measured from atomic absorption spectroscopy and $[K^+]$ is determined from electrostatic probe measurements. Hence, the right-hand side (RHS) of this expression is known and may be plotted as a function of total K added, i.e., $[K]_c$. A log–log plot of RHS vs $[K]_c$ should be linear if the reaction between K and KY is at equilibrium. Also,

TABLE 5.21

Reactions Related to Negative Ion Formation in Flames

Reaction	Reaction enthalpy, ΔH_0 (kcal mol^{-1})	References
$K + H_2WO_4 = KHWO_4 + H$	-10 ± 10	Jensen and Miller (1970)
$e^- + H_2WO_4 = HWO_4^- + H$	10 ± 10	
$HWO_4^- + H = WO_3^- + H_2O$	-5 ± 10	
$K + H_2MoO_4 = KHMoO_4 + H$	-14 ± 14	Jensen and Miller (1969b)
$e^- + H_2MoO_4 = HMoO_4^- + H$	17 ± 14	
$HMoO_4^- + H = MoO_3^- + H_2O$	-7 ± 12	
$e^- + HCrO_3 = CrO_3^- + H$	30 ± 10	Miller (1972)
$CrO_3^- + H_2 = HCrO_3^- + H$	19 ± 10	
$Li + HBO_2 = LiBO_2 + H$	-5 ± 5	Jensen (1969b)
$Na + HBO_2 = NaBO_2 + H$	5 ± 5	
$K + HBO_2 = KBO_2 + H$	2 ± 5	
$HBO_2 + e^- = H + BO_2^-$	22 ± 5	

the slope of the line should indicate the ratio of number of moles of KY to K contained in the equilibrium reaction. In practice the slope was found to be 1.0, thereby indicating that KY contains only one K atom (Jensen and Miller, 1970).

TABLE 5.22

Electron Affinities from Flame Studies[a]

Species	kcal mol^{-1}	eV	References
WO_3^-	84 ± 10	3.7 ± 0.5	Jensen and Miller (1970)
HWO_4^-	101 ± 10	4.4 ± 0.5	
MoO_3^-	60 ± 12	2.6 ± 0.5	Jensen and Miller (1969b)
$HMoO_4^-$	93 ± 14	4.0 ± 0.6	
CrO_3^-	55 ± 12	2.4 ± 0.5	Miller (1972)
$HCrO_3^-$	93 ± 12	4.0 ± 0.5	
BO^-	$\gtrsim 58$	$\gtrsim 2.5$	Jensen and Miller (1969b)
BO_2^-	97 ± 5	4.2 ± 0.2	
	94 ± 5	4.1 ± 0.2	Jensen (1969b)
Cl^-	83 ± 0.5[b]	3.6 ± 0.02	Bulewicz and Padley (1969)
Br^-	78.5 ± 0.5[b]	3.4 ± 0.02	
CN^-	> 63.5	> 2.7	

[a] See also the summary of Miller (1973).

[b] The values for Cl and Br are in good agreement with the more accurate photodetachment results of Berry and Reimann (1963).

Table 5.21 indicates the negative ion reactions for metal additives and their enthalpy changes. Electron affinity data, derived from such reactions, are given in Table 5.22. The large errors indicated are mainly the result of uncertainties in free energy function estimates.

IV. Analytical Flame Spectroscopy

A. Introduction

The utility of 1-atm flames as a medium for decomposing inorganic or metal organic compounds to their elements for analytical purposes is well known, e.g., see Grove, (1972), Mavrodineanu (1970), L'vov (1970), Dean and Rains (1969), Pungor (1967), and Mavrodineanu and Boiteux (1965). However, in applying the techniques of flame spectroscopy to chemical analysis, the practitioner uses skills that have often been described as more art than science. The development of high temperature chemistry has, in recent years, given analytical flame spectroscopy a more scientific basis, but before elaborating on this basis it is necessary to discuss the underlying rationale for the analytical techniques used. Basically, the technique relies on the tendency for a hydrocarbon- or hydrogen-fueled flame to degrade materials to their atomic state, particularly if the sample is initially in an aerosol or gaseous form. Most of the analytical limitations that occur are the result of departures from this ideal degradative characteristic of flames.

It is desirable to achieve a free-atom state because the line spectra of atoms, as opposed to the band or continuous spectra of molecular species, provides more assurance of the chemical identity and also better sensitivity for quantitative determinations. However, in a few instances use is made of molecular spectra for analytical purposes. For example, HPO is used in emission for P analysis, and InCl is used similarly for Cl analysis. In the latter example, indium metal is added to flame as a halogen scavenger.

For the majority of cases, the observed atomic emission or absorption intensities are independent of the initial chemical or physical nature of the metalloid introduced to the flame. In the case of Na and Li, intensity variations of only 1%, or less, have been found where the metalloids were introduced in the form of nitrate, acid sulfate, chloride, bromide, iodide, phosphate, carbonate, and tartrate solutions.

From the maximum input allowed by nebulization of solutions containing metals, the upper limit concentrations of these elements in the flames is about 10^{-3} atm. Therefore, on thermodynamic grounds, a direct influence of the presence of metal vapors on the burnt-gas flame temperature and composition is not to be expected. Also, the heat required for vaporization and dissociation of the additive removes heat from the flame to the extent of only about 0.5% of the total heat content and is therefore negligible. Likewise, the radiation emitted by the metal species does not produce any significant energy loss from the system. In practice, the evaporation of the solvent, which is usually water, provides the greatest heat sink, and temperature losses of from 10 to 600 °C are possible.

In multielement solutions the interaction of the elements with each other *in the flame* is unlikely, being a factor of 10^6 less probable than with the major flame gases. However, an interaction with the flame gases to form metal oxide, hydroxide, or hydride species is possible (see Table 5.13). In order to minimize errors that may arise from these and other departures from the ideal behavior, it is desirable to calibrate the analytical technique by using a standard sample which is as close as possible in chemical composition to the actual unknown sample. Evidence of nonideal behavior is usually manifested by anomalous "curves of growth" which are also known as Van der Held dilution curves. That is, the intensity of atomic emission, or absorption, does not show a one-to-one correspondence with the elemental concentration in the analytical solution. A detailed analysis of the various chemical factors that lead to this nonideal condition has been given by Bulewicz and Padley (1973).

B. Comparison of Absorption with Emission

There are at present three distinct spectroscopic techniques available for the quantitative analysis of element in flames, namely emission, absorption, and fluorescence spectroscopy. The emission and fluorescence techniques are similar in that one observes the radiation produced by the decay of an excited electronic state, which is usually the resonance level, to a lower energy level—usually the ground state. In the former case the excited-state condition is achieved by thermal, or chemical, means, whereas in the latter case a strong light source is used for species excitation. The absorption technique also utilizes a light source and in this case the frequency—and amount—of the absorbed incident radiation is used for the analysis. As the relatively new, and less general, technique of atomic fluorescence flame spectroscopy has common essential features with those for the absorption and emission methods, it will not be given further consideration. A detailed recent discussion may, however, be found in the article by Vickers and Winefordner (1972).

Both the absorption and emission techniques have a different dependence on the flame properties, and it is of interest to consider these factors from a high temperature chemistry point of view. The absorption analytical technique (e.g., see Willis, 1970) tends to be more specific than emission (e.g., see Alkemade, 1970), since the probability of a wavelength coincidence is about 50 times lower for absorption. As the most intense absorptions, such as resonance absorptions, arise from transitions out of the ground state, and as the thermal population of excited states is only about 10^{-4}–10^{-12} relative to the ground state, then, in principle, atomic absorption should be a more sensitive technique than emission. However, in general practice the analytical sensitivity is comparable for both absorption and emission, being in the part per billion to part per million range. With the absorption technique one is dealing with small signals that are the difference between the large incident and transmitted intensities. The limit of detection is then determined primarily by the noise of the light source and also any additional noise associated with concentration fluctuations of the absorbing flame species. Similarly, in emission one observes a small signal in the presence of a large and, to some extent, fluctuating flame background. As will become evident in the discussion to follow, an increased emission intensity can arise from chemical, as opposed to thermal, excitation processes. The absorption technique is particularly useful for those elements which are difficult to observe in emission. It suffers mainly from light-source availability, and stability, problems.

1. Absorption

The degree of absorption is given by the intensity of transmitted light, i.e.,

$$I = I_0 \, e^{-k_\nu l},$$

where I_0 is the intensity of the incident beam of light, k_ν is the absorption coefficient at the frequency ν, and l is the path length through the absorber, e.g., see L'vov (1970).

The nonchemical factors that determine the lower limit to the detection of atoms by atomic absorption may be represented by the expression

$$Nm = \frac{53.5 \, \Delta\nu_\mathrm{D} B(T) x (\Delta f)^{1/2}}{L \lambda_0^2 g_\mathrm{u} A_t \delta},$$

where

Nm is the number of atoms of analytical interest per cm^3 of flame gases at the limit of detection,

$\Delta\nu_D$ the Doppler broadening in sec^{-1},
$B(T)$ the partition function of the atom,
L the optical path length through the flame in cm,
λ_0 the wavelength of the transition in cm,
g_u the statistical weight of the upper state,
A_t the atomic transition probability in sec^{-1},
δ a line broadening factor other than that accounted for by the Doppler width,
x a fraction indicating the degree of fluctuation in the excitation source, in $sec^{1/2}$, and
Δf the frequency interval over which the amplifier readout system responds in sec^{-1}.

This relationship was suggested by Winefordner and Vickers (1964) and has recently been tested by Parsons and McElfresh (1972) on more than 50 elements. The comparison of calculated and experimental limits of detection is reasonable for air–acetylene flame data and for elements where the free-atom fraction, i.e., the ratio of free-atom concentration to the total atom concentration of a particular element, is expected to be near unity. Less satisfactory agreement is found for data obtained using a nitrous oxide–acetylene flame. This was explained by Parsons and McElfresh (1972) as being due to a departure from local equilibrium in this type of flame. This seems reasonable as the flame is a relatively faster burning one, and therefore the residence time of species in the flame is reduced.

2. *Emission*

Where measurements are made under equilibrium conditions, such as may exist in the burnt-gas region, the emission intensity is given by the expression

$$I = \frac{hc}{1.51}\frac{f}{\lambda^3}Nme^{-h\nu/kT}.$$

It is evident from this relationship that a high flame temperature is desirable. This contrasts with atomic absorption which is not particularly temperature sensitive, though the increased dissociation of molecular species to yield free atoms also sometimes warrants the use of high temperatures for absorption

With emission, the analytical sensitivity between elements can vary from 10^{-5} to 10^2 ppm and a precision of about 1% is obtainable (see Alkemade, 1970). This wide range of sensitivity is related to the specific nature of the energetics involved in chemiluminescence excitation processes.

Practically all of the reaction schemes suggested to account for the observed chemiluminescence are postulates and remain experimentally unproven. Two of the more likely reactions are thought to be

$$H + H + M \rightarrow H_2 + M^*,$$

and

$$C + MO \rightarrow CO + M^*.$$

A probable exception to these reactions is the source of excitation of HPO*, in the green, for which several chemiluminescent schemes have been suggested, e.g.,

$$PO + H_2 + OH \rightarrow HPO^* + H_2O,$$

which, for excitation in the green spectral region, would be almost thermoneutral—see the review of Gilbert (1970). It should be noted that the more practical HPO-emitting flames are relatively cool. For the case of PO*, the source of excitation is thought to be

$$PO + H + OH \rightarrow PO^* + H_2O$$

or

$$HPO^* + H \rightarrow PO^* + H_2,$$

and

$$HPO^* + OH \rightarrow PO^* + H_2O.$$

A dependence of overexcitation intensity on position in the flame is exemplified by the observed intensity profiles of Fig. 5.16. Padley and Sugden (1959a) have observed that in H_2–O_2–N_2 flames, the degree of metal-atom overexcitation increases rather monotonically with the energy

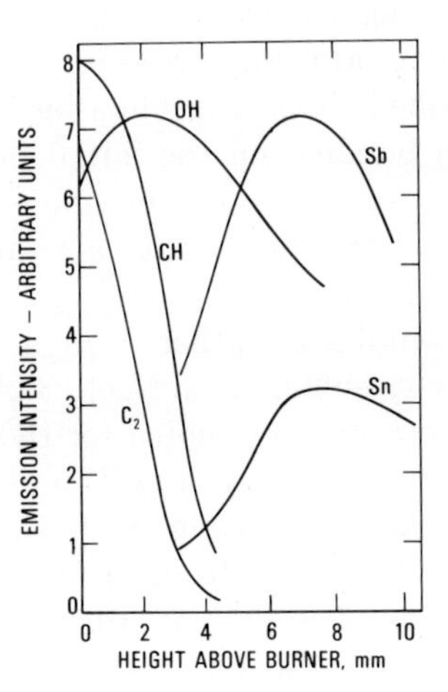

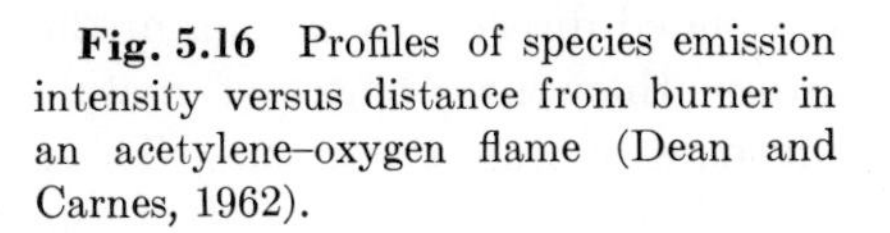

Fig. 5.16 Profiles of species emission intensity versus distance from burner in an acetylene–oxygen flame (Dean and Carnes, 1962).

of the emission over the range of 2–4.5 eV, and the increase is particularly pronounced in the region of 3.5–4.5 eV. Also, the presence of hydrocarbons leads to an increased degree of overexcitation, and this appears to be correlated with the region of maximum CH emission, i.e., the primary reaction zone.

As the degree of chemically produced emission is related to the concentration of flame radicals, such as H and OH, interferences can be expected for analytes containing elements that catalyze radical reactions, as discussed by Bulewicz and Padley (1973). We can expect such radical related interference effects to be particularly pronounced for metal concentrations of greater than 10^{-7} atm and for relatively cool flames, e.g., 2000 K or less.

C. Solution Effects on Sensitivity

The view has frequently been taken that the sensitivity of analytical flame spectroscopy is largely independent of the initial nature of the metal compound. Notable exceptions occur when a second metal is also present in the original solution and the possibility of matrix—and chemical—interaction arises during the desolvation–vaporization process. Formation of nonvolatile compounds in the desolvation process, or of occlusions in a refractory matrix, are responsible for many of the observed interference effects. That the volatilization process may be controlled by kinetic factors is evident from the observation that $CaCl_2$ analyte solutions give greater thermal emission signals than $CaSO_4$ solutions, even though both compounds are thermodynamically unstable at the flame conditions used.

Although, as we shall see in Section D to follow, the role of metal oxide species in affecting free-atom sensitivity is at least semiquantitatively established, little consideration has been given to the effect of metal–oxygen bonding on the initial processes of atom formation. Sastri *et al.* (1969) have therefore determined the effect of the solute character on the release of free atoms into air– and nitrous oxide–acetylene flames. They observed that if the metals were present in solution as metallocenes or fluorocomplexes, rather than as the acetylacetonates or oxysalts, then sensitivity enhancement factors of between 1.3 and 7.7 resulted. The higher enhancements were noted for those metals having higher metal oxide dissociation energies such as Ti and Ta. From these observations it was suggested that metal compounds which are bonded directly to oxygen in the initial solution would yield fewer free atoms than those where no metal–oxygen bond is present, such as metallocenes or fluorocomplexes. These observations also imply that chemical equilibrium does not exist at the point of observation, which seems reasonable since the measurements were made rather close to the burner top, i.e., 0.2 cm distance. Such observations are therefore not necessarily inconsistent with those of other workers where measurements made at later flame positions are less dependent on the initial chemistry of the solute.

Another example of "solute memory effects", which persist even in the burnt-gas region, is given by the positive ion studies of Kelly and Padley (1969b). The nitrates of Fe, Co, Ni, Mn, and Cr gave a higher total ionization than the corresponding chlorides and sulfates. Also, synergistic ionization effects between some of these metals and the alkali metals were observed in these studies.

The significance of kinetic effects in the sample degradative process is further demonstrated by the observed delay of about 0.2 msec in the release of Na-atoms from Na_2SO_4 solutions as compared with solutions of NaCl, NaOH, or $Na_2S_2O_3$ (Padley and Sugden, 1958).

It can be shown that the droplets of solution entering an analytical flame are desolvated after a period of several microseconds, which is short compared with the species lifetime in the flame of several milliseconds. A satisfactory theory has been developed for the mechanism of desolvation of sample droplets (Clampitt and Hieftje, 1972). Heat conduction from the flame to the droplet surface is the main rate-determining factor. Hence, the desolvation rates will depend on the flame temperature and composition and also on the thermal conductivity of the solvent vapor.

The details of the vaporization of the desolvated material are less well known. Under typical analytical conditions, in an air–acetylene flame at about 2100 °C temperature, it would take about 0.4 sec to vaporize Al_2O_3(c). This time scale is relatively long, and it is therefore desirable to prevent the formation of stable solids such as Al_2O_3 during the desolvation process. Experimental checks are sometimes made to determine whether complete vaporization of the sample has occurred in the flame. It is found that flame composition affects the volatilization of the sample. Also, particle size, in addition to the saturation pressure, has an important effect on the rate of vaporization. The rate-limiting step in the vaporization process is considered to be the rate of diffusion of vapor away from the particle surface.

1. Examples of Solution Interference Effects

A number of specific examples where chemical interferences occur, and techniques for reducing such effects, are given as follows.

- The sensitivity for systems containing metals such as Al and Mg, which form very refractory oxides resulting in low free-atom fractions, can be enhanced if the metals are introduced in the form of an organic complex,

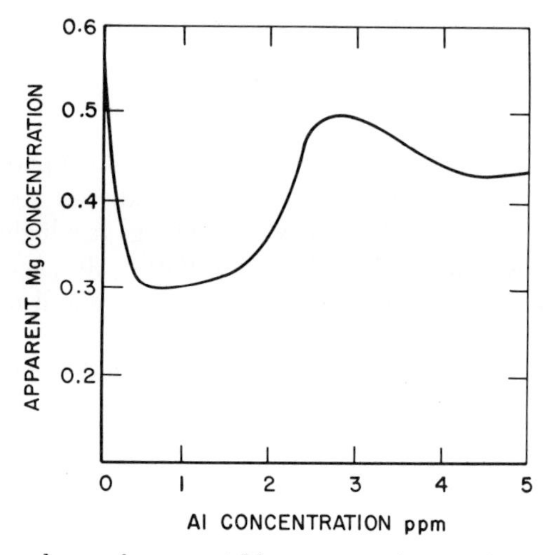

Fig. 5.17 Dependence of apparent Mg concentration on the presence of Al, in a flame spectrophotometric analysis. (Reprinted with permission from A. C. Menzies, *Anal. Chem.* **32,** 898. Copyright ©1960 by the American Chemical Society.)

such as the 8-hydroxyquinolate. This enhances the initial volatilization of the complexed metal which then readily decomposes in the gas phase to produce the free atom.

• Addition of ammonium chloride enhances the concentration of rare earth elements in the vapor phase, and this is presumably due to a reduction in the formation of condensed oxychlorides during the volatilization process.

• The interference of Al in the atomic absorption determination of Ca or Mg has been attributed to the formation of spinels, e.g., $MgO \cdot Al_2O_3(c)$, which are more stable than the component oxides. The addition of elements, such as La and Sr, which form higher stability spinels has the effect of competing more effectively than Mg for the Al to form $MO \cdot Al_2O_3(c)$ (Mansell, 1968). The more recent use of the higher temperature nitrous oxide–acetylene flames allows the thermal dissociation of the Mg spinel, thereby eliminating the interference of Al.

• For the case where Mg is present in aqueous solution with Al as the chloride, an unusual dependence of apparent Mg concentration on the amount of added Al is found, as shown in Fig. 5.17 (Menzies, 1960). This behavior is thought (L'vov, 1970) to be due to hydrolysis of $AlCl_3$ in solution to form $Al(OH)_3$ (c) which decomposes to $Al_2O_3(c)$ in the flame. The Mg component forms a matrix with Al_2O_3 and is therefore not readily available to the flame. As the amount of $AlCl_3$ is increased, the hydrolysis effect is reduced and the proportion of bound Mg decreases concomitant with an increase in sensitivity.

- Contrary to the above examples, the addition of small amounts (~ppm) of Al to fuel-rich air–acetylene flames can enhance the atomic absorption detectability of the elements Fe, Co, Ni, and Cr (Ottaway *et al.*, 1972). No mechanistic explanation has been given for this effect.
- The production of low-volatility compounds, such as $Sr_3(PO_4)_2$ and $MgSiO_3$, in the desolvation process can lead to a reduced analytical sensitivity for Sr and Mg.

Some of these sampling problems which result from the use of aerosols could be removed by a direct nebulization of metal samples using a sputtering technique (Winge *et al.*, 1971).

D. Free-Atom Fractions

The predominant factor that influences the availability of free atoms in flames for analytical absorption and emission spectroscopy is the tendency for metals to form stable oxide vapor species. Accordingly, the free-atom fraction should be particularly sensitive to changes in flame chemistry as, for example, may be achieved by adjusting the fuel content of the flame

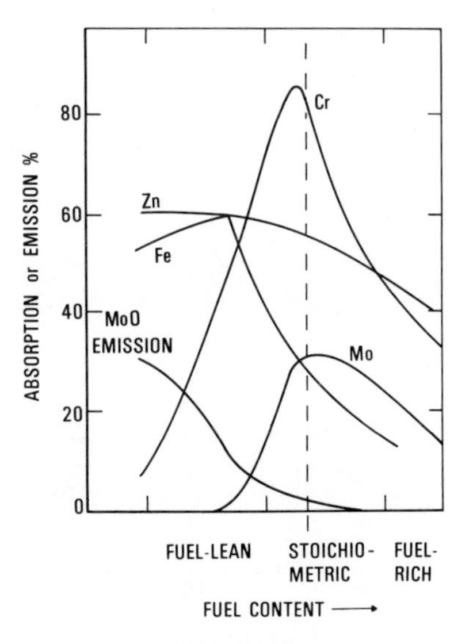

Fig. 5.18 Effect of flame composition on the concentration of free atoms, as determined by atomic absorption and emission (MoO only) in air–acetylene flames at 11 mm distance from burner (Coker *et al.*, 1971).

or, to a lesser extent, by changing the point of observation in the flame. The effect of flame composition on the formation of free atoms is experimentally demonstrated in the Fig. 5.18. Also shown is a correspondence between the reduction in the level of MoO, as determined by emission, and the increase in the number of Mo-atoms, as determined by absorption. Note also that the influence of changes in flame composition is greatest for those elements which form the more stable MO species, such as Cr and Mo where D CrO = 120 and D MoO = 116 kcal mol^{-1}, respectively; FeO is of intermediate stability and ZnO of low stability, i.e., D ZnO $\simeq$ 65 kcal mol^{-1}. A secondary effect of temperature is also present in these profile data.

1. *Thermodynamic Predictions*

If one assumes that chemical equilibrium exists at the point of observation in these analytical flames, then it is possible to make thermodynamic predictions about the influence of flame composition on the free-atom fractions. For example, using the free energy minimization technique for multicomponent systems, Chester *et al.* (1970) have made computations for various air– and nitrous oxide–acetylene flames containing Si and Al. Typical results are shown in Fig. 5.19. Note that an increase in fuel content,

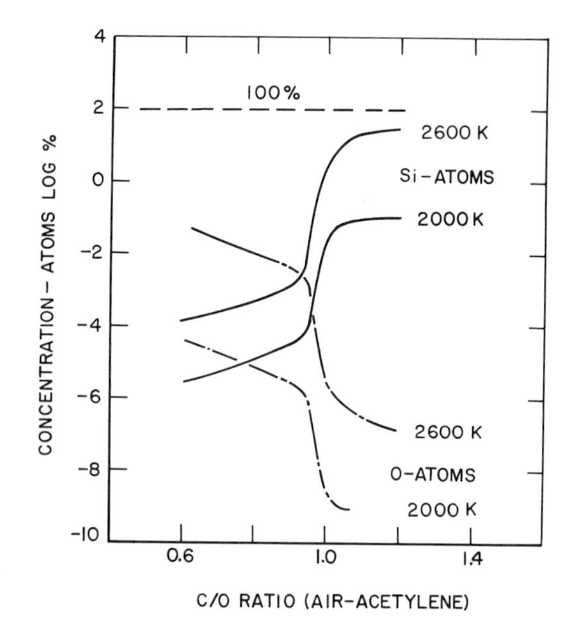

Fig. 5.19 Calculated equilibrium concentrations of Si- and O-atoms as a function of air–acetylene flame composition (Chester *et al.*, 1970).

or alternatively in temperature, has a pronounced effect on the enhancement of the free-Si-atom concentration, which occurs at the expense of SiO. The equilibrium concentration of O-atoms is also shown, for comparison. Dagnall and Taylor (1971) have extended calculations of this type to test mixed element systems where the formation of spinels may further decrease the availability of free atoms.

The validity of the data obtained by such calculations depends primarily on the selection of species which are thought to be present and the accuracy of their thermodynamic functions. Chester *et al.* (1970) considered the Si-containing species to be Si, SiO, SiO_2 (g, *l*), Si_2, Si_3, and SiC; and the Al-containing species to be Al, AlH, AlO, AlC, AlN, Al_2O, AlOH, AlO_2H, and Al_2O_2. The recent report of $Al(OH)_2$ as a dominant species under certain flame conditions (see Section II.K) serves to demonstrate the possible pitfalls associated with the a priori selection of species. Similar arguments also apply to the recently reported calculations of species concentrations for tungsten and molybdenum in acetylene–nitrous oxide and other flames (Reigle *et al.*, 1973). The results of these various calculations depend on the JANAF (1971, or earlier editions) selection of species and free energy functions which is not necessarily all inclusive.

Another source of uncertainty with the equilibrium description of free-atom fractions results from the flame radicals H, OH, and O often being present in excess of their equilibrium concentrations. The "equilibrium" condition of molecular species, such as MO, may then depend on the reactions of interest. That is, equilibrium may exist for a particular reaction but not necessarily for all reactions involving the molecular species of interest. For example, the reaction

$$LiOH + H \rightleftarrows Li + H_2O$$

is known to be at equilibrium, since the rates are fast in both directions. However, the reaction

$$Li + OH = LiOH$$

is not likely to be at equilibrium, since the highly exothermic forward reaction requires a third body to remove excess energy and is necessarily slow. Likewise, the reverse reaction is slow on account of its high endothermicity or activation energy.

Values of the free-atom fraction β have been measured many times over the past few years and an indication of the variation in the data between various workers may be found in the tabulation of Smyly *et al.* (1971; see their Table 5). Variations, in β, of up to a factor of 5 are found for supposedly the same system. However, the gross features are consistent, and there seems to be a consensus of opinion that the low β values correlate with the stability of diatomic molecular species in the flame—see the review of Alkemade (1970). Thus the disparity between workers in the β values for elements such as Mg, Mn, Si, and Sr, which form stable oxides or hydroxides, is thought to be due to the unusually strong dependence of β on flame composition, as may be thermodynamically predicted. As an example of the magnitude of this dependence, the results of Fig. 5.19 indicate that the concentration of Si atoms can vary by as much as an order of magnitude for a 2% change in the fuel-to-oxidant ratio. Consequently, the experimental control of combustion gases introduced to the flame is rather stringent, and an example of the procedures required may be found in the work of Smyly *et al.* (1971).

As flame characterization techniques have developed, the number of stable molecular species known to be present in flames has increased, e.g., see Table 5.13, and simple diatomic MO species can no longer be considered as the sole source of free-atom depletion. For example, in the case of Li, H_2-containing flames have a very low free-atom fraction of Li in the flame reaction zone due to the process

$$Li + H_2O \rightleftarrows LiOH + H,$$

and this decreases further as the excess concentration of H-atoms is depleted in the burnt-gas region.

A more complex example of the importance of molecular species formation in flames is given by the low β values found for Ca, Sr, and Ba which could be quantitatively accounted for using the following relative concentration data:

	Ca	Sr	Ba
MO/M	6.7	12.5	7.6×10^2
MOH/M	11	3.7	7.6×10^2
$M(OH)_2/M$	1	0.8	0.1×10^2

(Zeegers *et al.*, 1969). In the same study, using measured β values and assuming the formation of MO to be the only source of metal atom losses, except for Mg where the species $Mg(OH)_2$ was also taken into consideration, new values for the MO dissociation energies, given in eV, were derived as follows:

MgO	3.8 ± 0.1, (3.4 ± 0.3)
CrO	4.9 ± 0.1, (4.8 ± 0.4)
FeO	4.4 ± 0.1, (4.1 ± 0.2)
MnO	4.2 ± 0.1, (4.1 ± 0.3)

The values given in parentheses are the literature data as evaluated and averaged by Brewer and Rosenblatt (1969). The good agreement, with the possible exception of Mg where formation of $Mg(OH)_2$ is a complicating factor, seems to verify the assumptions of the β method. The flame used for these studies was acetylene–air ($T = 2410$ K).

Elements such as Ti, Si, Be, and Al are undetectable in an air–acetylene flame, and this has frequently been correlated with the high stability of their monoxides. However, in the hotter nitrous oxide–acetylene flame, i.e., ~3000 K as compared with ~2400 K, these elements are detectable by atomic absorption spectrometry and an enhancement in the free-atom fraction of between 10^2 and 10^4 is found over the air–acetylene case (Coker and Ottaway, 1971). If we consider the equilibrium

$$\mathrm{MO} \rightleftarrows \mathrm{M} + \mathrm{O}$$

then the enhancement of free atoms resulting from a temperature increase of from 2400 to 3000 K can be predicted from the thermodynamic functions of the above species. Coker and Ottaway (1971) note that the free-atom enhancement in the nitrous oxide–acetylene flame is greater than that predicted by the above thermodynamic argument. These workers suggest that the production of additional free atoms may occur via a reaction

$$\mathrm{MO} + \mathrm{CX} \rightarrow \mathrm{M} + \mathrm{CO} + \mathrm{X},$$

where CX is a species having a dissociation energy in the region of 126 kcal mol^{-1}, and C_2 and CO, necessarily in electronically excited states, are suggested as the most likely candidates. The view is taken, therefore, that the important property of the flames used for atomic absorption spectrometry is not the flame temperature but the chemical environment. It would be of interest to provide a systematic test of this suggestion by using carbon-free flames, such as H_2–O_2–N_2, of varying temperatures and stoichiometries. However, recent studies by Coker *et al.* (1971) suggest that the reducing power of air–hydrogen flames is even less than for air–acetylene flames.

As a means of avoiding the loss of atoms by metal hydroxide formation, it is possible to use flames without hydrogen content, such as CO–O_2 or $(CN)_2$–O_2. This latter flame has the highest temperature of any of the known practical flames (but see Table 5.5), excluding the impractical use of metal combustion, and for this reason is also of interest for systems where metal oxide stability is normally a limitation. However, the advantage of higher temperatures tends to be offset by an increase in thermal ionization. For example, in air– or nitrous oxide–acetylene flames, at temperatures of 2500–3200 K, the alkali—and most alkaline earth—metals are thermally ionized to the extent of 50–90%. Elements of high ionization potential such as Zn, Pt, and Be form only about 0.1% ion concentration (Woodward, 1971). In practice it is possible to add an electron source, such as an alkali metal, which is different from those requiring analysis, for suppressing the ionization of the element of interest. This follows from the equilibrium established between atoms, ions, and electrons. In this context, however, one should note the possible formation of molecular species such as K_2CrO_4 or K_2MoO_4, e.g., see Table 5.13, or of molecular ion species such as HUO_3^+, e.g., see Table 5.19.

Since, in many instances, the two main factors determining the free-atom fraction are formation of MO species and metal ionization, then it is possible to observe a very good correlation of β with the ionization potential *IP* of the metal and the dissociation energy of the metal oxide D_{MO} as follows:

$$\beta = b_1 + b_2 \cdot IP + b_3/D_{MO},$$

where b_1, b_2, and b_3 are constants (Parsons and McElfresh, 1972).

E. Distribution of Additive Atoms and Molecules in Flames

From atomic absorption atom-distribution studies, it has been possible to locate flame positions where maximum free-atom concentrations occur. Frequently, these positions are rather localized, and hence no benefit is gained by allowing the absorption beam to pass through an extended flame region. By taking measurements only from the region of maximum concentration the analytical sensitivity may be optimized. From the review of Hambly and Rann (1969), the distribution patterns of the free atoms indicate a more rapid depletion across the flame than along the flame, and the authors have suggested that the shape of the absorption beam should be elongated to take advantage of this effect.

It is important to recognize that the apparent distributions of free atoms, as determined by absorption and by emission,do not necessarily coincide as different chemical reactions may be involved in the generation of electronically excited atoms, as opposed to ground-state atoms. This is demonstrated by Fig. 5.20 where the maximum ground-state Fe-atom concentration, as determined by absorption, occurs well downstream of the Fe-emission maximum. Atomic emission flame profiles also tend to be sensitive to the chemistry of the initial solution. This is exemplified by the curves of Fig. 5.21. In this example it is likely that the aqueous solution allows initial oxidation of the metal to form PbO and hence slows the production of free Pb atoms.

Roos (1971) has observed that the emission profiles for metal atoms reach a maximum earlier in air–acetylene flames, with both lean and rich conditions, than those of the diatomic metal oxide species. Similarly, Koirtyohann and Pickett (1971) found that, by varying the flame stoichio-

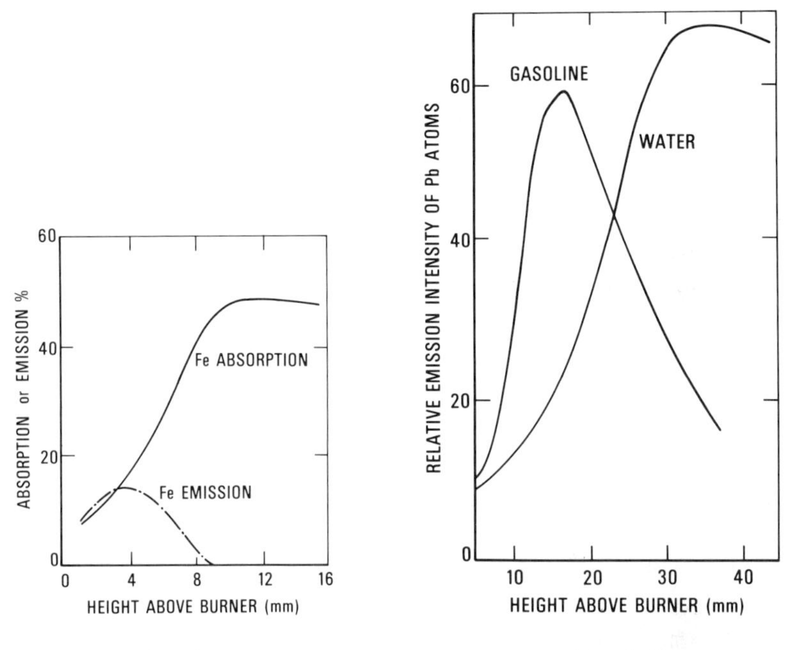

Fig. 5.20 **Fig. 5.21**

Fig. 5.20 Comparison of apparent distribution of Fe-atoms, as determined by emission and absorption, in a stoichiometric air–acetylene flame (Coker *et al.*, 1971).

Fig. 5.21 Distribution of excited Pb-atoms, in a premixed flame, as determined by emission, and using gasoline or water solutions. (Reprinted with permission from B. E. Buell, *Anal. Chem.* **34,** 635. Copyright ©1962 by the American Chemical Society.)

metry, a composition could be reached where an abrupt reduction in monoxide emission correlated rather well with an abrupt increase in atomic absorption. The species used were the oxides and elements of Al, Ti, Ce, La, and Tm.

A special burner technique has been devised to allow a separation of flame zones for the purpose of obtaining emission spectra separately under conditions which are progressively overexciting, reducing, and oxidizing as one moves downstream from the primary reaction zone. Such flames are well exemplified in the figures (i.e., 12-13 and 12-14) given by Mavrodineanu and Boiteux (1965). These authors have presented spectra for some 62 chemical species in these separated flames. As an example of these results, the spectra of flames which contain Sn or Sb show the presence of overexcited MO species in the inner-cone region, very little excitation, i.e., a chemical equilibrium condition, in the reducing interconal region, and excess excitation again in the oxidizing outer-cone region.

F. New Developments in Analytical Flame Spectroscopy

The development of more versatile radiation sources, in the form of lasers, should provide an impetus to atomic spectroscopy that may parallel that already given to Raman spectroscopy as an analytical tool.

Using an intercavity broadband dye-laser absorption technique, Thrash *et al.* (1971) demonstrated qualitatively the presence of Ba and Sr in an acetylene–air flame. They suggested that the technique should provide an improvement in sensitivity over conventional methods of at least two orders of magnitude.

Lasers may also be used as an excitation source for atomic resonance fluorescence, e.g., see Fraser and Winefordner (1972) and earlier cited work. Tunable dye lasers are particularly suited for this application because of their wide wavelength tunability, narrow emission linewidth, and high peak power.

Several other notable recent developments, in the field of analytical atomic absorption spectrophotometry, are the use of a plasma jet and also a high temperature graphite furnace or cuvette as flame substitutes—see L'vov (1970).

V. Chemical Inhibition of Flames

A. Introduction

The use of chemical additives to influence the performance of flames, explosives, rocket fuels, internal combustion engines, and other combustion systems is of considerable practical importance, and the science of high temperature vapors is particularly pertinent to this area of materials technology.

It is now recognized that the inhibition or extinguishment of flames can be more effectively achieved by chemical, rather than purely physical means. Most probably, practical fire retardancy situations involve both physical and chemical effects. However, for the present discussion we will be concerned only with the chemical aspects of flame inhibition, as reviewed by Hastie (1973a), Lovachev *et al.* (1973), Fristrom (1972), Creitz (1972), Friedman (1971), Fristrom and Sawyer (1971), Creitz (1970), McHale (1969), and Friedman (1961). Discussions of the relationship of flame inhibition chemistry to the overall consideration of fire extinguishment or fire proofing have been given by Pumpelly (1969), Hindersinn and Wagner (1967), Thiery (1970), Lyons (1970), and Hilado (1973).

The influence of additives on burning velocity is particularly revealing, in that only a small amount of additive is required to strongly affect flame propagation and also flammability limits. This suggests that the inhibitory function of the additives is catalytic.

In practical systems the manner in which the additives are released to the flame—or potential flame—system is of importance. One of the empirically based theories of flame retardancy requires the additive to be "at the right place at the right time." Thus the protection of a fire-retarded flammable polymer substrate requires that the vapor release properties of the inhibitor be concordant with the fuel-release properties of the flammable substrate. Likewise, spatial and time constraints exist in the external application of fire extinguishants. For the case of an in situ fire-retardant system one may cite the application of antimony oxide–organic halogen formulations in reducing the flammability of organic polymers, where the antimony–halogen component is released at about the same time as the polymer decomposes (Pitts, 1972).

In addition to the in situ function of chemical additivies for reducing the fire hazard of flammable substrates, flame-inhibiting compounds are also used as externally introduced fire extinguishants. Various halocarbons and alkali metal salt powders are common commercial extinguishants. It appears likely that the mechanistic action of fire retardants and fire extinguishants are similar and evidence of this similarity is given in the ensuing discussion. The considerable variety of chemical systems used in imparting fire retardancy to materials has been documented by Lyons (1970).

A satisfactory theory of flame inhibition has not yet been derived, although it is generally agreed that the additives interfere with the concentration of the flame propagating radicals H, OH, O, and perhaps CH_3 and HO_2. It is also apparent that the inhibiting action of additives, such as halogens, cannot be explained merely in terms of a reduction of the equilibrium concentration of radicals due to hydrogen halide formation (see Fristrom and Sawyer, 1971). That is, flame inhibition is primarily a kinetic phenomenon.

B. Rating of Flame Inhibitors

The degree of effectiveness of a flame inhibitor can simply be determined from the quantity of inhibitor needed to extinguish a flame. This quantity will vary somewhat with the flame type, i.e., if premixed or diffusion, and also the concentration ratio of fuel to oxidant. However, a similar ranking of inhibitors is usually found in these various flames. As a typical example of the amount of inhibitor required to extinguish a flame, about 2% of the commercial extinguishant CF_3Br is needed to extinguish a propane–air diffusion flame. If N_2 was used as the extinguishant in this case, the amount required would be about 15 times greater.

If the addition of an inhibitor is made to a premixed laminar-flow hydrocarbon–oxygen flame, where the reaction zone is both visually stable and luminous, one can readily determine the effect of the inhibitor (in the absence of extinction) by noting a shift of the reaction zone to a position further downstream from the burner opening. This shift results from a decrease in the burning velocity of the flame due to the presence of the inhibiting agent. As an example of this effect, the addition of about 10^{-3} mole fraction of HBr to an atmospheric lean CH_4–O_2–Ar flame would decrease the burning velocity by $\sim$2 cm sec^{-1}, and this would result in an increased flame area of about 6% (Levy *et al.*, 1962).

Burning velocity measurements on premixed laminar-flow flames provide a useful basis for the screening of flame-inhibiting agents. Flame-inhibition measurements are less frequently made on diffusion flames due to difficulty in defining a fundamental flame strength parameter, such as burning velocity, and also the strong geometrical influences on diffusion flame propagation. However, the recent use of an opposed jet burner has provided a means of measuring inhibitor effectiveness in terms of the flow conditions required to produce a hole in the diffusion flame, e.g., see Milne *et al.* (1970).

A more practical, though at present less fundamentally useful, test of the degree of effectiveness for gaseous flame inhibitors makes use of the so-called limiting-oxygen-index test. In this case the amount of inhibitor needed to extinguish a diffusion flame can be determined, e.g., see Petrella and Sellers (1970).

As the primary purpose of determining the relative efficiency of flame inhibitors relates to their potential application in the area of fire extinguishment and fire prevention, it is pertinent to examine the usefulness of data obtained primarily on laboratory premixed laminar-flow flames. Friedman, in particular, has considered this point on several occasions (Friedman, 1961, 1971). The propagation of real—as opposed to laboratory—fires frequently involves diffusion and turbulent flame conditions. However, there is evidence for a correspondence between the speed of turbulent flame propagation and the laminar-flow burning velocity of premixed flames. Also it is found that a diffusion flame contains a region of premixing at the base, and it is possible that the inhibitor functions in this region. Indeed, the fact that a similar amount of inhibitor (e.g., CH_3Br) can extinguish both a diffusion and a premixed flame strongly suggests the premixed region to be the relevant area of inhibition (Simmons and Wolfhard, 1956). Furthermore, a significant entrainment of air into the fuel region has been noted for several experimental diffusion flames, e.g., see the discussion given by Hastie (1973a). This suggests that in practice diffusion flames—and perhaps real fires—can be considered as resembling fuel-rich premixed flames. In fact, the ensuing discussion will reveal that the catalytic effect of chemical flame inhibitors can only operate under conditions of nonequilibrium radical concentrations, i.e., under strictly nondiffusion flame conditions.

As has been mentioned elsewhere (Section II), given the basic kinetic and transport properties for the flame reactions, it is possible to quantitatively calculate burning velocities. However, to date, this can only be done for the H_2–O_2 flame system as the basic data and mechanistic understanding are lacking for hydrocarbon flames. Similarly, if the flame kinetics of an additive were known, it would be possible to predict from basic principles the magnitude of the reduction in burning velocity due to inhibition. As such a determination is not feasible at the present time, it is necessary to develop less sophisticated mathematical models for flame inhibition.

The need for theoretical models has recently been emphasized by Fristrom and Sawyer (1971). These authors have developed a model which allows the definition of new parameters for evaluating flame inhibitors in terms of measurable macroscopic variables such as O_2 concentration, inhibitor concentration, and the change in burning velocity. The model has as a basis the notion that the inhibitor removes H-atoms from the system via reactions such as

$$H + HCl \rightarrow H_2 + Cl,$$

which compete with the chain-branching flame-propagating reaction:

$$H + O_2 \rightarrow OH + O.$$

The supporting evidence for the validity of such a model for flame inhibition is outlined in the following discussion on halogen inhibitors (Section C).

From the mathematical model a figure of merit for an inhibitor, using burning velocity data, may be defined as

$$\phi_v = ([O_2]/[I])(\delta V/V_0),$$

where

ϕ_v is the figure of merit for the inhibitor,
$[O_2]$ the oxygen concentration,

$[I]$ the inhibitor concentration,
V_0 the burning velocity in the absence of any inhibitor, and
δV the change in burning velocity due to the presence of the inhibitor.

It is assumed that the flame is slightly fuel rich, and hence the H atom is the dominant radical.

Sawyer *et al.* (1971) report that, in practice, ϕ_v correlates with other measures of inhibitor effectiveness, such as blow-off limits and extinction limits for halogen inhibitors. It also follows from the definition of ϕ_v that the degree of inhibition should be proportional to the amount of inhibitor added to the flame. Experimental observations, e.g., see Lask and Wagner (1962), show that this is approximately valid for a variety of homogeneous inhibitors.

Values of ϕ_v, as calculated from the data of Lask and Wagner (1962) and unpublished data of Wagner (cited by Morrison and Scheller, 1972) for *n*-hexane-fueled flames, and also from the data of Miller *et al.* (1963) for H_2-fueled flames, are listed in Table 5.23. These data are given in approximate increasing order of inhibitor effectiveness, and will be referred to again later when specific inhibitor types are discussed. At this point it is sufficient to note that only for the case of CO_2 is the degree of inhibition explainable in terms of a physical effect such as cooling. The magnitude of ϕ_v greater than unity for the other inhibitors can only be explained in terms of chemical effects. Some of these flame inhibitors have also been found to be effective in diborane–oxygen flames. For instance ϕ_v values in the range of 10–20 have been found for PCl_3, $SnCl_4$, and CF_3Br (Skinner and Snyder, 1964).

Common commercial halocarbon extinguishants have ϕ_v values similar to the value of 8.4 given for Br_2. It is pertinent to note that the effectiveness of these halocarbons, relative to that for an inert diluent such as CO_2 or N_2, is similar for both these premixed flame observations and the propane–air diffusion flame extinguishment studies of Petrella and Sellers (1970). Likewise, the relative effectiveness of $POCl_3:Br_2:CCl_4$ as inhibitors is about 7:2:1, respectively, both for the premixed *n*-hexane–air flame considered in Table 5.23 and for a methane–air counterflow diffusion flame with the inhibitor added to the fuel (Ibiricu and Gaydon, 1964). These comparisons provide further evidence for the validity of using burning velocity measurements in inhibited premixed flames as a means of rating inhibitors for practical fire applications—where diffusion is more significant.

Of the two flame types considered in Table 5.23, the *n*-hexane–air system more closely resembles the chemical conditions likely to occur in practical fire situations.

TABLE 5.23

Relative Effectiveness, ϕ_v, of Selected Flame Inhibitors

	Flame type (1 atm)	
Inhibitor[a]	*n*-hexane/air[b,c] ϕ_v	H_2–air[d] ϕ_v
CO_2	0.86	
Cl_2	1.8	−0.26[e]
$Si(CH_3)_4$	3.9	
CCl_4	4.2	
Br_2	8.4	
$SiCl_4$	10.5	3.5
$(CH_3)_3PO_4$	23	
$SbCl_3$	26	
$SbCl_5$		10.2
$TiCl_4$	30	10
$SnCl_4$	31	12.9
$POCl_3$	31	7.2
PCl_3	39	4.5
PBr_3	39	
CrO_2Cl_2	$\gtrsim$244	
$Fe(CO)_5$	356	19
$Pb(C_2H_5)_4$	390	3.4

[a] Amount of inhibitor used varied from 0.015% to several volume percent.

[b] From data given by Lask and Wagner (1962) for a stoichiometric mixture.

[c] A similar order of effectiveness has been noted in low pressure (0.01 atm) CH_4–O_2 premixed and diffusion flames (Miller, 1969).

[d] From data given by Miller *et al.* (1963) for a mixture with 1.75 fuel equivalence ratio.

[e] Negative sign indicates burning velocity increase rather than decrease.

C. Halogen Inhibitors

1. *Background*

Halogen compounds are among the most widely used commercial reagents for fire prevention and extinguishment. As examples one can cite:

- the early use of CCl_4 as a portable fire extinguishant;
- the current use of halocarbons such as CF_3Br as extinguishants,

particularly in connection with fuel fires resulting from aircraft mishaps, e.g., see Botteri (1971) and Fiala (1971), and also for the protection of electronic equipment;

- the incorporation of phosphorous–halogen or antimony–halogen formulations in materials such as natural and synthetic fabrics, furniture materials, paints, wood, paper, etc., e.g., see Lyons (1970).

It should be recognized that in some instances the halogen treatments of flammable materials act in the condensed state, altering the course of substrate decomposition such that less-flammable products are produced during the pyrolysis or oxidation processes. This mode of action is not considered in the following discussion, which is concerned only with fire retardancy aspects where vapor-phase chemistry is involved in the inhibition process.

In practical fire-retardent systems the halogen species can be introduced into the gas phase by mechanical means—as with halocarbon protection systems, or by chemical means—as with the release of HCl from decomposing polyvinylchloride, or as phosphorous chlorides or antimony halides, formed during the decomposition of a polymer substrate.

2. *Flame-Inhibition Mechanism*

Halogen systems have, in general, received the most attention by workers interested in the basic mechanisms of flame inhibition. As is evident from the high ϕ_v values given in Table 5.23, certain halogen species rank as chemical inhibitors and must therefore interfere with the chemistry associated with flame propagation. Furthermore, thermodynamic calculations show that flame cooling effects due to thermal dissociation of halogen species, such as Br_2, are negligible compared with the observed chemical effects (Simmons and Wolfhard, 1955). However for the case of Cl_2, the degree of inhibition observed in hydrocarbon flames is low enough to be consistent with thermal dissociation as a cooling and slightly flame-inhibiting process.

Some clues as to the basic cause of halogen-produced flame inhibition are provided from macroscopic observations such as:

- the effectiveness of halogens as retardants increases in the order, fluorides $\ll$ chlorides $<$ bromides $\lesssim$ iodides;
- CF_3Br is about four times more effective than CF_4 in preventing combustion of *n*-hexane–air mixtures;
- CF_3Br is a good inhibitor in hydrocarbon–air flames, but is mediocre in the faster burning H_2–air flames;

- Br_2 has little inhibiting effect on CO–O_2 flames but is very effective when small amounts of H_2 are introduced;
- halogen inhibitors are more effective in fuel-rich than fuel-lean premixed flames;
- with Br-substituted hydrocarbons and fluorocarbons, the inhibitor effectiveness increases with the number of Br-atom substitutions and, in certain instances, can be directly proportional to this number; for instance, in a premixed stoichiometric methane–air flame, the following ϕ_v values are found—Br_2 (24), CH_3Br (12), HBr (11), and CF_3Br (17).

From these macroscopic observations, it can be concluded that the flame-inhibition mechanism must involve an interaction between species containing either Cl or Br and a flame-propagating radical. Furthermore, either the halide species or the flame radical, or both, must contain a H-atom. Thus, reactions between HCl and O, OH or H, or between Cl and OH or H are possible.

The relatively high degree of flame retardation provided by CF_3Br, as compared with HBr or Br_2, has led to speculation that the CF_3 radical may also be an inhibiting moiety. However, in flames, it is known that CF_3 is readily converted to HF (Hastie, 1972) and, in some instances, COF_2 (Biordi *et al.*, 1973). It is difficult to conceive of a catalytic process involving the regeneration of CF_3 from such products.

The observation that inhibition is more effective in cooler than in hotter, faster burning flames may be due to the higher absolute concentration of radicals in the hotter flames, e.g., see Fig. 5.8. Alternatively, one could argue from the fact that in the hotter flames the excess in radical concentration over the equilibrium level is not nearly as great as for cooler flames. Hence, from the kinetic nonequilibrium nature of the inhibition process, one could expect a better degree of inhibition in the cooler flames. As an additional factor, the high temperature dissociation of potentially inhibiting species, such as HBr, could reduce their effectiveness in the hotter flames.

A more detailed description of the likely flame-inhibiting mechanism has been derived from a variety of experiments carried out at the molecular level in premixed flames. These studies include mass spectrometric microprobe sampling experiments, as described by Wilson *et al.* (1969) and Pownall and Simmons (1971), and molecular beam mass spectrometric experiments, such as those of Biordi *et al.* (1973) and Hastie (1973c).

a. Various Proposed Mechanisms

Before discussing the current status of the mechanistic understanding for flame inhibition by halogens, we should note that a considerable variety of mechanisms has been proposed in recent years. For example, Mills (1968) suggested that the inhibiting function of halogen species may involve electron attachment followed by dissociation to generate active inhibitor species, e.g.,

$$CF_3Br + e^- \rightarrow CF_3Br^- \rightarrow Br^- + CF_3$$

and

$$Br^- + H \rightarrow HBr + e^-.$$

He cites the observation of negative halogen ions in flames as support for this theory. However, the concentration of such ions is now known to be many orders of magnitude less than for the neutral species, and their effect on the flame-propagating radicals is most likely negligible (see Section III). In hydrocarbon flames the difference between radical and ion concentrations is greater than a factor of 10^3, and for H_2 flames this difference is about a factor of 10^{10}.

Creitz (1970, 1972) has suggested that halogenated extinguishing compounds act as catalytic agents for the recombination of oxygen atoms. The participation of OX, where X is Cl, Br, or I, as a catalytic intermediate is suggested by reactions such as,

$$O + X + M \rightarrow OX + M$$

or

$$O + X_2 \rightarrow OX + X$$

and

$$O + OX \rightarrow O_2 + X.$$

It is perhaps pertinent to this argument that ClO has been identified in the reaction zone of a low pressure, i.e., 15 Torr, premixed methane–perchloric acid flame (Williams and Wilkins, 1973). However, such species were not observed in premixed hydrocarbon–oxygen flames containing a source of Br or Cl, either at low pressure (Biordi *et al.*, 1973) or at 1 atm (Hastie, 1973c), and under combustion conditions varying from fuel lean to fuel rich, respectively. On the other hand, BrO has been observed in the region of the reaction zone for the case of a CH_3Br–O_2 diffusion flame, where the halogen concentration is necessarily unusually high (Simmons and Wolfhard, 1956). Thus the evidence for or against this model is inconclusive, though it will become evident in the discussion to follow that the evidence for other alternative models is more substantial.

Other workers have found that the presence of halogens in flames reduces the concentration of OH. For instance, Rosser *et al.* (1959) observed a decrease in the emission intensity of OH and an increase in that for CH due to the presence of CH_3Br or HBr in hydrocarbon flames. The observed radiation was not from a very well-defined part of the flame—both reaction zone and postflame regions being included. A similar decrease in OH emission intensity was found in a number of atmospheric counterflow fuel–air diffusion flames where the fuel was C_2H_4, CH_4, or H_2 and the inhibitors were Cl_2, Br_2, CH_3Br, CCl_4, and $POCl_3$ (Ibiricu and Gaydon, 1964). In time-resolved spectroscopic studies on the low pressure explosive combustion of styrene–oxygen mixtures, Petrella (1971) observed that the production of OH was delayed in the presence of HBr. Levy *et al.* (1962) suggest the inhibiting action of HBr in lean methane flames to be

$$HBr + OH \rightarrow H_2O + Br.$$

This is analogous to the suggestion of Wilson (1965) that methyl bromide reacts as follows:

$$CH_3Br + OH \rightarrow CH_2Br + H_2O.$$

However, more recently, Wilson and co-workers (1969) have suggested an alternative mechanism involving H-atom, rather than OH, reactions.

The general situation with regard to the action of halogens on OH in flames remains unsettled, and some of the apparent difficulty may be related to the use of different flames for the various inhibition studies. In particular, Wilson *et al.* (1969) have obtained indirect evidence for an increase in the maximum OH concentration level in low pressure lean CH_4–O_2 flames containing HBr. Similarly, an increase in OH emission intensity resulted from the addition of HCl to a low pressure acetylene–oxygen flame (Panah *et al.*, 1972). However, using 1-atm lean propane-fueled flames, Pownall and Simmons (1971) provide indirect evidence for a reduction in the maximum OH concentration in the presence of HBr.

b. H-Atom Mechanisms

The remaining principal flame-propagating radical to be considered, namely the H-atom, is favored by many workers as the species which is involved directly with halogen-controlled flame-inhibiting reactions. From the reviews of Fristrom and Sawyer (1971), Hastie (1973a), and Lovachev *et al.* (1973), the current status of halogen inhibition is as follows.

The low pressure flame ($\sim$0.05 atm) mass spectrometric studies reported by Wilson *et al.* (1969), indicate that the introduction of bromide or chloride halogen species into a premixed CH_4–O_2 flame leads to the production of the hydrogen halide early in the flame, i.e., in the preflame region. Also, in the presence of an inhibitor, the preflame zone is extended and the reaction zone narrows. Inhibition was therefore considered to occur primarily in the preflame region where the chain-branching reaction is slow and radical recombination is relatively competitive. In this model the inhibitor reduces the concentration level of radicals in the preflame region, but at the reaction zone the concentration may even be increased over that for an uninhibited flame. In effect, the radical concentration versus distance profiles shift downstream to a higher temperature region, where radical propagation is more competitive with the inhibitor reactions.

It was also observed in these low pressure flame studies that the formation of H_2CO is inhibited and the production of H_2 enhanced by the presence of halogen species. This provides indirect evidence for a removal of H-atoms from the flame and the predominant reaction is considered to be

$$H + HBr \rightarrow H_2 + Br.$$

This is known to be a fast reaction under flame conditions, and Fristrom and Sawyer (1971) have demonstrated that, in principle, it can effectively compete with the chain-branching reaction in the preflame region. They also show that the observed degree of inhibition is greater than that which would be predicted by considering the above reaction to reach equilibrium. This mode of inhibition for HBr is also supported by the work of Pownall and Simmons (1971) on atmospheric pressure lean propane–oxygen flames.

Further evidence for the inhibiting action of HBr in the preflame region is provided by the observations of Levy *et al.* (1962), where the effect of HBr was to decrease the rate of CH_4 consumption ahead of the reaction zone and also to *increase* the thickness of the reaction zone. Very little effect on the CO concentration profiles was found. This is most probably related to the fact that CO does not form until late in the flame system, i.e., close to the reaction zone. Thus, HBr inhibits only the early stage of methane oxidation.

The mechanism by which the Br-atom is removed from the system and HBr regenerated has not been established although, under relatively cool flame conditions, the reaction

$$HO_2 + Br \rightarrow HBr + O_2$$

could serve to replenish the inhibitor source. Similarly, the reaction

$$Br + H + M \rightarrow HBr + M^*$$

is possible (Day *et al.*, 1971). The reaction

$$Br + RH \rightarrow HBr + R,$$

where R is a hydrocarbon group, has also been suggested. Such a reaction relies on the presence of fuel, RH, and hence can only function in the preflame mixture. The production of HBr from other initial sources, such as RBr, can readily be achieved by a reaction of the type

$$H + RBr \rightarrow HBr + R,$$

and as R is most likely less reactive than H, this process is also flame inhibiting.

Note that this type of inhibition mechanism readily accounts for the non- or low-inhibiting properties of fluorides, since the high stability of HF provides an excessively high activation energy barrier for a reaction with H-atoms to result. The much lower effectiveness of chloride—as compared with bromide—inhibitors is probably due to the HCl reaction being very close to thermoneutral; this allows the reaction to proceed in the reverse direction and regenerate H-atoms. The probable importance of the back reaction is evident from the observation that Cl_2 actually promotes flame propagation in H_2–air flames, as shown by the negative ϕ_v value given in Table 5.23

This seems to be a very satisfactory—but not yet definitive—model for flame inhibition by halogens. More recent species concentration profile studies also support the model in that H-atoms are found to be present in relatively high concentration in the preflame region (Hastie, 1973b). The production of HBr in the preflame region of premixed halogen-containing flames at 1 atm has also been established (Hastie, 1973c). Similarly, the presence of HCl (as derived from CCl_4 as the initial additive) in the fuel, reaction-zone, and oxidant regions of acetylene–oxygen diffusion flames has been established (Combourieu *et al.*, 1971).

Another check on the model is provided by ignition-temperature studies. The inhibition model suggests that the flame reactions are shifted to a higher temperature region, hence the ignition temperature would be expected to be higher in the presence of the halogen inhibitor. Morrison and Scheller (1972) find the expected increase in ignition temperature for methyl halides, $SnCl_4$, and BBr_3. However, $SiCl_4$, $TiCl_4$, CrO_2Cl_2, and $Fe(CO)_5$ have no effect on the ignition temperature, and CCl_4 actually produces a decrease. Evidently, in this ignition context, the inhibition model is not of wide generality.

That inhibition is rather specifically related to the flame type is demonstrated by the observation of Tischer and Scheller (1969), where the tetrachlorides of C, Sn, Ge, and Si markedly increase the burning velocity of H_2–Cl_2 flames containing 0.6 mole fraction of H_2. This contrasts with the observed velocity decrease for flames where O_2 is the oxidant, e.g., see Table 5.23.

Several theoretical tests of the halogen inhibition model have been made in recent years, e.g., see Day *et al.* (1971), Lovachev and Gontkovskaya (1972), and Lovachev *et al.* (1973). Reactions between HBr and H, OH, and O are considered as sources of inhibition, and reactions between Br and HO_2, H, and Br are considered as routes for regenerating the active halogen species HBr. The relative importance of the individual HBr reactions with H, OH, and O is impossible to establish from these kinetic model calculations owing to the lack of relevant rate data for OH and O. However, these kinetic calculations do tend to support the model insofar as the considerable uncertainties in rate data permit. For example, the presence of 1% HBr in fuel-rich or fuel-lean H_2–O_2 flames, at a temperature of 1000 K, increases the time required to form 90% of the H_2O by at least three orders of magnitude.

Looking ahead to the time when sufficient basic data are at hand to allow rigorous kinetic modeling of flame inhibition, we can expect that the reaction rates of excited-state species will also need to be taken into consideration. A most pertinent example is given by the reaction

$$H + HCl \rightarrow H_2 + Cl,$$

where the rate for HCl in the first excited vibrational state is almost two orders of magnitude greater than for the ground or thermal equilibrium state (Arnoldi and Wolfrum, 1974).

Finally, we should recall that in the higher temperature reaction-zone region of flames, i.e., where $T > 1000$ K, reactions involving H, OH, and O are known to be balanced and a local equilibrium exists between these species (see Section II.I). Hence, under these conditions, arguments as to whether OH, O, or H are the inhibited species are not particularly critical, as the depletion of any of these radicals would serve to inhibit chain branching.

D. Metal-Type Inhibitors

1. Solid Inhibitors Containing Alkali Metals

Finely divided powders ($< 100\ \mu m$) of inorganic salts, particularly those containing an alkali metal such as Na or K, are known to be effective fire extinguishants. The compounds $KHCO_3$, $KHCO_3 \cdot$urea, and $K_2C_2O_4 \cdot H_2O$ are especially effective and are in widespread commercial use, ranging from small household or laboratory fires to fuel fires associated with aircraft mishaps.

Table 5.24 gives a comparison of the relative effectiveness of some of these alkali metal salts and other competitive extinguishants such as CF_3Br, $(NH_4)H_2PO_4$, and the chromate- and cryolite-type salts. In passing, it is of interest to note that the basic ammonium phosphate salts are used as aircraft-drop forest fire extinguishants. For each of the three laboratory test methods used, the data of Table 5.24 show the same general trend of decreasing effectiveness from the $KHCO_3 \cdot$urea complex to rock dust or silica.

TABLE 5.24

Comparison of Powder Inhibitors as Combustion Retardants

Retardant	Oxygen index[a]	Wt % for extinguishment[b]	50% decrease in blow-off velocity,[c] g/liter
KHO_3·urea	0.47	2.4	—
Na_2CrO_4	—	—	0.01
$(NH_4)_2CrO_4$	0.39	—	—
Na_3AlF_6	—	14	0.015
K_3AlF_6	0.36	—	0.012
$KHCO_3$	—	5.1	0.041
$NaHCO_3$	0.36	6.0	0.055
$(NH_4)H_2PO_4$	0.33	8–9	—
CF_3Br	—	19	0.03
Rock dust	—	48	—
SiO_2	—	—	$\gg 0.09$

[a] From Altman (1973), using premixed propane–oxygen–nitrogen flames; a high oxygen index indicates that a greater amount of O_2, relative to N_2, is necessary to support combustion in the presence of the inhibitor.

[b] From Grumer *et al.* (1973), using coal dust–air combustion mixtures; the wt % of additive necessary to prevent combustion is given.

[c] From Fiala and Winterfeld (1971); the concentration of extinguishant needed to decrease by 50% the blow-off velocity of a stoichiometric propane–air flame is given.

One can compare the effectiveness of these solids with the vapor-phase flame inhibitors considered in Table 5.23 by noting that about 10^{-3} mole fraction of a powdered alkali metal salt, such as K_2SO_4, Na_2CO_3, $KHCO_3$, or $NaHCO_3$, can reduce the burning velocity of a CH_4–air flame by 50%. Thus ϕ_v factors of about 100 are indicated. Powders of alkali metal salts are also found to be more effective on a weight basis than CF_3Br in a counter flow diffusion flame (Milne *et al.*, 1970). This result is basically in accord with the data shown in Table 5.24. The order of effectiveness for alkali metal salts usually follows the order Li < Na < K < Rb. Alkali metal carbonates are observed to be about twice as effective as the corresponding chlorides.

The mechanisms by which these powdered salts effect flame inhibition have not been resolved and the following discussion summarizes the current evidence. Flame temperature calculations (Dodding *et al.*, 1970) show that the effect of decomposing $NaHCO_3$ on flame cooling is small and that the mode of inhibition must therefore be chemical in nature. In practice, the smaller the initial particle size of the powder introduced to the flame, the greater the degree of inhibition found, e.g., see Dodding *et al.* (1970)

and Rosser *et al.* (1963). Friedman and Levy (1963) have observed that the introduction of elemental Na or K vapor to the fuel side of methane–air counter flow diffusion flames has no effect on the flame strength. Evidently this mode of introduction prevents formation of the functioning inhibiting species.

From observations such as these, arguments have been given in favor of both condensed- and vapor-phase inhibition mechanisms. Rosser *et al.* (1963) favored a gas-phase mechanism, since they calculated that under flame conditions an appreciable vaporization and dissociation of the solid powders should occur. This view is also supported by the studies of Vanpee *et al.* (1964) where the use of a long residence time (~80 msec) combustion system almost certainly ensures the complete vaporization of the alkali metal salts. The more recent study of Birchall (1970), using town gas–air flames, further supports this view. His observation of a relatively high efficiency for the alkali metal oxalates was accounted for by a model where the reactions

$$K_2C_2O_4 \cdot H_2O(s) \rightarrow K_2C_2O_4(s) + H_2O$$

and

$$K_2C_2O_4(s) \rightarrow K_2CO_3(s) + CO,$$

resulted in the production of submicron size carbonate particles early in the flame. These particles would then readily vaporize and decompose to yield the active inhibitor species. The following order for the effect of the anion component of the initial compound on the alkali efficiency was indicated:

$$\text{oxalate} > \text{cyanate} > \text{carbonate} > \text{iodide} >$$

$$\text{bromide} > \text{chloride} > \text{sulfate} > \text{phosphate}.$$

This order apparently represents the ease with which the alkali metal can be released to form the active species. It was also found that the presence of Cl_2 gas retards the efficiency of the oxalate and cyanate inhibition.

Proponents of a vapor-phase mode of inhibition mostly agree that the active gas-phase species would be a hydroxide, such as KOH. Under equilibrium conditions this species is much more stable than the oxides or elemental K, although with very lean flame conditions the oxides may also be important (Kaskan, 1965). The likely KOH-forming reaction

$$K + OH + M \rightarrow KOH + M$$

is considered to be kinetically more favorable than the alternative endothermic process

$$KOH + OH \rightarrow H_2O + KO.$$

Flame inhibition then most likely results from the reaction

$$KOH + H \rightarrow H_2O + K.$$

The observed poisoning effect of halogens, present in suitably large concentration (~1%), can be rationalized by the known stability of the alkali halides in flames resulting in a loss of KOH.

2. *Other Metal-Type Inhibitors*

It is apparent from the classification of inhibitor effectiveness given in Table 5.23, that metal-containing compounds such as CrO_2Cl_2, $Pb(C_2H_5)_4$, and $Fe(CO)_5$ are one to two orders of magnitude more effective than halogen inhibitors and several times more effective than the alkali metal salts. However, systems of this type have not yet been utilized in practice, with the possible exception of the recent commercial introduction of ferrocene as a flame inhibitor and smoke suppressant. It is probably relevant that lead tetraethyl is also an effective modifier of the preignition knocking phenomenon of internal combustion systems. One can reasonably expect that the modes of action in flame inhibition and knock prevention may be related. Unfortunately, despite considerable experimental effort, the mechanism of knock inhibition has not been definitely established, but see Section VII.

The metal and phosphorous halides, as indicated in Table 5.23, also show a degree of inhibition which is in considerable excess of what can be accounted for in terms of their halogen content. This is particularly evident in the H_2–air flame where Cl_2 itself does not provide any flame inhibition. It is clear then that the metals—and phosphorous—can themselves lead to flame inhibition and, more importantly, to a much greater degree than the halogen inhibitors. For instance, $POCl_3$ is about ten times more effective than the equivalent amount of chlorine. The two order-of-magnitude difference in ϕ_v between $Si(CH_3)_4$ and $Pb(C_2H_5)_4$ suggests that the inhibition mechanism is particularly sensitive to the properties of the metal itself. In fact, some metals show no inhibition at all. For example, the addition of several percent of Al_2Cl_6 vapor to premixed fuel-rich CH_4–O_2 flames produced a reduction in burning velocity that could be accounted for entirely by the amount of halogen present (Friedman and Levy, 1958).

A clue to the possible function of phosphorous additives as flame inhibitors is given by the observations of Fenimore and Jones (1964) on low pressure fuel-rich H_2 flames containing phosphorous. The species HPO was spectroscopically identified in the post-combustion gases by its characteristic green chemiluminescence. From indirect mass-action-type considerations the species P_2 was also believed to be a major P-containing species.

More recent mass spectrometric sampling experiments have verified the presence of P_2, PO, and other P-containing species in H_2-containing flames (Hastie, 1973a). Reactions such as

$$H + PO + M \rightarrow HPO + M,$$

$$HPO + H = H_2 + PO,$$

$$P_2 + O = P + PO,$$

$$P + OH = PO + H,$$

could conceivably lead to a catalytic loss of H-atoms from the combustion system.

Metals such as Fe, Cr, and Ti, and their oxides, have low vapor pressures under normal flame conditions, and their introduction to flames can result in the formation of condensed particles. These particles are highly luminous and are readily observed in the region of the reaction zone and the postflame gases. It has been argued, therefore, that these metals perturb the flame chemistry via heterogeneous rather than homogeneous reactions. A clue to the possible effectiveness of a heterogeneous, as opposed to a homogeneous, mode of inhibition is provided by the observation of Jost *et al.* (1961), where the flame-inhibiting effect of $Fe(CO)_5$ reached a limiting upper concentration level. At the higher concentration levels, in excess of the saturation pressure for iron-containing species, condensed particles should form, and the decreased rate of effectiveness above the limiting concentration level suggests that heterogeneous inhibition is less effective than for homogeneous reactions.

Evidence for heterogeneous processes can result from the fact that radical recombinations on solid particles should lead to a temperature rise at the particle. This has been observed for Cr, at concentrations of greater than 10^{15} cm^{-3} in H_2–O_2–N_2 flames, and occurs as a result of the reaction

$$Cr_2O_3(s) + H + H \rightarrow Cr_2O_3(s)^* + H_2.$$

However, at lower Cr concentrations it was found that the recombination of H-atoms occurred via a homogeneous reaction, as discussed elsewhere (see Section II.L). Similarly, the observation of a particle temperature apparently greater than the adiabatic flame value for uranium oxide particles led to the conclusion that radical recombination was occurring at the particle surface (Bulewicz and Padley, 1970b). However, Tischer and Scheller (1970) argue that the excess particle heating, if any, could well be due to gas–solid reactions and that radical recombination occurs primarily in the gas phase.

In these particle studies, observations were made downstream of the reaction zone, and the question arises as to whether the additives have an opportunity to form solids in the preflame region. Very little has been done to answer such a question. However, from the light scattering experiments of Cotton and Jenkins (1971), it is evident that for the case of Ba, at concentrations of 10^{-5}–10^{-6} mole fraction in a 1600 K H_2–O_2–N_2 flame, solids do form prior to the reaction zone. These flames did not show the streakiness which is commonly observed when solid particles are present. However, in spite of the presence of solids, the observed catalytic effect of alkaline earth metals on the recombination of H-atoms could be explained solely in terms of a homogeneous mechanism (see Section II.L).

The post-reaction-zone radical recombination studies such as those of Bulewicz and Padley (1970a), made in the presence of metal catalysts, have a significant bearing on the likely flame-inhibiting function of metal-containing species, e.g., see Section II.L. The results of these studies have also dispelled the often relied upon notion that the presence of metallic elements at ppm concentration levels has no effect on the flame chemistry. The survey experiments of Bulewicz and Padley (1970a) indicated that the elements Mg, Cr, Mn, Sn, U, and Ba had a pronounced effect on the recombination of H-atoms, whereas Na, Co, Ni, Cu, V, Zn, Ga, Th, Ce, and La were found to be ineffective as catalysts for H-atom recombination. These results were interpreted in terms of the model

$$\mathrm{MO + H + X \rightarrow MOH + X^*}$$

and

$$\mathrm{MOH + H\ (or\ OH) \rightarrow MO + H_2\ (or\ H_2O)},$$

i.e., the diatomic metal oxide species catalyzes the recombination of H or OH radicals via metal hydroxide intermediates. Many of these oxides and hydroxides are known to be stable under typical flame conditions (see Table 5.13). Unfortunately, only Sn and Cr are available for comparison with the flame speed measurements represented by Table 5.23. However, both of these elements show high ϕ_v values, and they are also among the most effective catalysts for H-atom recombination. Hence, it is reasonable to conclude that the role of the metallic flame inhibitors is somewhat analogous to that of the halogens, in that the overall effect results from catalysis of H-atom recombination. It is not inconceivable that P-containing species behave in a similar manner, particularly as PO and HPO are well-known flame species as discussed earlier.

It is apparent that the oxide–hydroxide mechanism indicated above relies on a rather special set of energetics. In particular, MOH must be stable under flame conditions but must readily react with a H-atom when the opportunity arises. Ideally, this last step should be exothermic to reduce the probability of a reverse reaction. A more critical set of energetic conditions is revealed by the case of Sn, where the initial step in the catalytic process is

$$SnO + H \rightarrow SnOH^*.$$

Here SnOH is produced in an excited electronic state which crosses to another state before undergoing further reaction with a second H-atom to produce H_2 and ground-state SnO (Bulewicz and Padley, 1971a). Such subtleties in energetics readily account for the striking variance in the ability of metals to inhibit flames or at least catalyze radical recombination. Further discussion along these lines has been given elsewhere (see Section II.L).

The poisoning effect of Cl_2 noted for alkali metals in flame inhibition can also be expected to occur for some of the other metal inhibitors. In particular, flame systems containing metals such as Ba, Cr, and Ca, and a halogen source can be expected to form stable MCl and MCl_2 species. This would reduce the effectiveness of these metals, as they would not be available to participate in the MO–MOH inhibition sequence. On the other hand, metals such as Sb and Sn have relatively weak metal–halogen bonds, and the halogen poisoning effect would be minimal in these cases.

Finally, we should note that the low ϕ_v value for $Pb(C_2H_5)_4$ in a fuel-rich H_2–air flame, i.e., see Table 5.23, indicates that lead is relatively ineffective in this flame relative to that in a stoichiometric *n*-hexane–air flame. This result is quite compatible with the general inhibition model for metals where the monoxide species, i.e., PbO in this case, is presumed to serve as a catalytic intermediate for H-atom recombination, the reason being that lead additives exist only as Pb-atoms in fuel-rich flames, whereas PbO is a dominant species under stoichiometric or lean conditions (Friswell and Jenkins, 1972).

E. Synergistic Systems: The Antimony Oxide–Halogen Example

Among practicing fire retardancy chemists, synergism is one of the more desirable goals. The achievement of a synergism allows the utilization of flame-inhibiting additives at lower concentrations, and the antimony oxide–halogen combination represents one of the more important synergistic fire-proofing formulations. The synergistic nature of this system is exemplified by the fact that a fire-proofed epoxy substrate containing a source of 15% Br, such as a halogenated paraffin, can be replaced by a combination of 5% Br and 3% $Sb_4O_6(s)$. Thus the quantity of halogen needed is significantly reduced by the presence of $Sb_4O_6(s)$. Also as $Sb_4O_6(s)$ is ineffective in the absence of the halogen, the combination is synergistic.

This fire retardant system has been in use for more than 30 years. However, as recently as 1967 no satisfactory theory had been suggested to explain the synergism between halogen and antimony compounds in imparting fire retardancy to polymer compositions (Hindersinn and Wagner, 1967).

Analysis of polymer substrates before and after pyrolysis, or charring, reveals that most of the antimony, and halogen, content is depleted (Fenimore and Jones, 1966). Pyrolysis usually occurs in the 300–500 °C temperature region and the low vapor pressure of Sb_2O_3 precludes the loss of antimony by a direct vaporization of the oxide. Chemical (Learmonth and Thwaite, 1970) and mass spectrometric (Hastie, 1973a) analyses indicate that the antimony is released to the vapor phase in the form of $SbCl_3$.

It is well established that the interactions leading to the release of $SbCl_3$ to the vapor phase do not appreciably alter the mode of decomposition of the polymer substrate, e.g., see Pitts (1972) and Fenimore and Jones (1966). Also, limiting oxygen index flammability rating tests in a nitrous oxide atmosphere showed very little flame inhibition as compared with those carried out in an oxygen atmosphere. Flames supported by nitrous oxide do not rely on chain branching for propagation, and this test is used as supporting evidence for the observed retardancy occuring in the flame rather than the substrate.

One could argue that the role of antimony in this system is merely to provide a source of halogen in the gas phase. However, this does not explain the synergism as the organic halide readily provides HCl to the flame in the absence of antimony. From thermogravimetric (Pitts, 1972) and mass spectrometric (Hastie, 1973a) studies made over the 200–500 °C temperature interval, it was found that the release of $SbCl_3$ occurs over an unusually wide temperature range, i.e., 250–400 °C. The mechanism for this moderated rate of release is attributed to the formation of several intermediate solid oxychlorides, as discussed elsewhere (Pitts, 1972). Mass spectrometric studies (Hastie, 1973a) show that this process is, in part, kinetically controlled, and it is not surprising, therefore, that the addition of a third substance to the substrate can modify and improve the process. For example, a partial replacement of Sb_2O_3 with $2ZnO \cdot 3B_2O_3 \cdot 3.5H_2O$, also known commerically as Firebrake ZB, can provide an improved retardancy effect (Woods and Bower, 1970). Thus it could be argued that the synergism is due to the moderating effect of antimony in allowing the release of halogen to the vapor phase over a wide temperature interval. This effect has a parallel in the use of substrate free radical initiators such as dicumyl peroxide, which are believed to delay the loss of halogen from the decomposing polymer (Eichhorn, 1964).

There is evidence to suggest that, given the opportunity to enter the vapor phase, antimony in the absence of a halogen can provide a useful degree of flame inhibition. In particular, triphenyl stibine shows all the characteristics of an effective vapor-phase flame inhibitor (Fenimore and Martin, 1972). Furthermore, the high ϕ_v values for $SbCl_3$ and $SbCl_5$, as compared with Cl_2 or CCl_4 (see Table 5.23), are indicative of antimony itself being involved in the flame-inhibiting process. In fact, the ϕ_v rating for $SbCl_3$ is not very different from that for $SnCl_4$, and it is reasonable to consider the possibility of an oxide–hydroxide inhibition process analogous to that suggested earlier for Sn.

From the results of recent mass spectrometric analyses of $SbCl_3$ and $SbBr_3$ in 1-atm fuel-rich CH_4–air flames, one can suggest a very plausible mechanism for the role of antimony trihalides as flame inhibitors (Hastie, 1973b). The results of the analysis for flames containing $SbBr_3$ are summarized by the species concentration profiles in Fig. 5.22. The appearance of CH_3Br and HBr in the preflame region is in accord with the observations of Wilson *et al.* (1969) for halogen additions in low pressure flames. Their model for halogen inhibition also provides the best explanation for the halogen component of this system. The role of antimony as an inhibitor most likely involves the species SbO and Sb which are shown to be the major antimony species in the region of the reaction zone. It should be noted that under the present conditions, i.e., antimony halide mole fractions in the range of 10^{-4}–10^{-3}, no condensation to form solids or liquids is possible and heterogeneous mechanisms do not require consideration.

A set of likely reactions involving antimony halides in these flames is given in Table 5.25. The reactions 4(a)–4(e) are analogous to those suggested for Sn flame inhibition. It would appear from the profile data of

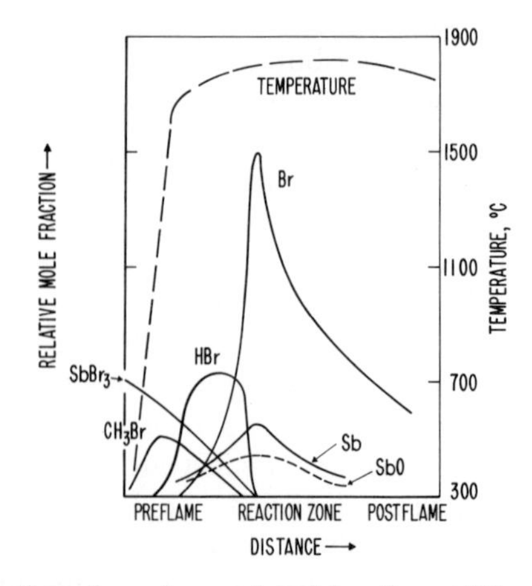

Fig. 5.22 Distribution of species produced by the addition of $SbBr_3$ vapor to a fuel-rich premixed CH_4–O_2 flame (Hastie, 1973b).

TABLE 5.25

Probable Reactions Involving Antimony Trihalides [a,b] in CH_4–O_2 Flames

1.	(a)	$SbX_3 + H \rightarrow HX + SbX_2$
	(b)	$SbX_3 \rightarrow X + SbX_2$
	(c)	$SbX_3 + CH_3 \rightarrow CH_3X + SbX_2$
2.	(a)	$SbX_2 + H \rightarrow HX + SbX$
	(b)	$SbX_2 + CH_3 \rightarrow CH_3X + SbX$
3.	(a)	$SbX + H \rightarrow Sb + HX$
	(b)	$SbX + CH_3 \rightarrow Sb + CH_3X$
4.	(a)	$Sb + O + M$[c] $\rightarrow SbO + M^*$
	(b)	$Sb + OH + M \rightarrow SbOH + M^*$
	(c)[d]	$SbOH + H \rightleftarrows SbO + H_2$
	(d)[d]	$SbO + H \rightarrow SbOH^*$
	(e)	$Sb + H_2O \rightleftarrows SbO + H_2$
	(f)[d]	$Sb + OH \rightleftarrows SbO + H$
	(g)	$Sb + H_2O \rightleftarrows SbOH + H$
5.	(a)	$X + X + M \rightarrow X_2 + M^*$
	(b)	$X_2 + CH_3 \rightarrow CH_3X + X$
	(c)	$X + CH_3 + M \rightarrow CH_3X + M^*$
6.	(a)[d]	$HX + H \rightarrow H_2 + X$
	(b)	$HX + CH_3 \rightarrow CH_4 + X$
7.	(a)	$X + HO_2 \rightarrow HX + O_2$
8.	(a)	$CH_3X + H_2 \rightarrow CH_4 + HX$
	(b)	$CH_3X + H \rightarrow CH_4 + X$ or $CH_3 + HBr$
	(c)	$CH_3X \rightarrow CH_3X^* \rightarrow CH_3 + X$

[a] That is, chlorides or bromides.
[b] * denotes an excited-state condition.
[c] M is a major gas component—third body.
[d] Reactions leading directly to flame inhibition.

Fig. 5.22 that inhibition involving SbO and Sb species would occur primarily in the region of the reaction zone. This contrasts with the suggested role of HBr where preflame processes are considered important. Thus, we are left with the not unreasonable conclusion that different flame inhibitors may operate in quite different flame regions. Furthermore, as the metal-type inhibitors appear to be more effective than the halogens, it is suggested that inhibition at the reaction zone is more desirable than in the preflame region. We recall that the main source of H-atoms in the preflame region is from diffusion out of the reaction zone, so that inhibition at this zone could be expected to have a delay effect on the flame reaction s'milar to that suggested by the halogen-inhibition model.

F. Conclusions

It is evident from the foregoing discussion that inorganic systems containing halogens, other than F, or certain metals can be effective flame inhibitors. It also appears that high temperature species—such as metal halides, oxides, and hydroxides—play an important part in the flame-inhibition processes. These processes are kinetically controlled and rely rather stringently on a special set of energetics, which remains to be determined. The basic feature of each of the flame-inhibiting systems considered is their interaction with H-atoms and, in some instances, OH species as the mode of inhibition. It also appears likely that the inhibition of H-atoms can be effectively achieved in both the preflame and reaction-zone flame regions.

We should note that hydrocarbons are themselves strong flame inhibitors. This is particularly evident when hydrocarbons are introduced into H_2–air flames (Miller *et al.*, 1963). The inhibiting effect is thought to be due to reactions such as

$$H + CH_4 \rightarrow H_2 + CH_3$$

and

$$H + C_2H_4 \rightarrow C_2H_5,$$

where the product radicals are far less reactive than the H-atom.

From the dependence of burning velocity on flame composition in the presence of inhibitors, as shown in Fig. 5.23, it is evident that the inhibitors, $Fe(CO)_5$ and $SbCl_5$, have a greater degree of effectiveness in the more fuel-rich flame mixtures, i.e., at fuel equivalence ratios of greater than about 2. Such conditions favor H atoms rather than O or OH as the predominant radical species. Hence the suggested inhibition mechanisms, for halogen-

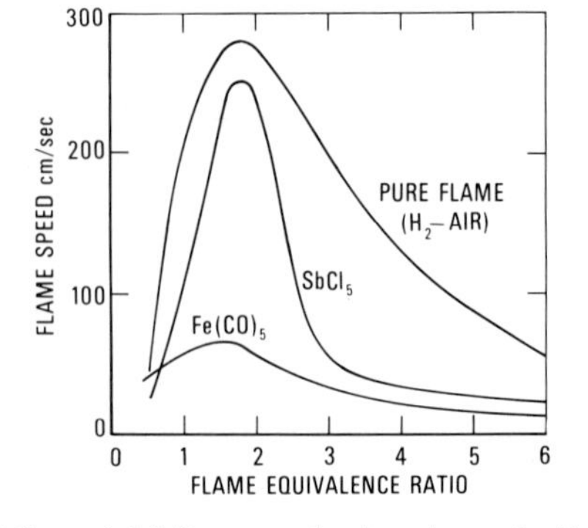

Fig. 5.23 Effect of flame inhibitors on the burning velocity of H_2–air flames and as a function of flame composition (Miller *et al.*, 1963).

and metal-type inhibitors, which basically involve catalysis of H-atom recombination, are further supported by these macroscopic observations.

The most reasonable model for the inhibiting effect of metals is one involving participation of metal oxide and hydroxide species. This effect is believed to be important primarily in the region of the reaction zone and also in the afterburning region of fuel-rich systems (Vanpee *et al.*, 1964). From the high metal–halogen bond energies for additives such as $TiCl_4$ and $SiCl_4$, it is most unlikely that any appreciable formation of the monoxide or hydroxide would occur prior to the reaction zone. For the low concentrations of metal additives used, it is not unreasonable that the onset of condensation and particle formation should lead to a reduction in the rate of inhibition with additive concentration, as the formation of particulates would tend to lower the collision probability of radicals with the additive.

As an example of the complexity involved in assessing, from basic principles, the potential of metals as flame inhibitors we may consider the Cr system. From the flame speed measurements it is known that Cr has an overall flame-inhibiting effect. Similarly, the catalytic enhancement of H-atom recombination is also indicative of possible flame inhibition. However, the second stage of hydrocarbon combustion in flames, namely, the oxidation of CO to CO_2, is actually accelerated by the presence of Cr at the ppm concentration level, e.g., see Matsuda and Gutman (1971). A similar enhancement is found with Ni and Fe additions, but Pb and Te show a retarding effect on this oxidation.

It is apparent that the various suggested mechanisms for flame inhibition do not satisfy all of the macroscopic observations and questions such as the following arise: why do PCl_3 and PBr_3 have the same degree of effectiveness in *n*-hexane–air mixtures when Br_2 and HBr are substantially superior to Cl_2 or HCl? Why is the inhibiting action of halogens more pronounced for fuels with low hydrogen content, e.g., C_6H_6, than for H_2 or CH_4? Why is $Fe(CO)_5$ much less effective in H_2-fueled flames? Similarly, why is $Fe(CO)_5$ less effective in a CH_4–O_2 flame than a CH_4–air stoichiometric flame (Lask and Wagner, 1962)?

It is apparent that various test methods indicate different degrees of effectiveness of metals as inhibitors of gas-phase combustion. For example, Fe and Pb appear to be very effective in reducing the flame speed of *n*-hexane–air flames but only moderately effective in H_2–air flames and ineffective in reducing the afterburning of fuel-rich ethylene–oxygen–nitrogen mixtures, as shown by Vanpee *et al.* (1964).

Another aspect involving "apparent" differences in inhibition effectiveness, dependent upon the method of observation, involves the H-atom recombination results of Bulewicz and Padley (1970a), which were discussed in Section II.L. While the catalytic effects of Sn and Cr were consistent with the observed reduction in burning velocity, the apparent non-catalytic effects of Na and K *appear* to discredit the suggested model for flame inhibition by these elements. With this model, the reaction

$$KOH + H \rightarrow K + H_2O$$

is involved in a catalytic loss of H-atoms with resulting flame inhibition. There are several possible explanations for this apparent anomaly. First, the H_2–O_2–N_2 flame used by Bulewicz and Padley (1970a) has a very high H-atom concentration in the reaction zone, namely about 3×10^{-2} atm. It follows from this, and the likely balanced character of the above reaction, that the concentration ratio of KOH to K is only about 0.006 in this particular flame. Also, the total amount of potassium added to the flame was equivalent to a species partial pressure of about 10^{-6} atm. Hence, at the reaction zone, the concentration of KOH would be less than 10^{-8} atm, that is about six orders of magnitude less than for the H-atom. Under such conditions it is not too surprising that no catalytic effect was observed. Flame speed reduction and extinguishment measurements, on the other hand, are made at significantly higher additive concentrations of about 10^{-4}–10^{-3} atm.

A second factor in this radical recombination versus flame speed "anomaly" involves a fundamental difference between the hydrogen and hydrocarbon flames, used for H-atom recombination studies and macroscopic flame-inhibition tests, respectively. For the particular hydrogen flame in question, the H-atom concentration at the reaction zone is two orders of magnitude in excess of the equilibrium level, whereas for hydrocarbon flames having a similar fuel-rich character the corresponding excess is only one order of magnitude, or less. Thus, in the latter case, the ratio of [KOH]/[K] would be greater by at least a factor of 10, and a more noticeable degree of inhibition could then be expected. It is suggested that such "apparent" discrepancies between various inhibition-related experiments would vanish if the complete chemical-kinetic character of the various sets of conditions is taken into account.

Finally, as much of the experimental evidence for the present understanding of flame inhibition has been derived from molecular studies on low pressure flames, it is important, for scaling purposes (e.g., see Section III.E), to consider the effect of pressure on inhibition. Bonne *et al.* (1962) found that as the flame pressure was lowered, the inhibiting action of $Fe(CO)_5$ decreased and had little effect on the OH concentration or the flame temperature. Inhibition was found to be about a factor of 3 greater at 1 atm than at $\frac{1}{2}$ atm. Similar results were obtained for the inhibition effect of trimethyl phosphate (TMP) on H_2–O_2–Ar flames (Fenimore and Jones, 1964). Flames at 0.13 atm pressure were unaffected by additions of

10^{-3} mole fraction TMP, whereas at 1-atm-pressure inhibition was observed. A factor-of-2 reduction in the H-atom concentration was also observed when 10^{-3} mole fraction of TMP was added to the 1-atm flame. Fenimore and Jones (1964) suggested that the lack of an observable inhibition effect for the low pressure flames may be the result of the much greater excess concentration of H-atoms present in low pressure flames. The flame speed is also greater at reduced pressure for most flames.

More recently, Homann and Poss (1972) have found that the degree of inhibition, as represented by a reduction in burning velocity for an ethylene–air flame, is significantly less at about 0.17 atm than at 1 atm for the additives $Fe(CO)_5$, CH_3I, CH_3Br, and CH_2Br_2, but not for CH_2Cl_2. In particular, the $Fe(CO)_5$ inhibitor was the system most affected by a pressure change. The authors suggest the possibility that this effect is indicative of the role of termolecular reactions—in addition to the normally considered bimolecular steps—in the inhibition process. Such an argument is in keeping with both the proposed model for metal oxide inhibition via the hydroxide intermediate and the model for halogen inhibition with three-body regeneration of the hydrogen halide species.

While the degree of inhibition appears to be pressure dependent, the relative effectiveness of most flame inhibitors is, nevertheless, remarkably similar for both 1- and 0.01-atm-pressure conditions, and also for both diffusion and premixed flames (Miller, 1969).

We may conclude from these few observations that there are qualitative similarities between the low pressure and 1-atm-pressure inhibition results, but that one should proceed with caution in attempting to extrapolate the results of inhibition studies on low pressure flames to the "real-life" application of atmospheric pressure fire systems.

VI. Smoke Formation and Suppression

A. Introduction

The importance of carbon formation in combustion processes is twofold—there being instances where carbon formation is desirable and others where it is undesirable. The preparation of carbon-black and the generation of highly luminous flames for lighting, e.g., in candles, and heat transfer purposes, e.g., in open hearth furnaces, represent examples where a controlled formation of carbon in flames is desirable. Undesirable aspects of carbon formation include the smoky exhausts of jet aircraft engines (Linden and Heywood, 1971), carbon formation and deposition in internal combustion engines, smoke emission from diesel engines, and smoking flames in gas-burning equipment (Singer and Grumer, 1962).

Smoke production can be controlled, to some degree, by the presence of inorganic additives in the combustion system. These additives are converted to high temperature species in the flame, and their chemical role in influencing smoke levels provides yet another application for high temperature vapors in combustion.

B. Smoke Formation

The phenomenological aspects of carbon formation, such as the fuel-type and the gas-dynamic conditions under which carbon formation occurs, are well known, e.g., see Street and Thomas (1955). Many different routes by which fuel molecules are converted to carbon particles have been suggested. However, the identity of the reaction intermediates involved is not firmly established but most reviewers favor the scheme suggested by Homann and co-workers where polyacetylene species ($C_{2n}H_2$) are the key intermediates, e.g., see Homann (1972) and cited earlier work. Species that are often considered as important reaction intermediates include:

$$C_4H_2,\quad C_2H,\quad C_2H_3,\quad C_2H_2,\quad H,\quad C_2,\quad C_4, \text{ and } C_6.$$

Unidentified charged species may also be important in the nucleation process—e.g., see the discussions of Gaydon and Wolfhard (1970) and Lawton and Weinberg (1969).

According to Homann (1972), the following reactions are likely steps leading to the formation of carbon in premixed flames:

$$O + C_2H_2 \rightarrow CH_2 + CO,$$

$$CH_2 + C_2H_2 \rightarrow HC_2CH_3 \text{ (methylacetylene)},$$

$$HC_2CH_3 + CH_2 \rightarrow CH_3CCCH_3 \text{ (dimethyl acetylene)},$$

or

$$C_2H + C_2H_2 \rightarrow C_4H_2 + H,$$

and

$$C_2H + C_4H_2 \rightarrow \underset{\substack{|\\ C\\ |||\\ C\\ |\\ H}}{HC = C}{-}C{\equiv}C{-}H$$

These branched hydrocarbon radicals may then form five- or six-membered rings with side chains available for further ring closure. Such rings are already present in flames using aromatic fuels, and this type of flame is known to be more susceptible to carbon formation.

Recent reviews of the chemistry of carbon formation have been given by Minkoff and Tipper (1962), Singer and Grumer (1962), Fenimore (1964), Bonne *et al.* (1965), Fristrom (1966), Homann (1967), Homann and Wagner (1967, 1968), Feugier (1969), Bradley (1969), Lawton and Weinberg (1969), Gaydon and Wolfhard (1970), Homann (1972), and Palmer and Seery (1973). In view of this extended coverage, additional detailed discussion of carbon formation appears unwarranted.

C. Smoke Suppression

The suppression of carbon in combustion systems can be achieved by,

- adjustment of the operational parameters such as fuel-to-oxidant ratio and aerodynamic flow conditions—as has been suggested for smoke reduction in jet engines (Linden and Heywood, 1971);
- use of electrical techniques which take advantage of the charged character of flame particulates—as has been suggested, for example, by Hardesty and Weinberg (1973);
- the use of chemical additives which either catalyze the oxidation of carbon within the flame, or reduce the concentration of charged condensation nuclei; the commercial use of Ba additives in diesel fuel to suppress exhaust smoke is an example of the chemical inhibitor approach (Golothan, 1967).

In considering the role of metal-containing chemical additives in smoke suppression, one is faced with the problem of setting up criteria for defining their degree of effectiveness. Cotton *et al.* (1971) chose the threshold value of the fuel-to-oxidant equivalence ratio at which soot was produced, for correlation with the degree of effectiveness of metal additives as smoke suppressants.

The smoke-inhibition mechanism suggested by Cotton *et al.* (1971) is particularly interesting, as it involves another aspect of the metal oxide–hydroxide catalytic process on H-atoms which was suggested as a mechanism for flame inhibition (see Section V). It also serves to demonstrate the subtle dependence of the function of flame additives on varying flame conditions. Their survey involved the effect of some 40 metals on the smoke point of propane–oxygen diffusion flames at atmospheric pressure. These flames, in essence, had cylindrical symmetry and were nonturbulent. Temperatures were typically in the region of 2100 K. The metals were introduced into the fuel stream as atomized aqueous solutions with N_2

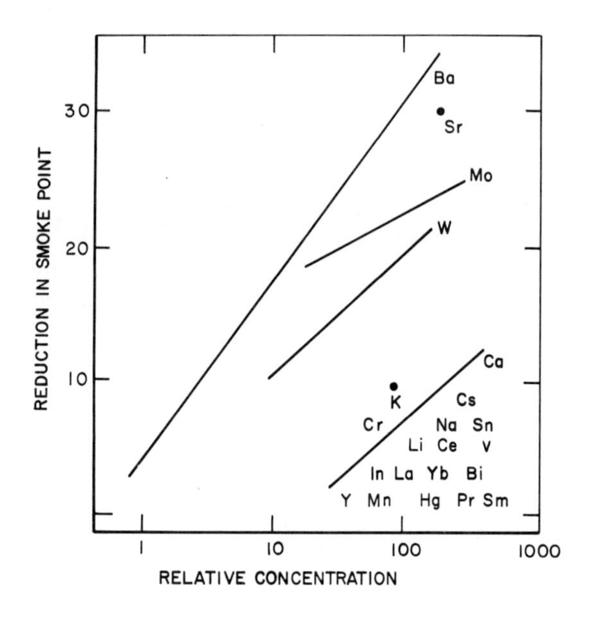

Fig. 5.24 Reduction in smoke point (relative to Ba) versus concentration of additive, in propane–oxygen diffusion flames (Cotton *et al.*, 1971).

as a carrier gas. For this study the metal concentration levels in the flame were typically in the region of 10^{-4} to 10^{-5} mole fraction. The efficiencies of the various metals, relative to Ba, as soot suppressants are represented in the Fig. 5.24. The other metals studied, and not shown in this figure, had essentially no effect relative to Ba and they include:

Ag, Al, As, Be, Cd, Co, Cu, Fe, Mg,

Ni, Pb, Si, Th, Ti, Tl, U, and Zn.

Notably, several of these metals, namely Fe, Pb, Si, and Ti, are known to be very good flame inhibitors—as discussed elsewhere (see Section V). From the Fig. 5.24, two groups of elements may be distinguished, with the group of Ba, Sr, Mo, and perhaps W, representing the highly effective smoke inhibitors.

Before discussing the significance of these observations, it is pertinent to note that Fe, which has a zero ranking in these studies, is actually used as a commercial smoke inhibitor in the form of ferrocene. Also, Spengler and Haupt (1969) found that the addition of Fe, in the form of the pentacarbonyl or as ferrocene, as well as Mn, in the form of methyl cyclopentadienylmanganese tricarbonyl, had a beneficial effect on smoke reduction in several diffusion flames and also a diesel engine. The reason for this apparent discrepancy between various studies is most probably related to the Fe additive being introduced to the flame as an aqueous solution on the one hand and as a premixed vapor on the other. It is conceivable that the aqueous solution form leads to iron oxide particle formation in the desolvation process, e.g., see Section IV. As will become evident in the discussion to follow, this form of iron (or indeed any metal) is not the most effective one for a catalytic action on smoke reduction.

Cotton *et al.* (1971) proposed a common mechanism for the soot-suppression activity of Ba, Sr, Ca, NO, and SO_2. Basically, the model is as follows. Under the fuel-rich conditions necessary for soot production, the active radical concentrations are low and, in fact, may even be below the equilibrium level. This follows from the well-known self-inhibiting effect of hydrocarbon flames, i.e.,

$$C_nH_m + H \rightarrow C_nH_{m+1},$$

and from the balanced reactions occurring between H, OH, and O, as discussed elsewhere (Section II.I). This condition of low radical concentrations should be contrasted with that found in faster burning flames, where premixing—and less fuel-rich—conditions exist, with the result that the reaction-zone radical concentration levels are well in excess of the equilibrium values. Under these latter conditions it is found that Ba and other alkali earth metals catalyze the recombination of H-atoms. The essence of the smoke-suppression mechanism involves a reversal of this recombination process under conditions of low radical concentration. That is, the metals are involved in a gas-phase catalysis of the decomposition of molecular hydrogen or water vapor to yield H-atoms.

The individual steps in the mechanism suggested by Cotton *et al.* (1971) are as follows:

$$MO + H_2 \rightarrow MOH + H, \tag{1}$$

$$MOH + H_2O \rightarrow M(OH)_2 + H, \tag{2}$$

$$M(OH)_2 + (X) \rightarrow MO + H_2O + (X), \tag{3}$$

$$H + H_2O \rightleftarrows OH + H_2, \tag{4}$$

and

$$OH + C(s) \rightarrow CO + H, \tag{5}$$

where M is a metal and X is any major flame component such as H_2O, CO_2, or N_2. The presence of MO species in flames, where M is Ba, Sr, or Ca, is well established, as is the formation of the metal hydroxide intermediates (see Section II.J). Also, the balanced nature of reaction (4) is well established, at least for the hotter flame regions. Thus, the net effect of these processes is to increase the production of the radicals H and OH

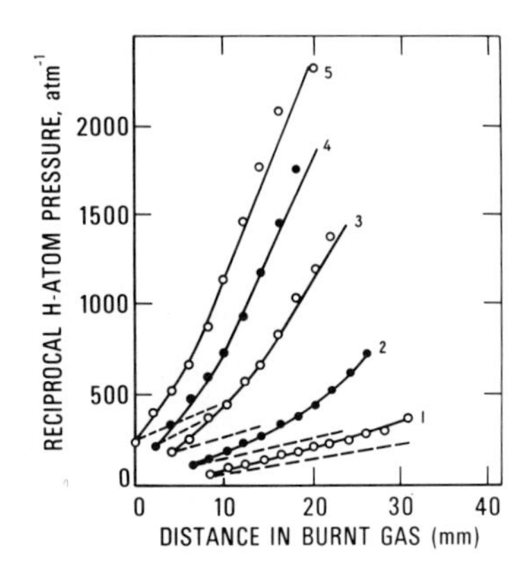

Fig. 5.25 Curves of reciprocal H-atom concentration versus distance from reaction zone, showing the catalytic effect of Ba as a function of flame composition; successive plots are displaced 2 mm for clarity; the broken curves are for flames without Ba present; the premixed flames represented have H_2–O_2–N_2 compositions of (1) 6–1–3, (2) 5–1–4, (3) 3.5–1–5.5, (4) 3–1–6, (5) 2.9–1–6.3 (Cotton *et al.*, 1971).

(and also O) and hence the oxidation of solid carbon.† Independent experimental support for the role of OH in carbon oxidation is given by the simultaneous observation of halogen-induced smoke formation and a depletion of OH species in hydrocarbon–air diffusion flames (Ibiricu and and Gaydon, 1964).

There is some evidence that O-atoms are also effective species for the oxidation of carbon, e.g., see the review of Palmer and Seery (1973). In this case the above mechanism would probably still hold true, since it is likely that the flame radicals H, OH, and O are each linked by balanced reactions such that an increase in the concentration of H or OH would also lead to an increased production of O-atoms (see Section II.I).

Note that the reaction (2)—leading to the production of the metal dihydroxide—is endothermic for the alkaline earth metals, and this suggests that the metal-catalyzed suppression of smoke will be much less effective as the temperature is decreased. We should point out, however, that in real fires the degree of fuel-oxidant mixing will also be an important factor in determining the effectiveness of metallic inhibitors.

From the catalytic model it can be argued that it is not possible to utilize metals for the *simultaneous* reduction of smoke and flame propagation. This is also suggested by the data shown in Fig. 5.25 where, as the flames

† This discussion of carbon oxidation is probably a simplification of the real situation since soot has a composition CH_x ($x < 1$), but the general mechanistic features are probably still valid.

become more fuel rich and deficient in radicals, the catalytic effect on radical recombination is substantially diminished. At the same time, the tendency for radical production by the catalytic reduction of H_2O and H_2 increases. It is therefore not at all surprising that the apparent effectiveness of fire and/or smoke retardant treatments may vary according to the test conditions used. It is also evident that the presence of solid carbon in flames is associated with an inhibiting effect on flame propagation. We may recall that the use of halogens as flame inhibitors often involves an increase in smoke production, e.g., see Petrella and Sellers (1970) and also Ibiricu and Gaydon (1964). Thus smoke production and flame propagation are intimately related phenomena and the catalytic use of additives to exclusively suppress one of these, without affecting the other, does not appear to be possible.

In conclusion, mention should be made of a recently suggested alternative—or perhaps supplementary—mechanism for smoke suppression, e.g., see the reviews of Miller (1973) and of Palmer and Seery (1973). Ions of the type $C_nH_m^+$ may serve as nucleation centers for smoke production. Hence the role of metal additives as smoke suppressants may well be one of charge exchange, i.e.:

$$C_nH_m^+ + M \rightarrow M^+ + C_nH_m,$$

which destroys the nucleation site. Such reactions are known to be rapid for

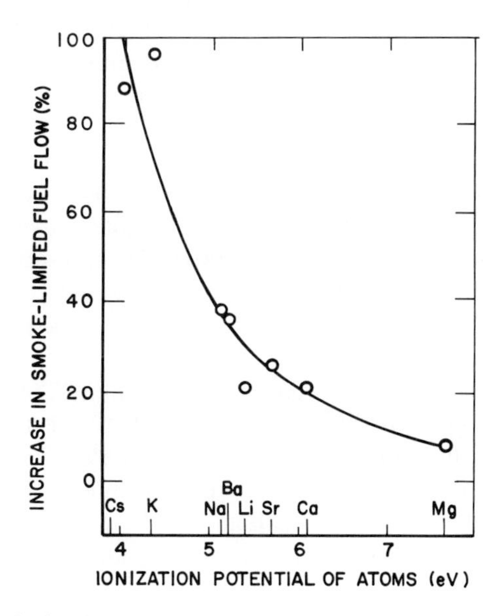

Fig. 5.26 Correlation between increase in smoke-limited fuel flow, i.e., smoke reduction, and decreasing ionization potential of the metal additives (Salooja, 1972).

H_3O^+–M charge exchange (see Section III.E). For this model, one can reasonably expect that those M species having the lowest ionization potentials would be the most effective smoke suppressants. This is borne out by the observations of Salooja (1972), as given in Fig. 5.26. In this study, contrary to the implication of Fig. 5.26, the metal additives are not necessarily present as metal atoms in the flames. If stable molecular species are formed, such as BaOH, then the molecular ionization potential may become the relevant parameter to correlate with smoke suppression. The ionization potentials of BaOH, SrOH, and CaOH are known to be slightly less than those of the metal atoms (see Section III, E).

VII. Combustion Knock

The phenomenon of "knock" in internal combustion engines is well known and metal-containing fuel additives, particularly lead tetraethyl (TEL), have been used for many years as knock suppressants.

According to the early review of Lewis and Von Elbe (1951), TEL is the most effective antiknock agent known, but iron carbonyl and nickel carbonyl are also reasonably adequate.† Compounds such as tin tetraethyl, stannic chloride, and titanium tetrachloride are almost two orders of magnitude less effective than TEL. This wide variation in the "antiknock" efficiency of various metals is in keeping with the similar behavior noted earlier in this chapter for flame inhibitors and smoke suppressants.

The mode of action of TEL as an antiknock agent appears to be an unsettled question (Bradley, 1969). It is generally agreed that some derivative of TEL is involved in radical scavenging but there are proponents of both homogeneous (Norrish, 1959, 1965; Agnew, 1960) and heterogeneous mechanisms (Kuppu-Rao and Prasad, 1972, and cited references). The TEL is generally believed to undergo rapid decomposition and oxidation to form PbO in the combustion chamber. However, the two schools of thought differ in their assignment of the state that PbO achieves, i.e., vapor or particulate.

Norrish (1959, 1965) has attempted to simulate the temperature and time scale—but not pressure—conditions of the combustion process. Spectroscopically, he observes the formation of PbO vapor species and a delay in the ignition process associated with knock. He argues that the expected temperature increase to about 850 °C precludes the presence of condensed PbO, which has a vapor pressure of about 0.25 Torr. However,

† An unpublished report has also indicated Ce—in an organometallic form—to be a very effective antiknock agent.

typical total PbO concentrations of the order of 10^{-4} mole fraction would be present under actual internal combustion conditions, and the dew point of PbO could be exceeded. Therefore, both particulate and vapor PbO are likely to coexist. However, the PbO content in the vapor phase is certainly sufficient to allow chemical reaction to occur between flame radicals and PbO vapor species. This is amply demonstrated by the known effect of lead additives on flame speed, and under less than saturation conditions with respect to the additive, e.g., see Section V.B.

The following homogeneous radical-scavenging reactions have been suggested (see the summary of Minkoff and Tipper, 1962):

$$PbO + R \rightarrow Pb + RO, \tag{1}$$

$$Pb + OH \rightarrow PbOH, \tag{2}$$

$$PbOH + OH \rightarrow PbO + H_2O, \tag{3}$$

and

$$PbO + OH \rightarrow PbO(OH), \tag{4}$$

$$PbO(OH) + OH \rightarrow PbO_2 + H_2O, \tag{5}$$

or

$$PbO_2 + R \rightarrow PbO + RO, \tag{6}$$

where R is a hydrocarbon group. Thus, a catalytic recombination of radicals could occur via reactions (1)–(3) or (4)–(6). Note the analogy of this model with those already considered in connection with flame inhibition and smoke suppression. Of the various Pb-containing species suggested in this model only PbO and Pb have been observed in laboratory flames, e.g., see Friswell and Jenkins (1972). The stability of the postulated lead hydroxide species can be estimated, i.e., see Section II.K.5, and it can be shown that they will usually be present in very minor steady-state concentrations relative to Pb and PbO. However, this does not necessarily detract from their possible utility as reactive intermediates for the proposed catalytic schemes as the formation-type reactions, i.e., (2) and (4), are likely to be slow relative to the loss-type reactions, i.e., (3) and (5). This would necessarily lead to a low steady-state PbOH—or PbO(OH)—concentration in a nonequilibrium regime.

Evidence for an alternative, or perhaps supplementary, heterogeneous mechanism involving radical recombination on particulate PbO(c) has recently been reviewed and extended by Kuppu-Rao and Prasad (1972). One might expect that if the heterogeneous mechanism were the sole mode of action, other solids would have been found with a similar degree of effectiveness. We can more reasonably argue that the antiknock action of iron carbonyl is probably heterogeneous owing to the nonvolatility of its oxides under any temperature regime of the Otto cycle.

One can only conclude that it appears likely that both homogeneous and heterogeneous processes contribute to the "antiknock" mechanism, which has yet to be established.

VIII. Combustion of Metals

A. Introduction

Metal combustion can be considered as an extreme form of oxidation, where the oxidant may be oxygen, nitrogen, water vapor, a halogen, etc. The basic processes of oxidation and combustion differ mainly in the time scale of their occurrence. Several examples of metal combustion have already been considered elsewhere in connection with gas–solid reactions (see Chapter 2, Section IX) and the $Mg–O_2$ diffusion flame (see Section II.C).

The final state of a metal combustion process, i.e., the temperature and product composition, is determined by the enthalpies and free energies of the initial- and final-state components. For the case of metal combustion in an oxygen atmosphere, the more pyrophoric metals are those with the higher heat-of-formation oxide products. Thus, Ta and Al can be expected to be highly pyrophoric, whereas Cu and Ag are barely reactive. On the other hand, the initiation and progress of the combustion are determined by reaction kinetic and transport factors which are less well understood than the thermodynamics, e.g., see Glassman *et al.* (1970) and Wolfhard *et al.*, (1964b).

The present discussion is restricted to those aspects of metal combustion that are of practical utility, with an emphasis on vapor-phase processes. Primarily, the combustion of metals is of interest in connection with metal-dust explosions, metal fires, the production of intense light emission, the attainment of high temperatures, and as a source of energy for propulsion.

B. Exploding Wires

The so-called exploding wire technique basically involves the rapid discharge, e.g., < 0.1 msec, of a very high electrical current, e.g., $\sim 10^5$ amp, through a thin metal conductor which is then rapidly vaporized at the generated high temperature. Exploding wires have found application as explosive ignition devices on rockets and satellites, as light sources for high-speed photography and flash photolysis experiments, and as a high temperature source for producing molecular species of spectroscopic interest or for the synthesis of a variety of inorganic compounds.

The utilization of exploding wires as a source of very high temperatures is well known, e.g., see Joncich (1969), Bennett (1968), and Cassidy *et al.* (1968), and until recently it was believed that the temperatures attained may be sufficient to produce nuclear fusion. However, the upper limit temperature provided by the exploding wire technique is about 10^6 K, whereas the temperature to produce fusion is thought to be one or two orders of magnitude greater (see Chapter 6, Section IV).

Exploding wires have also been used in a reactive environment for the synthesis of metal fluorides (Johnson and Siegel, 1969; Cook and Siegel, 1967; Mahieux, 1963a, b), oxides (Karioris and Fish, 1962), nitrides (Joncich *et al.*, 1966), and carbides (Cook and Siegel, 1968). This synthetic application of exploding wires has been reviewed by Joncich (1969). Some degree of control over the reaction can be obtained if a rapid product-quenching technique, such as an adiabatic expansion into a low pressure chamber, is utilized (Cook and Siegel, 1969).

Examples of compounds prepared by this technique include: $PtXeF_6$, AgF_3, ZrF_4, AgF, PtF_3, Al_2O_3, Cu_2O, CuO, ThO_2, U_3O_8, β-UO_2, LaC_2, TiC, ZrC, NbC, Nb_2C, Ta_2C, MoC, Mo_2C, W_2C, and the nitrides of Mg, Ti, Zr, Ta, Zn and Al. In some respects, the use of this technique for chemical-synthetic purposes is similar to the cocondensation synthetic technique discussed elsewhere (Chapter 3, Section V). Very little is known about the mechanisms by which such compounds are produced. However, some insight into the initial stages of reaction has been gained by time-resolved spectroscopic studies, where molecular reaction intermediates such as AlO and TiO have been observed (Cassidy, 1968).

C. Flash Lamps

The explosive combustion of metals forms the basis of commercial photoflash lamps. Until recently the aluminum–oxygen sytem was used for this purpose. However, spectroscopic and other studies on this system—e.g., see Rautenberg and Johnson (1960)—suggested that the maximum color temperature of $\sim$3800 K could not be significantly improved upon without changing the chemical system. As the light output is primarily blackbody radiation, which is derived from the heat of the combustion reaction, a more exothermic system should provide a more intense light output. However, at the high temperatures realized by these lamps, the maximum attainable temperature is limited by the thermal stability of the reaction products. Usually this temperature corresponds to the boiling point of the metal oxide—above which endothermic dissociation processes become more significant. As the blackbody radiative flux is proportional to the fourth power of the temperature, only modest increases in reaction temperature are needed to provide a substantially brighter light source.

The oxidation of zirconium by oxygen, or by fluorides such as tetrafluorohydrazine or nitrogen trifluoride, is energetically more favorable than for the aluminum–oxygen system and the zirconium–oxygen reaction is now used commercially in flash cubes (Nijland and Schroder, 1969). As these flash lamps are normally synchronized with the operation of camera shutters, the combustion process is usually designed to occur rapidly, and it is not likely that thermodynamic equilibrium conditions exist in such lamps. It is therefore necessary to obtain kinetic data for the oxidation of metals at high temperatures in order to develop theoretical models of flash lamps.

From Nijland and Schroder (1969), the present qualitative understanding of the operation of $Zr–O_2$ flash lamps is as follows. The ignition, using an external source of energy—such as the discharge of electric current—results initially in the generation of droplets of Zr metal which are dispersed in the bulb. These droplets react with the oxygen environment, which is initially at a pressure of about 20 atm, to produce a cloud of liquid ZrO_2 particles. This reaction proceeds until the oxygen pressure decreases to below 1 atm. The presence of particulates during the combustion reasonably accounts for the blackbody nature of the emitted radiation. An upper limit temperature results due to the endothermic formation of ZrO at the expense of ZrO_2 vapor species.

More detailed mechanistic models of the zirconium droplet–oxygen reaction have been developed. According to the model given by Maloney (1971), and extended by Gupta and Maloney (1973), the primary reaction zone is homogeneous—involving reactions such as

$$\mathrm{Zr} + \mathrm{O} = \mathrm{ZrO}, \qquad \mathrm{ZrO} + \mathrm{O} = \mathrm{ZrO_2}.$$

Condensation of these reaction products occurs at a later stage, i.e., at a greater distance from the metal droplet surface.

D. Combustion of Metals in Solid Rocket Propellants

1. *Background*

The combustion of metals is, in principle, a more effective energy source than conventional hydrogen- or hydrocarbon-fueled systems. This is apparent from the following comparison of specific exothermicities, given in kcal gm^{-1}, for various combustion products:

H_2O (3.2), CO_2 (2.1)

$BeO(s)$ (5.7), $B_2O_3(s)$ (4.4), $Al_2O_3(s)$ (3.9)

$BeF_2(s)$ (5.1), $LiF(s)$ (5.6).

Note that the light elements Li, Be, B, and also Al form very stable condensed-phase oxides and fluorides and are therefore prime candidates for supplementing the specific energy content of solid propellant fuels based on organic polymers. However, the relatively high oxygen requirements for boron combustion detracts from its use in solid fuels except for the case of air-breathing engines where the supply of oxidant is unlimited. Also, the high toxicity of Be has prevented its use in rockets to date. Rocket systems based on a lithium–fluorine combination are energetically more favorable than those of other metals, and they have been investigated both experimentally and theoretically but are not in routine use. Aluminum, then, is the most frequently used metal in practical solid propellant systems.

In addition to high combustion energy as a determining factor in the selection of a propellant system, the fuel density may also be a consideration, and in this case a high density metal such as Zr finds application as a fuel supplement.

As a general rule, the attainable combustion temperatures are limited by the boiling point of the metal oxide product. Beyond this temperature condition the absence of a condensed phase, and the increased importance of endothermic dissociative processes, reduces the translational energy content of the system. Typical boiling points range from 2316 K for B_2O_3 to 4548 K for ZrO_2 (see JANAF, 1971). Flammability limits for metals, in the form of fine particles, are around 10-mole % metal, relative to O_2 in air. This is a similar magnitude to that found for the more common fuels such as hydrocarbons. In practice, about 20 wt % of the metal is incorporated into a solid composite of oxidizer (e.g., ammonium perchlorate) and organic polymer fuel (e.g., polyethylene). The developed thrust per unit mass flow (i.e., specific impulse—see Section IX) of such a composite propellant is then about 250–280 lb/lb/sec.

From a consideration of the heats of formation of metal oxides and nitrides, and autoignition experiments, it has been shown that the metals Be, Li, Al, and Mg are the optimum fuels in an oxidizing atmosphere of CO_2 or N_2 (Rhein, 1965). Hence, these systems could conceivably be used for propulsion in the atmosphere of the planets Mars and Venus, respectively.

Comprehensive sources of information regarding the chemical aspects of metal-containing propellants include the references of Marsh and

Hutchison (1972), Holzmann (1969), Eringen *et al.* (1967), Siegel and Schieler (1964), and Huggett (1956). Additional discussion of rocket-related chemistry is also given in the following Section IX.

2. Mechanisms

In recent years much research has been directed to obtaining an understanding of the combustion mechanisms for propellant metals, e.g., see Prentice (1972), Mohan and Williams (1972), Edelman *et al.* (1971), Levine (1971), Bouriannes and Manson (1971), and cited earlier work. Studies on aluminum combustion in rocket engines show that the combustion occurs in the vapor phase (Cheung and Cohen, 1965). On the other hand, model studies of Be combustion indicate both condensed and vapor-phase processes to be important (Prentice, 1972). As was mentioned elsewhere (Chapter 4, Section III.C) the combustion or oxidation of this element can be particularly affected by water vapor owing to the stability of the hydroxide vapor species $Be(OH)_2$.

In general, combustion mechanisms of various metals differ considerably. These differences result from the changing stabilities of condensed and vapor species for different elements and are reflected in the wide boiling-point range for the metal oxide products. Even within the same group of the periodic table, the relative vapor pressures of oxides and metals can show a reversal, as is the case for Al and B. With Al, the metal has a higher vapor pressure than the oxide and vice versa for B.

A model for the combustion of B in air, as proposed recently by Mohan and Williams (1972), consists of both a low and a high temperature stage. The chemical basis of this model is demonstrated in Fig. 5.27. The low temperature stage is considered to be important at temperatures of less than about 2300 K, i.e., below the boiling point of B_2O_3. Note that for the low temperature stage, the combustion is controlled by the rate of O_2 diffusion through a liquid B_2O_3 barrier to the solid B fuel. Also, at this stage the overall reaction is

$$2B(s) + \tfrac{3}{2}O_2 \rightarrow B_2O_3.$$

For the high temperature stage, the rate-controlling steps are represented by the formal, i.e., nonelementary, reactions

$$B(l) + B_2O_3 \rightarrow \tfrac{3}{2}B_2O_2,$$

and

$$2B_2O_2 + O_2 \rightarrow 2B_2O_3.$$

This two-stage model satisfactorily accounts for the observed flammability limit and the burning rates.

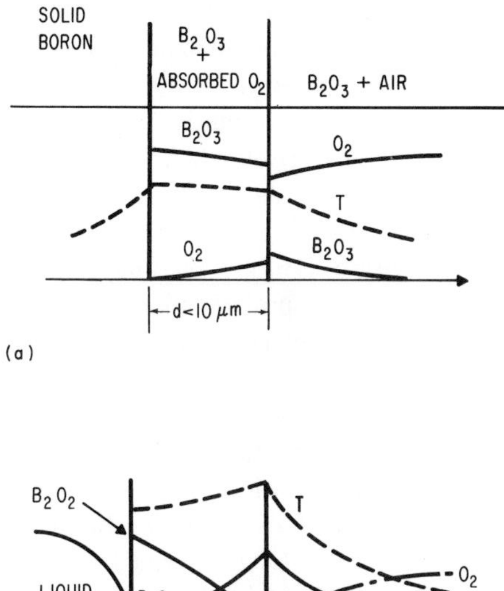

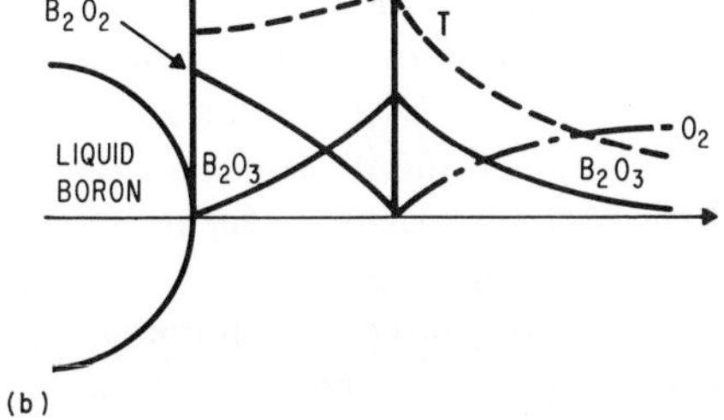

Fig. 5.27 Model for boron combustion in air showing, (a) a low temperature stage; (b) a high temperature stage (Mohan and Williams, 1972).

The frequent production of liquid or solid combustion products, which may hinder the further reaction of oxidant with the fuel, or alternatively introduce additional corrosion problems, represents one of the major difficulties associated with this application of metal combustion. The problem may be reduced, in part, through the use of volatile metal organic fuels, such as trimethylaluminum (Vanpee *et al.*, 1965). Fuels of this type lead to a two-stage combustion process, involving hydrocarbon oxidation in the first stage and metal oxidation in the second.

IX. Theoretical Performance of Chemical Rocket Engines

A. Introduction

The theoretical evaluation of rocket engine performance parameters represents one of the major developments in the application of high temperature thermodynamics, kinetics, and gas dynamics to practical high temperature processes. It can be said that the practical problems of rocket design, and choice of optimum propellant combinations, have been instrumental in the rapid accumulation—during the past 15 years—of measured and estimated thermochemical data on high temperature species, e.g., see the JANAF Thermochemical Tables (JANAF, 1971). In addition, the development of computer programs for dealing with multicomponent dynamic chemical systems, has also been in response to these problems e.g., see Gordon and McBride (1971). We may, in passing, also make reference to the significant developments in cryogenic, free radical, synthetic, boron, and noble gas chemistry that resulted from the search for more energetic propellant systems. A recent review discussion of such advanced propellant ingredients may be found in Holzmann (1969). It is apparent from the rocket literature, however, that relatively few esoteric propellant systems are used in practice. For instance, the Apollo rocket systems utilize the following fuel–oxidizer combinations: H_2–O_2, hydrocarbon–O_2, and various hydrazine–N_2O_4 mixtures.

In considering the theoretical merit of various propellant systems one should keep in mind the fact that the oxidizer usually comprises 70–80% of the total fuel–oxidizer propellant composition. The variety of thermochemically efficient oxidizers is limited to compounds of oxygen, fluorine, and chlorine. In practice, the fuel component is usually rich in hydrogen, e.g., H_2, N_2H_4, and "kerosene" although, as already mentioned (Section VIII), several of the light element metals may be used as supplemental additives. Metal hydrides are also considered to be potentially effective fuels. The theoretical basis for this limited choice of fuel–oxidizer chemistry will become evident in the discussion to follow.

Naturally, the urgent development of a new macrotechnology, namely rocketry, has led to a correspondingly prolific literature. The following citation of literature is therefore restricted to either general texts or to articles having particular pertinence to the high temperature vapor-phase aspects of rocket propulsion.

The application of thermodynamics to the calculation of rocket performance has been described in detail by Holzmann (1969) and by Wilkins (1963), among others. A related topic of rocket reentry thermodynamics has been considered in the articles edited by Loh (1968), and Ladacki (1972) has discussed the chemical aspects of ablation. A useful textbook discussion of rocket propulsion, given in connection with flame and combustion phenomena, may be found in Bradley (1969) and a brief introduction to the subject of chemical rocket propellants may be obtained from the articles of Lawrence and Bowman (1971a, b). Specialized discussions on the computation of rocket performance are contained in the various conference proceedings of the Western States Section of the Combustion

Institute (Bahn, 1960, 1963, 1968, 1969). More recent developments in the computation of kinetic and thermodynamic properties for high temperature gases are indicated in the Lewis Research Center conference proceedings (NASA, 1970). Some additional recent literature related to the kinetics and thermochemistry of chemical rocket propulsion is as follows: Jensen (1972), Eriksson (1971), Arbit *et al.* (1970), Miller (1970), Prothero (1969), Spengler and Buchner (1969), Lo (1969), Schadow (1969), Channapragada *et al.* (1969), Williams *et al.* (1969), Macek and Semple (1969), Barrere (1969), Lepie *et al.* (1968), Hammitt (1968), Hermsen (1968), Vanpee (1968), Gordon *et al.* (1959–1961), and White *et al.* (1958). Details on the general chemistry of rocket propellants may be found in Sarner (1966) and in Kit and Evered (1960); discussions of solid propellants may be conveniently located in Summerfield (1960), and Marsh and Hutchison (1972).

B. Performance Rating by Specific Impulse, I_{sp}

1. *Definition and Significance of* I_{sp}

Basically, a chemical rocket engine consists of a combustion chamber and an expansion nozzle, these being connected by a throat. In the combustion chamber, pressures in the range of about 10–100 atm, temperatures of the order of 4000 K, subsonic random gas velocities, and a relatively long residence time of about 10 msec are usual. The exit of the combustion chamber is constricted to form a throat where the velocity of the gaseous combustion products increases to about a sonic level. The gas then rapidly expands through an expansion nozzle, with a residence time of about 1 msec, where an appreciable cooling of the gas results, to about one-half or less of the combustion temperature, and an increase in gas flow velocity to supersonic levels occurs. This expansion process usually leads to a pressure reduction of a factor of 10 or more, and at the expansion nozzle exit the pressure approaches that of the ambient atmosphere.

According to Newton's third law, the force developed by the expulsion of high velocity gases from the expansion nozzle into the surrounding environment serves to propel the rocket device in the opposite direction to the gas flow. Ultimately, the force or thrust developed is related to the chemical energy released by the combustion process. From the fluid mechanical principle of conservation of momentum it can be shown (e.g., see Kit and Evered, 1960) that the force F may be expressed as

$$F = wV_e/g + A_c(P_e - P_a) \text{ lb},$$

where

- w is the propellant flow rate, lb sec^{-1};
- V_e the effective exhaust velocity, ft sec^{-1};
- g the gravitational acceleration constant, ft sec^{-2};
- A_e the nozzle exhaust exit area, ft^2;
- P_e the nozzle exhaust pressure, psi; and
- P_a the ambient pressure, psi.

a. Definition of I_{sp}

For the evaluation of the relative merit of various propellant systems the important parameter is the specific impulse, which is the impulse, i.e., force for unit time, developed per unit mass of propellant used and is usually given in units of seconds. The usefulness of this parameter becomes apparent when one recognizes that it can be determined both by measurement and from the application of basic thermodynamic and/or kinetic theory.

The second term of the above force equation is determined primarily by the nozzle design, and it is therefore an aerodynamic quantity. In discussions of the chemical aspects of propulsion it becomes unnecessary to consider this term which, in any case, is relatively small and in fact vanishes for the not uncommon condition of exhaust pressure equal to ambient pressure. One should note, however, that if the ambient pressure decreases with time, such as would be the case for a rocket leaving the earth's atmosphere, a corresponding increase in specific impulse results. This increase is usually about 10–20% of the total specific impulse. For the condition of exhaust pressure equal to ambient pressure, the specific impulse may then be given as

$$I_{sp} = V_e/g \text{ sec.}$$

A typical rocket system would have a specific impulse value in the range of 300–400 sec.

In addition to the specific impulse there are several other related parameters which one can calculate. These are the nozzle area ratio—i.e., the area ratio for the exit and the throat—and the characteristic velocity, which is given by the ratio of the exhaust velocity and the thrust coefficient. The interested reader may follow up the significance of these, and other, auxiliary parameters by referring to the report of Gordon and McBride (1971).

The primary variables for a particular propellant system that can be optimized by considering their effect on I_{sp} are, the nozzle area ratio and the fuel-to-oxidant concentration ratio. Plots of I_{sp} versus these parameters will vary smoothly and indicate an optimum ratio of nozzle areas or fuel-to-oxidant concentration. Clearly, the determination of such optimized conditions by actual rocket firing experiments is an undesirable, and most expensive, enterprise. The discussion to follow indicates that it is possible to determine I_{sp} by calculation without recourse to experiment or an experimentally based empiricism.

It can be shown that the directed exhaust velocity V_e, and hence I_{sp}, is proportional to the random gas velocity in the combustion chamber, which in turn is directly related to the gas kinetic term $(T_c/M)^{1/2}$, where T_c is the gas combustion temperature and M is the average molecular weight of gases. The combustion temperature is also approximately proportional to the heat of combustion.

If an isentropic expansion of the combustion gases is considered, i.e., a condition of constant entropy with reducing pressure, then terms involving heat capacities are introduced into the relationship between specific impulse, reaction temperature, and molecular weight. Nozzle and throat dimensions are also considered in this case, e.g., see Wilkins (1963). Without going into the details of derivation, a rigorous expression for I_{sp} is given by

$$I_{sp} = 6.940(H_c - H_e)^{1/2} \qquad \text{or} \qquad I_{sp} = 9.330(H_c - H_e)^{1/2},$$

depending on whether English or metric units are used, respectively, and where

H_c = enthalpy of chamber gases, units Btu/lb or cal/gm;
H_e = enthalpy of exhaust gases, units Btu/lb or cal/gm; and
I_{sp} = lb thrust/(lb mass/sec), i.e., sec.

The gas enthalpy is actually a sum total of the enthalpy for each individual species, i.e.,

$$H = \sum_i x_i h_i,$$

where, for species i, x_i is the mole fraction concentration, and h_i is the enthalpy.

It is evident, then, that an optimum rocket system requires reactions of high exothermicity that is, the generation of very stable products from relatively low stability reactants. It also follows that one should use a low average molecular weight system for maximum impulse. In practice, this low molecular weight requirement often offsets that of attaining a maximum combustion enthalpy change. For example, with H_2-fueled systems it is advantageous to carry out the combustion using fuel-rich mixtures rather than at the higher enthalpy stoichiometric condition. However, as a practical matter, the propellant density may also be an important consideration in the choice of propellant. It is then convenient to consider the product of I_{sp} and density ρ, known as the density specific impulse, as a figure of merit for a propellant system. The low molecular weight advantage of H_2 then tends to be offset by its low density as a liquid fuel, and other fuels become more competitive. For instance with H_2–O_2, I_{sp} is 391 sec and $I_{sp} \cdot \rho$ is 103 gm sec cm^{-3}, whereas for kerosene–O_2, I_{sp} is 301 sec but $I_{sp} \cdot \rho$ has a value of 322 gm sec cm^{-3} (Bradley, 1969).

b. Factors Affecting I_{sp} and Rocket Performance

A limitation in the use of high heat-of-combustion systems results from the tendency of reaction products to dissociate thermally if the temperature is too high. This results in the loss of kinetically available heat from the system. For example, at a typical pressure of 34 atm, and at a temperature of 3300 °C, the extent of dissociation of CO_2 is 40%, of H_2O is 12%, of H_2 is 6%, and of HF is 1%. Under these conditions CO and N_2 have negligible dissociation pressures. It can be shown that a practical upper limit temperature for chemical propulsion occurs at about 4000 °C (Silver, 1962). The potential advantage of utilizing higher pressures, where dissociation effects are reduced, is offset by the increased load requirements of higher pressure combustion vessels.

A further limiting feature of the use of extremely high temperatures results from the thermal population of the internal molecular modes, namely, the rotational and vibrational degrees of freedom. The distribution of energy in these forms is inefficient, as only kinetic energy contributes to the rocket thrust. An appreciation of these energy distribution problems may be gained from the fact that for H_2, at a relatively low temperature of 2000 K, the energy is partitioned equally between translation and dissociation and is about a factor of 2 less for rotational states. It follows that the formation of polyatomic products, with their many modes of internal degrees of freedom, leads to a relatively inefficient use of chemical energy.

As the initial combustion products are most likely populated in excess of the thermal distribution, a further inefficiency could result if the post-combustion conditions do not provide an ample opportunity for collisional relaxation of the excess internal energy to occur. However, the presence of water molecules as a combustion product in many systems has a desirable catalytic effect on the relaxation rates of internally excited products and, in view of the long combustion chamber residence times, one can normally expect to achieve complete relaxation—i.e., equilibrium—within the combustion chamber.

Another factor that influences the achievement of an optimum rocket performance is the dew point or condensation temperature of combustion products, or, if a kinetic condition is considered, rates of nucleation. It is desirable to maintain a gaseous state during the combustion and expansion process since the conversion of heat energy to kinetic energy during the expansion process cannot be achieved by condensed species. Particulate formation is also undesirable from a materials aspect since it promotes erosion of the expansion throat.

From specific impulse considerations, H_2–O_2 and H_2–F_2 are among the most effective chemical bipropellant systems with values of I_{sp} in the region of 400 sec. For most systems fluorine is a more energetic oxidizer than oxygen. With the use of a tripropellant, particularly where two-phase combustion is possible, a much higher I_{sp} can be obtained, as was demonstrated for the lithium–fluorine–hydrogen system, where an experimental value of $I_{sp} = 523$ sec was found (Arbit *et al.*, 1970). Similarly, one can show, from the thermodynamics, that a Be–O_2–H_2 system should yield a high specific impulse relative to a binary combination. Such systems benefit from the energy released by condensation in the expansion process [e.g., to form LiF(s) or BeO(s)] which is transferable to the gaseous product species (e.g., HF or H_2O).

More exotic systems that are potentially high energy propellants include trapped free radicals, particularly H-atoms, a trapped proton–electron combination, and noble gas compound oxidizers. In the latter case, considerable effort has been given to the unsuccessful attempted synthesis of light-element noble gas fluorides such as HeF_2 and NeF_4 where, in combination with H_2 as a fuel, specific impulses in the region of 500 sec are theoretically possible. Metal hydride–fluorine combinations have also been given consideration as high energy propellants. Of these, BeH_2 and LiH lead to the highest predicted specific impulse.

In order to achieve substantial improvements in rocket performance over that provided by the most favorable chemical combinations, it is necessary to utilize systems where bond breaking and formation by chemical means are not the primary source of energy release. Order-of-magnitude greater specific impulses are possible, i.e., ~5,000–15,000 sec range, using nuclear, arc-jet, ion-jet, or plasma propulsion systems (Silver, 1962). For these latter systems the potential for a high mass flow rate is less than for chemical systems, so that the total thrust is not necessarily any more favorable. However, electromagnetic plasma accelerators are potential candidates for rocket propulsion during deep space missions, e.g., see Bruckner and Jahn (1972) and Nerheim and Kelly (1968). Plasma temperatures of about 15,000 K are formed in such magnetoplasma dynamic accelerators, and the chemistry is mainly one of positive ion production and their recombination with electrons.

Finally, we should recognize that in the practical application of rockets, factors other than energy release are also important. These include propellant density, handling and storage convenience, burning rates, and corrosion properties of the combustion products, in addition to the strength, weight, and thermal transport characteristics of the structural materials.

2. *Thermodynamic Calculations of* I_{sp}

We have shown that the specific impulse is related to the enthalpy difference between the gases in the combustion chamber and those in the nozzle exit region. These enthalpy terms may be calculated if the gas composition, expressed as species concentrations, is known at these various rocket locations. The gas composition can be calculated, for conditions of thermodynamic equilibrium, using the multicomponent equilibria free energy minimization procedure, e.g., see the computer program of Gordon and McBride (1971) or the chemical engineering text of Balzhiser *et al.* (1972). This type of calculation requires a knowledge of what species are likely to be present and their basic enthalpy and entropy properties, such as may be found in the JANAF Thermochemical Tables (JANAF, 1971). The adiabatic flame temperature is also readily calculated from the same thermal data.

Given the high temperatures, and hence fast reaction rates, and also the long residence time of gas in the combustion chamber, the assumption of an equilibrium condition for this region is most reasonable. However, the rapid reduction of pressure and temperature that occurs during the gas expansion through the rocket nozzle may be too fast to allow the equilibrated combustion chamber gases to adjust to these kinetically slower conditions. Whether this is the case could be established if the relevant elementary reaction rate data were available. However, as the rate data are usually not known, one assumes either of two extreme rate conditions, namely, infinitely slow or infinitely fast. The former condition is termed frozen equilibrium, i.e., the gas species concentrations do not change between the combustion chamber and the rocket exit. The latter condition is a local equilibrium one where the gas composition rapidly adjusts to the changing pressure and temperature conditions of the expansion process. Clearly, the real condition of finite reaction rates would lead to a gas composition, and hence a calculated I_{sp} value, between those for the frozen and the equilibrium assumption. A few cases where the real kinetic conditions are considered in the calculation of I_{sp} will be mentioned in Section 3, which follows.

a. *Sources of Error*

The calculated difference between I_{sp}-equilibrium and I_{sp}-frozen is usually between 1 and 10%. As the actual value should fall between these two extremes, the uncertainty in I_{sp} using these model assumptions can be only a few percent. However, for practical applications this is still a serious uncertainty. For instance, the range of a missile can be altered by 20% for a 3% change in I_{sp} (see Holzmann, 1969).

Some additional simplifying assumptions, and sources of error, in these thermodynamic methods are as follows. As has been revealed many times throughout this book, the basic thermochemical data for high temperature species are often not well known, e.g., see the discussion of dissociation energies from flame studies (Section II.K) where discrepancies of more than 10 kcal mol^{-1} in bond dissociation energies are indicated. These uncertainties can lead to errors of one or two orders of magnitude in the calculated species concentrations. Whether or not this will have a significant effect on I_{sp} will depend on the importance of the species in question relative to the total gas composition. Thus for propellants containing predominantly compounds of C, H, O, or N, and for which the combustion products are the thermodynamically well-characterized species of H_2O, CO, CO_2, N_2, and H_2, the computational errors are insignificant. On the other hand, the metallized propellants will give a combustion gas mixture which contains species such as BOF, AlCl, AlOH, and AlO for which the uncertainties in heats of formation range from about 6–30 kcal mol^{-1}. The resulting uncertainty in I_{sp} could, depending on the actual initial propellant composition, far exceed the few percent error arising from the assumptions of frozen or local equilibrium. Clearly, for these circumstances, the consideration of a sophisticated kinetic reaction scheme would be of little value.

Since one requires heat capacity or entropy data at combustion temperatures as high as 4500 °C, a considerable extrapolation of the experimental data—which is obtained usually at 2000 °C or less—is necessary. This extrapolation can be made from well-known statistical mechanical relationships, provided the basic energy level—i.e., spectroscopic—data are available for each molecular species present in the combustion gas. In many instances this data has been estimated, and the calculated rocket performance is therefore dependent on the validity of the estimation process, e.g., see Jones (1963).

Several other assumptions made in these thermodynamic calculations of rocket performance concern the ideality of the gases, the absence of condensation, a constant steady-state pressure condition in the combustion chamber, and an isenthalpic combuston chamber—i.e., an absence of heat loss at the walls. The error in I_{sp} associated with the assumption of gas ideality is usually 1% or less, but for pressures of greater than about 100-atm, real-gas properties should be used. The efficiency of combustion is usually found to be in the range of 90–97%, and corrections for the heat losses may be included in the calculation of I_{sp}.

In view of these various uncertainties involved in the performance calculations, the view has often been expressed that an experimental observation of the composition of an operating rocket engine would be invaluable. Such a difficult experimental task has apparently not yet been realized, although some measurements are in progress (Houseman and Young, 1972).

b. Examples of Results

Some examples of calculated I_{sp} data for systems where high temperature inorganic species play a predominant role are indicated in Table 5.26. As these data result from a frozen equilibrium calculation, they can be considered as lower limit values. It is apparent from these I_{sp} data that the light metal hydrides are competitive with H_2 as rocket fuels.

A comparison of data obtained using both local, i.e. shifting, and frozen equilibrium assumptions is given by the Li–F_2 example in Table 5.27. Note that in this example I_{sp} frozen is about 17% less than I_{sp} local. This difference is also reflected by the compositional changes that can occur

TABLE 5.26

Theoretical Specific Impulse For Frozen Equilibrium[a]

Fuel	Oxidizer[b]	I_{sp} (sec)	T_c (K)	P_c (atm)	P_e (atm)
H_2	O_2 (78)	391	2769	68	1.0
H_2	OF_2	411	—	68	1.0
B_5H_9	OF_2 (80)	367	5169	68	1.0
B_5H_9	OF_2 (80)	422	5169	68	0.136
MgH_2	OF_2 (79)	305	4516	68	1.0
MgH_2	OF_2 (69)	352	4404	68	0.136
BeH_2	OF_2 (67)	383	4347	68	1.0
BeH_2	OF_2 (67)	445	4347	68	0.136
AlH_3	OF_2 (59)	333	4406	68	1.0
AlH_3	OF_2 (62)	386	4447	68	0.136
LiH	OF_2 (77)	337	4093	68	1.0
LiH	OF_2 (77)	386	4093	68	0.136

[a] Summarized from the more extensive tables given by Wilkins (1963); T_c and P_c are the calculated combustion temperature and designated combustion chamber pressure, respectively; P_e is the designated rocket exit gas pressure. Note that the effects of P_c and P_e on I_{sp} have been considered in these calculations.

[b] Values in parenthesis represent wt % oxidizer for an optimum I_{sp}.

TABLE 5.27

Calculated Performance for a Stoichiometric Li–F_2 Propellant[a]

Composition (mole fraction)		
	Chamber	Exit (local equilibrium)
e	0.00284	0
F	0.21128	0.12105
F^-	0.00496	0
F_2	0.00002	0
Li	0.12093	0
Li^+	0.00780	0
LiF	0.65079	0.67708
Li_2	0.00028	0
Li_2F_2	0.00109	0.1941
Li_3F_3	0	0.00777
	Local equilibrium	Frozen equilibrium
T exit (K)	1301	195
I_{sp} (sec)	498.8	412.8

[a] Data obtained from Gordon and McBride (1971); assigned combustion conditions are $P_c = 68.046$ atm, $P_e = 0.0002$ atm, and a Li/F_2 molar ratio of 2.0; the calculated adiabatic combustion temperature is $T_c = 5690$ K.

between the chamber and exit localities if the equilibrium is allowed to shift in accord with the dramatic temperature change. The importance of charged species in the combustion chamber and of polymeric inorganic species at the nozzle exit are revealed in these calculations, i.e., see Table 5.27. Note also the extreme condition of supersaturation imposed by the frozen equilibrium condition, that is, the presence of high temperature species at a temperature of only 195 K.

In concluding this thermodynamic discussion, it is useful to note that a convenient form of thermodynamic data for the performance evaluation of rocket and other high temperature gas systems is represented by so-called Mollier enthalpy versus entropy diagrams. Examples of such diagrams, and their utility, may be found in Vale (1963) and the Rand reports of Krieger, e.g., see Krieger (1970) and earlier cited reports. Both isentropic and isobaric work can be followed on such diagrams. Each diagram is constructed for only a single specified reaction mixture. For more general combustion problems where the mixture composition is also variable, enthalpy charts of the type suggested by Lutz (1957) may be utilized.

Tables of calculated combustion properties for the important H_2–O_2 propellant system have been given by Svehla (1964). Also, as this system is free of significant uncertainty in its basic thermodynamic data and is comprised of a relatively few simple species, it has been worthwhile to compute the gas transport properties from the equilibrium concentration and enthalpy data.

3. *Chemical Kinetic Factors in Rocket Performance*

The essential features involved in the calculation of rocket performance under nonequilibrium conditions differ from the equilibrium methods only in the assumption of finite reaction rates. The calculation of a kinetically controlled rocket system is then similar to that for any reacting gas mixture, for which the theoretical and computational procedures are well known, e.g., see Bittker and Scullin (1972) and also Chapter 1, (Section V).

Briefly, the basic procedure involves the selection of a set of elementary reactions which constitutes the kinetic model for the system. The forward and reverse rate constants are, hopefully, known quantities for each of these reactions and at each temperature of interest. Species continuity equations can then be derived. For instance, consider the hypothetical kinetic scheme of

$$\mathrm{H + H + M = H_2 + M}, \tag{1}$$

$$\mathrm{BO + HF = BOF + H}. \tag{2}$$

The net production rate of H-atoms at any time is then

$$\frac{d[\mathrm{H}]}{dt} = k_{-1}[\mathrm{H_2}][\mathrm{M}] - k_1[\mathrm{H}]^2[\mathrm{M}] + k_2[\mathrm{BO}][\mathrm{HF}] - k_{-2}[\mathrm{BOF}][\mathrm{H}],$$

where k_1 and k_{-1} are the rate constants for the forward and reverse reactions, respectively, and likewise for k_2 and k_{-2}. Similar equations may be written for each species and, in combination with the momentum and energy continuity equations, together with the gas equation of state, they constitute a set of differential equations whose solution yields the gas composition for any particular reaction time in the rocket gas-expansion region. Hence, the gas enthalpy H_e and I_{sp} can also be determined at any time of the gas-expansion history.

A kinetic system which is reasonably well understood is the H_2–F_2 combustion mixture and the reaction mechanism is considered to be

$$H + F_2 \rightarrow HF + F,$$

$$F + H_2 \rightarrow HF + H,$$

$$H + H + M \rightarrow H_2 + M,$$

$$F + F + M \rightarrow F_2 + M,$$

$$H + F + M \rightarrow HF + M,$$

where M is a third-body species. Calculations show that in this case, and for relatively high pressure ($>$ 20 atm) and high H_2-content conditions, the system closely follows local equilibrium conditions in the expansion process, i.e., see Sarner (1966).

In general, higher pressures allow a closer approach to equilibrium. For instance with the N_2H_4–N_2O_4 combustion sytem at P_c = 68 atm, the calculated I_{sp} data are almost identical for both the kinetic, i.e., finite rate, and shifting equilibrium, i.e., infinitely fast rate, computational assumptions. However, when the combustion chamber pressure is reduced to P_c = 4.08 atm, the value of I_{sp}-kinetic lags that of I_{sp}-equilibrium by about 11% and is, in fact, closer to the I_{sp}-frozen value (Dipprey, 1972).

a. The OF_2–B_2H_6 Example

An example of the accuracy that can be obtained for I_{sp}-kinetic, if sufficient basic data is available, is given by the comparison of predicted and actual performance for the developmental OF_2–B_2H_6 propellant system, as shown in the Fig. 5.28 (Bittker, 1970). Very good agreement is found when the nonequilibrium, i.e., kinetic, processes are accounted for.

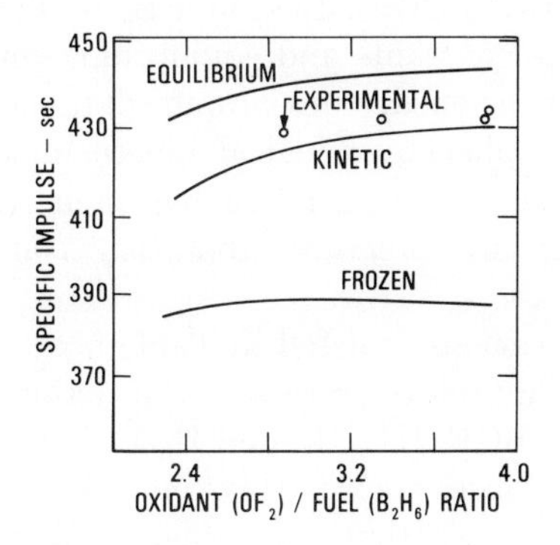

Fig. 5.28 Specific impulse data as a function of OF_2–B_2H_6 ratio, as separately obtained by experiment or thermodynamic (equilibrium or frozen) and kinetic calculation (Bittker, 1970).

TABLE 5.28

Elementary Reactions in the OF_2–B_2H_6 Rocket Propellant System (Bittker, 1970)[a]

Reaction[c]	Energy release rate (%)[b]
$H + H + M \leftrightarrows H_2 + M$	53.1
$BO + HF \rightleftarrows BOF + H$	13.9
$H + F + M \rightleftarrows HF + M$	11.1
$H + OH + M \rightleftarrows H_2O + M$	9.8
$BF + OH \rightleftarrows BOF + H$	3.5
$F + H_2 \rightleftarrows HF + H$	2.3
$H_2 + OH \rightleftarrows H_2O + H$	2.3
$BF + O \rightleftarrows BO + F$	1.2
$H + BF_2 \rightleftarrows HF + BF$	1.0
$BO + F + M \rightleftarrows BOF + M$	0.7
$BF + O + M \rightleftarrows BOF + M$	0.6
$O + H + M \rightleftarrows OH + M$	0.5

[a] Data obtained from Bittker (1970); the mole ratio of B_2H_6/OF_2 is 0.279, i.e., the mixture is fuel rich.

[b] Energy release rate is defined as the product of the net reaction rate and the heat of reaction.

[c] As usual, M denotes a major gas component third body, and is usually a hypothetical composite of species such as H_2O and H_2.

For these calculations the kinetic reaction steps were assumed to be those listed in Table 5.28.

A general description of the kinetic computation procedure used may be found in the report of Bittker and Scullin (1972). High temperature species included in their computer data bank include: BF, BF_2, BF_3, BH, BH_2, BH_3, BO, BOF, BOF_2, BO_2, and B_2O_3. The critical input data consist of reaction rate constants and equilibrium constants. For species such as those just mentioned, these input data are mainly estimated quantities. An overall phenomenological kinetic model, i.e., elementary steps not elucidated, has been proposed by Rhein (1971) for this combustion system under low pressure laboratory conditions, i.e., $\sim$1–70 Torr range.

The first four reaction steps listed in Table 5.28 accounted for almost 90% of the total energy-release rate and the major steady-state species were determined to be BOF, HF, H_2, and H_2O. Fortunately, the rate data for the first, third, and fourth steps in this kinetic scheme are relatively well known, and it can be shown that only order of magnitude estimates are required for the remaining reactions to yield reasonably accurate I_{sp}-kinetic data. The estimation of the rate data is also facilitated by the fact that the reactions are either close to thermoneutral or they are exothermic, and the probably small activation energies resulting will have little influence on the high temperature rate constants.

b. Sources of Error

An indication of the effects of uncertain rate data in rocket combustion systems may be obtained from the study of Jensen (1972). A major uncertainty in most radical recombination reactions is the relative efficiency of the various third-body M species.

It is also apparent from the above example that radical recombination reactions account for most of the heat release (see Table 5.28). Therefore, unless these radicals can recombine in the expansion chamber, a significant loss of heat and kinetic energy results. The fundamental difficulty is that radical recombination processes, which are normally highly exothermic, require a third body for removal of the excess energy generated. Such termolecular processes are kinetically slow and are particularly sensitive to the rapidly decreasing pressure conditions in the expansion chamber. Interest has therefore been generated in the prospect of catalyzing these recombinations with suitable additives (Lordi *et al.*, 1968). In principle, the catalyst would allow the recombination to occur via a series of rapid bimolecular steps rather than the inherently slow termolecular process. Kinetic studies in 1-atm H_2–O_2–N_2 flames have shown that certain metals, particularly Cr and Sn, at the ppm concentration level, dramatically increase the recombination rates of H-atoms (see Section II.L). However, the effect of these observations on specific impulse calculations has apparently not yet been determined.

Other kinetic factors of possible importance, but neglected on account of the absence of rate data, include relaxation of internal to translational energy in the expansion regime and nucleation and condensation rates for condensed species. The recent interest in combustion systems as potential laser sources and the importance of relaxation processes in such systems can be expected to catalyze the generation of some of this lacking data in the future.

While the emphasis on rocket performance calculations has involved steady-state conditions, there exists a number of equally important transient phenomena involving ignition and combustion instability. The need for kinetic data to deal with these types of problems is of paramount importance.

It is appropriate to conclude this discussion with a quote from Bradley (1969) concerning rocket performance: "The basic requirements for all future improvements, however, are more extensive and reliable thermodynamic and rate process data."

6

Energy Systems

I. Introduction

The projections of an imminent energy shortage, e.g., see Krieger (1972), together with the need to control air pollution from energy sources, point to the necessity of improving and optimizing the efficient use of energy systems which are at present no more than about 40% efficient.

Hottel and Howard (1971) have recently surveyed present and future possible energy sources, and have made a number of assessments with regard to research requirements and priorities in this area.† A more recent discussion of present and future energy systems has been given by Hammond *et al.* (1973). More advanced, i.e., higher temperature, energy conversion systems are being considered, or developed, in connection with energy conversion in space systems (De Groff *et al.*, 1967). Szergo (1971)

† A bibliography on energy-related research is available—"Energy: A Special Bibliography with Indexes," NASA SP-7042 (1974), NASA, Washington, D.C.

has discussed, in some detail, the economic aspects of the various development options with respect to the energy problem.

For practically all of the viable energy systems the common optimization parameter is the temperature. Generally, the higher the temperature, for example in a gas turbine or a steam boiler, the greater the efficiency in converting the source of energy to useful work. This is a direct consequence of the Carnot efficiency relation, which may be expressed as

$$E_{max} = (T_{source} - T_{sink})/T_{source},$$

where E_{max} is the maximum efficiency attainable for a given source and sink temperature, e.g., see Weller and Bagby (1967). It is apparent, then, that a high source temperature is desirable for energy systems.

In practice, suitable high source temperatures are attainable—for instance, in coal or oil combustors or in nuclear fission fuel elements where temperatures of 2000 °C, or greater, are readily achieved. The basic problem lies in the containment of such high temperature reactive fuel systems and of the working fluid used to transfer the heat to an energy conversion unit. This is essentially a materials problem involving, for the most part, either mechanical or solid-state chemical phenomena. However, some of the materials problems arise from high temperature gas–solid corrosion reactions and, in some instances, an undesirable vapor transport of material to another part of the system can result. In other cases, vapor-phase transport in the materials-energy system may even be desirable as, for example, in the preparation of refractory coatings or in fuel processing. We have therefore selected for discussion a number of energy systems where, to a varying degree, high temperature vapor-phase processes either play an important role in the present technology, or may do so in anticipated future developments.

The energy systems to be considered represent a wide range of temperature and pressure conditions as shown schematically in Fig. 6.1. In addition to these systems, one should be aware of several other energy sources where high temperature processes play a role. For instance, there is considerable interest in hydrogen as an alternative to fossil fuels, and high temperature catalytic processes for its production from H_2O have been proposed, e.g., see Wentorf and Hanneman (1974) and Hammond *et al.* (1973). These processes have some pertinence to the present discussion in that they involve gas–solid interactions and metal–halide dissociations at temperatures of about 500–700 °C. Similarly, the increased attention given to natural or induced geothermal power sources (Barnea, 1972) gives rise to questions concerning the interaction of steam with minerals at subterranean levels. The utilization of subterranean nuclear explosions (e.g.,

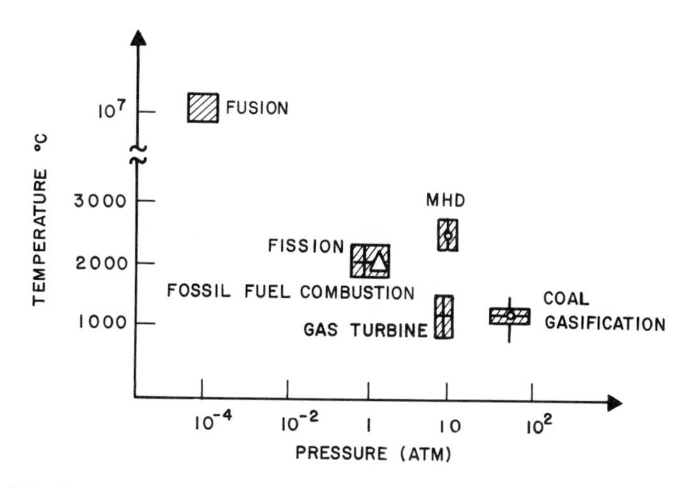

Fig. 6.1 Approximate temperature and pressure regimes for energy systems.

Operation Plowshare) for releasing fossil fuels poses similar questions of a high temperature chemical nature.

The discussions of previous chapters on chemical transport and vapor deposition (Chapter 3), corrosion (Chapter 4), and combustion (Chapter 5) are clearly relevant background material to a discussion of energy systems. In addition to this basic material, which applies primarily to the static or material transport phenomena likely to be found in energy systems, it is appropriate to consider, as part of the following discussion, some basic aspects of energy transport—particularly with regard to the role of high temperature species in heat transfer.

II. Thermal Energy Transport

A. Introduction

As was indicated in the introductory chapter (Chapter 1, Section V.B), energy—or heat—transport, mass transport, and chemical reaction are all interdependent phenomena.† It follows that, in addition to the already demonstrated role of high temperature species in mass transport systems (e.g., see Chapter 3), such species may also affect the transport and distribution of thermal energy in high temperature systems, particularly

† More detailed discussion of this point may be found in the basic texts of Frank-Kamenetskii (1969), Williams (1965), and Geiger and Poirier (1973).

where chemical reactions are present. A brief discussion of this application of high temperature species follows.

A recent interest in energy transport has resulted from the need to model rocket exhausts and the boundary layers of hypersonic vehicles. However, the basic results and modeling procedures developed can be applied equally well to other combustion-supported energy systems, such as magnetohydrodynamic generators and gas turbines, in addition to nuclear fission and fusion devices.

The observable heat-related macroscopic properties of high temperature gas-phase systems, such as rate of heat release from flames (see Fristrom and Westenberg, 1965), thermal conductivity (see Brokaw, 1960, 1961a, b, 1967; Barsukov and Belov, 1969), and ablation and corrosion (see Mills *et al.*, 1971), can, in principle, be rigorously related to basic molecular properties such as enthalpies, partition functions, reaction rates, and collision cross sections (see Chapter 1, Section V).

Thermal conductivities can be calculated, to a good approximation, from viscosity and density data using the "modified Eucken factor," i.e.,

$$\lambda/C_{\mathrm{v}}\mu = 15R/4C_{\mathrm{v}}M + (\rho D/\mu)(1 - 3R/2C_{\mathrm{v}}M),$$

where λ is the thermal conductivity, μ the viscosity, ρ the density, R the gas constant, C_{v} the heat capacity at constant volume, D the diffusivity, and M the molecular weight, e.g., see Ferron (1968). Explicit expressions for the heat conductivity of nonreacting polyatomic and polar gases have been given by Mason and Monchick (1962) and more recently by Monchick *et al.* (1965). Thermal transport in high temperature gases, including those where chemical reaction is also occurring, has been considered in detail by Brokaw (1961b). It is apparent that the methods for calculating transport phenomena are fairly well developed, and the main limitation is the lack of basic kinetic and collision-diameter data.

B. Collision Diameters

Some methods for the estimation of collision diameters are as follows (Brokaw, 1960, 1961a, b). The collision diameter between atoms a and b is empirically approximated by

$$\sigma_{\mathrm{ab}} = \bar{r}_{\mathrm{a}} + \bar{r}_{\mathrm{b}} + 1.8\ \text{Å},$$

where $\bar{r}$ are the mean Slater radii of the outermost electrons as given by

$$\bar{r} = [n(2n + 1)/2(Z - S)]a$$

where Z is the atomic number, n the effective principal quantum number, S the screening constant, and a the Bohr radius. The collision diameter for homonuclear diatomic molecules is approximated by

$$\sigma_{\text{mol}} = (1.849\sigma^2_{\text{atom}} - 1.528\sigma_{\text{atom}})^{1/2}.$$

These expressions can also be used for species in excited electronic states where the collision diameters can be appreciably larger.

Considerable uncertainty exists in dealing with estimations for polar molecular species, because of their ability to interchange translational and rotational energy over large distances; i.e., the collisions are nonadiabatic.

C. Heat Conduction in Reacting Gases

Many of the high temperature systems of practical interest are gases undergoing chemical reactions and the enthalpy changes associated with these reactions strongly affect heat transport. Heat from a chemical reaction may be transported as molecular enthalpy—that is, translational and internal energy—and by diffusion across concentration-temperature gradients. An endothermic reaction, such as molecular dissociation, can transport heat to a lower temperature region where heat is released by recombination processes. For example, the reaction

$$Al_2Cl_6 = 2AlCl_3,$$

gives Al_2Cl_6 a high thermal conductivity; in fact, this chloride vapor has been suggested as a power plant working fluid (see Section III.D.1). Similarly, heat may be removed from a hot metal surface where molecular dissociation occurs followed by diffusion of the fragments away from the hot surface with subsequent recombination.

Expressions for the thermal conductivity of reacting gas mixtures in chemical equilibrium have been given by Brokaw (1960) and Butler and Brokaw (1957). For the more general case, where a finite reaction rate leads to a departure from chemical equilibrium, see Brokaw (1961a).

For gas mixtures in local chemical equilibrium the contribution of a single chemical reaction to the coefficient of thermal conductivity is given by

$$\lambda_r = \frac{\Delta H^2/RT^2}{\sum_{k=1}^{u-1} \sum_{l=k+1}^{u} \left(\dfrac{RT}{D_{kl}P}\right)(X_k X_l)\left(\dfrac{n_k}{X_k} - \dfrac{n_l}{X_l}\right)^2},$$

where

ΔH is the heat of the reaction,
P the pressure,

X_k, X_l are mole fractions of components k and l,
n_k, n_l are the stoichiometric coefficients for species k and l in the reaction, and
D_{kl} is the binary diffusion coefficient between species k and l.

A more general expression for the case where a number of reactions are present has been given by Brokaw (1960, 1961b). The heat flux may be calculated from λ_r using the expressions given in the introductory chapter (Chapter 1, Section V.B).

Very good agreement between calculated and measured thermal conductivities for HF gas is obtained when the equilibria between monomer–dimer–hexamer are taken into account. If the gas had been assumed to be nonreacting, i.e., only monomeric species present, then the calculated thermal conductivity would have been about a factor of 33 lower. This example serves to demonstrate the importance to thermal transport of a complete accounting of the chemical species present in a multicomponent gas-phase system. Good agreement between experiment and theory has also been demonstrated, with the assumption of local thermodynamic equilibrium, for the thermal conductivity of PCl_5 which undergoes dissociation to PCl_3 and Cl_2 (Chakraborti, 1963).

For the case of Br_2 or F_2 dissociation, finite reaction rates preclude the assumption of equilibrium for the calculation of thermal conductivity. In such cases, the thermal conductivity is related to the ratio of a diffusion time and a relaxation time.

The application of thermal transport calculations to rocket combustion systems has been discussed by Taylor and Petrozzi (1968). The calculations are most sensitive to the Lennard-Jones potential function parameters σ and ϵ/k which for high temperature species are usually estimated. As examples of potential function parameters, for AlF_3, $\sigma = 5.198$ Å and $\epsilon/k = 1846$ K, whereas for the permanent gas CO_2, $\sigma = 3.996$ Å and $\epsilon/k = 190$ K. These authors have given similar estimates for a number of other high temperature species, such as

B_2O_2, B_2O_3, HBO_2, LiOH, LiO, Li_2O, BOH,

AlOCl, $AlCl_3$, $AlCl_2$, AlO, BCl, and BF.

D. Thermal Energy Storage

Energy may be stored by the heating, melting, or vaporization of a thermodynamically reversible material. This stored energy may then be released as heat by a reversal of these endothermic processes.

The use of the latent heat associated with a phase change for energy storage does not require a temperature change. As was pointed out by Hottel and Howard (1971), the high heat of fusion for salts such as ferric chloride, i.e., 20,000 Btu ft^{-3} at 305 °C m.p., could be of use in smoothing electric space-heating loads. A review of this type of energy storage has been given by Altman (1967).

Another potentially useful aspect of the energy-storage principle involves the application of multiple fluid cycles. It is evident that multiple fluid cycles, where one of the fluids is a high temperature vapor and the other is steam, can provide an increase in thermal efficiency, e.g., see Hougen *et al.* (1959). The heat of condensation of the high temperature fluid is used as a heat source for the vaporization of the lower temperature fluid. To date, only the mercury-steam cycle has been used commercially.

III. Nuclear Fission Systems

There exists a multitude of materials problems associated with the operation and development of nuclear fission systems as energy sources. They cover such diverse areas as neutron damage to the alloy of fuel containers (i.e., cladding)—which results in swelling and creep of the metal, gas–solid corrosion, and vapor-phase transport of fissile material. Some of these problems are the result of high temperature vapor-phase phenomena, and these are given emphasis in the discussion which follows.

A. Reactors in General

A general discussion of the role of nuclear fission as a source of electric power has recently been given as part of a symposium on "Energy for the Future—Problems and Prospects" (Benedict, 1971). Hottel and Howard (1971) have discussed the prospects for nuclear fission energy systems, in comparison with other available energy sources, and a cost evaluation and technical comparison of various reactor concepts has also been made recently by the EEI Reactor Assessment Panel (1971).

The fundamental aspects of various reactor types have been outlined by Smith (1967). Basically, the process of nuclear fission results in the release of fission atoms, neutrons, and photons. About 85% of the energy is carried off in the form of kinetic energy which can be converted to heat by collision processes. Most fission reactors obtain energy from the slow neutron fission of the low abundance (0.7%) isotope ^{235}U; however, the future use of fast breeders would make use of the more abundant (99.3%)

^{238}U isotope. The isotope ^{235}U is the only natural isotope that fissions readily and a fission of this nucleus releases about 200 MeV total kinetic energy. About 10^{10} fissions sec^{-1} are required to produce 1 watt of electric power. The low abundance of natural nuclear fuel is offset by the fact that the nuclear fission process releases about 10^6 times more energy than a chemical process. Existing commercial reactors use a factor of about 10^{-4} less fuel than would be required by a coal-fueled power plant.

In order for a nuclear fission chain reaction to occur, a critical mass must be present. This allows the production of neutrons to outweigh their loss by nonfission capture processes. The cross section for the fission of ^{235}U is several orders of magnitude greater with slow, or thermal, neutrons than for fast neutrons. Hence moderators, such as water, graphite, Be, BeO, and BeC are used to thermalize the fast neutrons in the system.

Reactor control can be achieved by the use of control rods made of neutron absorbing materials such as Cd, B_4C, and certain rare earth elements. Unproductive neutron losses are reduced by the use of neutron reflector shields. These are usually constructed of high density elements such as Fe, Bi, or Pb.

A typical reactor fuel assembly is comprised of an array of metal rods, known as fuel pins, which contain the fissionable fuel material. Rods containing a moderator material may also be dispersed among the fuel pins. These pins are situated in a coolant bath which serves to control the fuel temperature and prevent a catastrophic melting of the fuel. The coolant also acts as a heat transfer medium to a conventional power plant such as a steam turbine electric generator. The choice of coolant is dictated by the need for low neutron absorption and high heat transfer efficiency. Coolants used to date include: liquid metals such as Na, molten fluoride salts, light water (H_2O), heavy water (D_2O), and He or CO_2 gases.

1. Reactor Types

A comparison of various reactor types has been given by Smith (1967). These include the following:

- Light Water Reactors (LWR),
- Heavy Water Reactors (HWR),
- High Temperature Gas Reactors (HTGR),
- the Molten Salt Reactor Experiment (MSRE), and the breeder systems:
- Liquid Metal Fast Breeder Reactor (LMFBR),
- Gas Cooled Fast Breeder Reactor (GCFBR),
- Molten Salt Breeder Reactor (MSBR).

The reactor terminology used refers primarily to the type of coolant employed.

Reactor fuels used are mostly solid oxides although, more recently, carbide fuels have been developed, e.g., for HTGR. Some early reactors used aqueous solutions of uranyl sulfate or the nitrate. Experimental reactors have used liquid metal fuels of U in Bi, or Pu in Fe, and molten fluoride fuels. At present, chloride salts are unacceptable because of the high neutron absorption cross section of the ^{35}Cl component, for both thermal and fast neutrons, coupled with a high cost for separating chlorine isotopes.

Most reactors in the U.S.A. are light water reactors, i.e., H_2O coolant fluid is used, and the water may be in the form of liquid under pressure or steam. The water is used in a primary loop which removes heat from the reactor and then transfers heat to a separate secondary water loop in which steam is generated and used to drive a turbine as part of an electric current generator. The reactor contains rods of UO_2 enriched in ^{235}U (3%) and sealed in zirconium alloy tubes which are contained in a steel pressure vessel. The water surrounding this vessel typically has temperatures of around 280–320 °C and a pressure of 185 atm in the pressurized water system. This primary water also contains boric acid which serves as a neutron moderator. These reactors have a thermal efficiency of about 32%.

The high temperature helium-gas-cooled reactor has a higher efficiency than either the light or heavy water reactors. This is a practical demonstration of the significance of the Carnot efficiency principle in energy systems. The boiling water reactor has a fuel core input water temperature of 190 °C and an outlet temperature of 286 °C. On the other hand the HTGR utilizes He which enters at 405 °C and leaves at 777 °C. The fuel type for this reactor is also novel, consisting of particles of uranium and thorium dicarbide, i.e.:

$$5.03\%\,^{235}U + 94.6\%\,^{232}Th + 0.37\%\,^{238}U,$$

in a pyrolytic carbon matrix and a silicon carbide seal to contain fission products. The higher thermal conductivity of the carbide fuel element, as opposed to the oxides, allows use of high average power densities which increases the efficiency. Also the higher density of carbides gives a greater number of neutron absorptions by the fuel.

Mention should also be made of the more advanced nuclear fission systems developed for small scale use in space systems, e.g., see the review of Pittman and LaFleur (1967). These systems operate at relatively high temperatures to achieve good thermal efficiency, and hence ceramic fuels and moderators are used. For instance, a hydrided U–Zr alloy serves as an intimate mixture of fuel and moderator. Future fuel candidates include uranium carbides and nitrides.

B. Breeder Reactors

The breeder reactor concept utilizes either ^{238}U or ^{232}Th as initial fuel sources and produces, or breeds, ^{233}U and ^{239}Pu, respectively, as additional new fuel. This production of new fuel is also accompanied by the release of neutrons which may be used, simultaneously, as a source of heat for a steam driven electric power plant. The breeding of new fuel usually takes place in a blanket of primary fuel around the core of an active fuel pin element.

1. LMFBR

Of the various breeder reactor concepts, the liquid metal fast breeder reactor is being given top development priority in the U.S.A.—see Shaw (1967), Seaborg and Bloom (1970), and Culler and Harms (1972). Fast gas and molten salt breeder reactors are being considered as alternative systems but at lower priority. In fast breeder systems the water coolant is substituted by liquid Na ($\sim$1 atm and 620 °C) which does not appreciably slow the neutrons; that is, Na has a hard neutron spectrum. The fuel usually consists of a mixture of the oxides, or carbides, of plutonium and uranium. A number of developmental plants have been operated with an electric power output of up to 66 MW. Commercial systems with 1000-MW output are planned, e.g., see Seaborg and Bloom (1970).

As the Na coolant becomes radioactive during operation, it is desirable to incorporate a secondary coolant system—or loop. Heat is exchanged between the radioactive primary Na coolant loop and the secondary loop which remains inactive. The secondary loop then exchanges heat to a third loop containing high pressure steam ($\sim$160 atm) for use in a conventional turbine system. This arrangement minimizes the prospect of an accidental interaction of radioactive Na, which may also contain fuel and fission product impurities, with the steam loop.

It is expected that in these reactors the oxide ceramic fuel temperatures will be at least 2200 °C, the cladding temperature may reach 760 °C, and the sodium outlet temperature will be about 590–650 °C. Some consideration, in the form of research and development, is also being given to plutonium–uranium nitrides, sulfides, and phosphides as fuels for breeder systems. A discussion of such ceramic nuclear fuels may be found in the review of Kendall and Riley (1965).

As one of the objectives of these reactors is to attain economically advantageous high burn-up rates, considerable quantities of fission products will also be present under actual operating conditions. Burn ups in excess of 10 at. % of heavy metal atoms are required for an economical fuel cycle. This would require continuous operation for a period of about 3 years and under a high neutron flux of 10^{23} neutron cm^{-2}. The ability of fuel pin cladding alloys to withstand such an excessive radiation level is one of the major current concerns in reactor technology. Cladding materials under consideration include Ni-base alloys, austentite stainless steel, and refractory metal alloys.

2. *GCFBR*

Proposed gas cooled breeder reactors would use He at temperatures of 316–650 °C and at pressures of 85 atm. Steam, at a temperature of 560 °C, would be used to drive the turbines. This relatively high temperature steam allows an improved thermal efficiency of around 40%, which is better than existing water-cooled slow reactor systems and is comparable with advanced fossil-fueled power plants. However, it should be noted that the poor heat transfer characteristics of He gas is a disadvantage over the liquid-Na coolant system.

Feasibility studies on GCFBR systems, e.g., see Whitman *et al.* (1967), have indicated both technical practicality and economic attractiveness, particularly for large systems. The harder neutron spectrum available with gas—as opposed to liquid-Na—cooling should allow a higher breeding ratio and hence lower fuel cycle costs. The gas most frequently considered in the U.S.A. is He; in fact, He-cooled reactors are now beginning to come into commercial use. However, in the United Kingdom, the less expensive gas CO_2 is favored and operating pressures of the order of 140 atm are contemplated (Moore and Fawcett, 1967).

3. *MSBR and MSRE*

The molten salt breeder reactor, which uses as a fuel source a molten mixture of U, Th, Be, and ^{7}Li fluorides appears to compare favorably with other proposed breeder reactor systems (EEI Reactor Assessment Panel, 1971), but is currently being given a very low development priority relative to the LMFBR and GCFBR systems (Hammond *et al.*, 1973). A number of reviews have recently appeared on the molten salt reactor concept and also the results of the molten salt reactor experiment which convincingly demonstrated the feasibility of MSBR; e.g., see Rosenthal *et al.* (1971a), Grenon and Geist (1971), Opstad and Schafer (1970), Rosenthal *et al.* (1970), Haubenreich and Engel (1970), Grimes (1970), McCoy *et al.* (1970), and Whatley *et al.* (1970), in addition to the annual ORNL reports—e.g., see Rosenthal *et al.* (1971b).

Operating in a nonbreeding mode, the MSRE generated a thermal energy output of about 8 MW for several years and was terminated during December 1969. The essential materials features of this experimental reactor were

- the molten salt fuel, i.e., $^{7}LiF(65) + BeF_2(29.1) + ZrF_4(5) + UF_4$ (0.9, and 33% ^{235}U) where the mole % composition is given in parentheses and $T = 650\ °C$,
- the molten salt coolant—initially $^{7}LiF(66) + BeF_2(34)$ and later $NaBF_4(92) + NaF(8)$, at $T = 454\ °C$,
- the unclad graphite moderators—i.e., in direct contact with the fluoride fuel bath, and
- the Hastalloy-N container material.

The vapor pressure of the fuel mixture is less than 0.1 Torr and a low pressure, ~0.3 atm, gas blanket of He was in contact with the fuel.

The molten salt breeder concept allows either ^{239}Pu, ^{233}U, or ^{235}U to be used as fuels, and a potential fuel doubling time of about 20 years has been estimated. When operating as a breeder system, ThF_4 would be added to the fuel as a source of fissile material with ^{233}U being produced. This breeding process was demonstrated experimentally in later stages of the MSRE.

A special requirement of the MSBR involves the need for rapid in situ fuel reprocessing, i.e., a complete cycling about every 3 days. This results from the high neutron capture cross sections of the fission products—particularly protoactinium, i.e., ^{233}Pa, and rare earth elements. A good breeding performance then depends on new fuel reprocessing techniques such as xenon stripping, fluoride volatility, and liquid metal extraction. The assessment of Hottel and Howard (1971) indicated that research and development is still needed on the chemical processing problems encountered in separating ^{233}U, ^{232}Th, and rare earth fission products in a single salt stream. However, Rosenthal *et al.* (1971a, b) have suggested that the development of a new Bi metal solvent extraction technique has overcome many of the fuel reprocessing problems.

The main advantages of molten fluoride fuels are as follows (Smith, 1967):

- a high thermal efficiency, and low salt vapor pressure,
- minimal radiation damage to the fuel, and
- good solubility of fuel and the fission products.

The main disadvantages are

- high melting points, i.e., ~449 °C for the fuel and 385 °C for the coolant, and
- a less favorable breeding ratio than for alternative breeder reactors.

A further problem, common to all high neutron flux breeder systems, is the neutron damage incurred by the alloy container material which for Hastalloy-N results in embrittlement.

It is perhaps pertinent to note the comment of Grimes (1970) that—"the mechanisms by which the Mo, Nb, Te, and Ru isotopes appear in the gas phase are still not fully understood." This observation refers to the presence of these isotopes in the He cover gas where they form an aerosol. Under the redox potentials of the MSRE fuel salt bath, which is intentionally kept slightly reducing by the presence of UF_3, it is unlikely that the volatile high valence fluorides of these elements would form. The vaporization process may involve unidentified lower valence fluorides or, as has been suggested by Grimes (1970), even the metals themselves but in a colloidal-aerosol state. The presence of ^{110}Ag and Pd is also noted in the purge gas samples. These elements do not form known volatile fluorides. However, it is reasonable to speculate that perhaps complex vapor species, such as those considered elsewhere (Chapter 3, Section IV), are involved in these "vaporization" processes. Further discussion of reactor-related vaporization phenomena may be found in the subsections to follow.

C. Some Reactor Materials Limitations

The combined effects of high temperatures, a corrosive atmosphere, and the high flux of energetic neutrons, place severe demands on reactor materials. In fact it has been contended that if the development of breeder reactors had been attempted 20 years ago, the materials problems would have been insurmountable at that time.

Smith (1967) and Nevitt and Duffield (1970) have discussed the general role of materials in reactors. Some recent research report literature, which deals extensively with the characteristics of advanced reactor materials and fuels, include: Patriarca and Rucker (1971), Keller (1969), Keller and Chubb (1969), and a series of AEC R and D reports, e.g., Public Service Company of Colorado 330—MW(E) "High Temperature Gas Cooled Reactor Research and Development Program," Rep. No. GA-9130 (1969), and earlier cited reports.

The effects of radiation on reactor materials, as discussed by Smith (1967), Bowen and Clarke (1970), and Kangilaski (1970), primarily involve the formation of defects in solids. In nonmetallic components, ionization can also be important, resulting in the formation of color centers in glasses and ionic crystals. Materials such as glass, transistor semiconductors, and polymers are adversely affected by neutron fluxes of the order of 10^{14} neutron cm^{-2}, whereas for stainless steel and metal alloys the corresponding figure is about 10^{21} neutron cm^{-2}. Despite the susceptibility of nonmetallic materials to neutron damage, the need to use increasingly higher temperatures in reactors has elevated the importance of refractory ceramics in reactor technology. In addition to the use of oxide and carbide ceramic fuels, ceramics have found use in moderators (BeO), insulators (Al_2O_3), control rods (B_4C), canning (graphite and SiC), and other structural materials (graphite), e.g., see Bowen and Clarke (1970). These materials frequently are in contact with gases or vapors and, as was emphasized by Nightingale (1966) for the case of graphite, more basic information is needed on the processes of material transport and compatibility.

For the case of fast breeder reactors, the high neutron fluxes cause swelling of stainless-steel components (Hafeele *et al.*, 1970). Cladding alloys used in Na-cooled systems are composed primarily of Fe, Cr, Ni, and Mo. Corrosion and radiation damage tests on a number of alloys have been described by Nettley *et al.* (1967). Dry steam was initially considered as a potential primary loop coolant material; however, corrosion was found to be a problem at the high temperatures required. Also, steam-cooled reactors require high-Ni-content steels to reduce corrosion and the swelling which results from the high neutron flux is even more pronounced with these alloys. Hence, steam cooling is not of current interest. The corrosion of stainless steel by sodium is controlled by minimizing the O_2 impurity level to less than 50 ppm. A cladding wall maximum temperature of about 700 °C is usually acceptable in fast breeder reactors.

D. High Temperature Vapor-Phase Processes Associated with Nuclear Fission Reactors

Recognizing that most of the materials requirements and problems associated with reactor operation and development involve the solid or liquid states, there remain a number of materials applications and processes that are basically high temperature vapor-phase phenomena. These applications of the science of high temperature vapors involve primarily: fuel and fission product vaporization, corrosion, reactor safety, fuel processing, the development of gaseous core reactors, and energy transport by dissociating vapors.

1. Dissociating Gases as a New Class of Coolant

The concept of utilizing dissociating gases as coolants and working substances for power plants has recently been discussed by Krasin and Nesterenko (1971). It is claimed that N_2O_4 and Al_2Cl_6 can provide about a 5 and 10% efficiency increase (i.e., to ~40%), respectively, over steam when used separately as coolants in fast neutron reactors, i.e., breeders. Binary circuits, i.e., two separate loops, each containing a dissociating gas such as Al_2Br_6 and N_2O_4, can lead to even greater absolute efficiencies in the region of 60%. Also in magnetohydrodynamic power plants (see Section VII), the improved efficiency over H_2O as a working fluid would be about 10%.

These potential coolants rely on the ability of the molecular dimers to dissociate to monomers at high temperature with the absorption of energy and, conversely, to recombine at a lower temperature with the release of energy to the system. Thus, the heat—or enthalpy—of dissociation is of fundamental importance here. We may recall that from the relationships between heat flux, coefficients of thermal conductivity, and reaction enthalpy (see Chapter 1, V.B, and Chapter 6, II.C), the heat flux varies as the square of the reaction enthalpy. For N_2O_4 the dissociation enthalpy is 13.7 kcal mol^{-1}, whereas for Al_2Cl_6 the corresponding energy is 29 kcal mol^{-1}, and hence, Al_2Cl_6 should be a far more effective working fluid. The other aluminum halides also have dissociation enthalpy values similar to Al_2Cl_6. The higher heats of condensation for Al_2Cl_6 and Al_2Br_6, relative to H_2O, are also favorable for heat transport applications. Furthermore, the increase in the number of gas molecules resulting from dissociation is advantageous to gas turbine operation. For Al_2Cl_6 as a coolant, the likely operating conditions would be 370–570 °C temperature and 90 atm pressure.

To summarize, the main advantages of dissociating high temperature vapors as coolants are—higher heat transfer coefficients, the need for only a single loop in fast neutron reactor operation, and the replacement of sodium which allows a harder neutron spectrum and a shorter fuel doubling time to be obtained in a breeder reactor.

2. Application of Vapor Deposition to Nuclear Fuel Preparation

Modern reactors of the HTGR type use encapsulated fuel particles to protect the fuel and to retain fission products; chemical vapor deposition processes are used to prepare such particles (see Chapter 3, Section II). For instance, a high temperature gas cooled reactor has used carbon-coated uranium–thorium dicarbide particles at temperatures up to 1500 °C. The coating can be applied by pyrolysis of hydrocarbons, such as methane or propane, at temperatures of about 1400 °C. Oxide coatings, e.g., BeO and Al_2O_3, have been prepared by the hydrolysis of metal halides in fluidized bed reactors. The deposition temperature is important since, if too low, intermediates such as solid particles, oxyhalides, and hydroxides may form and, if too high, rapid grain growth may reduce structural strength, e.g., see Powell *et al.* (1966).

A recent discussion of the preparation and properties of nuclear fuel pellet coatings has been given by Ford *et al.* (1972). The mechanical and thermal properties of the coating are currently of more concern than their chemical stability.

3. Gas-Phase Composition and Material Transport in the Fuel Core

a. Actinide Redistribution

It is well documented that a substantial redistribution of the actinides occurs during the operation of a $(PuU)O_2$ fuel element, e.g., see Bober *et al.* (1973) and cited literature. This effect can result in a temperature increase of several hundred degrees centigrade at the fuel pin center and possibly can lead to melting with an even more dramatic separation of the actinide components. To sustain an efficient and controllable fuel pin operation it is desirable that the distribution of the fuel components remains uniform.

In practice, the redistribution of UO_2 and PuO_2 in the fuel mixture is found to vary according to the initial stoichiometry and also the irradiation time and intensity. Attempts to develop fundamental chemical models that explain these dependencies, and extend understanding of the phenomena, have been made. Under the large temperature gradients associated with fuel pin operation, i.e., a change from ~2200 to ~500 °C across several millimeters, the possible modes of transport are

- vapor transport via cracks and voids,
- thermal diffusion through the lattice,
- solidus–liquidus separation, and
- electrotransport.

It is believed that the first listed mode of transport is the predominant redistribution factor at temperatures of about 1700–2000 °C and the thermal diffusion mode is predominant at temperatures of greater than about 2000 °C (Bober *et al.*, 1973).

The vapor transport mode is of particular interest to the present discussion and, according to the model described by Olander (1973a, b), gives a satisfactory explanation for the observed actinide redistribution where the cooler fuel pin regions become enriched in UO_2. The driving force for material transport is provided by the high concentration gradient for vapor species along cracks or in mobile pores, as determined by the variation of partial pressures across the temperature gradient. This force allows molecular diffusion to occur through the gas space with the result that the more volatile component is transported to the cooler outer region of the fuel pin. As is usual, the species flux J may be described by Fick's law, i.e.,

$$J = -\frac{D}{kT}\frac{\partial p}{\partial z},$$

where p is the species partial pressure—which is typically $\sim$0.1 atm or less, z is the direction of the temperature gradient and transport flow, and D is the binary diffusion coefficient of the species in the inert carrier gas. Under the assumed model condition of hyperstoichiometric $(UPu)O_2$ mixtures, i.e., $O/(U + Pu) > 2.0$, the only significant vapor species is known, from basic equilibrium thermodynamic data, to be UO_3. In extending the model to hypostoichiometric compositions one would need to consider also the species UO, UO_2, PuO, and PuO_2, and the corresponding diffusion coefficients, together with the changing reaction equilibria for these species along a temperature gradient. As has been alluded to elsewhere (Chapter 1, Section V.B), the diffusion coefficient for a particular species may be estimated from kinetic theory. The key estimate is for the effective collision diameter (see Chapter 6, Section II.B), e.g., for UO_2 Oldfield and Brown (1970) estimate a molecular collision diameter of 5.47 Å. For the conditions considered by Olander, i.e., UO_3 in 10-atm Xe with a temperature of $\sim$2000 K, the kinetic theory estimate for the UO_3 diffusion coefficient is determined to be $D_{UO_3} = 0.2$ cm^2 sec^{-1}. This value may be compared with those given elsewhere for other high temperature species, (see Chapter 3, Section III.C).

One can reasonably expect that the presence of fission product impurities may also influence the fuel redistribution process and, in principle, future extended models could incorporate this additional complexity provided the vapor pressure and transport data were available for each impurity vapor species.

b. Fission Product Distribution

In addition to the above mentioned possible influence of fission products on fuel redistribution, the development of models describing the release of fission products from a faulty fuel pin requires a knowledge of the gas-phase composition. This is difficult to predict because the nonstoichiometric nature of solid uranium oxide fuel has a pronounced influence on the oxidation state of the fission products.

Assuming a given composition for the condensed phase, multicomponent equilibria calculations have been made for the uranium–oxygen–strontium model system (Fontana and Bailey, 1968). The results depend on an estimation of Henry's law coefficients and are therefore only approximate. For a fuel composition given by U = 1.000, Sr = 0.005, and O = 2.005, and a temperature of about 2000 °C, the partial pressures of UO_2, UO_3, and Sr are determined to be of the order of 10^{-6} atm and of SrO about 10^{-7} atm. Further calculations of fission product gas pressures have been made by Paulson and Springborn (1968).

For the case of fuels containing polonium, it is likely that vapor transport occurs as the metal rather than as the oxide PoO_2. In the pure state, PoO_2 has a vapor pressure of several Torr at 900 °C, though the exact nature of the vapor state has not yet been determined—see Steinmeyer and Kershner (1971). In this connection it is useful to recall the complex vaporization behavior of the chemically similar TeO_2 system where many different species were found (Muenow *et al.*, 1969).

The fission of ^{235}U can proceed via at least 30 different routes, yielding more than 60 distinct nuclei. In a flux of 10^{13} neutron cm^{-2} the relative yield of these nuclei peaks in the 90–100 and 130–145 amu ranges, i.e., the second row transition elements and the third long row including the early members of the rare earths (Bowen and Clarke, 1970). Elements such as Zr, Mo, Nd, Ce, Cs, and Sr form the major fraction of the metallic fission products.

For a 6% combustion level of $(UPu)O_2$, products with a 10^4 ppm order-of-magnitude concentration include

Sr, Zr, Mo, Ru, Pd, Xe, Cs, Ce, and Nd;

of 10^3 ppm magnitude:

Kr, Rb, Y, Nb, Tc, Rh, Te, I, Ba, La, Pr, Pm, and Sm;

and of 10^2 ppm magnitude:

Se, Br, In, Sb, Cd, Sn, Eu, and Gd—

see Halachmy (1971). Note that the even-numbered isotopes are favored as fission products.

The fission of $^{235}UO_2$ results in a decrease in average metal valence of from 4 to 3.5. Thus oxygen is made available for oxidation of some of the fission products, though not all. Electron probe microanalysis suggests that Mo, Tc, Ru, and Re are precipitated in the metallic state, and Ba, Sr, Nd, and Zr form the oxides.

Some of the fission products may have a considerably higher volatility than the fuel components. Halachmy (1971) has made measurements on the vaporization of these fission products from the irradiated fuel (UPu)O_2 over a range of temperature and neutron dosage conditions. The vapors, generated in a tungsten Knudsen effusion cell, were trapped and analyzed with the various radioactive constituents being characterized by their in situ radiochemical spectra. The vaporization rate was primarily a function of the oxygen potential of the fuel matrix and the period of irradiation. Another study on the vaporization of fission products from irradiated uranium indicated that at temperatures between 1250 and 1900 °C iodine was lost as a uranium iodide under a He atmosphere. However, in air only elemental iodine vaporized (Castleman and Tang, 1969).

Chemical analyses of spent (UPu)O_2 fuel pins indicate that the fission products Cs, I, Te, Mo, and Ru, in addition to oxygen, are preferentially transported to the container walls of the fuel pin. The chemical form in which these elements are transported is not yet known though volatile oxides, halides, oxyhalides, and mixed metal complexes, of the type mentioned throughout Chapters 2 and 3, are likely candidates.

In some instances it is desirable to extract the short-lived nuclides continuously during irradiation. If UF_4 is the fuel, then it is possible to vaporize and separate the nuclides, Mo, Tc, Sb, Te, I, and Zr by a suitable selection of vaporization and condensation temperatures (Weber *et al.*, 1971b). Where several nuclides vaporize at similar temperatures, a temperature gradient in the condenser allows their separation and in situ characterization, using γ-ray spectroscopy.

4. *Advanced Nuclear Gaseous Core Reactors*

According to Barthelemy (1971), the feasibility of a gaseous core nuclear system has been established. In principle, the reactor core would be a flow system of uranium vapor and hydrogen. This particular core type could be used to heat H_2 as a source of energy for aerospace propulsion systems, e.g., see Rothman (1968) and Lanzo (1970). A specific impulse (see Chapter 5, Section IX) in the range of 1500–2500 sec appears possible with this concept. Typical reactor conditions would be, a total gas pressure of 20 atm and a uranium metal partial pressure of 1 atm. Barthelemy (1971) also considers this type of reactor to have potential as a power source, in combination with a magnetohydrodynamic generator. However, the corrosion of the graphite moderator by the H_2 gas–uranium metal mixture is a known problem with this reactor.

An attempted development of a low altitude nuclear ramjet air-breathing missile system (Project Pluto) encountered serious vapor transport related problems, namely, the corrosion of BeO ceramic components by water vapor and the increased volatility of uranium under oxidizing conditions (Rothman, 1968).

5. *Corrosion in High Temperature Reactors*

An a priori indication of the likely corrosion effects of gas impurities on reactor components may be obtained from thermodynamic arguments. Hales and Pearce (1970) have made such an evaluation of the possible effects of impurities in the He-gas coolant on the various metallic components in the so-called MK III-type reactor. This reactor would operate at a gas temperature of 800 °C and 50 atm pressure. Neglecting the possible importance of kinetic factors, free energy considerations for reactions such as

$$M(s) + H_2O \leftrightarrows MO(s) + H_2,$$

$$M(s) + CO_2 \rightleftarrows MO(s) + CO,$$

and

$$M(s) + 2CO \leftrightarrows MC(s) + CO_2,$$

$$M(s) + CH_4 \rightleftarrows MC(s) + 2H_2,$$

indicate that the significant parameters for oxidation are the impurity partial pressure ratios of H_2O/H_2 and CO_2/CO, and for carburization the ratios $CH_4/(H_2)^2$ and $(CO)^2/CO_2$. Typical major impurity levels in He-cooled reactors are H_2(30), CO(30), CO_2(10), CH_4(~0.2), and H_2O (~0.1), given in ppm. These concentration levels are equivalent to partial pressures of up to several Torr and are therefore significant. The metal activity in alloys also has an important effect on the stability of corrosion products. Thus, for unit activity Cr, one could expect $Cr_{23}C_6$ to be stable, but at a factor of 10 lower activity this compound would not be stable.

The additional effect of radiation on the corrosion properties of materials, such as Zircaloy, has been of particular concern, e.g., see Cerrai *et al.* (1972). In the case of aqueous solutions, radiolysis of H_2O readily occurs with the formation of corrosive peroxides. However, the corrosive properties of liquid metals and molten fluorides are virtually unaffected by radiation.

6. *Fuel-Coolant Interactions*

An ongoing concern regarding reactor safety and the fortunate lack of empirical data derived from major nuclear excursions (i.e., accidents) has resulted in an extensive effort to model the possible chemical interactions that could arise during the initial stages of a nuclear mishap, e.g., see Ybarrondo *et al.* (1972). Of particular concern is the prediction of the thermal and hydraulic response of commercial nuclear power reactors to a loss-of-coolant accident and the possible release of radioactive material to the environment.

a. Water-Coolant Interactions

The modeling problem is primarily one involving heat transfer. It is predicted that as a result of a loss of coolant in the pressurized water reactor (LWR), the fuel cladding temperature will rise to more than 900 °C in the first 6 sec of the excursion. Under these conditions additional heat will be generated by reaction of the cladding, e.g., Zircaloy, with steam and this heat contributes to the total heat flux of the system. In addition to this, the possibility exists for an interaction of the steam coolant with the uranium oxide fuel and the associated fission products. Such an interaction would be possible, for example, in the not uncommon event of a leak in the fuel cladding.

Reactions of uranium oxides with liquid water and oxygen at high temperatures have been reported by Habashi and Thurston (1967) and by Markowitz and Clayton (1970). It was found that UO_2 had a lower resistance to water corrosion than mixtures such as ZrO_2–CaO–UO_2, ZrO_2–UO_2, or ThO_2–UO_2. Also, it appeared that the dissolution was favored by the presence of oxygen which allowed U^{IV} to convert to U^{VI}. One of the mechanisms suggested to explain the dissolution of UO_2 involves the production of hydroxide species in solution, such as $UO(OH)_2$ and $UO_2(OH)_2$, and it is not inconceivable that a similar mode of dissolution could occur in high temperature, high pressure steam (see Chapter 2, Section VIII.E). Such an argument tends to be supported by the fact that the observed corrosion effects appear to be similar in either aqueous or steam media (Markowitz and Clayton, 1970).

The dissolution and transport of the fission products by steam is also of concern. Therefore, the known enhanced volatility of oxides such as TeO_2 in the presence of water vapor (see Chapter 2, Section VIII) is important in determining the migration of fission products from a defective fuel pin—see Malinauskas *et al.* (1970).

b. Na-Coolant Interactions

In Na-cooled fast reactors it becomes necessary to vent gaseous fission products, i.e., Kr and Xe, from the fuel in order to prevent an excessive pressure buildup. As discussed by Whitman *et al.* (1967), venting can be achieved either internally to an end plenum or externally to the coolant and the latter procedure is favored. This process provides an opportunity for interaction of sodium vapor with the fuel, and also the possible gas transport of fission products into the sodium. However, under normal operation, these interactions do not appear to be serious problems. A more drastic interaction between Na and fuel is possible in the event of a fuel pin cladding failure. It is thought that such an interaction would lead to the formation of a two-phase (vapor + liquid) Na bubble which would move to a secondary containment region and possibly carry a plutonium oxide aerosol to the environment. A model involving primarily gas dynamic and heat transfer considerations suggests that about 2–10% of the fuel may be transported by the Na bubble, mostly during the first 2 min of the Na-fuel interaction (Kennedy and Reynolds, 1973).

A Na–(UPu)O_2 interaction may also lead to a swelling of the fuel, and hence further cladding failure, owing to the formation of a compound Na_3MO_4 (M is U and Pu) which is stable in liquid Na—e.g., see Smith (1973) and cited literature.

In addition to Na-fuel interactions, a failure in the liquid-Na–steam-heat exchanger is possible and has in fact already occurred with serious results. The reaction of liquid Na and steam is vigorous, leading to the generation of heat and hydrogen, and hence a high potential fire and explosion hazard results.

The compatibility of other prospective fast reactor fuels, such as nitride and nitride–carbide systems, with the coolant environment, whether it be sodium or steam, needs to be established.

c. Molten Salt Coolant Interactions

With the MSBR concept, an accidental fuel-coolant interaction would involve a mixing of fluoride salts with the decomposition of $NaBF_4$ to yield BF_3 gas. However, in this instance, the solubility of BF_3 in the molten fuel is sufficiently high to cause a 20% reduction in reactor activity, i.e., to a relatively harmless subcritical level (Cantor, 1973).

E. Fuel Processing and Vaporization Processes

1. Introduction

The need for fuel reprocessing, as opposed to discarding the spent fuel, arises from the low proportion of fuel, i.e., ~1–3%, that is actually used in reactor operation. In practice the reprocessing of spent fuel, and also the extraction of uranium from its ore, is primarily achieved by aqueous processes, e.g., see Bruce (1961). However, given an increased need to operate reactors with a high fuel burn-up requirement, particularly in connection with fast breeders, such as the LMFBR, the development of alternative processing techniques appears to be warranted (Schmets, 1970). Even a decade ago some consideration was given to alternative processes involving vaporization, such as distillation, fluoride volatility, chloride volatility, and pyrometallurgical procedures (Bruce, 1961). Details of the more recent status of these particular process types may be found in the articles edited by Chiotti (1969). A discussion of pyrometallurgical processing has also been given by Chiotti and Voigt (1961). The present status of these high temperature methods has been summarized by Schmets (1970), who comments that more research and development is necessary to allow continuous, rather than batch type, operation.

2. *Fluoride Volatility Processes*

Fluoride volatility processing is considered to be one of the possible economical means for reprocessing spent fast reactor oxide fuels, and Shaw (1967) has emphasized the need for more research into such processes. The results of research in this area have recently been extensively reviewed by Schmets (1970)—see also Chiotti (1969).

For the case of a LMFBR PuO_2–UO_2 spent fuel mixture, containing fission products and Na impurities (from coolant leakage), a proposed fluorination process would involve a fluidized bed conversion of UO_2 to UF_6 using F_2 at a temperature of 350 °C and a similar second-stage fluorination at ~550 °C to yield PuF_6. Various distillation steps would be used to remove the volatile fluoride fission products from the UF_6 and PuF_6 fractions (Jonke *et al.*, 1969; Thompson *et al.*, 1969). One of the proposed methods of converting PuF_6 to the solid form of PuF_4 involves use of a H_2–F_2 flame as a reducing medium.

In addition to F_2, other fluorinating reagents used are the interhalogens such as BrF_5 and BrF_3. Alkali metal fluoride melts may also be used to oxidize uranium, plutonium, and their fission products according to the reaction

$$\mathrm{Me} + n\mathrm{RF} \rightleftarrows \mathrm{MeF}_n + n\mathrm{R},$$

where n is usually 4 or 6. Continuous removal of the alkali metal R by vaporization allows the reaction to proceed to the right. Another feasible molten salt process involves the dissolution of thorium–uranium oxide particles in a fluoride melt, such as LiF–NaF–ZrF_4, followed by a reaction with HF or F_2 to produce volatile UF_6, e.g., see Bannasch *et al.* (1969) and cited literature.

Of fundamental importance to fluoride volatility processing is the question of relative volatility for the various fluorides present. From Schmets (1970), the important volatile fluorides in fuel reprocessing, and their boiling points, are those listed in Table 6.1. Usually, the fluorides of Mo, Te, Nb, Sb, Ru, Rh, and Pd are the most abundant by-products. Under the normal temperature conditions of fluoride volatility processing, i.e., < 500 °C, a number of fission product elements form relatively nonvolatile fluorides. These include the rare earths, Y, Rb, Cs, Ag, Sr, Cd,

TABLE 6.1

Volatile Fluorides in Fuel Reprocessing[a]

Molecule	Boiling point (°C)	Molecule	Boiling point (°C)
UF_6	56.5	IF_5	100
NpF_6	55.2	SbF_5	142
PuF_6	62.2	RuF_5	227
TeF_6	−36	NbF_5	235
SbF_3	3.19		
IF_7	4.5	IOF_5	4.5
MoF_6	35	$TcOF_4$	178
TcF_6	55.3	$RuOF_4$	184
NpF_6	56.4	$MoOF_4$	184
RuF_6	62.2		

[a] Data obtained from the review of Schmets (1970).

Ba, Zr, and Sn. In addition to volatility, the dissociation equilibria between the hexafluorides and tetrafluorides of U and Pu are of basic importance.

Distillation of UF_6 and PuF_6 from the other metal fluorides has been experimentally evaluated, and it is believed that no azeotropic problems should exist for UF_6 distillations, with the probable exception of MoF_6. In general, the gross distribution of the actinides and fission products follows that expected from the fluoride volatilities, e.g., see Steindler *et al.* (1969). However, there are some uncertainties with regard to the actual identity of the volatile components, particularly for Ru.

Another aspect of the fluoride volatility separation concept involves the selective absorption and complexing of volatile fluorides by solid salts such as alkali metal fluorides. For the alkali metal salt systems the stability of the solid complex increases with increasing atomic number of the alkali metal. The vapor pressure of these complexes is relevant to the problems of storage where, over an extended period of time, material transport and contamination could occur.

A fused-salt volatility process (Cathers *et al.*, 1961) utilizes a fused-salt bath of 50% ZrF_4–50% NaF, at 600–700 °C, which is sparged with HF to dissolve zirconium–uranium oxide fuels as the nonvolatile tetrafluorides. Fluorination of the melt then releases uranium as the volatile UF_6. The alloy Hastalloy-N, with metal constituents of Ni, Mo, Cr, Fe, and Mn, provides suitable resistance to salt corrosion and may be used as a container material.

Under the practical conditions of fluoride volatility processing, the heterogeneous reactions, e.g., fluorination of the oxides, are kinetically rather than thermodynamically controlled. Hence the effect of temperature, particle size, porosity, etc., on the rate of fluorination is of interest, e.g., see Henrion and Leurs (1971).

3. *Production of UF_6 from Ore Concentrates*

The production of UF_6 from ore concentrates using fluoride volatility processing has been described by Lawroski *et al.* (1961). The initial step involves partial fluorination of the ore with HF, i.e.,

$$\text{ore} + \text{HF} \xrightarrow{450\ ^\circ\text{C}} \text{UF}_4(\text{s}) + \text{impurity fluorides.}$$

This is followed by a fluorination process, i.e.,

$$\text{UF}_4(\text{s}) + \text{F}_2 \xrightarrow{>200\ ^\circ\text{C}} \text{UF}_6,$$

which has an activation energy of about 20 kcal mol^{-1}. The lower cost of HF, as compared with F_2, is the reason why direct fluorination of the ore is not used. As the other ore constituents also form fluorides, this results in an inefficient use of the halogen. This fluorination process for refining uranium is particularly advantageous if the metal is to be isotopically enriched, which is practically always the case, since UF_6 is required for the gas diffusion isotope enrichment process.

The SiO_2 and B_2O_3 ore impurities are removed in this process as the volatile species SiF_4 and BF_3. Fluorides of Mo and V are troublesome in the fractional distillation of UF_6, and the crude UF_6 can contain impurities such as MoF_6, VOF_3, SbF_5, CrF_4, and BiF_5, which are listed in order of increasing boiling point, i.e., from 40 to 550 °C. The residual fluorinated gangue may contain solid fluorides such as NiF_2, PbF_2, NaF, CuF, CaF_2, AlF_3, CrF_3, MgF_2, and MnF_2. These solids lead to caking problems in stationary-column reactors, and this has lead to the use of a fluidized bed reactor for the gas–solid fluorination reaction.

In the final purification step, which involves distillation of a mixture of UF_6, MoF_6, and VOF_3, the optimum conditions are considered to be a pressure of about 2 atm and a temperature of 75 °C. In the fraction containing VOF_3 and UF_6 a vapor-phase enrichment of V, relative to U, was found. The reason for this enrichment is apparently not known—see Lawroski *et al.* (1961). However, it should be noted that at 75 °C the pure systems would have *calculated* vapor pressures, in Torr, of $VOF_3 \sim 150$, $UF_6 \sim 1600$, $MoF_6 \sim 3000$, and $VF_5 \sim 3000$. At such low temperatures and high pressures a vapor-phase association between VOF_3 and UF_6 seems likely and this would explain the observed enrichment—see Example 4 in Chapter 3, Section IV.D.

4. Fluoride Low Pressure Distillation Processes

As part of the MSBR development, experience with the molten salt reactor experiment has shown that the extraction of rare earth fission products from the fluoride fuel bath is relatively difficult owing to the high stability of the rare earth fluorides. Several types of separation procedure have been tested, namely, a high temperature distillation process and a lower temperature counter-current solvent extraction process which utilizes the solvent properties of liquid-Bi–metal, e.g., see the review of Grimes (1970). The current status of these separation procedures is that the Bi extraction technique is feasible but suffers from the corrosive nature of liquid Bi; with regard to vacuum distillation it has been stated by Grimes (1970) that more work is needed on recovery schemes for Li, Be, and Th.

As the molten fluoride mixtures form solutions that are far from being ideal, in the thermodynamic solution sense, then the relative volatilities cannot be determined simply from a knowledge of the individual pure salt vapor pressures and the composition of the mixture. That is, Raoult's law does not apply here. It is necessary, therefore, to determine experimentally the relative volatilities of individual components for each mixture of interest. This has been done recently for molten solutions of LiF and $LiF–BeF_2$, containing fluorides of Ce, La, Pr, Nd, Sm, Eu, Ba, Sr, Y, and Zr (Hightower and McNeese, 1972).

A summary of thermodynamic activity and free energy data pertinent to the molten salt mixtures of interest to the MSBR program may be found in the review of Baes (1969). These data illustrate the strong dependence in the vapor pressures of fission-product species, such as RuF_5 and MoF_6, on the concentration ratio of UF_4/UF_3 in the molten fuel mixture. For the most probable MSBR operating conditions this ratio will be in the range of 10–100 and the reducing character of this amount of UF_3 is sufficient to preclude the formation of WF_6, MoF_6, and RuF_5 as volatile fission products. The more stable lower valence species are generally less volatile but very little is known about their identity or thermochemistry.

Laboratory tests on the separation of NdF_3, as a representative fission product, from the more volatile $LiF–BeF_2$-ZrF_4 carrier salt have been carried out in the temperature region of 1000 °C (Hightower and McNeese, 1971). Evidence for entrainment and a concentration polarization phenomenon was found. The latter effect was proposed as an explanation for a gradual decrease in distillation effectiveness due to an increase in the concentration of NdF_3 at the surface of the still-pot liquid relative to the bulk liquid. Such an increase would allow a more than normal amount of NdF_3 to enter the vapor phase. Alternatively, one might suspect that formation of a vapor-phase complex, such as $LiNdF_4$, may be a contributing factor (see Chapter 3, Section IV). No thermodynamic analysis of such a possibility has been made, however.

5. *Chloride Volatility Processes*

Chlorination processes, somewhat analogous to fluoride volatility processes, have also been developed on a laboratory scale in recent years, e.g., see Chiotti, (1969), Reinhard *et al.* (1972), and the related discussion of Chapter 3, Section VII.B. Particular emphasis has been given to fuel mixtures containing thorium since its fluoride-volatility processing is restricted by the low volatility of ThF_4, whereas the chloride is relatively volatile (see Fischer *et al.*, 1969). In addition to this, the initial separation of fuel from the Zircaloy fuel pin reactor elements can also utilize a chlorination procedure, and it is therefore advantageous if a similar processing chemistry can be used for reprocessing the fuel itself.

In essence, the chlorination process involves a carbothermic chlorination of the oxide fuel-fission product mixture, e.g.,

$$ThO_2(s) + 2C(s) + 2Cl_2 = ThCl_4 + 2CO.$$

If carbon is not already present, i.e., as a fuel pellet coating, then a mixture of CCl_4 and Cl_2 may be similarly utilized. The separation conditions are determined to a large extent by the relative volatilities of the metal chlorides. To achieve reasonable pressures and reaction rates, temperatures of about 500–800 °C are required. The metal chloride vapor pressures for this temperature interval can be classified as follows:

- high volatility for $ZrCl_4$, $MoCl_5$, $NbCl_5$, $InCl_3$, UCl_6, UCl_5 and $PaCl_5$;
- medium volatility for CsCl, $RuCl_3$, $ThCl_4$ and UCl_4; and
- low volatility for the rare earth chlorides, $BaCl_2$, $SrCl_2$, $RhCl_3$ and $PuCl_3$.

Actual vapor pressure data may be found in thermodynamic texts such as Kubaschewski and Evans (1956). Condensed-phase thermodynamic activities will also affect the relative volatilities and data for relatively simple, i.e., binary or ternary, mixtures may be found in Lumsden (1966).

Apart from the separation achieved from the difference in volatility, a further degree of separation can be obtained by selective absorption of the metal chloride vapors by solids such as NaCl or $BaCl_2$. An effective degree of absorption results from the formation of stable condensed-phase compounds, such as $UCl_4 \cdot 2NaCl$, $UCl_5 \cdot NaCl$, or $UCl_4 \cdot 2BaCl_2$, which, at a later stage in the separation process, may be thermally decomposed to yield uranium chlorides, e.g., see Hariharan *et al.* (1969) and Hirano and Ishihara (1969).

6. *Salt Transport Processes*

A number of salt transport processes have been suggested or demonstrated, particularly in connection with the reprocessing of fast breeder spent fuel, e.g., see Steunenberg *et al.* (1969) and Knighton *et al.* (1969). These processes basically involve molten salt or liquid-metal extraction techniques with some distillation stages. Hence, the important basic parameters are solubility and liquid distribution coefficients. However, it is relevant to the present discussion of high temperature vapors to note that these salt transport processes involve temperatures in the region of 600–800 °C and that they utilize chloride solvent, or carrier, salt mixtures such as $MgCl_2$ (50 mole %)-NaCl (30)-KCl(20). Many of the fission product elements, in addition to U and Pu, will be present in a chloride form in these melts. Thus the possibility exists for mixed chloride enhanced volatility processes to occur—leading to an unwanted vapor-phase transport of complex metal chlorides of the type considered in Chapter 3, Section IV. No theoretical analysis has been made for such effects which, from the high separation yields reported, cannot be of major consequence but may ultimately impose a limit on the process efficiency.

IV. Fusion Power

A. General Considerations

Most projections on the future availability of fossil fuel, and even nuclear fuel for fission reactors, indicate the eventual need for an alternative energy source—perhaps in this century. Nuclear fusion, utilizing deuterium extracted from sea water as a primary fuel source, is most frequently suggested as the energy source of the future. The several basic schemes proposed for using nuclear fusion as an energy source contain high temperature materials problems of a gas- or vapor-phase character. It is desirable, therefore, to consider briefly the present status and future outlook for the controlled fusion process. A number of reviews have appeared on this subject in the last few years, e.g., see Post (1970, 1971), Seif (1971), Rose (1971), Yoshikawa (1971), Coppi and Rem (1972), Wood and Nuckolls (1972), and Post (1973). Discussions of the chemical aspects of fusion technology have also recently appeared (Gruen, 1972).

The basic fusion reaction most frequently considered is

$$^{2}D + {}^{3}T \rightarrow {}^{4}He + n + 17.6 \text{ MeV},$$

with the neutron carrying off about 14.1 MeV of the excess energy. There exists a considerable activation energy barrier to this process, owing to the electrostatic repulsion of the deuterium and tritium particles, i.e., ions. It is generally agreed that a temperature in the range of 10^7–10^8 °C, equivalent to a particle energy of about 10 keV, is necessary to overcome this barrier and allow a self-sustained fusion to proceed. It is also believed that for this reaction to occur the density and time scale must be such that the fuel plasma is above a number density–time product of 10^{14} sec cm^{-3} (Seif, 1971). In terms of ion densities a range of 10^{14}–10^{17} ions cm^{-3} is expected. Energy output is proportional to the square of plasma density. Thus, in the suggested low pressure steady-state reactor, a fuel pressure of 10^{-4}–10^{-6} atm and a confinement time of 10^{-2}–1 sec would be required. The fundamental problem, then, is to contain the combustion system for a sufficient period of time, and "magnetic bottle" effects have been pursued for this purpose. The variety of magnetic bottle techniques applied to the confinement of plasmas for the purpose of attaining fusion temperatures is exemplified by the account of project Sherwood, e.g., see Bishop (1958). The so-called Tokamak magnetic plasma confinement scheme is considered to show the most immediate promise, and an attainment of fusion plasma densities has recently been reported in the popular press.

Post (1970) considers the following high temperature plasma technique to show promise for a steady-state-type operation. Energetic neutral atomic beams are generated by passing ions into a chamber containing a gas or metal vapor where charge transfer occurs and the ions are converted to energetic neutrals. This neutral beam can cross the magnetic field and enter the plasma where it is reionized.

In order for the energy carrying neutrons to reach a moderator, such as Li or graphite, a thin-walled vacuum chamber will be needed to contain the plasma system. The necessary passage of highly energetic neutrons through such a wall is expected to lead to serious structural damage, every metal atom being displaced at least once per day. Radiation damage and wall interactions with radially diffusing ions lost from the plasma are also likely. Suggested materials include vacuum chamber walls of niobium, vanadium, or molybdenum and liquid lithium or a lithium salt such as Li_2BeF_4 as a coolant.

As commercial power plants are required to operate for periods of at least several years, even minor gas impurities, i.e., at the ppm concentration level, may create serious long-term materials corrosion problems. It is expected that the reactor walls may have to sustain temperatures of 500–

900 °C which are sufficiently high for gas–solid reactions to occur between metals and common impurities, such as O_2, CO, N_2, and H_2O. Furthermore, these interactions could well be promoted by the high energy neutron flux present, although very little is known about these radiation effects.

A Li coolant also serves the purpose of tritium fuel regeneration, according to the reactions

$$^{7}\text{Li} + \text{fast neutron} \rightarrow \text{T} + {}^{4}\text{He} + \text{slow neutron} - 2.5\ \text{MeV}$$

and

$$^{6}\text{Li} + \text{slow neutron} \rightarrow \text{T} + {}^{4}\text{He} + 4.8\ \text{MeV}.$$

For economic and environmental reasons, a complete recovery of tritium is required. Also, as only about 5% of the fuel is expected to react for a single pass through the plasma, then fuel reprocessing—most likely by distillation—is an important aspect of fusion reactor technology. Where an absence of radioactivity is desirable, and fuel breeding is not required, the following reaction is of interest:

$$\text{p} + {}^{6}\text{Li} \rightarrow {}^{3}\text{He} + {}^{4}\text{He} + 4.0\ \text{MeV}.$$

B. Laser-Induced Fusion

The use of high power lasers provides an alternative energy source to the use of magnetic fields for fuel activation, e.g., see Lubin and Fraas (1971), Fraas (1971), Boyer (1973), and Hammond *et al.* (1973). Lasers also allow the use of fuel pellets without the need for magnetic field plasma confinement. This latter aspect bypasses the central problem of confinement that has prevented the demonstration of a viable fusion reactor to date. While the laser fusion concept does not require magnetic containment there is nevertheless an interest in using laser generated plasmas as feed material for a magnetically confined system (Andryukhina *et al.*, 1971).

An immediate challenge to laser technology is presented by the requirement of a high ignition temperature for fusion. Depending on the wavelength used, laser pulse ($\sim 10^{-7}$ sec) power outputs of the order of 10^5–10^6 joules appear to be necessary. This is about two or three orders of magnitude greater than the power of presently available lasers, although Solon (1971) predicts a rapid upward trend in laser power capability during the next several years. Indeed, laser-induced nuclear fusion has already been demonstrated experimentally, e.g., see Basov *et al.* (1968). However, the energy output is far below the greater than breakeven point representing the generation of more energy than the input level. It is likely that chemical laser systems will provide the best opportunity for a laser-induced commercially viable thermonuclear reactor device.

One of the suggested laser-controlled thermonuclear reactor systems involves laser ignition of a frozen, compressed (i.e., density $\sim$1000 gm cm^{-3}), deuterium–tritium pellet as it falls through a molten Li-vortex cavity (Lubin and Fraas, 1971). Time scales are particularly important to this concept. Energy transfer from the oscillating electric field of the laser beam relies on the heating of free electrons followed by an electron–ion energy transfer interaction. This latter collisional process, which should occur on a time scale of 10^{-11} sec, appears to be rate determining in the transfer of energy to the plasma. Laser pulses of 10^{-10} sec appear to be suitable for energy transfer to the plasma fuel. Theoretically, it appears necessary to provide a symmetrical interaction between laser beam and fuel pellet such that the pellet becomes compressed or imploded.

The propagation of a shock wave through the molten Li bath to the container walls is believed to be a potential problem, and the introduction of gas bubbles throughout the Li melt has been suggested as a means of absorbing this shock energy. In the region of the reaction zone, the absorption of alpha particles and x rays by lithium would result in its vaporization and expansion, leading to an additional exerted pressure on the reaction vessel. The vessel must therefore be capable of containing molten Li and sustaining pressure bursts. Materials such as alloys of chrome–molybdenum–steel or niobium–zirconium are likely candidates for construction of the reaction vessel. The latter material would sustain wall temperatures of about 1000 °C and allow the use of potassium vapor to transfer heat from the walls for use in a Rankine cycle having a thermal efficiency of about 58%.

V. Coal Gasification

The reaction of coal with steam to yield a relatively low Btu gas, containing primarily H_2 and CO as fuel sources, is a well-known commercial coal gasification process. However, at the present time, there are a number of developmental processes where the production of the higher Btu fuel CH_4 is the primary objective. As CH_4 is also a major component of natural gas, these developmental gasification processes could be used to supplement—or eventually substitute for—natural gas. The most frequently considered processes of this type are those known as the steam-iron, Hygas, CSG, Bureau of Mines, BCR, and Kellogg processes, e.g., see the summary of Henry and Louks (1971).

Thermodynamic predictions indicate that an endothermic reaction of steam with solid carbon or coal should be favorable for the production of CO and H_2 at temperatures of about 700 °C, and greater. Also, an exothermic reaction of carbon with hydrogen to yield CH_4 is thermodynamically feasible at temperatures below about 500 °C. The various coal gasification stages rely on both of these reaction conditions for the production of fuel gas. The proposed conditions of temperature range from about 430 to 1500 °C and of pressure from about 100–1 atm. These high temperature conditions, together with the reactive nature of the gas phase, require the use of refractory ceramic liners in the reaction vessel.

At present, it is difficult to identify major areas of application of the science of high temperature inorganic vapors to the rapidly developing technology of coal gasification.† However, the processes under development are at high temperature and the possible presence of molten inorganic media, such as carbonates and hydroxides, and the heterogeneous nature of the gasification processes, warrant a very brief mention of some of the concepts of coal gasification (see also Gould, 1967; Zahradnik and Glenn, 1971). It is reasonable to assume that as the theoretical modeling of the various processes develops, increased attention will be given to the chemical history of the volatile coal impurities, especially sulfur, e.g., see Sinha and Walker (1972) and Karn *et al.* (1972). Also, we can reasonably expect that some materials limitations will result from the stability of liquid- and vapor-phase products of reaction between the ceramic walls and slag impurities such as FeO, SiO_2, and V_2O_3 together with SO_2 and CO_2—see the discussions given elsewhere on corrosion (Chapter 4, Sections II.C and III.C).

The essential feature of the Kellogg process is that it uses molten sodium carbonate (~1200 °C) both as a catalyst and also to supply heat for a carbon–steam gasification reaction. The products of this reaction are CO, CO_2, CH_4, and H_2.

The BCR process would utilize a fluidized bed of coal at 800–1000 °C with steam and oxygen at 20–30 atm for the production of coal gas. A yield of about 8% CH_4 has been obtained at 8 atm. A complete conversion of the carbon in coal to hydrocarbon gas has, in fact, been demonstrated for the higher pressure conditions of 500 atm H_2 and a temperature of about 900 °C (Moseley and Paterson, 1967; Zahradnik and Glenn, 1971).

It is possible to remove CO_2 continuously from the reaction products by means of an acceptor such as dolomite (MgO·CaO) which forms a

† Much of the recent research in this area is contained in U.S. Government R and D reports, available from the National Technical Information Service (NTIS); several hundred such reports are indexed under coal gasification.

stable carbonate at about 540 °C and which can be reactivated by heating to about 1070 °C. The key reactions in this acceptor coal gasification process are considered to be

$$CaO(s) + CO_2 \rightarrow CaCO_3(s),$$

$$C(s) + H_2O \rightarrow CO + H_2,$$

$$C(s) + 2H_2 \rightarrow CH_4,$$

and

$$CO + H_2O \rightarrow CO_2 + H_2.$$

In this process it is important to keep the steam pressure below about 13 atm, because of the undesirable formation of low melting eutectics of $Ca(OH)_2$–$CaCO_3$ and CaO–$Ca(OH)_2$–$CaCO_3$.

VI. High Temperature Vapor-Phase Processes Associated with Gas Turbine Systems

The relationship of efficiency to gas inlet temperature requires operation of gas turbine systems at the highest temperature allowed by the materials of construction.

Existing gas turbines, such as the JT9D which powers the Boeing 747, operate at gas inlet temperatures of up to 1260 °C and have a lifetime of up to 30,000 hours (Bradley and Donachie, 1970). It is expected that in the next decade, engines with gas temperatures of 1650 °C will be employed. Much of the structural material in use is Ti- or Ni-base superalloy. In the future, materials such as Haynes 188 alloy, containing Co and La, are expected to be used. Use of vapor-deposited oxidation-resistant coatings, such as Al_2O_3 and SiO_2, and composites is also anticipated. A discussion of the relative merits of different super-alloy types has been given by Goward (1970). Recent developments in cooling techniques for turbine blades and protective coatings for the super alloys used for construction of the blades were discussed at an AGARD Conference, e.g., see Esgar and Reynolds (1971).

Many of the current materials problems associated with gas turbine systems involve stress and creep rather than loss of blade material by chemical attack (Barnes, 1968). A notable exception is the pronounced corrosion by combustion gases containing sodium, sulfur, and vanadium impurities such as may be found in operating marine turbine systems (Bornstein *et al.*, 1971). A synergistic interaction involving sodium sulfate and vanadium oxide salts appears to cause the observed corrosion—see Chapter 4, Section II.C for further discussion of this type of corrosion.

There is a current interest in the use of ceramic rotor blades and stator vanes at temperatures of 1370 °C and greater. The prime candidates for this application are Si_3N_4 and SiC. A composition of lithium–aluminum–silicon oxides is considered to be a good candidate for heat exchanger parts. It is expected that the potential advantages of ceramics would be a reduced weight, higher operating efficiency, and lower cost.

In addition to ceramics, there is considerable interest in composites such as silicon carbide coated fibers of boron or carbon. Alumina-base coatings have been used for Co- or Ni-base alloys. Vapor deposition processes are frequently used to prepare these coatings.

We can expect that with the continued trend towards higher temperatures in turbines, vapor-phase processes will become more significant than heretofore.

VII. Magnetohydrodynamic Energy Systems

A. Introduction

The generation of electrical energy using combustion powered magnetohydrodynamic (MHD) devices has been the subject of considerable research effort in the past decade, e.g., see Heywood and Womack (1969). At present, the factors limiting the successful demonstration of such devices as viable power plants are primarily materials problems associated with the interaction of high temperature combustion products with the diffuser wall–electrode assembly of the generator, e.g., see Hottel and Howard (1971), Harris and Moore (1971), and Medin *et al.* (1972).

The MHD principle, as discussed in detail by Rosa (1968), involves an electrically conducting fluid which, on passage through a magnetic field, develops an induced electric field across the fluid along which a current can flow. These elements of force, field, and current are, from electromagnetic theory, mutually perpendicular. A direct electric current can pass along the induced field to electrodes which may be connected to an external load. Removal of energy from the system results in a slowing down—or reduction in kinetic energy— of the flowing conductor gas. Essentially, the system contains no moving mechanical parts and allows a direct conversion of gas kinetic energy to electric power. The source of kinetic energy most frequently considered is a hot gas generated by combustion. Alternative schemes involving gas heating and ionization using nuclear reactors have also been suggested, e.g., see Imani (1972) and Seikel and Nichols (1972).

An MHD generator is usually comprised of a sequentially connected combustion chamber, a supersonic expansion nozzle, and a gas flow channel which is sometimes known as a diffuser. This configuration is quite analogous to that of a rocket engine, and similar computational procedures apply in calculating performance parameters (see Chapter 5, Section IX). The magnetic field and current tapping is located along the gas flow channel.

The hot post-combustion gases normally have an insufficient free electron concentration to be effective current carriers, and thermally ionizable seed materials, such as K_2CO_3 or K_2SO_4, are added. Gas conductivity should be greater than 10 mho m^{-1}; that is, gas electron concentrations of about 10^{14} cm^{-3} are required. Thus for practical seed levels of about 1 mole % and typical pressure conditions of about 5 atm, equilibrium calculations indicate that gas temperatures of 2500–3000 K are necessary.

In order that the gas conductivity is maintained, the MHD generator can operate with only a few hundred degrees drop in temperature along the diffuser. Consequently, to conserve thermal efficiency, the hot exit gases must be further utilized, e.g., as an energy source for a steam turbine generator. Hence, an MHD power system is usually considered in conjunction with other energy systems and the combinations most commonly discussed are open cycle/fossil fueled–MHD–steam, and closed cycle/nuclear fueled–MHD–steam. For economic reasons, most of the research effort has been directed to the former system, with particular emphasis on coal as the fossil fuel source. In principle, such a system could operate with efficiencies of 50 to 60%, i.e., almost double that of conventional fossil-fueled steam-power generators. An advanced system has been suggested where a possible efficiency of greater than 60% could be realized (Barthelemy, 1971). This system involves closed-cycle operation of a hypothetical gaseous core nuclear reactor coupled to an MHD device. A complex vapor-phase chemistry involving interactions of nuclear fuel, radioactive fission products, MHD seed material, and materials of construction can be expected in such a system.

Recent evaluations of the status of MHD devices as viable and competitive power systems indicate that the problems of maintaining gas conductivity, seed recovery—which must be greater than 98% to be economical— and channel materials degradation, remain largely unsolved (Hottel and Howard, 1971; Harris and Moore, 1971). Also, at the high combustion temperatures used, relatively high levels of NO are generated and must be removed from the exhaust gases. This does not appear to be a serious problem, however (Hammond *et al.*, 1973).

Experimental plants have operated in the 1- to 30-MW range for short periods up to several minutes, and an MHD-topped steam station with a projected electrical output of 75 MW has been constructed in Russia. A comparison of the design and performance parameters for the various developmental MHD systems may be found in the review of Medin *et al.* (1972).

B. MHD Materials Problems

The high gas temperatures required for MHD power systems create a number of severe materials problems. The walls of the MHD channel must necessarily contain both electrically conducting and insulating components. If external cooling is provided, then metal electrodes may be used. However, if the walls are too cool then condensation of seed components and fuel impurities, such as slag from coal, increase corrosion and electrical arcing problems. The nature of the corrosion process is very temperature dependent. At relatively low temperatures, e.g., 1300 °C, condensed-phase interactions between the channel wall–electrode assembly and seed material, a seed and slag mixture, and slag, may occur—in sequence—with increasing temperature. At higher temperatures, e.g., $>$ 1700 °C, corrosion by vaporization may be predominant.

Many workers have favored the use of yttria- or ceria-stabilized zirconia and, more recently, lanthanum chromate for ceramic electrodes, W, Re, or Ta for cooled metal electrodes, and MgO or Al_2O_3 for insulation and wall materials. Tests of such materials under simulated operating conditions have been reported (Rekov, 1969). In the presence of the potassium seed, the loss of ZrO_2 from the walls was found to increase by about 10–20%.

The specific effect of the water vapor component of the combustion gases on these ceramics does not appear to have been particularly noted, but from previous discussion (e.g., see Chapter 2, Section VIII; Chapter 5, Section II.K) it appears likely that volatile hydroxides could be formed under typical MHD operating conditions. Indeed, it has been recently predicted, by thermodynamic calculation, that the vaporization recession of $LaCrO_3$ would be about a factor of 10 greater in the presence of 0.08 atm of H_2O—as may be derived from coal ($\sim CH_{0.7}$) combustion (Bowen *et al.*, 1974). This predicted enhanced loss is due to the formation of $CrO_2(OH)$ as a stable vapor species in the diffuser. At typical MHD temperatures, the Langmuir maximum free evaporation rate for a $LaCrO_3$ electrode is then calculated to be about 10^4 gm cm^{-2} for a reasonable practical operating time of 10,000 hrs.

In view of this predicted importance of $CrO_2(OH)$, and hence its thermodynamic enthalpy and free energy characteristics, from which the H_2O-enhanced vaporization rates are derived, it is pertinent to note that the evidence for this species is rather limited. The sole source of data is from an unpublished transpiration study by G. R. Belton and Y. W. Kim, as cited by Graham and Davis (1971). As was discussed elsewhere (Chapter 2, Section VIII) the indirect characterization of species, e.g., by transpiration, can lead to ambiguity regarding the species identity and also such methods do not readily recognize the presence of additional species. In view of these reservations, the conclusions regarding a predicted loss of $LaCrO_3$ as an MHD electrode material are rather tentative. Furthermore, the assumed absence of a diffusive boundary layer at the walls has not been established.

The high velocity of the gas flow in the MHD channel and the possible presence of fly-ash components are likely to function as abrasives and further enhance wall degradation.

The wall temperature to be used represents something of a compromise, since at lower temperatures, i.e., below about 1500 °C, the stability of mixed oxide compounds such as vanadates and zirconates would allow ash components to destructively interact with the zirconia, magnesia, and calcia components. Condensation of seed and slag also provides a medium for electrolytic wall corrosion. However, some workers have argued that a layer of condensed slag, which contains an electrically conducting iron oxide (nonstoichiometric) component, may be used as an integral part of the electrode system and would also serve as a protective thermal and oxidation barrier. Based on thermodynamic arguments, Bowen *et al.* (1974) have recently concluded that none of the candidate electrode materials will survive for long periods of time in the presence of slag—as derived from coal combustion. It is not yet known to what extent kinetic, i.e., nonequilibrium and mass transport, factors may modify this conclusion.

The prospect of replenishing wall losses by an appropriate seeding of the combustion gases has been considered, but this may increase the difficulty of alkali recovery and also increase the deposition of solids in downstream components such as heat exchangers.

It is evident that many of the existing materials and seed-recovery problems are derived from the presence of combustion impurities, particularly coal slag. The future prospects for a hydrogen fuel economy may provide a source of clean fuel for MHD systems and an experimental and theoretical evaluation of a H_2–O_2 MHD generator is under investigation (Marlin Smith *et al.*, 1974). A cost analysis indicates the H_2-fuel system to be comparable with other fuels and the reduced materials problems and low pollution produced make this an attractive system. However, the presence of a large H_2O concentration may prove to be detrimental if the high temperature materials are susceptible to the formation of hydroxide vapor species.

C. MHD Gas Chemistry

In order to successfully model, and thereby optimize, the design and operation of an MHD unit, it is essential to provide a rigorous accounting of the gas chemistry, particularly within the diffuser. Despite considerable effort in this direction, it is not presently possible to predict satisfactorily the electrical conductivity of seeded coal combustion gas.

1. *Electron Conductivity*

The basic factors involved in the calculation of MHD electrical conductivity are the species concentrations—including the free electron—and the electron mobility through the gas mixture. As was indicated to be the case with the modeling of rocket performance (Chapter 5, Section IX), it is not experimentally possible to determine species concentrations in these complex high temperature systems. If one can assume that the gas system is at thermodynamic equilibrium then, in principle, the composition can be calculated. The steady-state electron concentration is then defined, and its mobility through the gas medium may be determined from the momentum transfer cross section between electrons and the other species, which in turn is determined by the electron-species collision cross sections. These cross-section data are not known and can only be estimated very approximately, but as a rule the most polar species, such as H_2O and KOH, should have the highest cross sections.

In addition to an inadequate set of basic thermochemical and collision cross-section data, the troublesome factors in the modeling problem appear to be a complex gas composition which may not correspond to an equilibrium composition, and slag condensation in the MHD channel and its possible interaction with the seed material.

Electron producing ionization kinetics may also be an important factor in determining MHD conductivity, e.g., see Hottel and Howard (1971). In this connection, it is pertinent to recall the complex ion chemistry that results between metal additives and electrons in atmospheric laboratory flames, i.e., see Chapter 5, Section III.F. Indeed, the presence of excess concentrations of charged species in the reaction zone of hydrocarbon-containing flames has led to the suggestion of a nonequilibrium low pressure ($\sim$0.1 atm) MHD generator (Fells, 1967). The use of low gas pressures would allow the excess radical and ion concentrations generated by the combustion process to be maintained in the diffuser itself. Calculations show that under these conditions a 1500 K flame would give an ion yield equivalent to that for an equilibrium system at 2500 K.

The prospect of generating electron concentrations in excess of the equilibrium level, by external means, has also been of some interest (Coney,

1970). Using an artificial electron source, such as photoionization or an electron gun at the entrance of the MHD diffuser, it is possible to generate significantly more electrons than by thermal ionization. This offers an advantage, provided the residence time of the gases in the MHD duct is less than the lifetime for electron–positive ion attachment processes. The prospects for maintaining an excess electron concentration level appear to be more favorable for monatomic gases in a closed-cycle system than for an open-cycle combustion-supported MHD system where polyatomic gases of high electron attachment cross section would be present (Coney, 1970; Brogan, 1963).

2. *Seed Materials*

It is generally agreed that the optimum seed material is K_2CO_3 or K_2SO_4, the cesium salts being excluded solely on grounds of greater cost (Heywood and Womack, 1969; Moore, 1963). The next best candidates of Na and Ba yield almost an order of magnitude fewer electrons on thermal ionization. Electron scavengers—i.e., species having electron affinities of $\sim$2 eV, or greater, such as Cl atoms—have been found to have an undesirable effect of reducing the free electron levels and this precludes the use of alkali halide additives (Brogan, 1963).

The chemistry of the seed material and its recovery from the combustion gases has been dealt with in considerable detail elsewhere, e.g., see Heywood and Womack (1969). Factors such as rates of vaporization, diffusion, condensation, interaction with impurities—particularly slag—are each found to be important in characterizing the seed chemistry at different stages of the MHD system.

3. *Equilibrium Calculations*

If equilibrium is assumed in the MHD channel, one can, from a knowledge of all of the species present and their thermodynamic functions, calculate steady-state free electron concentrations for a variety of composition, temperature, and pressure conditions. Many such calculations have been made recently for potassium seeded systems, e.g., see Zielinski (1968), Park (1967), Kmonicek *et al.* (1971), Hirano (1970), Lanshov *et al.* (1969), Heywood and Womack (1969), Feldmann *et al.* (1970), and Spencer *et al.* (1973). Notably, these calculations are based essentially on the same primary thermochemical input data which is very uncertain for some of the species (i.e., see JANAF, 1971). The results also rely heavily on a correct choice of species which, as has been alluded to previously, is not always certain.

From thermochemical calculations of species concentrations, it is predicted that about half of the potassium seed is present in the form of the essentially unionizable species KOH and is therefore unavailable as an electron source. Typically, at a temperature of 2300 °C and with a slightly lean coal/air combustion system at 3 atm pressure, the major potassium-containing species are calculated to be KOH, K, KO, and K^+, given in decreasing order of abundance. Other significant vapor species include SiO_2, FeO, OH, SO_2, and NO, in addition to the major combustion gas components of N_2, H_2O, CO_2, and CO. The primary slag or fly-ash components include SiO_2, Al_2O_3, Fe_3O_4, TiO_2, MgO, CaO, Na_2O, and K_2O. For a coal-fired system the total slag concentration in the diffuser is believed to be of the order of 0.1%. Spencer *et al.* (1973) have shown, by thermodynamic calculation, that the presence of such slag impurities in a coal-fired MHD system could result in a 19% loss of electrical conductivity.

The combined uncertaintities in the heats of formation and the molecular constants used in computing partition functions for KOH (see JANAF, 1971) cause the concentration ratio of K to KOH to be uncertain by at least an order of magnitude. This degree of uncertainty does not allow one to make a reasonable judgment as to whether the presence of KOH is sufficiently high as to require restrictions on the amount of hydrogen present in the fuel. Clearly, the successful thermodynamic modeling of MHD gas chemistry requires better basic data than are presently available.

4. *Electron Scavengers*

The recent observation of highly electronegative metal oxide species in flames (see Chapter 5, Section III.F) suggests that impurities of these metals could significantly reduce the free electron concentration. In particular, the formation of the species AlO_2^-, BO_2^-, PO_2^-, and SO_2^- is predicted to reduce the free electron concentration by about 1–20% (Spencer *et al.*, 1973). These calculations are very uncertain owing to the experimental, or estimational, errors in the electron affinities of the species considered. For instance, using the various sets of experimental electron affinity data for AlO_2^- results in a 70% variation in the calculated electron concentrations.

Another source of considerable computational uncertainty concerns the species CO_2^- which would compete strongly, and in fact prohibitively so, for free electrons, if its electron affinity was as high as 0.5 eV—as suggested by the JANAF Thermochemical Tables (JANAF, 1971). However, as was noted by Spencer *et al.* (1973), more recent data suggest a lower electron affinity, i.e., $\sim$0.1 eV, and hence a negligible concentration of CO_2^-. This lower value, although still somewhat uncertain, tends to be supported by the experimentally proven absence of CO_2^- in flames where many other negative ions of low-electron-affinity species were found, e.g., see Miller (1968).

The recent development of laser spectroscopy techniques, such as the laser photoelectron spectroscopy of negative ions, provides a means for determining the necessary electron affinity data and with relatively high accuracy, that is, to less than $\pm$ 0.04-eV uncertainty, e.g., see Celotta *et al.* (1974).

For systems where H_2O is present in the combustion gases, OH^- is a stable species which also competes for free electrons. Its increasing importance at higher temperatures tends to offset the electron gains achieved by a temperature increase. In this connection it is of note that Jensen (1972) has shown the predicted conductivity of K-seeded rocket combustion systems to be particularly sensitive to the unknown rates of reactions, such as,

$$e + H_2O \rightleftarrows OH^- + H$$

and

$$e + OH + M \rightleftarrows OH^- + M.$$

7

Chemistry of High Temperature Species in Space

I. Introduction

Chemistry in space can involve extreme environmental conditions of high temperature, high pressure, high vacuum, and high radiation levels. The subject of high temperature species in the upper atmosphere, the outer solar system, and interstellar and stellar space, provides an example of the utility of the laboratory characterization of high temperature species for determining compositions and temperatures of remote regions by spectroscopic means.

The transmission of electromagnetic radiation through the earth's atmosphere is restricted owing to absorption by N_2, O_2, and O_3 in the 1–$10^{3.5}$ Å wavelength region—i.e., γ-ray, x-ray, and uv radiation—and by H_2O and CO_2 in the 10^5–10^8 Å region—i.e., infrared and far-infrared radiation. Further absorption occurs at wavelengths greater than about 10^{11} Å due to the ionosphere cutoff. Thus the utility of electromagnetic radiation as a chemical probe in the upper atmosphere, stellar, and interstellar space has, in the past, been restricted to an optical window around

10^4 Å and a radio window between about 10^8–10^{11} Å (Aller, 1963). However, more recently, this limitation has been eliminated by the inclusion of spectrometers in satellites and other spacecraft. In order to fully utilize this added capability for extraterrestrial observation, it is essential that basic laboratory-derived spectroscopic data be made available for the newly available spectral regions. Thus a new impetus exists for spectroscopic study of high temperature species in the uv and infrared spectral regions.

In addition to the utility of spectral data for species identification, the spectroscopically derived thermodynamic free energy functions of high temperature species may also be used to evaluate the equilibrium composition of stars.

Theories concerning the geochemical history of the moon also make use of basic thermodynamic volatility data for high temperature species. A notable recent example is the demonstration that the anomalously low abundance of Eu found in lunar rock and soil samples can be accounted for by the high volatility of europium oxide, under vacuum conditions, as compared with that for the other rare earth oxides (Nguyen *et al.*, 1973).

II. The Upper Atmosphere

A. Background

In the immediate vicinity of the earth's surface, a number of atmospheric regions are defined, namely,

- the troposphere at 0–15 km altitude,
- the stratosphere at 15–70 km,
- the D region at 70–100 km,
- the E region at 100–160 km,
- the F region at 160–1000 km, and
- the ionosphere at larger distances.

As a source of basic data on the upper atmosphere one should consult the handbook of Johnson (1965). The F region prevents passage of far-uv radiation, the E region x-rays, while cosmic rays are usually restricted by the stratosphere. Some attenuation of radio waves occurs below the E region and infrared radiation is partially absorbed by the troposphere. For the E and F regions, and beyond, high temperature conditions can exist, i.e., 1000–1800 K, particularly during sunspot activity. However most upper atmosphere chemical phenomena are nonthermal, owing to the influence of solar radiation which is primarily distributed in the visible and infrared spectral regions.

Much of the research involving the upper atmosphere is aimed at deriving a model of its composition in terms of individual species and the variations that may occur with altitude, solar conditions, and artificial disturbances such as aircraft emissions and nuclear detonations. A natural disturbance of interest is the optical aurora, e.g., see Omholt (1971).

As an indication of the approximate composition of the upper atmosphere, at an altitude of 260 km the number density of molecular N_2 is believed to be about 10^8 cm^{-3} (Reid, 1971). Typical oxygen atom concentrations are about 10^{11} and 10^8 cm^{-3} at 120 and 350 km, respectively. The other likely radical species, the H atom, is relatively insignificant at these altitudes, but becomes a predominant species at altitudes of greater than about 900 km. The chemistry at these upper altitudes is greatly influenced by the large mean free path required for molecular collisions. This varies from about 10 cm at 90 km to about 10^7 cm at 800 km.

Free electrons and positive and negative ions are also important species beyond the stratosphere. The major positively charged species are, O^+, NO^+, O_2^+, N^+, and H^+ and the maximum species number density is about 10^4–10^5 cm^{-3} and occurs at altitudes of between 150 and 1000 km, e.g., see Kasha (1969).

B. Natural Disturbances

Natural disturbances in the upper atmosphere, such as the optical aurora and air-glow or twilight, are manifested as light emission and are therefore amenable to spectroscopic examination. Air-glow or twilight results from upper atmosphere, i.e., 100–300 km altitude, chemiluminescence reactions, of which the Na–D line chemiexcitation is an important contributor. The excitation is believed to result from reactions involving species such as NaO and NaH—for instance,

$$NaO + O \rightarrow O_2 + Na^*$$

$$NaH + H \rightarrow H_2 + Na^*,$$

and

$$NaH + O \rightarrow OH + Na^*$$

(Shukla and Srivastava, 1970; Ingham, 1972). From a consideration of all the likely formation- and loss-reactions for these species, it is possible to calculate their steady-state concentrations from basic rate data. Such a calculation is basically similar to those already mentioned; e.g., in connection with plasmas (Chapter 3, Section VI.B), flames (Chapter 5, Section II), and rocket exhausts (Chapter 5, Section IX). Shukla and Srivastava (1970) calculate steady-state concentrations of NaO $\sim 10^3$ cm^{-3} and NaH $\sim 10^{-1}$ cm^{-3} at altitudes in the region of 100 km. These concentrations may be compared with those of the other reactive species at this altitude, i.e., Na $\sim 10^3$, OH $\sim 10^7$, H $\sim 10^{10}$, N $\sim 10^{10}$, and O $\sim$ 10^{12} cm^{-3}. From the calculations, significant variations in the NaO and NaH concentrations, and hence the air-glow, are predicted for various altitudes and seasons.

Another reaction which is believed to contribute to air-glow is the recombination process

$$O + O + O \rightarrow O_2 + O(^1S).$$

It has been suggested that the emission from $O(^1S)$ could be used, in conjunction with the appropriate reaction rate constant, to determine the concentration of O-atoms in the region of 80–120 km. The air-glow spectrum also contains emissions from Mg^+, Ca^+, Li, and K.

C. Chemical Releases

Artificial disturbances, in the form of chemical releases by rockets, have been used to probe the magnetic fields and wind patterns of the upper atmosphere. Very recently, the proposed commercial use of the stratosphere by supersonic aircraft, such as the SST, has created concern over the influence of engine exhaust releases on the environmental balance.

The subject of chemical releases at high altitudes has been reviewed by Rosenberg (1966); Harang (1969) has reviewed the spectroscopic aspects. Some of these releases simulate the air-glow phenomenon and the release of sodium vapor from rockets has been extensively used to determine wind profiles in addition to diffusion rates and temperatures. In this case a visible boundary is observed around 90 km because of the predominance of yellow Na-atom vapor above, and white sodium oxide smoke below, this altitude.

Other metal containing emissions frequently used as upper atmosphere probes include Li, K, Cs, Mg, Be, Sr, Ba, and Al. Formation of oxide species is often an important function of such releases. For example, the AlO species provides an intense luminescence to aluminum releases. The responsible reactions are thought to be

$$Al + O \rightarrow AlO + h\nu$$

and

$$AlO + O \rightarrow AlO_2 + h\nu,$$

e.g., see Woodbridge (1961). Also, the AlO resonance emission spectrum has utility as a temperature indicator, as does BaO, e.g., see Harang (1969). Elemental potassium is also used as a temperature probe.

Barium releases in the upper atmosphere have been used primarily for mapping magnetic fields, e.g., see Haerendel *et al.* (1967). In order to understand how such releases lead to the formation of ion clouds, laboratory studies of Ba releases, produced by the combustion of Ba–CuO–NaN_3 and followed by an expansion into a vacuum, have been made by Batalli-Cosmovici and Michel (1971). They suggest that excited and ionized Ba-atoms are important only in the initial period of expansion. For the most part, barium is present in the atomic ground state and the ionization observed in rocket releases is attributable to photo- and chemi-ionization processes.

Another aspect to metal combustion in the upper atmosphere involves the effect of nuclear detonations on the environment and on radar transmission. Attempts are being made to model these effects in terms of elementary process chemical kinetics, e.g., see Bortner and Bauner (1972). Similarly, the release of metals as impurities from jet turbine fuels into the stratosphere is of environmental concern. For example, if exhaust gases engaged in a catalytic destruction of the ozone layer, a damaging increase could occur in the amount of uv radiation reaching the earth's surface.

The artificial generation of "electron clouds" in the E and F regions is of interest for the scattering and reflection of radio-frequency signals, e.g., see Rosenberg and Golomb (1963). Free electrons are conveniently generated by an explosion, or rapid combustion, of Cs-containing mixtures such as $CsNO_3$–Al explosive (e.g., TNT). The explosion yields free Cs atoms, together with condensed Al_2O_3. At the high temperatures generated, the thermal ionization of Cs results in the production of free electrons. However, the steady-state concentration level of electrons is found to be only about 10^{-3} of the theoretical maximum due to recombination and electron attachment processes occurring during the initial expansion of explosion products. Hence, the lifetime of electron clouds is sensitive to particle density and therefore altitude. A maximum lifetime was observed experimentally at about 100 km in the experiments of Rosenberg and Golomb (1963).

In addition to man-made chemical releases it is expected that species such as MgO, which may be derived from meteor debris, should exist in the upper atmosphere though no identification has yet resulted (Wentink and Brown, 1971).

III. The Outer Solar System

The formation of satellite bodies, such as meteorites, in the outer solar system has been described by equilibrium condensation models involving the cooling and aggregation of nebular gases, e.g., see Lewis (1973). In the initial stages of this process the temperatures are moderately high, i.e., ~800 K, and gas–solid reactions, such as

$$Fe(s) + H_2S \rightarrow FeS(s) + H_2$$

and

$$Fe(s) + H_2O \rightarrow FeO(s) + H_2,$$

are believed to be of importance.

For the particular case of chondritic meteorites, it is possible to combine a knowledge of their elemental and mineral composition with basic high temperature vapor pressure and solubility data to yield information on the temperatures and pressures under which the chondritic materials accreted; e.g., see the review of Larimer (1973). These conditions have thus been shown to fall in the pressure and temperature ranges of 10^{-4}–10^{-6} atm and 350–500 K. Key volatile trace elements used for these determinations include Pb, Bi, In, and Tl. Similar arguments made for the more refractory components indicate a temperature condition of greater than 1300 K for the inner part of the nebula.

IV. The Interstellar Region

The average particle density of matter in the interstellar region is less than 1 cm^{-3}. However, there are galactic cloud regions, such as nebulae, where densities of the order of 10^{10} cm^{-3} are observed.

In the past several years the rate of discovery of new interstellar molecules has increased dramatically. Most of these species have been organic in nature, for example, CH_3CN, HNC, HNCO, CH_3C_2H, and CS—see the review of Snyder (1972). One notable exception is the recent observation of the well-known high temperature species SiO in the gas cloud Sagittarius B2, located near the center of the Milky Way Galaxy (Wilson *et al.*, 1971). Species such as these are identified by their radio emission spectra and the availability of corresponding laboratory data is imperative to their detection in interstellar media. For the case of SiO, the line emission of the $J = 3 \rightarrow 2$ rotational transition was observed at 130246 MHz and a column number density of 4×10^{13} cm^{-2} was estimated. Based on a lack of detectability of the $J = 1 \rightarrow 2$ transition, a column density for species in this state of less than 10^{12} cm^{-2} has been estimated (Dickinson and Gottlieb, 1971).

V. Stellar Atmospheres

Stellar atmospheres include primarily the stars and, to a lesser extent, comets. The astrophysical terminology (Harvard System) for star types is as follows:

O B A F G K M S N,

also known as "Oh be a fine girl kiss me so nicely." This listing is in order of decreasing surface temperature over a range 80,000–3000 K. The sun is an example of a G-type star, having a temperature at the photosphere, i.e., near the surface, in the region 4000–5700 K. The atmosphere or chromosphere of the sun is at a much higher temperature, i.e., in the range 4×10^3–5×10^6 K, and the corona has a temperature range of 5×10^5–3×10^6 K.

These star-type temperature extremes are also known as hot or blue-white, and cool or red, stars; the former radiate mostly in the far-uv and the latter in the infrared. The composition of these stars varies from being rich in light elements, such as He and H for O and B stars, and metals for the yellow stars of F, G, K, M, to rich in heavy metals for S stars. White stars such as Sirius exhibit strong H-atom spectra, yellow stars such as Capella indicate line spectra of metals—particularly Na, Mg, Ca, Ti, V, and Fe—and red stars such as α Herculis and 19 Piscium show molecular band spectra, for instance, TiO in M-type and ZrO in S-type stars (Merrill, 1963).

Swings (1969) has recently reviewed the role played by free radicals and high temperature species, such as TiO, AlO, AlH, ZrO, VO, LaO, YO, SiC_2, C_3, and C_2, in astronomy. The intensity of molecular spectra for species such as these may be used to calculate elemental abundances of stars, based on an assumption of local thermodynamic equilibrium. Such an approximation is reasonable for the high density environment of stars but is not likely to be valid in low density comet tails and interstellar media.

At local thermodynamic equilibrium the Kirchoff–Planck relation holds, i.e.,

$$B_\nu(T) = (2h\nu^3/c^2)[\exp(h\nu/kT) - 1]^{-1},$$

where

$B_\nu(T)$ is the specific intensity, at temperature T, of blackbody radiation,
h Planck's constant,
ν the spectral frequency,
c the velocity of light, and
k the Boltzmann constant.

This relationship, which is known to hold to a reasonable degree of approximation, particularly in the solar atmosphere and the lower chromosphere, is used to define the temperature in a stellar atmosphere (Greenstein, 1960). It can be shown that collisional excitation, and hence a departure from local thermodynamic equilibrium, is negligible in the solar photosphere but is important in hot stars for transitions in the visible region.

Free electrons practically follow a Maxwellian energy distribution in stellar atmospheres. However, deviations from equilibrium ionization and excitation occur in most stellar atmospheres owing to the large number of radiative processes present, as compared with collisional processes.

An important quantity in the utilization of spectral intensities for composition determinations is the oscillator strength or f parameter of the observed electromagnetic transition. As has been mentioned elsewhere (Chapter 5, Section II.F.3), such data are very sparse for high temperature species and only recently have laboratory data become available for the important diatomic metal oxide species MgO, SiO, and AlO, e.g., see Vanpee *et al.* (1970). In determining absolute f numbers it is necessary to know the species number density; that is, the partial pressure and the temperature of the laboratory experiment. Thus, the considerable uncertainties that one usually associates with partial pressure measurements of high temperature species are also transferred to f data and hence determinations of stellar composition.

The application of thermodynamic calculations of multicomponent equilibria have been particularly useful for describing the chemistry of stellar atmospheres, e.g., see Aller (1963), Tsuji (1964), and Lord (1965). While much of the basic thermochemical data has been modified in recent years, the original composition plots given by Aller (1963) and Tsuji (1964) are still indicative of the qualitative chemistry of the various star types. As an example, Fig. 7.1 shows the calculated composition-temperature profiles for the atmospheres of giant stars with an assumed equal carbon and oxygen abundance. These calculations are based on the following dissociation energy values (given in electron volts): TiO (6.9), ZrO (7.8), and AlO (5.8). The results also depend intimately on the initial choice of species. Lord (1965) has included perhaps the greatest variety of species in his analysis of the composition for M5 dwarf- and giant-stellar atmospheres. The most significant (i.e., partial pressure $> 10^{-9}$ atm) high temperature molecular species in this instance (for 0.5 atm H_2 and 2000 K) were determined to be

$$\begin{gathered} SiO, \quad SiS, \quad AlH, \quad MgH, \quad Al_2O, \quad TiO, \\ NaH, \quad NaOH, \quad VO, \quad TiO_2, \quad AlO(OH), \\ MgS, \quad FeO, \quad AlO, \quad \text{and} \quad CaO. \end{gathered}$$

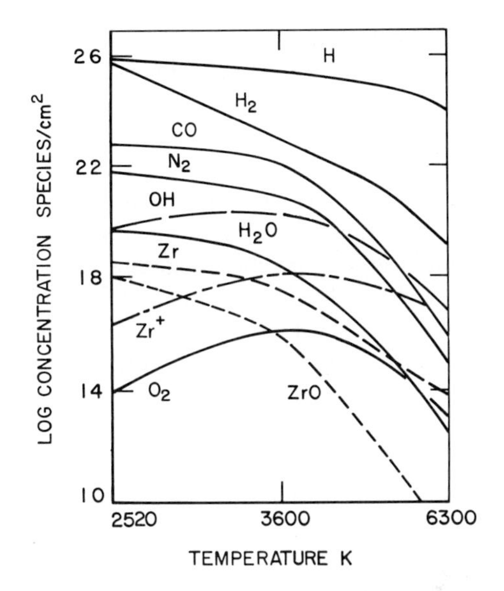

Fig. 7.1 Calculated composition as a function of temperature for a giant star atmosphere with equal total carbon and oxygen abundance; Ti-containing species are also present in similar amount to those indicated for Zr but are omitted from this plot. (Reprinted with permission from L. H. Aller, "Astrophysics—The Atmospheres of the Sun and Stars," 2nd ed. Copyright ©1963 by The Ronald Press Company, New York.)

TABLE 7.1

High Temperature Species Spectroscopically Identified in Stellar Atmospheres[a]

Species detected	Conditions
BH, MgH, SiH, CaH, TiO, MgO ScO, AlO, ZrO, YO, MgF, SrF	Sun (G 2), 6000 K[b]
TiO, MgH, SiH, AlH, CaH ZrO, ScO, YO, CrO, AlO BO, C_2, SiF, SiN	M star, β Pegasi ~3000 K
ZrO, YO, LaO, SiH, AlH, AlO	S star ~3000 K
C_2, C_3, SiC_2	C star ~3000 K, e.g., R. Y. Draconis
CaOH	M-dwarf stars of late-type, absorption spectra, <3000 K[c]

[a] See the reviews of Swings (1969) and Weltner (1967).
[b] Also see Babcock (1945).
[c] See Pesch (1972).

Thermodynamic calculations also predict the formation of condensed species, such as titanates and aluminates, in some of the cooler atmospheres, i.e., at 1700–2000 K (Lord, 1965).

Some of the high temperature species spectroscopically identified as stellar molecules are listed in Table 7.1. We can expect that in cool stars a number of additional as yet unidentified polyatomic species may be present and observable in the infrared. It is of note that an unaccounted-for-depletion of Si in the later C-type stars suggests the existence of as yet unidentified species such as HSiN and Si_2H_2 (Tsuji, 1964). One must also include for speculation the more recently discovered (in the laboratory) species SiN, Si_2N, and SiCN (Muenow and Margrave, 1970).

References[*]

Ackermann, R. J., Thorn, R. J., and Winslow, G. H. (1961). *Symp. Chem. Thermodynamic Properties High Temp., 18th Int. Congr. Pure Appl. Chem., Montreal* p. 270.

AGARD (North Atlantic Treaty Organization Advisory Group for Aerospace Research and Development) 84-71 (1971). *AGARD Symp. Aircraft, Fuels, Lubricants, Fire Safety, 37th* The Hague.

AGARD (North Atlantic Treaty Organization Advisory Group for Aerospace Research and Development) *Conf. Proc. No. 52* (1970). "Reactions between gases and solids." AD 702 657.

Agnew, W. G. (1960). *Combust. Flame* **4,** 29.

Alcock, C. B., and Hooper, G. W. (1960). *Proc. Roy. Soc. London* **A254,** 551.

Alcock, C. B., and Zador, S. (1971). *In* "Ceramics in Severe Environments" (Mater. Sci. Res. Vol. 5) (W. W. Kriegel and H. Palmour, III, eds.), p. 1. Plenum Press, New York.

Alexander, C. A., Ogden, J. S., and Levy, A. (1963). *J. Chem. Phys.* **39,** 3057.

Alexander, P. A. (1963). *In* "The Mechanism of Corrosion by Fuel Impurities" (H. R. Johnson and D. J. Littler, eds.), p. 571. Butterworths, London and Washington, D.C.

* Reports cited may be obtained from the U.S. Government Clearinghouse for Scientific and Technical Information, Springfield, Virginia.

Alkemade, C. Th. J. (1970). *In* "Analytical Flame Spectroscopy" (R. Mavrodineanu, ed.), p. 1. Springer-Verlag, Berlin and New York.
Allen, B. C. (1970a). *Reactor Mater.* **13,** 47.
Allen, B. C. (1970b). *Reactor Mater.* **13,** 139.
Aller, L. H. (1963). "Astrophysics: The Atmospheres of the Sun and Stars," 2nd ed. Ronald Press, New York.
Almer, F. H. R., and Wiedijk, P. (1971). *Z. Anorg. Allg. Chem.* **385,** 312.
Altman, M. (1967). *In* DeGroff *et al.* (1967), p. 135.
Altman, R. L. (1973). Laboratory evaluation of dry chemical fire extinguishants. Fall meeting Western States sect., Combust. Inst., Los Angeles.
Anderson, J. S. (1963). *Int. Symp. High Temp. Technol., Asilomar, Stanford Res. Inst.* p. 254.
Anderson, R. C. (1957). *In* "Combustion Researches and Reviews 1957" (AGARDograph no. 15), p. 17. Butterworths, London and Washington, D.C.
Andrews, G. E., and Bradley, D. (1972). *Combust. Flame* **19,** 275.
Andryukhina, E. D., *et al.* (1971). *JETP Lett.* **14,** 210.
Anon. (1971a). *Chem. Eng. News* **29,** 31.
Anon. (1971b). *J. Iron Steel Inst.* **209,** 25.
Antill, J. E., and Warburton, J. B. (1970). *AGARD Conf. Proc. No. 52,* Paper 10 (see AGARD, 1970).
Arbit, H. A., Clapp, S. D., and Nagai, C. K. (1970). *J. Spacecr. Rockets* **7,** 1221.
Argent, B. B., and Birks, N. (1968). *In* "High Temperature Materials: The Controlling Physical Processes" (A. J. Kennedy, ed.), p. 36. Oliver and Boyd, London.
Arnoldi, D., and Wolfrum, J. (1974). *Chem. Phys. Lett.* **24,** 234.
Arshadi, M., Yamdagni, R., and Kebarle, P. (1970). *J. Phys. Chem.* **74,** 1475.
Arthur, J. R. (1968). *In* "The Structure and Chemistry of Solid Surfaces" (G. A. Somorjai, ed.), Paper 46. Wiley, New York.
Ashton, A. F., and Hayhurst, A. N. (1973). *Combust. Flame* **21,** 69.
Asilomar (1960). *Proc. Int. Symp. High Temp. Technol., 1st, Asilomar, California, 1959.* McGraw-Hill, New York.
Asilomar (1964). *Proc. Int. Symp. High Temp. Technol., 2nd, Asilomar, California, 1963.* Butterworths, London and Washington, D.C.
Asilomar (1969). *Proc. Int. Symp. High Temp. Technol., 3rd, Asilomar, California, 1967.* Butterworths, London and Washington, D.C.
Audsley, A., and Bayliss, R. K. (1969). *J. Appl. Chem.* **19,** 33.
Babcock, H. D. (1945). *Astrophys. J.* **102,** 154.
Baddour, R. F., and Timmins, R. S. (eds.) (1967). "The Application of Plasmas to Chemical Processing." MIT Press, Cambridge, Massachusetts.
Baes, C. F., Jr. (1969). *In* Chiotti (1969), p. 617.
Bahn, G. S. (ed.) (1963). "Kinetics Equilibria and Performance of High Temperature Systems" (*Proc. Conf. Western States Sect., 2nd, Combustion Inst. 1962*). Gordon and Breach, New York.
Bahn, G. S. (ed.) (1968). "The Performance of High Temperature Systems" (*Proc. Conf. Western States Sect., 3rd, Combust. Inst.,* Pasadena, *California, 1964*), Vol. 1. Gordon and Breach, New York.
Bahn, G. S. (ed.) (1969). "The Performance of High Temperature Systems" (*Proc. Conf. Western States Sect., 3rd, Combust. Inst., Pasadena, California, 1964*), Vol. 2. Gordon and Breach, New York.
Bahn, G. S., and Zukoski, E. E. (eds.) (1960). "Kinetics, Equilibria and Performance of

High Temperature Systems" (*Proc. Conf. Western States Sect., 1st, Combust. Inst., 1959*). Butterworths, London and Washington, D.C.
Baker, L. A., and Ward, R. G. (1967). *J. Iron Steel Inst.* **205,** 714.
Baker, R. (1967). *J. Iron Steel Inst.* **205,** 637.
Bakish, R., Gellar, C. A., and Marinow, I. (1962). *J. Metals* **14,** 770.
Ballard, W. E. (1963). "Metal Spraying and the Flame Deposition of Ceramics and Plastics." Griffin, London.
Balwanz, W. W. (1965). *Symp. (Int.) Combust., 10th* p. 685. Combust. Inst., Pittsburgh, Pennsylvania.
Balzhiser, R. E., Samuels, M. R., and Eliassen, J. D. (1972). "Chemical Engineering Thermodynamics." Prentice-Hall, Englewood Cliffs, New Jersey.
Ban, V. S. (1971). *J. Electrochem. Soc.* **118,** 1473.
Ban, V. S. (1972a). *J. Electrochem. Soc.: Solid-State Sci. Technol.* **119,** 761.
Ban, V. S. (1972b). *J. Cryst. Growth* **17,** 19.
Banerjee, B. R. (ed.) (1968). *Proc. Nat. Symp. Heterogeneous Catal. Contr. Air Pollut., 1st.* Nat. Air Pollut. Contr. Admin.
Bannasch, W., Jonas, H., and Podschus, E. (1969). *In* Chiotti (1969), p. 279.
Barksdale, J. (1966). "Titanium," 2nd ed. Ronald Press, New York.
Barnea, J. (1972). *Sci. Amer.* **226,** 70.
Barnes, H. L. (ed.) (1967). "Geochemistry of Hydrothermal Ore Deposits." Holt, New York.
Barnes, J. F. (1968). *Combustion* **39,** 23.
Barr, J. (1955). *In* "Combustion Researches and Reviews," AGARDograph No. 9, p. 1. Butterworths, London and Washington, D.C.
Barrere, M. (1969). Recent Experimental Results on the Combustion of Aluminum and Other Metals. N70-13861 (ONERA-TP-735).
Barrett, M. F., Havard, D. C., Sayce, I. G., Selton, B., and Wilson, R. (1973). Active Aluminas Prepared by Plasma Vaporization. Paper 13 at Faraday Symp. No. 8, High Temp. Stud. Chem., London.
Barsukov, V. I., and Belov, V. A. (1969). *High Temp. USSR* **7,** 388.
Barthelemy, R. R. (1971). *In* "Technology Utilization Ideas for the 70s and Beyond" (AAS, Sci. and Technol. Ser. Vol. 26) (F. W. Forbes and P. Dergarabedian, eds.), p. 1. AAS, Tarzana, California.
Bartlett, R. W. (1968). *In* "The Performance of High Temperature Systems" (G. S. Bahn, ed.), Vol. 1, p. 79. Gordon and Breach, New York.
Barton, P. B., Jr., Bethke, P. M., and Toulmin, P. (1963). Miner. Soc. Amer. Spec. Pap. 1, 171.
Basco, N., and Morse, R. D. (1971). *Proc. Roy. Soc. London* **A321,** 129.
Bascombe, K. N., Green, J. A., and Sugden, T. M. (1962). *Advan. Mass Spectrom.* **2,** 66.
Basov, N. G., Kriukov, P. G., Zakharov, S. D., Senatsky, Y. V., and Tchekalin, S. V. (1968). *IEEE J. Quantum Electron.* **4,** 864.
Batalli-Cosmovici, C., and Michel, K. W. (1971). *Z. Naturforsch.* **A26,** 1147.
Batty, J. C., and Stickney, R. E. (1969). *In* "Rarefied Gas Dynamics" (*Proc. Int. Symp., 6th*) (L. Trilling and H. Y. Wachman, eds.), Vol. II, p. 1165. Academic Press, New York.
Batty, J. C., and Stickney, R. E. (1971). *Oxid. Metals* **3,** 331.
Bauer, S. H., and Porter, R. F. (1964). *In* "Molten Salt Chemistry" (M. Blander, ed.), p. 607. Wiley (Interscience), New York.
Baulch, D. L., Drysdale, D. D., Horne, D. G., and Lloyd, A. C. (1972). "Evaluated

Kinetic Data for High Temperature Reactions," Vol. I, Homogeneous Gas Phase Reactions of the H_2–O_2 System. Butterworths, London and Washington, D.C.
Beguin, C. P., Kanaan, A. S., and Margrave, J. L. (1964). *Endeavour* **23,** 55.
Behrndt, K. H. (1968). *In* "Techniques of Metals Research" (R. F. Bunshah, ed.), Vol. I, p. 1226. Wiley (Interscience), New York.
Bell, W. E., Merten, U., Tagami, K., and Garrison, M. C. (1961). *Symp. Chem. Thermodynam. Properties High Temp., Int. Congr. Pure Appl. Chem., 18th, Montreal* p. 82.
Belouet, C. (1972). *J. Cryst. Growth* **13/14,** 342.
Belousov, V. I., Sidorov, L. N., and Akishin, P. A. (1970). *Zh. Fiz. Khim.* **44,** 263.
Belton, G. R., and Jordan, A. S. (1967). *J. Phys. Chem.* **71,** 4114.
Belton, G. R., and McCarron, R. L. (1964). *J. Phys. Chem.* **68,** 1852.
Belton, G. R., and Worrell, W. L. (eds.) (1970). "Heterogeneous Kinetics at Elevated Temperatures." Plenum Press, New York.
Belyaev, I. N., Kesarev, V. V., Borchenko, G. V. (1972). *Zh. Neorg. Khim.* **17,** 833.
Benedict, M. (1971). *Proc. Nat. Acad. Sci. U.S.* **68,** 1923.
Bennett, F. D. (1968). *Progr. High Temp. Phys. Chem.* **2,** 1.
Bennett, J. E., Mile, B., and Summers, R. (1970). *Nature (London)* **225,** 932.
Berkowitz, J. (1962). *Proc. Symp. Thermodynam. Nucl. Mater., IAEA, Vienna, 1962* p. 505.
Berkowitz, J., Meschi, D. J., and Chupka, W. A. (1960). *J. Chem. Phys.* **33,** 533.
Berry, R. S., and Reimann, C. W. (1963). *J. Chem. Phys.* **38,** 1540.
Berry, W. E. (1970). *Reactor Mater.* **13,** 44.
Beus, A. A. (1958). *Geokhimiya* No. 4, 307.
Beus, A. A. (1963). *In* "Problems in Post Magmatic Ore Deposition" (J. Kutina, ed.), Vol. I, p. 105. Czech. Acad. Sci., Prague.
Biordi, J. C., Lazzara, C. P., and Papp, J. F. (1973). *Symp. (Int.) Combust. 14th,* p. 367. Combust. Inst., Pittsburgh, Pennsylvania.
Birchall, J. D. (1970). *Combust. Flame* **14,** 85.
Bird, R. B., Stewart, W. E., and Lightfoot, E. N. (1960). "Transport Phenomena." Wiley, New York.
Bishara, M. N., Fisher, S. S., Kuhlthau, A. R., and Scott, J. E., Jr. (1968). *In* "The Structure and Chemistry of Solid Surfaces" (G. A. Somorjai, ed.), paper 47. Wiley, New York.
Bishop, A. S. (1958). "Project Sherwood. The US Program in Controlled Fusion." Addison-Wesley, Reading, Massachusetts.
Bittker, D. A. (1970). Kinetics and Thermodynamics in High Temperature Gases. NASA SP-239.
Bittker, D. A., and Scullin, V. J. (1972). General Chemical Kinetics Computer Program for Static and Flow Reactions with Application to Combustion and Shock Tube Kinetics. NASA TN-D-6586.
Blander, M., ed. (1964). "Molten Salt Chemistry." Wiley (Interscience), New York.
Bleicher, M. (1972). *J. Electrochem. Soc.: Solid State Sci. Technol.* **119,** 613.
Bober, M., Schumacher, G., and Geithoff, D. (1973). *J. Nucl. Mater.* **47,** 187.
Bocek, P., and Janak, J. (1971). *Chromatogr. Rev.* **15,** 111.
Bockris, J. O'M., White, J. L., and MacKenzie, J. D. (eds.) (1959). "Physicochemical Measurements at High Temperatures." Academic Press, New York.
Bol'shakov, K. A., Galchenko, I. E., Dorfman, V. F., and Kislyakov, I. P. (1971). *Izv. Akad. Nauk. SSSR, Neorg. Mater.* **7,** 5.
Bonne, U., Jost, W., and Wagner, H. G. (1962). *Fire Res. Abstr. Rev.* **4,** 6.

Bonne, U., Homann, K. H., and Wagner, H. G. (1965). *Symp. (Int.) Combust., 10th* p. 503. Combust. Inst., Pittsburgh, Pennsylvania.

Boow, J. (1972). *Fuel* **51,** 170.

Borgianni, C., Capitelli, M., Cramarossa, F., Triolo, L., and Molinari, E. (1969). *Combust. Flame* **13,** 181.

Bornstein, N. S., DeCrescente, M. A., and Roth, H. A. (1971). Effect of Vanadium and Sodium Compounds on Accelerated Oxidation of Nickel-Base Alloys. Annu. rep. for Office Naval Res., Contract N00014-70-C-0234.

Bortner, M. H., and Bauner, T. (eds.) (1972). "Defence Nuclear Agency Reaction Rate Handbook," 2nd ed. DOD Nucl. Informat. Analy. Center, GE, Santa Barbara, California.

Botteri, B. P. (1971). *In* AGARD (1971) 84–71, Paper 13.

Bougnot, G., Chevrier, J., Etienne, D., and Bohe, C. (1971). *Mater. Res. Bull.* **6,** 137.

Boumans, P. W. J. M. (1972). *In* "Analytical Emission Spectroscopy" (E. L. Grove, ed.), Part II, p. 1. Dekker, New York.

Bouriannes, R., and Manson, N. (1971). *C.R. Acad. Sci. Paris Ser. C* **272,** 561.

Bowen, D. H., and Clarke, F. J. P. (1970). *In* "Chemical and Mechanical Behavior of Inorganic Materials" (A. W. Searcy, D. V. Ragone, and U. Colombo, eds.), p. 585. Wiley (Interscience), New York.

Bowen, H. K., Halloran, J. W., Petuskey, W. T., See, J. B., and Lynch, D. (1974). Chemical stability and degradation of MHD electrodes, *In Symp. Eng. Aspects Magnetohydrodynam. 14th, April 8, 1974, Tullahoma, Tennessee* p. IV.1.1.

Bowles, P. J., Ramsted, H. F., and Richardson, F. D. (1964). *J. Iron Steel Inst.* **202,** 113.

Bowser, R. J., and Weinberg, F. J. (1972). *Combust. Flame* **18,** 296.

Boyer, K. (1973). *Astronaut. Aeronaut.* **11,** 44.

Bradley, E. F., and Donachie, M. J., Jr. (1970). *J. Metals* **22,** 25.

Bradley, J. N. (1969). "Flame and Combustion Phenomena." Methuen, London.

Bradshaw, A. V. (1970). *Inst. Mining Met. Trans. Sect. C* **79,** 281.

Brady, E. L. (1953). *J. Phys. Chem.* **57,** 706.

Braunstein, J., Mamantov, G., Smith, G. P., eds. (1971). "Advances in Molten Salt Chemistry," Vol. I. Plenum Press, New York.

Braunstein, M., and Kikuchi, R. (1969). Condensation and Nucleation Processes of Single Crystal Thin Films. Tech. Rep. AFML-TR-69-149.

Brecher, L. E. (1970). *In* "Chemical Vapor Deposition" (*Int. Chem. Vapor Deposition Conf., 2nd*), p. 37. Electrochem. Soc., New York.

Brewer, L. (1950). *In* "Chemistry and Metallurgy of Miscellaneous Materials: Thermodynamics" (L. L. Quill, ed.), p. 193. McGraw-Hill, New York.

Brewer, L. (1957). *In* "Experientia Supplementum," Vol. VII, p. 227. Birkhauser Verlag, Stuttgart.

Brewer, L. (1962). Principles of High Temperature Chemistry, *Proc. R. A. Welch Foundation Conf. Chem. Res., Houston, Texas, 1962.*

Brewer, L. (1965). *In* "High Strength Materials" (V. F. Zackay, ed.), p. 12. Wiley, New York.

Brewer, L. (1968). *Science* **161,** 115.

Brewer, L. (1971). *J. Opt. Soc. Amer.* **61,** 1101.

Brewer, L. (1972). *J. Electrochem. Soc.* **119,** 5.

Brewer, L., and Lofgren, N. (1948). The Thermodynamics of Gaseous Cuprous Chloride Monomer and Trimer. AECD-1834.

Brewer, L., and Rosenblatt, G. M. (1969). *Advan. High Temp. Chem.* **2,** 1.
Brewer, L., and Searcy, A. W. (1956). *Annu. Rev. Phys. Chem.* **7,** 259.
Bril, J. (1968). *Spectrochim. Acta* **23B,** 375.
Brinckman, F. E., and Gordon, G. (1967). Energetic intermediates in inorganic synthesis characterization of transport species in electric discharge, *Proc. Int. Symp. Decomposition Organo-met. Compounds Refract. Ceram., Metals, Met. Alloys, Dayton, Ohio.*
Brittan, M. I. (1970). *J. S. Afr. Inst. Mining Met.* **70,** 278.
Brogan, T. R. (1963). *In* "Ionization in High Temperature Gases" (K. E. Shuler and J. B. Fenn, eds.), p. 319. Academic Press, New York.
Broida, H. P. (1954). *J. Chem. Phys.* **22,** 348.
Broida, H. P., and Heath, D. F. (1957). *J. Chem. Phys.* **26,** 223.
Brokaw, R. S. (1960). *J. Chem. Phys.* **32,** 1005.
Brokaw, R. S. (1961a). *J. Chem. Phys.* **35,** 1569.
Brokaw, R. S. (1961b). *Planet. Space Sci.* **3,** 238.
Brokaw, R. S. (1967). Transport Properties of High Temperature Gases. U.S. Govt. R. and D. Rep. N68-27697 (NASA TM X-52315).
Bronfin, B. R. (1967). *In* "The Applications of Plasmas to Chemical Processing" (R. F. Baddour and R. S. Timmins, eds.), p. 157. MIT Press, Cambridge, Massachusetts.
Bronfin, B. R. (1969). *In* "Chemical Reactions in Electrical Discharges" (R. F. Gould, ed.), Advan. Chem. Ser. 80, p. 423. Amer. Chem. Soc., Washington, D.C.
Brown, J., and Burns, G. (1970). *Can. J. Chem.* **48,** 3487.
Brown, R. A. S. (1971). *Can. Met. Quart.* **10,** 47.
Brown, R. L., Everest, D. A., Lewis, J. D., and Williams, A. (1968). *J. Inst. Fuel* **41,** 433.
Browne, W. G., Porter, R. P., Verlin, J. D., and Clark, A. H. (1969). *In Symp. (Int.) Combust., 12th* p. 1035. Combustion Inst., Pittsburgh, Pennsylvania.
Bruce, F. R. (1961). *In* "Process Chemistry" (F. R. Bruce, J. M. Fletcher, and H. H. Hyman, eds.), Vol. 3, p. 427. Pergamon, Oxford.
Bruckner, A. P., and Jahn, R. G. (1972). Spectroscopic Studies of the Exhaust Plume of a Quasisteady MPD Accelerator. NASA-CR-128317.
Brzustowski, T. A., and Glassman, I. (1964). *In* "Heterogeneous Combustion" (H. G. Wolfhard, I. Glassman, and L. Green, Jr., eds.), p. 117. Academic Press, New York.
Buchler, A., and Berkowitz-Mattuck, J. B. (1967). *Adv. High Temp. Chem.* **1,** 95.
Buchler, A., Stauffer, J. L., and Klemperer, W. (1967). *J. Chem. Phys.* **46,** 605.
Buell, B. E. (1962). *Anal. Chem.* **34,** 635.
Bulewicz, E. M., and Padley, P. J. (1969). *Trans. Faraday Soc.* **65,** 186.
Bulewicz, E. M., and Padley, P. J. (1970a). *Symp. (Int.) Combust., 13th* p. 73. Combust. Inst., Pittsburgh, Pennsylvania.
Bulewicz, E. M., and Padley, P. J. (1970b). *Combust. Flame* **15,** 203.
Bulewicz, E. M., and Padley, P. J. (1971a). *Trans. Faraday Soc.* **67,** 2337.
Bulewicz, E. M., and Padley, P. J. (1971b). *Proc. Roy. Soc. London* **A323,** 377.
Bulewicz, E. M., and Padley, P. J. (1973). *Spectrochim. Acta* **28B,** 125.
Bulewicz, E. M., and Sugden, T. M. (1956a). *Trans. Faraday Soc.* **52,** 1481.
Bulewicz, E. M., and Sugden, T. M. (1956b). *Trans. Faraday Soc.* **52,** 1475.
Bulewicz, E. M., and Sugden, T. M. (1958a). *Trans. Faraday Soc.* **54,** 830.
Bulewicz, E. M., and Sugden, T. M. (1958b). *Trans. Faraday Soc.* **54,** 1855.
Bulewicz, E. M., and Sugden, T. M. (1959). *Trans. Faraday Soc.* **55,** 720.
Bulewicz, E. M., James, C. G., and Sugden, T. M. (1956). *Proc. Roy. Soc. London* **A235,** 89.

Bulewicz, E. M., Phillips, L. F., and Sugden, T. M. (1961). *Trans. Faraday Soc.* **57,** 921.
Bulewicz, E. M., Padley, P. J., and Smith, R. E. (1970). *Proc. Roy. Soc. London* **A315,** 129.
Bulewicz, E. M., Padley, P. J., Cotton, D. H., and Jenkins, D. R. (1971). *Chem. Phys. Lett.* **9,** 467.
Bunsen, R. (1852). *J. Prakt. Chem.* **56,** 53.
Bunshah, R. F., and Douglass, D. L. (1971a). The Properties of Rare-Earth Metals and Alloys. Rep. UCLA Eng. 7161.
Bunshah, R. F., and Douglass, D. L. (1971b). New Techniques for the Synthesis of Metals and Alloys. Rep. UCLA Eng. 7112.
Burgess, D. D. (1972). *Space Sci. Rev.* **13,** 493.
Burke, R. R., and Miller, W. J. (1970). Study of Mass Spectrometric Ion Sampling Processes. AFCRL-70-0550, AeroChem TP-247.
Burmeister, J. (1971). *Mater. Res. Bull.* **6,** 219.
Burns, R. P., and Blaz, H. L. (1971). *Advan. Mass Spectrom.* **5,** 462.
Burrows, G. (1960). "Molecular Distillation." Oxford Univ. Press, London and New York.
Butler, J. N., and Brokaw, R. S. (1957). *J. Chem. Phys.* **26,** 1636.
Calcote, H. F. (1962). *Symp.* (*Int.*) *Combust., 8th* p. 181. Williams and Wilkins Co., Baltimore, Maryland.
Calcote, H. F. (1963). *In* "Ionization in High Temperature Gases" (K. E. Shuler and J. B. Fenn, eds.), Vol. 12, p. 107. Academic Press, New York.
Calcote, H. F., and Miller, W. J. (1971). *In* "Reactions under Plasma Conditions" (M. Venugopalan, ed.), Vol. II, p. 327. Wiley (Interscience), New York.
Calcote, H. F., and Reuter, J. L. (1963). *J. Chem. Phys.* **38,** 310.
Calcote, H. F., Kurzius, S. C., and Miller, W. J. (1965). *Symp.* (*Int.*) *Combustion, 10th* p. 605. Combust. Inst., Pittsburgh, Pennsylvania.
Campbell, R. J. (1969). *High Temp. Sci.* **1,** 303.
Cantor, S. (1973). *J. Nucl. Mater.* **47,** 177.
Capitelli, M., Ficocelli, V. E., and Molinari, E. (1970). Equilibrium Compositions and Thermodynamic Properties of Mixed Plasmas. Nat. Tech. Informat. Serv. NP 18772.
Caplan, D., and Cohen, M. (1961). *J. Electrochem. Soc.* **108,** 438.
Carlton, H. E., and Oxley, J. H. (1967). *AIChE J.* **13,** 571.
Carlton, H. E., Oxley, J. H., Hall, E. H., and Blocher, J. M., Jr. (1970). *In* "Chemical Vapor Deposition" (*Proc. Chem. Vapor Deposition Conf., 2nd*), p. 209. Electrochem. Soc., New York.
Cassidy, E. C. (1968). *Naturwissenschaften* **55,** 125.
Cassidy, E. C., Abramowitz, S., and Beckett, C. W. (1968). Investigations of the Exploding Wire Process as a Source for High Temperature Studies. Nat. Bur. Std. Monograph 109.
Castleman, A. W., Jr., and Tang, I. N. (1969). U.S. Govt. Rep. N70-21532 (BNL-13651).
Cathers, G. I., Carr, W. H., Lindauer, R. B., Milford, R. P., and Whatley, M. E. (1961). *In* "Process Chemistry" (F. R. Bruce, J. M. Fletcher, and H. H. Hyman, eds.), Vol. 3, p. 307. Pergamon, Oxford.
Celotta, R. J., Bennett, R. A., and Hall, J. L. (1974). *J. Chem. Phys.* **60,** 1740.
Cerrai, E., Gadda, F., and Scaroni, A. (1972). *Can. Met. Quart.* **11,** 21.
Chakraborti, P. K. (1963). *J. Chem. Phys.* **38,** 575.
Channapragada, R. S., Anderson, R., Tirumalesa, D., and Gopalakrishnan, A. (1969). *AIAA J.* **7,** 1581.

Chaudron, G., and Trombe, F. (eds.) (1973a). "Les Hautes Temperatures et leurs Utilizations en Physique et en Chimie," Vol. I, Realisation des Hautes Temperatures. Masson and Cie., Paris.
Chaudron, G., and Trombe, F. (eds.) (1973b). "Les Hautes Temperatures et leurs Utilizations en Physique et en Chimie," Vol. II, Mesures Physiques a Hautes Temperatures. Masson and Cie., Paris.
Chedaille, J., and Braud, Y. (1972). "Industrial Flames," Vol. I, Measurements in Flames. Arnold, London.
Chester, J. E., Dagnall, R. M., and Taylor, M. R. G. (1970). *Anal. Chim. Acta* **51,** 95.
Cheung, H., and Cohen, N. S. (1965). *AIAA J.* **3,** 250.
Chiotti, P. (ed.) (1969). "Nuclear Metallurgy," Vol. 15, Reprocessing of Nuclear Fuels. USAEC Conf.-690801.
Chiotti, P., and Voigt, A. F. (1961). *In* "Process Chemistry" (F. R. Bruce, J. M. Fletcher, and H. H. Hyman, eds.), Vol. 3, p. 340. Pergamon, Oxford.
Chopra, K. L. (1969). "Thin Film Phenomena." McGraw-Hill, New York.
Chupka, W. A., Berkowitz, J., Meschi, D. J., and Tasman, H. A. (1963). *Advan. Mass Spectrom.* **2,** 99.
Clampitt, N. C., and Hieftje, G. M. (1972). *Anal. Chem.* **44,** 1211.
Clarke, J. F., and Moss, J. B. (1970). *Combust. Sci. Technol.* **2,** 115.
Clarke, J. T. (1967). *In* "The Application of Plasmas to Chemical Processing" (R. F. Baddour and R. S. Timmins, eds.), p. 132. M.I.T. Press, Cambridge, Massachusetts.
Cochran, C. N., Sleppy, W. C., and Frank, W. B. (1970). *J. Metals* **22,** 54.
Cockayne, B., Filby, J. D., and Gasson, D. B. (1971). *J. Cryst. Growth* **9,** 340.
Coker, D. T., and Ottaway, J. M. (1971). *Nature (London)* **230,** 156.
Coker, D. T., Ottaway, J. M., and Pradhan, N. K. (1971). *Nature (London)* **233,** 69.
Colin, R., and Drowart, J. (1962). *J. Chem. Phys.* **37,** 1120.
Combourieu, J., Falinower, C., and Denis, G. (1971). *In* AGARD (1971) 84-71, paper no. 16.
Coney, M. W. E. (1970). *J. Phys.* **D3,** 1702.
Conway, J. B., Wilson, R. H., Jr., and Grosse, A. V. (1953). *J. Amer. Chem. Soc.* **75,** 499.
Cook, E., and Siegel, B. (1967). *J. Inorg. Nucl. Chem.* **29,** 2739.
Cook, E., and Siegel, B. (1968). *J. Inorg. Nucl. Chem.* **30,** 1699.
Cook, E., and Siegel, B. (1969). *High Temp. Sci.* **1,** 238.
Coppi, B., and Rem, J. (1972). *Sci. Amer.* **227,** 65.
Corey, R. C., ed. (1969). "Principles and Practices of Incineration." Wiley (Interscience), New York.
Cotton, D. H. (1970). *Sci. Progr.* **58,** 161.
Cotton, D. H., and Jenkins, D. R. (1968). *Trans. Faraday Soc.* **64,** 2988.
Cotton, D. H., and Jenkins, D. R. (1969). *Trans. Faraday Soc.* **65,** 376.
Cotton, D. H., and Jenkins, D. R. (1971). *Trans. Faraday Soc.* **67,** 730.
Cotton, D. H., Friswell, N. J., and Jenkins, D. R. (1971). *Combust. Flame* **17,** 87.
Coudurier, L., Wilkomirsky, I., and Morizot, G. (1970). *Inst. Mining Met. Trans. Sect. C* **79,** 34.
Coward, H. F., and Jones, G. W. (1952). U.S. Bur. Mines Bull. 503.
Crank, J. (1956). "The Mathematics of Diffusion." Oxford Univ. Press, London and New York.
Creitz, E. C. (1972). *Fire Technol.* **8,** 131.
Creitz, E. C. (1970). *J. Res. Nat. Bur. Std.* **74A,** 521.
Cruikshank, D. P., Morrison, D., and Lennon, K. (1973). *Science* **182,** 277.

Culler, F. L., Jr., and Harms, W. O. (1972). *Phys. Today* **25,** 28.

Cullis, C. F., and Mulcahy, M. F. R. (1972). *Combust. Flame* **18,** 225.

Cummings, G. A. McD. (1957). *In* "Combustion Researches and Reviews 1957," AGARDograph No. 15, p. 1. Butterworths, London and Washington, D.C.

Cuthbert, J. (1966). *Advan. Mass Spectrom.* **3,** 821.

Dagnall, R. M., and Taylor, M. R. G. (1971). *Spectrosc. Lett.* **4,** 147.

Day, M. J., Stamp, D. V., Thompson, K., and Dixon-Lewis, G. (1971). (*Int.*) *Combust., 13th* p. 705. Combust. Inst., Pittsburgh, Pennsylvania.

Dean, J. A., and Carnes, W. J. (1962). *Analyst* **87,** 743.

Dean, J. A., and Rains, T. C. (eds.) (1969). "Flame Emission and Atomic Absorption Spectrometry," Vol. I, Theory. Dekker, New York.

Decker, B. E. L., and Rao, B. H. G. (1970). *In* AGARD Conf. Proc. No. 52, Paper 26, see AGARD (1970).

De Groff, H. M., Hoglund, R. F., Fabri, J., Nagey, T. F., and Rumbaugh, M. E., Jr. (eds.) (1967). Combustion and Propulsion, *AGARD Colloq. 6th, Energy Sources and Energy Conversion.* Gordon and Breach, New York.

De Maria, G. (1970). *In* "Chemical and Mechanical Behavior of Inorganic Materials" (A. W. Searcy, D. V. Ragone, and U. Colombo, eds.), p. 81. Wiley (Interscience), New York.

Demerdache, A., and Sugden, T. M. (1963). *In* "The Mechanism of Corrosion by Fuel Impurities" (H. R. Johnson, and D. J. Littler, eds.), p. 12. Butterworths, London and Washington, D.C.

Dettingmeyer, J. H., Tillack, J., and Schafer, H. (1969). *Z. Anorg. Allg. Chem.* **369,** 161.

Dewing, E. W. (1967). *Nature* (*London*) **214,** 483.

Dewing, E. W. (1970). *Met. Trans.* **1,** 2169.

Dewing, E. W. (1971). *J. Phys. Chem.* **75,** 1260.

Diamant, R. M. E. (1971). "The Prevention of Corrosion." Business Books, London.

Dickinson, D. F., and Gottlieb, C. A. (1971). *Astrophys. Lett.* **7,** 205.

Dipprey, D. F. (1972). *In* "Chemistry in Space Research" (R. F. Landel and A. Rembaum, eds.), p. 465. Amer. Elsevier, New York.

Distin, P. A., and Whiteway, S. G. (1970). *Can. Met. Quart.* **9,** 419.

Dixon-Lewis, G. (1967). *Proc. Roy. Soc. London* **A298,** 495.

Dixon-Lewis, G. (1968). *Proc. Roy. Soc. London* **A307,** 111.

Dixon-Lewis, G. (1970a). *Proc. Roy. Soc. London* **A317,** 235.

Dixon-Lewis, G. (1970b). *Combust. Flame* **15,** 197.

Dixon-Lewis, G., and Williams, A. (1963). *Symp.* (*Int.*) *Combustion, 9th* p. 576. Academic Press, New York.

Dixon-Lewis, G., Sutton, M. M., and Williams, A. (1965). *In Symp.* (*Int.*) *Combust., 10th* p. 495. Combust. Inst., Pittsburgh, Pennsylvania.

Dixon-Lewis, G., Sutton, M. M., and Williams, A. (1970). *Proc. Roy. Soc. London* **A317,** 227.

Dodding, R. A., Simmons, R. F., and Stephens, A. (1970). *Combust. Flame* **15,** 313.

Dor, A. A. (ed.) (1972). "Nickel Segregation." Metallurgical Soc., AIME, New York.

Dougherty, G. J., Dunn, M. R., McEwan, M. J., and Phillips, L. F. (1971). *Chem. Phys. Lett.* **11,** 124.

Drowart, J., and Goldfinger, P. (1962). *Annu. Rev. Phys. Chem.* **13,** 459.

Drowart, J., Goldfinger, P., and Verhaegen, G. (1969). *In* "High Temperature Technology" (*Proc. Int. Symp. High Temp. Technol., 3rd, Asilomar, California, 1967*), p. 159. Butterworths, London and Washington, D.C.

Dudash, J. J., and Searcy, A. W. (1969). *High Temp. Sci.* **1,** 287.

Dunderdale, J., Durie, R. A., Mulcahy, M. F. R., and Schafer, H. N. S. (1963). *In* "The Mechanism of Corrosion by Fuel Impurities" (H. R. Johnson and D. J. Littler, eds.), p. 139. Butterworths, London and Washington, D.C.

Dutton, W. A., Janjua, M. B. I., Van den Steen, A. J., and Watkinson, A. P. (1971). *Can. Met. Quart.* **10,** 97.

Dzidic, L., and Kebarle, P. (1970). *J. Phys. Chem.* **74,** 1466.

Eatherly, W. P., and Piper, E. L. (1962). *In* "Nuclear Graphite" (R. E. Nightingale, ed.), p. 21. Academic Press, New York.

Eckert, E. R. G., and Drake, R. M., Jr. (1972). "Analysis of Heat and Mass Transfer." McGraw-Hill, New York.

Eckert, E. R. G., Sparrow, E. M., Ibele, W. E., Goldstein, R. J., and Scott, C. J. (1968). *Int. J. Heat Mass Transfer* **11,** 1421.

Edelman, R. B., Economos, C., and Boccio, J. (1971). *AIAA J.* **9,** 1935.

Edwards, A. B. (1956). *Proc. Aust. Inst. Mining Met.* **177,** 69.

Edwards, J. B. (1974). "Combustion Formation and Emission of Trace Species." Ann Arbor Science Publ., Ann Arbor, Michigan.

EEI Reactor Assessment Panel (1971). Combustion, June, 12.

Eichhorn, J. (1964). *J. Appl. Polym. Sci.* **8,** 2497.

Elliott, G. R. B. (1952). Gaseous Hydrated Oxides, Hydroxides, and Other Hydrated Molecules. UCRL-1831.

Elliott, J. F., Gleiser, M., and Ramakrishna, V. (1963). "Thermochemistry for Steelmaking," Vol. II. Addison-Wesley, Reading, Massachusetts.

Ellis, A. F., and Glover, J. (1971). *J. Iron Steel Inst. London* **209,** 593.

Ellis, A. J. (1957). *Amer. J. Sci.* **255,** 416.

Ellis, W. C. (1968). *In* "Techniques of Metals Research" (R. F. Bunshah, ed.), Vol. I, p. 1024. Wiley (Interscience), New York.

Eltenton, G. C. (1947). *J. Chem. Phys.* **15,** 455.

Emmenegger, F. P. (1972). *J. Cryst. Growth* **17,** 31.

Eriksson, G. (1971). *Acta Chem. Scand.* **25,** 2651.

Eringen, A. C., Leibowitz, H., Koh, S. L., and Crowley, J. M., eds. (1967). "Mechanics and Chemistry of Solid Propellants." Pergamon, Oxford.

Esgar, J. B., and Reynolds, R. A. (1971). Tech. Evaluation Rep. on AGARD Tech. Meeting No. 73 on High Temp. Turbines. Nat. Tech. Informat. Serv. AD 721857.

Evenson, K. M., Radford, H. E., and Moran, M. M., Jr. (1971). *Appl. Phys. Lett.* **18,** 426.

Everest, D. A., Sayce, I. G., and Selton, B. (1971). *J. Mater. Sci.* **6,** 218.

Eyring, L. (ed.) (1967, 1969, 1971, 1972). *Advan. High Temp. Chem.* **1, 2, 3, 4.**

Ezell, J. B., Thompson, J. C., and Margrave, J. L. (1967). *Advan. High Temp. Chem.* **1,** 219.

Faktor, M. M., and Garrett, I. (1971). *J. Chem. Soc. A* 934.

Faraday, M. (1957). "The Chemical History of a Candle." Crowell Co., New York.

Farber, M., and Srivastava, R. D. (1973a). *Combust. Flame* **20,** 33.

Farber, M., and Srivastava, R. D. (1973b). *Combust. Flame* **20,** 43.

Farber, M., Srivastava, R. D., Frisch, M. A., and Harris, S. P. (1973). Faraday Symp. 8, High Temperature Studies in Chemistry, London, 1973.

Farber, M., Harris, S. P., and Srivastava, R. D. (1974). *Combust. Flame* **22,** 191.

Faure, F. M., Mitchell, M. J., and Bartlett, R. W. (1972). High Temp. Sci. **4,** 181.

Feist, W. M., Steele, S. R., and Readey, D. W. (1969). *Phys. Thin Films* **5,** 237.

Feldmann, H. F., Leonards, M. A., Simons, W. H., and Bienstock, D. (1970). Thermo-

dynamic, Electrical, Physical, and Compositional Properties of Seeded Coal Combustion Products. U.S. Bureau of Mines Bull. 655.
Fells, I. (1967). *In* De Groff *et al.* (1967), p. 477.
Fenimore, C. P. (1964). "Chemistry in Premixed Flames." Pergamon, Oxford.
Fenimore, C. P., and Jones, G. W. (1958). *J. Phys. Chem.* **62,** 693.
Fenimore, C. P., and Jones, G. W. (1959). *J. Phys. Chem.* **63,** 1154.
Fenimore, C. P., and Jones, G. W. (1964). *Combust. Flame* **8,** 133.
Fenimore, C. P., and Jones, G. W. (1964b). *J. Chem. Phys.* **41,** 1887.
Fenimore, C. P., and Jones, G. W. (1966). *Combust. Flame* **10,** 295.
Fenimore, C. P., and Martin, F. J. (1972). *In* The Mechanisms of Pyrolysis, Oxidation and Burning of Organic Materials. Nat. Bur. Std. Spec. Publ. 357, p. 159.
Ferguson, F. A., and Phillips, R. C. (1962). *Advan. Chem. Eng.* **3,** 61.
Ferron, J. R. (1968). *In* "Extreme Conditions of Temperature and Pressure in Chemical Industry" (M. Simonetta, ed.), p. 17. Pergamon, Oxford.
Feugier, A. (1969). *Rev. Inst. Fr. Petrol. Ann. Combust. Liquides* **24,** 1374.
Feugier, A. (1970). *Rev. Gen. Therm.* **9,** 1045.
Feugier, A., and Van Tiggelen, A. (1965). *Symp. (Int.) Combust., 10th* p. 621. Combust. Inst., Pittsburgh, Pennsylvania.
Fiala, R. (1971). *In* AGARD (1971) 84-71.
Fiala, R., and Winterfeld, G. (1971). *In* AGARD (1971) 84-71.
Ficalora, P. J., Hastie, J. W., and Margrave, J. L. (1968a). *J. Phys. Chem.* **72,** 1660
Ficalora, P. J., Uy, O. M., Muenow, D. W., and Margrave, J. L. (1968b). *J. Amer. Ceram. Soc.* **51,** 574.
Fincham, C. J. B., and Richardson, F. D. (1954). *Proc. Roy. Soc.* **A223,** 40.
Fischer, E., Laser, M., and Merz, E. (1969). *In* Chiotti (1969), p. 645.
Fletcher, A. W. (1969). *Metals Mater.* **3**(1), 9.
Foner, S. N., and Hudson, R. L. (1953). *J. Chem. Phys.* **21,** 1374.
Fontana, M. G. (1971). *Corrosion* **27,** 129.
Fontana, M. G., and Bailey, R. E. (1968). *Trans. Amer. Nucl. Soc.* **11,** 376.
Ford, L. H., Hibbert, N. S., and Martin, D. G. (1972/73). *J. Nucl. Mater.* **45,** 139.
Fraas, A. P. (1971). The Blascon—an Exploding Pellet Fusion Reactor. ORNL-TM-3231.
Frank-Kamenetskii, D. A. (1969). "Diffusion and Heat Transfer in Chemical Kinetics" (transl. ed. by J. P. Appleton), 2nd ed. Plenum Press, New York.
Fraser, L. M., and Wineforner, J. D. (1972). Anal. Chem. **44,** 1444.
Freeman, E. S., and Anderson, D. A. (1966). *Combust. Flame* **10,** 337.
Freeman, M. P. (1969a). *Progr. High Temp. Phys. Chem.* **3,** 255.
Freeman, M. P. (1969b). *Advan. High Temp. Chem.* **2,** 151.
Friedman, R. (1961). *Fire Res. Abstr. Rev.* **3,** 128.
Friedman, R. (1971). *Fire Res. Abstr. Rev.* **13,** 187.
Friedman, R., and Levy, J. B. (1958). *Combust. Flame* **2,** 105.
Friedman, R., and Levy, J. B. (1963). *Combust. Flame* **7,** 195.
Frieser, R. G. (1968). *J. Electrochem. Soc.: Solid State Sci.* **115,** 401.
Fristrom, R. M. (1961). *In* "Experimental Methods in Combustion Research" (J. Surugue, ed.), p. 6. Pergamon, (for AGARD), Oxford.
Fristrom, R. M. (1963a). *Science* **140,** 297.
Fristrom, R. M. (1963b). *Symp. (Int.). Combust., 9th* p. 560. Academic Press, New York.
Fristrom, R. M. (1966). *Surv. Progr. Chem.* **3,** 55.

Fristrom, R. M. (1972). *In* The Mechanisms of Pyrolysis, Oxidation, and Burning of Organic Materials. Nat. Bur. Std. Spec. Publ. 357, p. 131.
Fristrom, R. M., and Sawyer, R. F. (1971). *Proc. AGARD Symp. Aircr. Fuels, Lubricants Fire Safety, 37th, The Hague.*
Fristrom, R. M., and Westenberg, A. A. (1957). *Combust. Flame* **1,** 217.
Fristrom, R. M., and Westenberg, A. A. (1965). "Flame Structure." McGraw-Hill, New York.
Friswell, N. J., and Jenkins, D. R. (1972). *Combust. Flame* **19,** 197.
Frolov, A. S., Trofimov, M. G., and Verenkova, E. M. (1971). Gas Flame Sprayings of ZrO_2 and Al_2O_3 Coatings with Aluminum Phosphate Additions. NASA Tech. Tr. TT F-13555.
Fromm, E., and Jehn, H. (1969). *Vacuum* **19,** 191.
Fruehan, R. J. (1972). *Met. Trans.* **3,** 2585.
Fuhs, A. E. (1959). *Ind. Eng. Chem.* **51,** 739.
Fujii, E., Nakamura, H., Haruna, K., and Koga, Y. (1972). *J. Electrochem. Soc.* **119,** 1106.
Gallo, C. F. (1971). *Appl. Opt.* **10,** 2517.
Ganeev, I. G. (1962). *Geokhimiya* 917.
Ganeev, I. G. (1963). *In* "Problems in Postmagmatic Ore Deposition" (J. Kutina, ed.), Vol. I Appendix, p. 141. Czech. Acad. Sci., Prague.
Gatz, C. R., Rosser, W. A., and Smith, F. T. (1961). Study of Radar Beam Attenuation in Rocket Exhaust Gases, Part 2. The Chemistry of Ionization in Rocket Exhausts. Tech. Rep. AFBMD-TR-61-39, Part 2 (AD 258777).
Gaydon, A. G. (1948). "Spectroscopy and Combustion Theory," 2nd ed. Chapman and Hall, London.
Gaydon, A. G. (1968). "Dissociation Energies and Spectra of Diatomic Molecules." Chapman and Hall, London.
Gaydon, A. G., and Wolfhard, H. G. (1970). "Flames, Their Structure, Radiation and Temperature," 3rd rev. ed. Chapman and Hall, London.
Geiger, G. H., and Poirier, D. R. (1973). "Transport Phenomena in Metallurgy." Addison-Wesley, Reading, Massachusetts.
Gen, M. Y., and Petrov, Y. I. (1969). *Russ. Chem. Rev.* **38,** 1007.
Gilbert, M. (1956). *Symp. (Int.) Combust., 6th* p. 74. Van Nostrand-Reinhold, Princeton, New Jersey.
Gilbert, P. T. (1970). *In* "Analytical Flame Spectroscopy" (R. Mavrodineanu, ed.), p. 238. Springer-Verlag, New York.
Gilles, P. W. (1961). *Annu. Rev. Phys. Chem.* **12,** 355.
Glassman, I., Mellor, A. M., Sullivan, H. F., and Laurendeau, N. M. (1970). AGARD Conf. Proc. No. 52, Paper 19 (See Agard, 1970).
Glemser, O., and Mueller, A. (1962). *Naturwissenschaften* **49,** 279.
Glemser, O., and Wendlandt, H. G. (1963). *Advan. Inorg. Chem. Radiochem.* **5,** 215.
Goes, C., and Kul, F. (1960). *Tonind.-Z. Keram. Runds.* **84,** 125.
Goldberger, W. M. (1966). *Chem. Eng. (N.Y.)* **73,** 173.
Gole, J. L., and Zare, R. N. (1972). *J. Chem. Phys.* **57,** 5331.
Golothan, D. W. (1967). *Soc. Auto Eng. Trans.* **76,** item 670092.
Gordon, J. S. (1967). Combustion Gas Emitted Radiation and Chemiluminescence Phenomena in Solid-Gas Reacting Systems. AFOSR 67-1795 (AD-656 426).
Gordon, S., and McBride, B. J. (1971). Computer Program for Calculation of Complex Chemical Equilibrium Compositions, Rocket Performance, Incident and Reflected Shocks, and Chapman-Jouguet Detonations. NASA SP-273.

Gordon, S., Zeleznik, F. J., and Huff, V. N. (1959–61). A General Method for Automatic Computation of Equilibrium Compositions and Theoretical Rocket Performance of Propellants. NASA Tech. Note D-132.

Gould, R. F. (ed.) (1967). "Fuel Gasification" (Advan. Chem. Ser. 69). Amer. Chem. Soc., Washington, D.C.

Goward, G. W. (1970). *J. Metals* **22,** 31.

Grabner, H., and Pilz, W. (1967). *Proc. Int. Ioniz. Phenomena Gases, 8th, Vienna, 1967* p. 228.

Graham, H. C. (1970). *In* AGARD Conf. Proc. No. 52, Paper 9. See AGARD (1970).

Graham, H. C., and Davis, H. H. (1971). *J. Amer. Ceram. Soc.* **54,** 88.

Greene, F. T., Randall, S. P., and Margrave, J. L. (1959). *In* "Thermodynamic and Transport Properties of Gases, Liquids and Solids," p. 222. Amer. Soc. Mech. Eng., New York.

Greene, F. T., Milne, T. A., Vandergrift, A. E., and Beachey, J. (1969). An Experimental Study of the Structure, Thermodynamics, and Kinetic Behavior of Water." U.S. Dept. of Interior R. and D. Progr. Rep. 493.

Greenstein, J. L. (ed.) (1960). "Stellar Atmospheres." Univ. of Chicago Press, Chicago, Illinois.

Grenon, M., and Geist, J. J. (1971). *Energ. Nucl.* **13,** 86.

Gretz, R. D. (1966). *In* "Vapor Deposition" (C. F. Powell, J. H. Oxley, and J. M. Blocher, Jr., eds.), p. 149. Wiley, New York.

Grimes, W. R. (1970). *Nucl. Appl. Technol.* **8,** 137.

Groh, G. (1968). *J. Appl. Phys.* **39,** 5804.

Gross, B., Grycz, B., and Miklossy, L. (1969). "Plasma Technology" (English transl., R. C. G. Leckey). Elsevier, New York.

Gross, P., and Lewin, R. H. (1973). *In* "High Temperature Studies in Chemistry," Faraday Symp. No. 8, Paper 14. Chem. Soc., London.

Gross, P., and Stuart, M. C. (1972). *In* "Metallurgical Chemistry Symp (O. Kubaschewski, ed.), p. 499. HM Stationary Office, London.

Gross, P., Levi, D. L., Dewing, E. W., and Wilson, G. L. (1959). *In* "Physical Chemistry of Process Metallurgy" (G. R. St. Pierre, ed.), Part 1, p. 403. Wiley (Interscience), New York.

Grove, E. L. (ed.) (1972). "Analytical Emission Spectroscopy," Part II. Dekker, New York.

Gruber, P. E. (1970). *In* "Chemical Vapor Deposition" (*Int. Chem. Vapor Deposition Conf., 2nd*), p. 25. Electrochem. Soc., New York.

Gruen, D. M. (ed.) (1972). "The Chemistry of Fusion Technology." Plenum Press, New York.

Gruen, D. M., and McBeth, R. L. (1969). *Inorg. Chem.* **8,** 2625.

Grumer, J., Miller, L. F., Bruszak, A. E., and Dalverny, L. E. (1973). Minimum Extinguishant and Maximum Oxygen Concentrations for Extinguishing Coal Dust-air Explosions. U.S. Bur. of Mines Rep. of Investigations 7782.

Gschneider, K. A., Jr. (ed.) (1964). "Metallurgy at High Pressures and High Temperatures." Gordon and Breach, New York.

Guiochon, G., and Pommier, C. (1973). "Gas Chromatography in Inorganics and Organometallics." Ann Arbor Science Publ., Ann Arbor, Michigan.

Gulbransen, E. A. (1966). *Nature* (*London*) **212,** 1420.

Gulbransen, E. A. (1970). *Corrosion* **26,** 19.

Gulbransen, E. A., and Jansson, S. A. (1970). *In* "Heterogeneous Kinetics at Elevated

Temperatures" (G. R. Belton and W. L. Worrell, eds.), p. 181. Plenum Press, New York.

Gulbransen, E. A., Andrew, K. F., and Brassart, F. A. (1969). *Proc. Int. Congr. Metallic Corrosion, 3rd* **4,** 16. Mir Publ., Moscow.

Gulbransen, E. A., Andrew, K. F., and Brassart, F. A. (1964). *In* "Heterogeneous Combustion" (H. G. Wolfhard, I. Glassman, and L. Green, Jr., eds.), p. 227. Academic Press, New York.

Gupta, S. K. (1971). *J. Phys. Chem.* **75,** 112.

Gupta, S. K., and Maloney, K. M. (1973). *J. Appl. Phys.* **44,** 3339.

Gurvich, L. V., and Ryabova, V. G. (1964a). *High Temp. USSR* **2,** 190.

Gurvich, L. V., and Ryabova, V. G. (1964b). *High Temp. USSR* **2,** 366.

Gurvich, L. V., and Ryabova, V. G. (1964c). *High Temp. USSR* **2,** 486.

Gurvich, L. V., Ryabova, V. G., Khitrov, A. N., and Starovoitov, E. M. (1971). *Teplofiz. Vys. Temp.* **9,** 290.

Gurvich, L. V., Ryabova, V. G., and Khitrov, A. N. (1973). *In* "High Temperature Studies in Chemistry," Faraday Symp. No. 8, Paper 8. Chem. Soc., London.

Gusarov, A. V., and Gorokhov, L. N. (1968). *Russ. J. Phys. Chem.* **42,** 449.

Gutmann, V., and Mayer, U. (1972). *In* "Structure and Bonding" (P. Hemmerich, C. K. Jorgensen, J. B. Neilands, R. S. Nyholm, D. Reinen and R. J. P. Williams, eds.), Vol. 10, p. 127. Springer-Verlag, New York.

Habashi, F. (1969). "Principles of Extractive Metallurgy," Vol. I, General Principles. Gordon and Breach, New York.

Habashi, F., and Thurston, G. A. (1967). *Energ. Nucl. Milan* **14,** 238.

Haerendel, G., Lust, R., and Rieger, E. (1967). *Space Res.* **7,** 77.

Hafeele, W., Faude, D., Fischer, E. A., and Laue, H. J. (1970). *Annu. Rev. Nucl. Sci.* **20,** 393.

Hager, J. P., and Hill, R. B. (1970). *Metall. Trans.* **1,** 2723.

Halachmy, M. (1971). Contribution a l'etude du comportement des produits de fission dans les combustibles nucleaires par la mesure des tensions de vapeur des constituents de $(UPu)O_2$ irradie dans un flux de neutrons rapides. Thesis, Univ. of Paris.

Hales, R., and Pearce, R. J. (1970). The Thermodynamics of Some Metal/Coolant Reactions in MKIII Reactors. Nat. Tech. Informa. Serv. RD-B-N-1816.

Halls, D. J., and Pungor, E. (1969). *Combust. Flame* **13,** 108.

Halpern, C., and Ruegg, F. W. (1958). *J. Res. Nat. Bur. Std.* **60,** 29.

Halstead, C. J., and Jenkins, D. R. (1967). *Combust. Flame* **11,** 362.

Halstead, W. D., and Raask, E. (1969). *J. Inst. Fuel* **42,** 344.

Hambly, A. N., and Rann, C. S. (1969). *In* "Flame Emission and Atomic Absorption Spectrometry" (J. A. Dean and T. C. Rains, eds.), p. 241. Dekker, New York.

Hammitt, A. G. (1968). *Astronaut. Acta* **14,** 57.

Hammond, A. L., Metz, W. D., and Maugh, T. H. II (1973). "Energy and the Future." Amer. Ass. Advan. Sci., Washington, D.C.

Hangos, I., Juhasz, I., and Verkonyi, L. (1972). *Magy. Kem. Foly.* **78,** 261 (Hung.).

Harang, O. (1969). *In* "Atmospheric Emissions" (B. M. McCormac and A. Omholt, eds.), p. 489. Van Nostrand-Reinhold, Princeton, New Jersey.

Hardesty, D. R., and Weinberg, F. J. (1973). *Symp. (Int.) Combust., 14th* p. 907. Combust. Inst., Pittsburgh, Pennsylvania.

Hariharan, A. V., Sood, S. P., Prasad, R., Sood, D. D., Rengan, K., Balakrishnan, P. V., and Ramaniah, M. V. (1969). *In* Chiotti (1969), p. 261.

Harris, L. P., and Moore, G. E. (1971). *IEEE Trans. Power App. Syst.* **90,** 2030.

Harrison, H., Hummer, D. G., and Fite, W. L. (1964). *J. Chem. Phys.* **41,** 2567.

Harvey, F. J. (1972). *Metall. Trans.* **3,** 2973.
Haskell, R. W., and Byrne, J. G. (1972). *In* "Treatise on Materials Science and Technology" (H. Herman, ed.), Vol. 1, p. 293. Academic Press, New York.
Hass, G., and Thun, R. E. (eds.) (1966). *Phys. Thin Films* **3,**
Hastie, J. W. (1971). *In* "Advances in Molten Salt Chemistry" (J. Braunstein, G. Mamantov, and G. P. Smith, eds.), Vol. I, p. 225. Plenum Press, New York.
Hastie, J. W. (1972). Unpublished observations.
Hastie, J. W. (1973a). *J. Res. Nat. Bur. Std.* **A77,** 733.
Hastie, J. W. (1973b). *Combust. Flame* **21,** 187.
Hastie, J. W. (1973c). *Combust. Flame* **21,** 49.
Hastie, J. W. (1974). *Chem. Phys. Lett.* **26,** 338.
Hastie, J. W. (1975). *Int. J. Mass Spectrom. Ion Phys.* **16,** 89.
Hastie, J. W., and Kaldor, A. (1972). Unpublished observations.
Hastie, J. W., and Margrave, J. L. (1968a) *Fluorine Chem. Rev.* **2,** 77.
Hastie, J. W., and Margrave, J. L. (1968b). *J. Chem. Eng. Data* **13,** 428.
Hastie, J. W., and Margrave, J. L. (1969). *J. Phys. Chem.* **73,** 1105.
Hastie, J. W., Hauge, R. H., and Margrave, J. L. (1971). *High Temp. Sci.* **3,** 257.
Hastie, J. W., Hauge, R. H., and Margrave, J. L. (1970). *Annu. Rev. Phys. Chem.* **21,** 475.
Hastie, J. W., Hauge, R. H., and Margrave, J. L. (1969a). *Inorg. Chim. Acta* **3,** 601.
Hastie, J. W., Hauge, R. H., and Margrave, J. L. (1969b). *J. Mol. Spectrosc.* **29,** 152.
Haubenreich, P. N., and Engel, J. R. (1970). *Nucl. Appl. Tech.* **8,** 118.
Hauge, R. H., and Margrave, J. L. (1972). *High Temp. Sci.* **4,** 170.
Havel, J. J., McGlinchey, M. J., and Skell, P. S. (1973). *Accounts Chem. Res.* **6,** 97.
Hayes, F. H. (1972). *In* Kubaschewski (1972), p. 581.
Hayhurst, A. N., and Kittelson, D. B. (1972a). *Combust. Flame* **19,** 306.
Hayhurst, A. N., and Kittelson, D. B. (1972b). *J. Chem. Soc. Chem. Commun.* p. 422.
Hayhurst, A. N., and Kittelson, D. B. (1972c). *Nature (London)* **235,** 136.
Hayhurst, A. N., and Sugden, T. M. (1966). *Proc. Roy. Soc. London* **A293,** 36.
Hayhurst, A. N., and Sugden, T. M. (1967). *Trans. Faraday Soc.* **63,** 1375.
Hayhurst, A. N., and Telford, N. R. (1970). *Trans. Faraday Soc.* **66,** 2784.
Hayhurst, A. N., and Telford, N. R. (1971). *Proc. Roy. Soc. London* **A322,** 483.
Hayhurst, A. N., Mitchell, F. R. G., and Telford, N. R. (1971). *Int. J. Mass Spectrom. Ion Phys.* **7,** 177.
Hecht, N. L. (1972). *In* "Techniques of Metals Research" (R. F. Bunshah, ed.), Vol. VII, p. 229. Wiley (Interscience), New York.
Helgeson, H. C. (1964). "Complexing and Hydrothermal Ore Deposition." Pergamon, Oxford.
Hellner, C., and Keller, R. A. (1972). *J. Air Pollut. Contr. Ass.* **22,** 959.
Henderson, A. W., Campbell, T. T., and Block, F. E. (1972). *Metall. Trans.* **3,** 2579.
Henderson, U. V., Jr., Woods, H. P., and Poplin, G. (1964). *In* "Heterogeneous Combustion" (H. G. Wolfhard, I. Glassman, and L. Green, Jr., eds.), p. 203. Academic Press, New York.
Henderson, W. A. (1964). *J. Metals AIME* **16,** 155.
Henley, E. J., and Rosen, E. M. (1969). "Material and Energy Balance Computations." Wiley, New York.
Henrion, P. N., and Leurs, A. (1971). *J. Nucl. Mater.* **41,** 1.
Henry, J. P., Jr., and Louks, B. M. (1971). *Chem. Tech.* (April) 236.
Herman, H. (ed.) (1972). "Treatise on Materials Science and Technology," Vol. I. Academic Press, New York.

Hermann, E. (1970). *Aluminum* **46,** 764.
Hermsen, R. W. (1968). Vapor-phase combustion of beryllium and aluminum. U.S. Govt. R. and D. Rep. AD 670 529.
Hessing, H. (1964). *Int. Symp. High Temp. Technol., Asilomar, Stanford Res. Inst.* p. 325. McGraw-Hill, New York.
Heywood, J. B., and Womack, G. J. (eds.) (1969). "Open-Cycle Magnetohydrodynamic Power Generation." Pergamon, Oxford.
High Temperature Institute (1971). USSR Academy of Sciences Significant Research Results for 1971. Nauka Press. Tech. Transl. FSTC-HT-23-1016-73, AD 765 753.
Hightower, J. R., Jr., and McNeese, L. E. (1971). "Low Pressure Distillation of Molten Fluoride Mixtures: Nonradioactive Tests for the MSRE Distillation Experiment." ORNL Tech. Rept. 4434.
Hightower, J. R., Jr., and McNeese, L. E. (1972). *J. Chem. Eng. Data* **17,** 342.
Hilado, C. J. (1973). *Fire Technol.* **9,** 198.
Hildenbrand, D. L. (1963). *In* "Kinetics, Equilibria and Performance of High Temperature Systems" (G. S. Bahn, ed.), p. 27. Gordon and Breach, New York.
Hills, A. W. D. (1970). *In* "Heterogeneous Kinetics at Elevated Temperatures" (G. R. Belton and W. L. Worrell, eds.), p. 449. Plenum Press, New York.
Hindersinn, R. R., and Wagner, G. M. (1967). *In Encycl. Polym. Sci. Technol.* **7,** 1.
Hinshelwood, C. N. (1926). "Kinetics of Chemical Changes." Oxford Univ. Press, London and New York.
Hirano, K., and Ishihara, T. (1969). *In* Chiotti (1969), p. 241.
Hirano, T. (1970). *Bull. Inst. Space Aeronaut. Sci. Univ. Tokyo* **6,** 26.
Hirschfelder, J. O., McClure, F. T., and Weeks, I. F. (1942). *J. Chem. Phys.* **10,** 201.
Hirschfelder, J. O., Curtiss, C. F., and Bird, R. B. (1954). "Molecular Theory of Gases and Liquids." Wiley, New York.
Hirth, J. P. (1972). *J. Cryst. Growth* **17,** 63.
Hirth, J. P., and Pound, G. M. (1963). "Condensation and Evaporation." MacMillan, New York.
Hoare, M. R., and Pal, P. (1972). *J. Cryst. Growth* **17,** 77.
Hollahan, J. R., and Wydaven, T. (1973). *J. Inorg. Nucl. Chem.* **35,** 1079.
Holland, N. H., and Rosborough, D. F. (1971). *J. Inst. Fuel* **44,** 300.
Hollander, T. J., Kalff, P. J., and Alkemade, C. T. (1963). *J. Chem. Phys.* **39,** 2558.
Holzl, R. A. (1968). *In* "Techniques of Metals Research" (R. F. Bunshah, ed.), Vol. I, p. 1377. Wiley (Interscience), New York.
Holzmann, R. T. (1969). "Chemical Rockets and Flame and Explosives Technology." Dekker, New York.
Homann, K. H. (1967a). *Combust. Flame* **11,** 265.
Homann, K. H. (1967b). *In* "Oxidation and Combustion Reviews" (C. F. H. Tipper, ed.), Vol. 2, p. 230. Elsevier, Amsterdam.
Homann, K. H. (1972). *In* The Mechanisms of Pyrolysis, Oxidation, and Burning of Organic Materials, Nat. Bur. Std. Spec. Publ. 357, p. 143.
Homann, K. H., and MacLean, D. I. (1971a). *J. Phys. Chem.* **75,** 3645.
Homann, K. H., and MacLean, D. I. (1971b). *Ber. Bunsenges. Phys. Chem.* **75,** 945.
Homann, K. H., and Poss, R. (1972). *Combust. Flame* **18,** 300.
Homann, K. H., and Wagner, H. G. (1967). *Symp. (Int.) Combust., 11th,* p. 371. Combust. Inst., Pittsburgh, Pennsylvania.
Homann, K. H., and Wagner, H. G. (1968). *Proc. Roy. Soc. London* **A307,** 141.
Homann, K. H., Mochizuki, M., and Wagner, H. G. (1963). *Z. Phys. Chem.* **37,** 299.

Horster, H., Kauer, E., and Lechner, W. (1971). *Phillips Tech. Rev.* **32,** 155.
Hottel, H. C., and Howard, J. B. (1971). "New Energy Technology—Some Facts and Assessments." MIT Press, Cambridge, Massachusetts.
Hougen, O. A., Watson, K. M., and Ragatz, R. A. (1959). "Chemical Process Principles. Part II. Thermodynamics," 2nd ed. Wiley, New York.
Houldcraft, P. T. (1967). "Welding Processes." Cambridge Univ. Press, London and New York.
Houseman, J., and Young, W. S. (1972). *In Mol. Beam Sampling Conf., Missouri, 1972* p. 70. Midwest Res. Inst., Missouri.
Hove, J. E., and Riley, W. C. (eds.) (1965). "Ceramics for Advanced Technologies." Wiley, New York.
Hudson, J. B. (1970). *J. Vac. Sci. Technol.* **7,** 53.
Huggett, C. (1956). *In* "Combustion Processes" (B. Lewis, R. N. Pease, and H. S. Taylor, eds.), p. 514. Oxford Univ. Press, London and New York.
Huggett, L. G., and Piper, L. (1966). *In* "Materials Technology in Steam Reforming Processes" (C. Edeleanu, ed.), p. 337. Pergamon, Oxford.
Hunt, L. P., and Sirtl, E. (1970). *In* "Chemical Vapor Deposition" (*Int. Chem. Vapor Deposition Conf., 2nd*), p. 3. Electrochem. Soc., New York.
Hunt, L. P., and Sirtl, E. (1972). *J. Electrochem. Soc.* **119,** 1741.
Ibberson, V. J. (1969). *High Temp.—High Press.* **1,** 243.
Ibberson, V. J., and Thring, M. W. (1969). *Ind. Eng. Chem.* **61,** 49.
Ibiricu, M. M., and Gaydon, A. G. (1964). *Combust. Flame* **8,** 51.
Imani, K. (1972). *Diss. Abstr. Int.* **B32,** 5835.
Ingham, M. F. (1972). *Sci. Amer.* **226,** 78.
Inghram, M., and Drowart, J. (1960). *Proc. High Temp. Technol. Conf., 1st, Asilomar, California, 1959* p. 219. McGraw-Hill, New York.
Ingraham, T. R., and Parsons, H. W. (1969). *Can. Met. Quart.* **8,** 291.
Ivanov, K. I., Lipshtein, R. A., and Chmovzh, V. E. (1973). *Therm. Eng.* (*USSR*) **20,** 85.
Iwasaki, I., Takahaski, Y., and Kahata, H. (1966). *Trans. Soc. Mining Eng. AIME* **235,** 308.
Jackson, D. D. (1971). Thermodynamics of the Gaseous Hydroxides, UCRL-51137.
Jacobs, P. W. M., and Powling, J. (1969). *Combust. Flame* **13,** 71.
Jacobs, P. W. M., and Russell-Jones, A. (1968). *J. Phys. Chem.* **72,** 202.
Jamrack, W. D. (1963). "Rare Metal Extraction by Chemical Engineering Techniques." MacMillan, New York.
JANAF (1971). Joint Army Navy Air Force Thermochemical Tables, 2nd ed., NSRDS-NBS 37. U.S. Govt. Printing Office, Washington, D.C.
Jansen, L. (1968). *In* "Molecular Processes on Solid Surfaces" (E. Drauglis, R. D. Gretz, and R. I. Jaffee, eds.), p. 49. McGraw-Hill, New York.
Jeffes, J. H. E., and Jacob, K. T. (1972). *In* Kubaschewski (1972), p. 513.
Jeffes, J. H. E., and Marples, T. N. R. (1972). *J. Cryst. Growth* **17,** 46.
Jenkins, D. R., and Sugden, T. M. (1969). *In* "Flame Emission and Atomic Absorption Spectrometry" (J. A. Dean and T. C. Rains, eds.), p. 151. Dekker, New York.
Jensen, D. E. (1968). *Combust. Flame* **12,** 261.
Jensen, D. E. (1969a). *J. Chem. Phys.* **51,** 4674.
Jensen, D. E. (1969b). *Trans. Faraday Soc.* **65,** 2123.
Jensen, D. E. (1972). *Combust. Flame* **18,** 217.
Jensen, D. E., and Jones, G. A. (1972). *Trans. Faraday Soc.* **68,** 259.

Jensen, D. E., and Jones, G. A. (1973). *J. Chem. Soc. Faraday Trans. I* **69,** 1448.
Jensen, D. E., and Miller, W. J. (1971). *Symp. (Int.) Combust., 13th* p. 363. Combust. Inst., Pittsburgh, Pennsylvania.
Jensen, D. E., and Miller, W. J. (1969a). Thermodynamic Studies in Metal Containing Flames. AeroChem Rep. TP-223.
Jensen, D. E., and Miller, W. J. (1969b). Electron Attachment and Compound Formation in Flames. IV. Negative Ion and Compound Formation in Flames Containing Potassium and Molybdenum. AeroChem Rep. TP-220.
Jensen, D. E., and Miller, W. J. (1970). *J. Chem. Phys.* **53,** 3287.
Jensen, D. E., and Padley, P. J. (1966a). *Trans. Faraday Soc.* **62,** 2140.
Jensen, D. E., and Padley, P. J. (1966b). *Trans. Faraday Soc.* **62,** 2132.
Jensen, D. E., and Padley, P. J. (1967). *Symp. (Int.) Combust., 11th* p. 351. Combust. Inst., Pittsburgh, Pennsylvania.
Jensen, D. E., and Travers, B. E. L. (1974). *IEEE Trans. Plasma Sci.* **PS-2,** 34.
Johnson, F. S. (1965). "Satellite Environment Handbook," 2nd ed. Stanford Univ. Press, Stanford, California.
Johnson, G. M., Matthews, C. J., Smith, M. Y., and Williams, D. J. (1970). *Combust. Flame* **15,** 211.
Johnson, H. R., and Littler, D. J., eds. (1963). "The Mechanism of Corrosion by Fuel Impurities." Butterworths, London and Washington, D.C.
Johnson, R. L., and Siegel, B. (1969). *J. Inorg. Nucl. Chem.* **31,** 955.
Jolly, W. L. (1969). *In* "Chemical Reactions in Electrical Discharges" (R. F. Gould, ed.), *Advan. Chem. Ser. 80* p. 156. Amer. Chem. Soc., Washington, D.C.
Jona, F., and Mandel, G. (1964). *J. Phys. Chem. Solids* **25,** 187.
Jona, F., and Mandel, G. (1963). *J. Chem. Phys.* **38,** 346.
Joncich, M. J. (1969). *Progr. High Temp. Phys. Chem.* **3,** 231.
Joncich, M. J., Vaughn, J. W., and Knutsen, B. J. (1966). *Can. J. Chem.* **44,** 137.
Jones, M. J. (ed.) (1972). "Advances in Extractive Metallurgy and Refining." Inst. Min. Met., London.
Jones, W. H. (1963). *In* "Kinetics, Equilibria and Performance of High Temperature Systems" (G. S. Bahn, ed.), p. 1. Gordon and Breach, New York.
Jones, W. H., Griffel, M., and Hochstim, A. R. (1963). Astronaut. Aerosp. Eng. Oct., p. 86.
Jonke, A. A., Levitz, N. M., and Steindler, M. J. (1969). *In* Chiotti (1969), p. 231.
Joshi, M. M., and Yamdagni, R. (1967). *Indian J. Phys.* **41,** 275.
Jost, W. (1960). "Diffusion in Solids, Liquids, Gases." Academic Press, New York.
Jost, W., Bonne, U., and Wagner, H. G. (1961). *Chem. Eng. News* **39,** 76.
Kaldis, E. (1971). *J. Cryst. Growth* **9,** 281.
Kaldis, E. (1972). *J. Cryst. Growth* **17,** 3.
Kalff, P. J., and Alkemade, C. T. J. (1970). *J. Chem. Phys.* **52,** 1006.
Kalff, P. J., and Alkemade, C. T. J. (1972). *Combust. Flame* **19,** 257.
Kallend, A. S. (1967). *Trans. Faraday Soc.* **63,** 2442.
Kallend, A. S. (1972). *Combust. Flame* **19,** 227.
Kana'an, A. S., Beguin, C. P., and Margrave, J. L. (1966). *Appl. Spectrosc.* **20,** 18.
Kangilaski, M. (1970). *Reactor Mater.* **13,** 21.
Karioris, F. G., and Fish, B. R. (1962). *J. Colloid Sci.* **17,** 155.
Karn, F. S., Friedel, R. A., and Sharkey, A. G., Jr. (1972). *Fuel* **51,** 113.
Karpenko, N. V. (1969a). *Vestn. Leningrad Univ. Fiz. Khim.* **24,** 77.
Karpenko, N. V. (1969b). *Vestn. Leningrad Univ. Fiz. Khim.* **24,** 114.
Karpenko, N. V. (1970). *Zh. Neorg. Khim.* **15,** 1378.

Karpenko, N. V., and Dogadina, G. V. (1971). *Zh. Neorg. Khim.* **16,** 818.
Karpenko, N. V., and Sevastyanova, T. N. (1967). *Vestn. Leningrad Univ. Fiz. Khim.* **22,** 109.
Kasha, M. A. (1969). "The Ionosphere and its Interaction with Satellites." Gordon and Breach, New York.
Kaskan, W. E. (1958). *Combust. Flame* **2,** 229.
Kaskan, W. E. (1965). *Symp. (Int.) Combust., 10th* p. 41. Combust. Inst., Pittsburgh, Pennsylvania.
Kaskan, W. E., and Millikan, R. C. (1962). *Symp. (Int.) Combust., 8th* p. 262. Williams and Wilkins, Baltimore, Maryland.
Kaufman, F. (1969). *In* "Chemical Reactions in Electrical Discharges" (R. F. Gould, ed.), *Advan. Chem. Ser. 80* p. 29. Amer. Chem. Soc., Washington, D.C.
Kaufman, M., Muenter, J., and Klemperer, W. (1967). *J. Chem. Phys.* **47,** 3365.
Kay, E. (1971). *Ann. Rev. Mater. Sci.* **1,** 289.
Keller, D. L. (1969). Progress on Development of Materials and Technology for Advanced Reactors. January–March, 1969. Rep. no. N70-21021 (BMI-1862).
Keller, D. L., and Chubb, W. (1969). Progress on High-Temperature Fuels Technology During August 1968–July 1969, Annu. rep. Rep. no. N70-21213 (BMI-1870).
Kelley, K. K. (1960). Bur. Mines Bull. 584.
Kelley, K. K., and King, E. G. (1961). Bur. Mines Bull. 592.
Kellogg, H. H. (1966). *Trans. Metall. Soc. AIME* **236,** 602.
Kellogg, H. H., and Basu, S. K. (1960). *Trans. Metall. Soc. AIME* **218,** 70.
Kelly, R., and Padley, P. J. (1969a). *Trans. Faraday Soc.* **65,** 355.
Kelly, R., and Padley, P. J. (1969b). *Trans. Faraday Soc.* **65,** 367.
Kelly, R., and Padley, P. J. (1970). *J. Chem. Soc. D Chem. Commun.* No. 23, p. 1606.
Kelly, R., and Padley, P. J. (1971a). *Trans. Faraday Soc.* **67,** 1384.
Kelly, R., and Padley, P. J. (1971b). *Trans. Faraday Soc.* **67,** 740.
Kendall, E. G. (1965). *In* "Ceramics for Advanced Technologies" (J. E. Hove and W. C. Riley, eds.), p. 364. Wiley, New York.
Kendall, E. G., and Riley, W. C. (1965). *In* "Ceramics for Advanced Technologies" (J. E. Hove and W. C. Riley, eds.), p. 251. Wiley, New York.
Keneshea, F. J., and Cubicciotti, D. (1964). *J. Chem. Phys.* **40,** 191.
Kennedy, A. J. (ed.) (1968). "High Temperature Materials: The Controlling Physical Processes." Oliver and Boyd, Edinburgh.
Kennedy, G. C., Wasserburg, G. J., Heard, H. C., and Newton, R. C. (1962). *Amer. J. Sci.* **260,** 501.
Kennedy, M. F., and Reynolds, A. B. (1973). *Nucl. Technol.* **20,** 149.
Kettani, M. Ali, and Hoyaux, M. F. (1973). "Plasma Engineering." Wiley, New York.
Khambatta, F. B., Gielisse, P. J., Wilson, M. P., Jr., Adamski, J. A., and Sahagian, Ch. (1972). *J. Cryst. Growth* **13,** 710.
Kikuchi, T., Kurosawa, T., and Yagihashi, T. (1969). *Nippon Kinzoku Gakkaishi* **33,** 305. (*Chem. Abstr.* **70,** 117248y).
King, I. R. (1963). *In* "Ionization in High Temperature Gases" (K. E. Shuler and J. B. Fenn, eds.), p. 197. Academic Press, New York.
Kirshenbaum, A. D., and Grosse, A. V. (1956). *J. Amer. Chem. Soc.* **78,** 2020.
Kit, B., and Evered, D. S. (1960). "Rocket Propellant Handbook." MacMillan, New York.
Kleinert, P. (1969). *In* "Reactivity of Solids," *Proc. Int. Symp. Reactivity Solids, 6th, Schenectady, New York, 1968* (J. W. Mitchel, R. C. DeVries, R. W. Roberts, and P. Cannon, eds.), p. 487. Wiley (Interscience), New York.

Kmonicek, V., Veis, S., and Hoffer, V. (1971). *Trans. Czech. Acad. Sci. Tech. Sci. Ser.* **81,** No. 4.

Knacke, O. (1972). *In* Kubaschewski (1972), p. 549.

Knewstubb, P. F. (1965). *Symp. (Int.) Combust., 10th* p. 623 (discussion). Combust. Inst., Pittsburgh, Pennsylvania.

Knewstubb, P. F., and Sugden, T. M. (1962). *Nature (London)* **196,** 1311.

Knighton, J. B., Johnson, I., and Steunenberg, R. K. (1969). *In* "Nuclear Metallurgy" (P. Chiotti, ed.), Vol. 15, Reprocessing of Nuclear Fuels, p. 337. USAEC Conf-690801.

Knuth, E. L. (1973). *In* "Engine Emissions Pollutant Formation and Measurement" (G. S. Springer and D. J. Patterson, eds.), p. 319. Plenum Press, New York.

Koenig, C. J., and Green, R. L. (1967). Water Vapor in High-Temperature Ceramic Processes. Ohio State Univ. Eng. Exp. Sta., Bull. 202, Ohio State Univ., Columbus, Ohio.

Kofstad, P. (1964). *Proc. Int. Symp. High Temp. Technol., 2nd, Asilomar* p. 176. Butterworths, London and Washington, D.C.

Kofstad, P. (1966). "High Temperature Oxidation of Metals." Wiley, New York.

Kofstad, P. (1970). *In* "Chemical and Mechanical Behavior of Inorganic Materials" (A. W. Searcy, D. V. Ragone, and U. Colombo, eds.), p. 245. Wiley (Interscience), New York.

Koirtyohann, S. R., and Pickett, E. E. (1971). *Spectrochim. Acta* **26B,** 349.

Kolosov, E. N., Sidorov, L. N., and Voronin, G. F. (1971). *Russ. J. Phys. Chem.* **45,** 1548.

Komarek, K. L. (1972). *In* Kubaschewski (1972), p. 75.

Kondratiev, V. N. (1972). Rate Constants of Gas Phase Reactions. COM-72-10014.

Kononyuk, I. F., and Kulikovskaya, N. P. (1970). *Izv. Akad. Nauk SSSR Metal.* **3,** 64.

Korb, L. J., and Crockett, L. K. (1970). *Metal Progr.* **97,** 99.

Krakowski, R. W., and Olander, D. R. (1968). *J. Chem. Phys.* **49,** 5027.

Krasin, A. K., and Nesterenko, V. B. (1971). *At. Energ. Rev.* **9,** 177.

Krause, H. H. (1959). *In* "Corrosion and Deposits in Coal- and Oil-Fired Boilers and Gas Turbines," p. 99. ASME, New York.

Krauskopf, K. B. (1957). *Econ. Geol.* **52,** 786.

Krauskopf, K. B. (1959). *In* "Researches in Geochemistry" (P. H. Abelson, ed.), p. 260. Wiley, New York.

Krauskopf, K. B. (1963). *In* "Problems of Postmagmatic Ore Deposition" (J. Kutina, ed.), Vol. I, p. 434. Czech. Acad. Sci., Prague.

Krauskopf, K. B. (1964). *Econ. Geol.* **59,** 22.

Krauskopf, K. B. (1965). *In* "Problems of Postmagmatic Ore Deposition" (M. Stemprok, ed.), Vol. II, p. 332. Czech. Acad. Sci., Prague.

Krauskopf, K. B. (1967a). "Introduction to Geochemistry." McGraw-Hill, New York.

Krauskopf, K. B. (1967b). *In* "Geochemistry of Hydrothermal Ore Deposits" (H. L. Barnes, ed.), p. 1. Holt, New York.

Krell, E., and Lumb, E. C. (1963). "Handbook of Laboratory Distillation." Elsevier, Amsterdam.

Krempl, H. (1971). *In* "Physical Chemistry, an Advanced Treatise" (W. Jost, ed.), Vol. I, Thermodynamics, p. 545. Academic Press, New York.

Krieger, F. J. (1970). U.S. Govt. R. and D. Rep. AD-706886.

Krieger, J. H. (1972). Chem. Eng. News, Nov. 13, p. 20.

Krupenie, P. H., Mason, E. A., and Vanderslice, J. T. (1963). *J. Chem. Phys.* **39,** 2399.

Krzhizhanovskaya, E. K., and Suvorov, A. V. (1971). *Russ. J. Inorg. Chem.* **16,** 1355.

Kubaschewski, O. (1968). *Naturwissenschaften* **55,** 525.

Kubaschewski, O., ed. (1972). *Metallurg. Chemi. Symp., 1971* H.M.Stationary Office, London.

Kubaschewski, O., and Evans, E. L. (1956). "Metallurgical Thermochemistry." Wiley, New York.

Kubaschewski, O., Cibula, A., and Moore, D. C. (1970). "Gases and Metals." Elsevier, Amsterdam.

Kumar, A., and Pandya, T. P. (1970). *Indian J. Pure Appl. Phys.* **8,** 42.

Kuppu–Rao, V., and Prasad, C. R. (1972). *Combust. Flame* **18,** 167.

Lachnitt, J. (1961). "Les Hautes Temperatures." Univ. of France Press, Paris.

Ladacki, M. (1972). *In* "Chemistry in Space Research" (R. F. Landel and A. Rembaum, eds.), p. 253. American Elsevier, New York.

Lamprey, H., and Ripley, R. L. (1962). *J. Electrochem. Soc.* **109,** 713.

Landsberg, A., and Block, F. E. (1965). "Study of the Chlorination Kinetics of Germanium, Silicon, Iron, Tungsten, Molybdenum, Columbium, and Tantalum. U.S. Bur. of Mines Rep. Invest. 6649.

Landsberg, A., and Hoatson, L. (1970). *J. Less-Common Metals* **22,** 327.

Landsberg, A., Hoatson, C. L., and Block, F. E. (1971). *J. Electrochem. Soc.* **118,** 1331.

Landsberg, A., Hoatson, C. L., and Block, F. E. (1972). *Metall. Trans.* **3,** 517.

Lanshov, V. N., Bazarov, G. P., and Kufa, E. N. (1969). *High Temp. USSR* **7,** 598.

Lanzo, C. D. (1970). *Nucl. Appl. Technol.* **8,** 6.

Larimer, J. W. (1973). *Space Sci. Rev.* **15,** 103.

Lascelles, K., and Schafer, H. (1971). *Z. Anorg. Allg. Chem.* **382,** 249.

Lask, G., and Wagner, H. G. (1962). *Symp. (Int.) Combust., 8th* p. 432. Williams and Wilkins, Baltimore, Maryland.

Lawn, B. R. (1974). *Mater. Sci. Eng.* **13,** 277.

Lawrence, R. M., and Bowman, W. H. (1971a). *J. Chem. Educ.* **48,** 335.

Lawrence, R. M., and Bowman, W. H. (1971b). *J. Chem. Educ.* **48,** 458.

Lawroski, S., *et al.* (1961). *In* "Process Chemistry" (F. R. Bruce, J. M. Fletcher, and H. H. Hyman, eds.), Vol. 3, p. 98. Pergamon, Oxford.

Lawton, J., and Weinberg, F. J. (1969). "Electrical Aspects of Combustion." Oxford Univ. Press (Clarendon), London and New York.

Laxton, J. W. (1963). *In* "The Mechanism of Corrosion by Fuel Impurities" (H. R. Johnson and D. J. Littler, eds.), p. 228. Butterworths, London and Washington, D.C.

Layne, G. S., Huml, J. O., Bangs, L. B., and Meserve, J. H. (1972). *Light Metal Age* **30** (3, 4), 8.

Lazzara, C. P., Biordi, J. C., and Papp, J. F. (1973). Radical Species Profiles for a Methane-oxygen-argon Flame. U.S. Bur. Mines RI 7766.

Learmonth, G. S., and Thwaite, D. G. (1970). *Brit. Polym. J.* **2,** 104.

Lebeau, P., and Trombe, F., eds. (1950). "Les Hautes Temperatures et leurs utilisations en Chimie," Vol. I. Masson, Paris.

Leeds, D. H. (1968). *In* "Ceramics for Advanced Technologies" (J. E. Hove and W. C. Riley, eds.), p. 197. Wiley, New York.

Lepie, A. H., Zimmer, M. F., and Baroody, E. E. (1968). *AIAA J.* **6,** 179.

Leslie, R. T. (1966). *Ann. N.Y. Acad. Sci.* **137,** 19.

Lever, R. F., and Mandel, G. (1962). *J. Phys. Chem. Solids* **23,** 599.

Levine, H. S. (1971). *High Temp. Sci.* **3,** 237.

Levy, A., and Merryman, E. L. (1965). *Trans. ASME J. Eng. Power* **87,** Ser. A, 374.

Levy, A., Droege, J. W., Tighe, J. J., and Foster, J. F. (1962). *Symp. (Int.) Combust., 8th* p. 524. Williams and Wilkins, Baltimore, Maryland.

Lewis, B., and Von Elbe, G. (1951). "Combustion, Flames and Explosions of Gases." Academic Press, New York.

Lewis, B., and Von Elbe, G. (1961). "Combustion, Flames and Explosions of Gases," 2nd ed. Academic Press, New York.

Lewis, J. S. (1973). *Space Sci. Rev.* **14,** 401.

Linden, L. H., and Heywood, J. B. (1971). *Combust. Sci. Technol.* **2,** 401.

Linevsky, M. J. (1971). Metal Oxide Studies; Iron Oxidation. Tech. Rep. RADC-TR-71-259.

Linville, W., and Spencer, J. D. (1973). Informat. Circ. 8612, U.S. Bur. Mines.

Livey, D. T., and Murray, P. (1959). *In* "Physico-chemical Measurements at High Temperatures" (J. O'M. Bockris, J. L. White, and J. D. Mackenzie, eds.), p. 87. Academic Press, New York.

Lo, R. E. (1969). "Theoretical Performance of the Tripropellant Rocket System F_2, O_2/LiH, Al/H_2 and Related Subsystems. U.S. Govt. Rep. N70-26436 (DLR-MITT-69-21).

Loh, W. H. T., ed. (1968). "Re-entry and Planetary Entry Physics and Technology. I. Dynamics, Physics, Radiation, Heat Transfer and Ablation." Springer-Verlag, New York.

Logani, R. C., and Smeltzer, W. W. (1971). *Can. Met. Quart.* **10,** 149.

Lord, H. C., III (1965). *Icarus* **4,** 279.

Lordi, J. A., Mates, R. E., and Hertzberg, A. (1968). *AIAA J.* **6,** 172.

Lorel, R. (1970). *Entropie* No. 36, 32.

Lovachev, L. A., and Gontkovskaya, V. T. (1972). *Dokl. Akad. Nauk SSSR* **204,** 379.

Lovachev, L. A., Babkin, V. S., Bunev, V. A., V'yun, A. V., Krivulin, V. N., and Baratov, A. N. (1973). *Combust. Flame* **20,** 259.

Lubin, M. J., and Fraas, A. P. (1971). *Sci. Amer.* **224,** 21.

Luikov, A. V., and Mikhailov, Yu. A. (1965). "Theory of Energy and Mass Transfer" (English ed.). Pergamon, Oxford.

Lumsden, J. (1966). "Thermodynamics of Molten Salt Mixtures." Academic Press, New York.

Lumsden, J. (1972). *In* Kubaschewski (1972), p. 533.

Lundin, C. E. (1969). NASA CR-1271.

Lustman, B., and Kerze, F., Jr. (eds.) (1955). "The Metallurgy of Zirconium." McGraw-Hill, New York.

Lutz, O. (1957). *In* "Combustion Researches and Reviews 1957," AGARDograph No. 15, p. 173. Butterworths, London and Washington, D.C.

L'vov, B. V. (1970). "Atomic Absorption Spectrochemical Analysis." American Elsevier, New York.

Lydtin, H. (1970). *In* "Chemical Vapor Deposition" (*Int. Chem. Vapor Deposition Conf.*), p. 71. Electrochem. Soc., New York.

Lynde, R. A., and Corbett, J. D. (1971). *Inorg. Chem.* **10,** 1746.

Lyons, J. W. (1970). "The Chemistry and Uses of Fire Retardants." Wiley (Interscience), New York.

Macek, A., and Semple, J. M. (1969). Combustion of Boron Particles at Atmospheric Pressure. AD-693055 (SQUID-TR-ARC-12-PU; N70-12511).

MacKenzie, K. J. D. (1970). *J. Inorg. Nucl. Chem.* **32,** 3731.

Madix, R. J., and Schwarz, J. A. (1971). *Surface Sci.* **24,** 264.

Madix, R. J., Parks, R., Susu, A. A., and Schwarz, J. A. (1971). *Surface Sci.* **24,** 288.

Mahe, R. (1973). *In* Chaudron and Trombe (1973), p. 139.

Mahieux, F. (1963a). *C. R. Acad. Sci. Paris* **257,** 1083.

Mahieux, F. (1963b). *C. R. Acad. Sci. Paris* **258,** 3497.
Maloney, K. M. (1971). *High Temp. Sci.* **3,** 445.
Maier, C. G. (1942). Sponge Chromium. U.S. Bur. Mines Bull. 436.
Maissel, L. I., and Glang, R. (1970). "Handbook of Thin Film Technology." McGraw-Hill, New York.
Malinauskas, A. P., Gooch, J. W., Jr., and Redman, J. (1970). *Nucl. Appl. Tech.* **8,** 52.
Mandel, G. (1962). *J. Chem. Phys.* **37,** 1177.
Manes, M. (1969). *In* "Chemical Reactions in Electrical Discharges" (R. F. Gould, ed.), Advan. Chem. Ser. 80, p. 133. Amer. Chem. Soc., Washington, D.C.
Mansell, R. E. (1968). *Appl. Spectrosc.* **22,** 790.
Margrave, J. L. (1957). *Ann. N.Y. Acad. Sci.* **67,** 619.
Margrave, J. L. (1962). *Science* **135,** 345.
Margrave, J. L. (ed.) (1967a). "The Characterization of High-Temperature Vapors." Wiley, New York.
Margrave, J. L. (chairman) (1967b). "High-Temperature Chemistry: Current and Future Problems" (*Proc. Conf. Houston, Texas, 1966*), Publ. No. 1470. Nat. Acad. Sci., Nat. Res. Council, Washington, D.C.
Markowitz, J. M., and Clayton, J. C. (1970). Corrosion of Oxide Nuclear Fuels in High Temperature Water. Nat. Tech. Informat. Serv. WAPD-TM-909.
Markstein, G. H. (1963). *Symp. (Int.) Combust., 9th* p. 137. Academic Press, New York.
Markstein, G. H. (1969). *Combust. Flame* **13,** 212.
Marlin Smith, J., Nichols, L. D., and Seikel, G. R. (1974). Nasa Lewis H_2-O_2 MHD Program, *In Symp. Eng. Aspects of Magnetohydrodynam., 14th April, 1974, Tullahoma, Tennessee* p. III 7.1.
Marrero, T. R., and Mason, E. A. (1972). *J. Phys. Chem. Ref. Data* **1,** 3.
Marsh, H. E., Jr., and Hutchison, J. J. (1972). *In* "Chemistry in Space Research" (R. F. Landel and A. Rembaum, eds.), p. 361. American Elsevier, New York.
Mash, D. R. (1962). *In* "Materials Science and Technology for Advanced Applications" (D. R. Mash, ed.), p. 656. Prentice-Hall, Englewood Cliffs, New Jersey.
Mason, E. A., and Monchick, L. (1962). *J. Chem. Phys.* **36,** 1622.
Matthews, C. J. (1968). *Rev. Pure Appl. Chem.* **18,** 311.
Matousek, J. (1972). *Silikaty* **16,** 1.
Matousek, J., and Hlavac, J. (1971). *Glass Technol.* **12,** 103.
Matsuda, S., and Gutman, D. (1971). *J. Phys. Chem.* **75,** 2402.
Matsumoto, O., Shirato, Y., and Miyazaki, M. (1968). *J. Electrochem. Soc. Japan* **36,** 219.
Mavrodineanu, R. (1970). *Develop. Appl. Spectrosc.* **8,** 18.
Mavrodineanu, R., and Boiteux, H. (1965). "Flame Spectroscopy." Wiley, New York.
Mavrodineanu, R., and Hughes, R. C. (1964). *Develop. Appl. Spectrosc.* **3,** 305.
Mayer, S. W., Schieler, L., and Johnston, H. S. (1967). *Symp. (Int.) Combust., 11th* p. 837. Combust. Inst., Pittsburgh, Pennsylvania.
McCarroll, B. (1967a). *J. Chem. Phys.* **47,** 5077.
McCarroll, B. (1967b). *J. Chem. Phys.* **46,** 863.
McCoy, H. E., *et al.* (1970). *Nucl. Appl. Tech.* **8,** 156.
McEwan, M. J., and Phillips, L. F. (1965). *Combust. Flame* **9,** 420.
McEwan, M. J., and Phillips, L. F. (1967). *Combust. Flame* **11,** 63.
McHale, E. T. (1969). *Fire Res. Abstr. Rev.* **11,** 90. Also N68-37380.
McKinley, J. D., Jr. (1966). *J. Chem. Phys.* **45,** 1690.
McKinley, J. D., Jr. (1969). *In* "Reactivity of Solids" (J. W. Mitchell, R. C. DeVries, R. W. Roberts, and P. Cannon, eds.), p. 345. Wiley (Interscience), New York.
McLain, H. A. (1969). *Nucl. Safety* **10,** 392.

McTaggart, F. K. (1967). "Plasma Chemistry in Electrical Discharges." Elsevier, Amsterdam.
McTaggart, F. K. (1969). *In* "Chemical Reactions in Electrical Discharges" (R. F. Gould, ed.), Advan. Chem. Ser. 80, p. 176. Amer. Chem. Soc., Washington, D.C.
Medin, S. A., Ovcharenko, V. A. and Shpil'rain, E. E. (1972). *High Temp. USSR* **10,** 390.
Meek, R. L., and Braun, R. H. (1972). *J. Electrochem. Soc.* **119,** 1538.
Meissner, H. P. (1971). "Processes and Systems in Industrial Chemistry." Prentice-Hall, Englewood Cliffs, New Jersey.
Menzies, A. C. (1960). *Anal. Chem.* **32,** 898.
Merrill, P. W. (1963). "Space Chemistry." Univ. Michigan Press, Ann Arbor, Michigan.
Messier, D. R., and Wong, P. (1971). *J. Electrochem. Soc.* **118,** 772.
Mezey, E. J. (1966). *In* "Vapor Deposition" (C. F. Powell, J. H. Oxley, and J. M. Blocher, Jr., eds.), p. 423. Wiley, New York.
Michels, H. H. (1972). *J. Chem. Phys.* **56,** 665.
Millan, G., and Da Riva, I. (1960). *Symp. (Int.) Combust., 8th* p. 398. Williams and Wilkins, Baltimore, Maryland.
Miller, C. L. (1970). "Basic Metal Ignition and Combustion Studies Relating to Air-augmented Combustion. Final report. AD-707860 (CETEC-FR 01700; AFOSR-70-1691TR).
Miller, D. R., Evers, R. L., and Skinner, G. B. (1963). *Combust. Flame* **7,** 137.
Miller, R. C., and Ayen, R. J. (1969). *Ind. Eng. Chem. Proc. Design Develop.* **8,** 370.
Miller, W. J. (1968). *Oxid. Combust. Rev.* **3,** 97.
Miller, W. J. (1969). *Combust. Flame* **13,** 210.
Miller, W. J. (1972). *J. Chem. Phys.* **57,** 2354.
Miller, W. J. (1973). *Symp. (Int.) Combust., 14th* p. 307. Combust. Inst., Pittsburgh, Pennsylvania.
Mills, A. F., Gomez, A. V., and Strouhal, G. (1971). *J. Spacecr. Rockets* **8,** 618.
Mills, R. M. (1968). *Combust. Flame* **12,** 513.
Milne, T. A., and Greene, F. T. (1966). *J. Chem. Phys.* **44,** 2444.
Milne, T. A., and Greene, F. T. (1969). *Advan. High Temp. Chem.* **2,** 107. Academic Press, New York.
Milne, T. A., Green, F. T., and Benson, D. K. (1970). *Combust. Flame* **15,** 255.
Minagawa, S., and Seki, H. (1972). *Jap. J. Appl. Phys.* **11,** 855.
Minkoff, G. J., and Tipper, C. F. H. (1962). "Chemistry of Combustion Reactions." Butterworths, London and Washington, D.C.
Miwa, T., Yoshimori, T., and Takeushi, T. (1964). *Kogyo Kagaku Zasshi* **67,** 2045.
Mohan, G., and Williams, F. A. (1972). *AIAA J.* **10,** 776.
Monchick, L., Pereira, A. N. G., and Mason, E. A. (1965). *J. Chem. Phys.* **42,** 3241.
Moore, G. E. (1963). *In* "Ionization in High Temperature Gases" (K. E. Shuler and J. B. Fenn, eds.), p. 347. Academic Press, New York.
Moore, J. H. (1958). *In* "Vacuum Metallurgy" (R. F. Bunshah, ed.), p. 435. Van Nostrand-Reinhold, Princeton, New Jersey.
Moore, R. V., and Fawcett, S. (1967). *In* "Fast Breeder Reactors" (P. V. Evans, ed.), p. 99. Pergamon, Oxford.
Morey, G. W. (1957). *Econ. Geol.* **52,** 225.
Morokuma, K., and Pedersen, L. (1968). *J. Chem. Phys.* **48,** 3275.
Morokuma, K., and Winick, J. R. (1970). *J. Chem. Phys.* **52,** 1301.
Morrison, M. E., and Scheller, K. (1972). *Combust. Flame* **18,** 3.
Moseley, F., and Paterson, D. (1967). *J. Inst. Fuel* **40,** 523.

Muenow, D. W. (1973). *J. Phys. Chem.* **77,** 970.
Muenow, D. W., and Margrave, J. L. (1970). *J. Phys. Chem.* **74,** 2577.
Muenow, D. W., Hastie, J. W., Hauge, R., Bautista, R., and Margrave, J. L. (1969). *Trans. Faraday Soc.* **65,** 3210.
Mukherjee, N. R., Fueno, T., Eyring, H., and Ree, T. (1962). *Symp.* (*Int.*) *Combust., 8th* p. 1. Williams and Wilkins, Baltimore, Maryland.
Murata, K. J. (1960). *Amer. J. Sci.* **258,** 769.
Murphy, J. E., Morrice, E., and Wong, M. M. (1970). A Comparison of Sublimation and Vaporization for Purification of Samarium Metal. Bur. Mines Rep. RI 7466.
Myers, J. W. (1971). *Ind. Eng. Chem. Prod. Res. Develop.* **10,** 200.
Naboko, S. I. (1959). *Bull. Volcanolog. Ser. II* **20,** 121.
Naegeli, D. W., and Palmer, H. B. (1968). *J. Mol. Spectrosc.* **26,** 277.
Nagasawa, K., Bando, Y., and Takada, T. (1972). *J. Cryst. Growth* **17,** 143.
NASA (1970). Kinetics and Thermodynamics in High-Temperature Gases, Conf. at Lewis Res. Center, 1970. SP-239.
Nazimova, N. A. (1967). *Zh. Prikl. Spektrosk.* **7,** 169.
Nelson, H. W. (1959). *In* "Corrosion and Deposits in Coal- and Oil-Fired Boilers and Gas Turbines," p. 7. ASME, New York.
Nelson, W., and Cain, C., Jr. (1960). *Trans. ASME J. Eng. Power* p. 194.
Nerheim, N. M., and Kelly, A. J. (1968). A Critical Review of the Magnetoplasmadynamic Thruster for Space Applications. NASA Tech. Rep. 32-1196.
Nettley, P. T., Bell, I. P., Bagley, K. Q., Harries, D. R., Thorley, A. W., and Tyzack, C. (1967). *In* "Fast Breeder Reactors" (P. V. Evans, ed.), p. 825. Pergamon, Oxford.
Neumann, G. M. (1973a). *Z. Metallkd.* **64,** 117.
Neumann, G. M. (1973b). *Z. Metallkd.* **64,** 26.
Nevitt, M. V., and Duffield, R. B. (1970). *In* "Chemical and Mechanical Behavior of Inorganic Materials" (A. W. Searcy, D. V. Ragone, and U. Colombo, eds.), p. 615. Wiley (Interscience), New York.
Newman, R. N., and Page, F. M. (1970). *Combust. Flame* **15,** 317.
Newman, R. N., and Page, F. M. (1971). *Combust. Flame* **17,** 149.
Nguyen, L. D., de Saint Simon, M., Puil, G., and Yokoyama, Y. (1973). *Geochim. Cosmochim. Acta Suppl. 4* **2,** 1415.
Nickl, J. J., and Reichle, M. (1971). *J. Less-Common Metals* **24,** 63.
Nightingale, R. E. (1966). *In* "High Temperature Nuclear Fuels" (*Metallurg. Soc. Conf.*), Vol. 42, p. 11. Gordon and Breach, New York.
Nijland, L. M., and Schroder, J. (1969). *Symp.* (*Int.*) *Combust., 12th* p. 1277. Combust. Inst., Pittsburgh, Pennsylvania.
Nisel'son, L. A., Petrusevich, I. V., Shamrai, F. I., and Fedorov, T. F. (1971). *In* "High Temperature Inorganic Compounds" (G. V. Samsonov, ed.), p. 477. Published for A.E.C. and Nat. Sci. Foundation by Indian Nat. Sci. Documentation Center, New Delhi (AEC-TR-6873).
Norrish, R. G. W. (1959). *Symp.* (*Int.*) *Combust., 7th* p. 203. Butterworths, London and Washington, D.C.
Norrish, R. G. W. (1965). *Symp.* (*Int.*) *Combust., 10th* p. 1. Combust. Inst., Pittsburgh, Pennsylvania.
Novikov, G. I., and Gavryuchenkov, F. G. (1967). *Russ. Chem. Rev.* **36,** 156.
Ogden, J. S., Hinchcliffe, A. J., and Anderson, J. S. (1970). *Nature* (*London*) **226,** 940.
O'Halloran, G. J., and Walker, L. W. (1964). Determination of Chemical Species Prevalent in a Plasma Jet. Tech. Document Rep. ASD TDR 62-644, Part II.

O'Halloran, G. J., Fluegge, R. A., Betts, J. F., and Everett, W. L. (1964). Determination of Chemical Species Prevalent in a Plasma Jet. Tech. Document Rep. ASD TDR 62-644, Part I.

Olander, D. R. (1967). *Ind. Eng. Chem. Fundam. Quart.* **6,** 178, 188.

Olander, D. R. (1968). *In* "The Structure and Chemistry of Solid Surfaces" (G. A. Somorjai, ed.), paper 45. Wiley, New York.

Olander, D. R. (1973a). *J. Nucl. Mater.* **49,** 21.

Olander, D. R. (1973b). *J. Nucl. Mater.* **49,** 35.

Olander, D. R., and Schofill, J. L. (1970). *Metall. Trans.* **1,** 2775.

Olander, D. R., Siekhaus, W., Jones, R. H., and Schwarz, J. A. (1972a). *J. Chem. Phys.* **57,** 408.

Olander, D. R., Jones, R. H., Schwarz, J. A., and Siekhaus, W. J. (1972b). *J. Chem Phys.* **57,** 421.

Oldfield, W., and Brown, J. B., Jr. (1970). *Mater. Sci. Eng.* **6,** 361.

Oldright, G. L. (1924). *Trans. AIME* **70,** 471.

Olette, M. (1961). *In* "Physical Chemistry of Process Metallurgy" (G. R. St. Pierre, ed.), Part 2, p. 1065. Wiley (Interscience), New York.

Omholt, A. (1971). "The Optical Aurora." Springer-Verlag, New York.

Oppermann, H. (1971). *Z. Anorg. Allg. Chem.* **383,** 285.

Opstad, M. T., and Schafer, K. (1970). Combustion, June, 30.

O'Reilly, A. J., Doig, I. D., and Ratcliffe, J. S. (1972). *J. Inorg. Nucl. Chem.* **34,** 2487.

Orlovskii, V. P., Schafer, H., Repko, V. P., Safronov, G. M., and Tananaev, I. V. (1971). *Izv. Akad. Nauk SSSR Neorg. Mater.* **7,** 971.

Oshcherin, B. N. (1971). *In* "High Temperature Inorganic Compounds" (G. V. Samsonov, ed.), p. 257. Published for AEC and Nat. Sci. Foundation by Indian Nat. Sci. Documentation Center, New Delhi (AEC-TR-6873).

Ottaway, J. M., Coker, D. T., and Singleton, B. (1972). *Talanta* **19,** 787.

Oxley, J. H. (1966). *In* "Vapor Deposition" (C. F. Powell, J. H. Oxley, and J. M. Blocher, Jr., eds.), p. 102. Wiley, New York.

Oxley, J. H., and Campbell, I. E. (1959). *J. Metals* **11,** 135.

Øye, H. A., and Gruen, D. M. (1969). *J. Amer. Chem. Soc.* **91,** 2229.

Padley, P. J. (1959). Ph.D. Thesis, Cambridge Univ., Cambridge.

Padley, P. J. (1969). *New Sci.* **41,** 23.

Padley, P. J., and Sugden, T. M. (1958). *Proc. Roy. Soc. London* **A248,** 248.

Padley, P. J., and Sugden, T. M. (1959a). *Symp. (Int.) Combust., 7th* p. 235. Butterworths, London and Washington, D.C.

Padley, P. J., and Sugden, T. M. (1959b). *Trans. Faraday Soc.* **55,** 2054.

Padley, P. J., and Sugden, T. M. (1962). *Symp. (Int.) Combust., 8th* p. 164. Williams and Wilkins, Baltimore, Maryland.

Palmer, H. B., and Seery, D. J. (1973). *Annu. Rev. Phys. Chem.* **24,** 235.

Palmer, K. N. (1973). "Dust Explosions and Fires." Chapman and Hall, London.

Panah, M., Mellottee, H., and Delbourgo, R. (1972). *C. R. Acad. Sci. Paris Ser. C* **274,** 1430.

Papatheodorou, G. N. (1973). *J. Phys. Chem.* **77,** 472.

Park, S. K. (1967). The Equilibrium Thermodynamic Properties of a High-temperature Nitrogen-alkali Metal Vapor Mixture. NASA TN D-4106.

Parker, T. A., and Heinsohn, R. J. (1968). *In* "The Performance of High Temperature Systems" (G. S. Bahn, ed.), Vol. 1, p. 209. Gordon and Breach, New York.

Parsons, M. L., and McElfresh, P. M. (1972). *Appl. Spectrosc.* **26,** 472.

Parsons, R. J. (1970). *J. Inst. Fuel* **43,** 524.
Patel, C. C., and Jere, G. V. (1960). *Trans. Metall. Soc. AIME* **218,** 219.
Paton, N. E., Robertson, W. M., and Mansfeld, F. (1973). *Metall. Trans.* **4,** 321.
Patriarca, P., and Rucker, D. J. (1971). Fuels and Materials Development Program Quarterly Report for Period Ending Sept. 30, 1970. ORNL-4630.
Paulson, D. L., and Oden, L. L. (1971). *Corrosion* **27,** 146.
Paulson, W. A., and Springborn, R. H. (1968). NASA TR D-4823.
Peeters, J., and Mahnen, G. (1973). *Symp. (Int.) Combust., 14th* p. 133. Combust. Inst., Pittsburgh, Pennsylvania.
Peeters, J., Vinckier, C., and Van Tiggelen, A. (1969). *Oxid. Combust. Rev.* **4,** 93.
Penner, S. S. (1955). "Introduction to the Study of Chemical Reactions in Flow Systems" (AGARDograph No. 7). Butterworths, London and Washington, D.C.
Pesch, P. (1972). *Astrophys. J.* **174,** L155.
Petrella, R. V. (1971). Studies of the Combustion of Hydrocarbons by Kinetic Spectroscopy. II. The Explosive Combustion of Styrene Inhibited by Halogen Compounds. Private communication.
Petrella, R. V., and Sellers, G. D. (1970). *Fire Technol.* **6,** 93.
Phillips, J. C. (1973). *In* "Treatise on Solid State Chemistry" (W. B. Hannay, ed.), Vol. I, The Chemical Structure of Solids, p. 1. Plenum Press, New York.
Phillips, L. F., and Sugden, T. M. (1960). *Can. J. Chem.* **38,** 1804.
Pilkington, L. A. B. (1969). *Proc. Roy. Soc.* **A314,** 1.
Pint, P., and Flengas, S. N. (1971). *Can. J. Chem.* **49,** 2885.
Pistorius, C. W. F. T., and Sharp, W. E. (1960). *Amer. J. Sci.* **258,** 757.
Pittman, F. K., and Lafleur, J. D., Jr. (1967). *In* DeGroff *et al.* (1967), p. 5.
Pitts, J. J. (1972). *J. Fire Flammability* **3,** 51.
Pitzer, K. S., and Brewer, L. (1961). "Thermodynamics," 2nd ed. McGraw-Hill, New York.
Polak, L. (1971). *In* "Reactions under Plasma Conditions" (M. Venugopalan, ed.), Vol. II, p. 142. Wiley (Interscience), New York.
Polanyi, M. (1932). "Atomic Reactions." Williams and Norgate, London.
Pollard, B., and Milner, D. R. (1971). *J. Iron Steel Inst.* **209,** 291.
Polyachenok, O. G., and Komshilova, O. N. (1972). *High Temp. USSR* **10,** 173.
Porter, R. F., and Zeller, E. E. (1960). *J. Chem. Phys.* **33,** 858.
Post, R. F. (1970). *Annu. Rev. Nucl. Sci.* **20,** 509.
Post, R. F. (1971). *Proc. Nat. Acad. Sci. U.S.* **68,** 1931.
Post, R. F. (1973). *Phys. Today* **26,** 30.
Powell, C. F. (1966). *In* "Vapor Deposition" (C. F. Powell, J. H. Oxley, and J. M. Blocher, Jr., eds.), p. 191. Wiley, New York.
Powell, C. F., Oxley, J. H., and Blocher, J. M., Jr., eds. (1966). "Vapor Deposition." Wiley, New York.
Pownall, C., and Simmons, R. F. (1971). *Symp. (Int.) Combust., 13th* p. 585. Combust. Inst., Pittsburgh, Pennsylvania.
Pratt, G. L. (1969). "Gas Kinetics." Wiley, New York.
Prentice, J. L. (1972). Beryllium Particle Combustion. Final Rep. N72-28924 (NASA-CR-127722; NWC-TP-5330).
Prescott, R., Hudson, R. L., Foner, S. N., and Avery, W. H. (1954). *J. Chem. Phys.* **22,** 145.
Pressley, H. (1970). *Trans. Brit. Ceram. Soc.* **69,** 205.
Prothero, A. (1969). *Combust. Flame* **13,** 399.

Pszonicki, L., and Minczewski, J. (1963). *Spectrochim. Acta* **18,** 1325.
Pumpelly, C. T. (1966). *In* "Bromine and its Compounds" (Z. E. Jolles, ed.), p. 657. Academic Press, New York.
Pungor, E. (1967). "Flame Photometry Theory." Van Nostrand-Reinhold, Princeton, New Jersey.
Quill, L. L. (ed.) (1950). "The Chemistry and Metallurgy of Miscellaneous Materials." McGraw-Hill, New York.
Rabenau, A. (1967). *Angew. Chem. Int. Ed.* **6,** 68.
Rains, R. K., and Kadlec, R. H. (1970). *Metall. Trans.* **1,** 1501.
Ralston, O. C. (1924). *Trans. AIME* **70,** 447.
Rao, D. B. (1970). *High Temp. Sci.* **2,** 381.
Rapp, R. A. (1970). *In* AGARD (1970), paper 2.
Rapp, D., and Johnston, H. S. (1960). *J. Chem. Phys.* **33,** 695.
Rat'kovskii, I. A. (1972). *Vestsi Akad. Nauk Belarus. SSR, Ser. Khim. Navuk* No. 2, 112. (*Chem. Abstr.* **77,** 11075u).
Rautenberg, T. H., Jr., and Johnson, P. D. (1960). *J. Opt. Soc. Amer.* **50,** 602.
Recker, K., and Leckebusch, R. (1969). *J. Cryst. Growth* **5,** 125.
Reed, T. B. (1967). *Advan. High Temp. Chem.* **1,** 259.
Regnier, P. R., and Taran, J. P. E. (1973). *Appl. Phys. Lett.* **23,** 240.
Reid, R. H. G. (1971). *Planet. Space Sci.* **19,** 801.
Reid, R. W., and Sugden, T. M. (1962). *Discuss. Faraday Soc.* **33,** 213.
Reid, W. T. (1971b). "External Corrosion and Deposits: Boilers and Gas Turbines." American Elsevier, New York.
Reigle, L. L., McCarthy, W. J., and Ling, A. C. (1973). *J. Chem. Eng. Data* **18,** 79.
Reinhard, G., Stahlberg, R., and Edelmann, B. U. (1972). *Kernenergie* **15,** 23.
Reisman, A. (1970). "Phase Equilibria." Academic Press, New York.
Reisman, A., and Alyanakyan, S. A. (1964). *J. Electrochem. Soc.* **111,** 1154.
Reisman, A., and Berkenblit, M. (1966). *J. Electrochem. Soc.* **113,** 146.
Reisman, A., and Landstein, J. E. (1971). *J. Electrochem. Soc.* **118,** 1479.
Rekov, A. I. (ed.) (1969). "Materials for Magneto-hydrodynamic Generator Channel." Nauka, Moscow (English transl., 1971, JPRS-53939).
Remirez, R. (1968). *Chem. Eng.* **75,** 114.
Rhein, R. A. (1965). *Pyrodynamics* **3,** 161.
Rhein, R. A. (1971). *AIAA J.* **9,** 353.
Rich, R. (1965). "Periodic Correlations." Benjamin, New York.
Rigg, T. (1972). *In* Jones (1972), p. 153.
Rigg, T. (1973). *Can. J. Chem. Eng.* **51,** 714.
Riley, W. C. (1968). *In* "Ceramics for Advanced Technologies" (J. E. Hove and W. C. Riley, eds.), p. 14. Wiley, New York.
Robertson, D. G. C., and Jenkins, A. E. (1970). *In* "Heterogeneous Kinetics at Elevated Temperatures" (G. R. Belton and W. L. Worrell, eds.), p. 393. Plenum Press, New York.
Rolsten, R. F. (1961). "Iodide Metals and Metal Iodides." Wiley, New York.
Rolsten, R. F. (1959). *J. Electrochem. Soc.* **106,** 975.
Roos, J. T. H. (1971). *Spectrochim. Acta* **26B,** 285.
Rosa, R. J. (1968). "Magnetohydrodynamic Energy Conversion." McGraw-Hill, New York.
Rose, D. J. (1971). *Science* **172,** 797.
Rosenberg, N. W. (1966). *Science* **152,** 1017.

Rosenberg, N. W., and Golomb, D. (1963). *In* Shuler and Fenn (1963), p. 395.
Rosenblatt, G. M. (1970). *In* "Heterogeneous Kinetics at Elevated Temperatures" (*Proc. Int. Conf. Met. Mater. Sci.* (G. R. Belton, ed.), p. 209. Plenum Press, New York.
Rosenburg, H. S. (1970). *Reactor Mater.* **13,** 59.
Rosenthal, M. W., Kasten, P. R., and Briggs, R. B. (1970). *Nucl. Appl. Tech.* **8,** 107.
Rosenthal, M. W., Haubenreich, P. N., McCoy, H. E., and McNeese, L. E. (1971a). *At. Energy Rev.* **9,** 601.
Rosenthal, M. W., Briggs, R. B., and Haubenreich, P. N. (1971b). Molten Salt Reactor Program semi-annu. rep., ORNL Rep. 4622.
Rosner, D. E. (1972a). *In* "Annual Review of Materials Science" (R. A. Huggins, R. H. Bube, and R. W. Roberts, eds.), Vol. 2, p. 573. Annual Reviews, Palo Alto, California.
Rosner, D. E. (1972b). *Oxid. Metals* **4,** 1.
Rosner, D. E., and Allendorf, H. D. (1970). *J. Phys. Chem.* **74,** 1829.
Rosner, D. E., and Allendorf, H. D. (1971). *J. Phys. Chem.* **75,** 308.
Rosser, W. A., Wise, H., and Miller, J. (1959). *Symp. (Int.) Combust., 7th* p. 175. Butterworths, London and Washington, D.C.
Rosser, W. A., Jr., Inami, S. H., and Wise, H. (1963). *Combust. Flame* **7,** 107.
Rothman, A. J. (1968). *In* "Ceramics for Advanced Technologies" (J. E. Hove and W. C. Riley, eds.), pp. 107, 306. Wiley, New York.
Rovner, L. H., Drowart, A., Degreve, F., and Drowart, J. (1968). Mass Spectrometric Studies of the Vaporization of Refractory Metals in Oxygen at Low Pressures. AFML-TR-68-200; AD 675319.
Rozsa, J. T. (1972). *In* "Analytical Emission Spectroscopy" (E. L. Grove, ed.), Part II, p. 508. Dekker, New York.
Runnalls, O. J. C., and Pidgeon, L. M. (1952). *AIME J. Metals* **4,** 843.
Ryabchenko, E. V., Arzamasov, B. N., and Prokoshkin, D. A. (1969). *In* "Protective Coatings on Metals" (G. V. Samsonov, ed.), Vol. I, p. 70. Consultants Bureau, New York.
Ryabova, V. G., and Gurvich, L. V. (1964). *High Temp. USSR* **2,** 834.
Ryabova, V. G., Gurvich, L. V., and Khitrov, A. N. (1971). *Teplofiz. Vys. Temp.* **9,** 755.
Rybakov, B. N., Moskvicheva, A. F., and Beregovaya, G. D. (1969). *Russ. J. Inorg. Chem.* **14,** 1531.
Sale, F. R. (1971). *Miner. Sci. Eng.* **3,** 3.
Salooja, K. C. (1972). *Nature (London)* **240,** 350.
Saltsburg, H. (1973). *Annu. Rev. Phys. Chem.* **24,** 493.
Samuel, R. L. (1958). A Survey of the Factors Controlling Metallic Diffusion from the Gas Phase. Murex Rev., Vol. I.
Sandulova, A. V., and Ostrovskii, P. I. (1971). *Izv. Akad. Nauk SSSR Neorg. Mater.* **7,** 10.
Sarner, S. F. (1966). "Propellant Chemistry." Van Nostrand-Reinhold, Princeton, New Jersey.
Sastri, V. S., Chakraborti, C. L., and Willis, D. E. (1969). *Can. J. Chem.* **47,** 587.
Sawyer, R. F., Fristrom, R. M., and Williamson, R. B. (1971). *Proc. 1971 Spring Meeting, Western States Sect.* Combust. Inst.
Sayce, I. G. (1972). *In* Jones (1972), p. 241.
Schaaf, D. W., and Gregory, N. W. (1971). *J. Phys. Chem.* **75,** 3028.
Schadow, K. (1969). *AIAA J.* **7,** 1870.
Schafer, H. (1964). "Chemical Transport Reactions." Academic Press, New York.
Schafer, H. (1970). *Z. Anorg. Allg. Chem.* **376,** 11.

Schafer, H. (1971). *J. Cryst. Growth* **9,** 17.
Schafer, H. (1972a). Preparation of Oxides and Related Compounds by Chemical Transport. Nat. Bur. Std. Spec. Publ. 364, p. 413.
Schafer, H. (1972b). *In* "Preparative Methods in Solid State Chemistry" (P. Hagenmuller, ed.), p. 251. Academic Press, New York.
Schafer, H. (1973). *In* "Crystal Growth: an Introduction" (P. Hartman, ed.), p. 143. North-Holland Publ., Amsterdam.
Schafer, H., and Fuhr, W. (1962). *Z. Anorg. Chem.* **319,** 52.
Schafer, H., and Wiese, U. (1971). *J. Less-Common Metals* **24,** 55.
Schaffer, P. S. (1965). *J. Amer. Chem. Soc.* **48,** 509.
Schick, H. L. (1966). "Thermodynamics of Certain Refractory Compounds," Vol. 2. Academic Press, New York.
Schirmer, H., and Seehawer, J. (1967). *Proc. Int. Conf. Ioniz. Phenomena Gases, 8th, Vienna* p. 229.
Schissel, P. O., and Trulson, O. C. (1965). *J. Chem. Phys.* **43,** 737.
Schlechten, A. W. (1968). *In* "High Temperature Refractory Metals" (W. A. Krivsky, ed.), p. 347. Gordon and Breach, New York.
Schmets, J. J. (1970). *At. Energy Rev.* **8,** 3.
Schofield, K., and Broida, H. P. (1968). *In* "Atomic Interactions" (B. Bederson and W. L. Fite, eds.), Vol. 7B, p. 189. Academic Press, New York.
Schofield, K., and Sugden, T. M. (1971). *Trans. Faraday Soc.* **67,** 1054.
Schofield, K., and Sugden, T. M. (1965). *Symp. (Int.) Combust., 10th* p. 589. Combust. Inst., Pittsburgh, Pennsylvania.
Schoonmaker, R. C., and Porter, R. F. (1959). *J. Chem. Phys.* **30,** 283.
Schwar, M. J. R., and Weinberg, F. J. (1969). *Combust. Flame* **13,** 335.
Schwarz, J. A., and Madix, R. J. (1968). *J. Catal.* **12,** 140.
Schwerdtfeger, K., and Turkdogan, E. T. (1970). *In* "Techniques of Metals Research" (R. A. Rapp, ed.), Vol. IV, p. 321. Wiley (Interscience), New York.
Seaborg, G. T., and Bloom, J. L. (1970). *Sci. Amer.* **223,** 13.
Searcy, A. W. (1962). *Progr. Inorg. Chem.* **3,** 49.
Searcy, A. W. (1964). *Int. Symp. High Temp. Technol. IUPAC, Asilomar, 1963* p. 105. Butterworths, London and Washington, D.C.
Searcy, A. W., Ragone, D. V., and Colombo. U. (eds.) (1970). "Chemical and Mechanical Behavior of Inorganic Materials." Wiley (Interscience), New York.
Seif, M. (1971). *Science* **173,** 802.
Seikel, G. R., and Nichols, L. D. (1972). *J. Spacecr. Rockets* **9,** 322.
Seki, I., and Minagawa, S. (1972). *Jap. J. Appl. Phys.* **11,** 850.
Semenenko, K. N., Naumova, T. N., Gorokhov, L. N., and Novoselova, A. V. (1964a). *Dokl. Akad. Nauk SSSR* **154,** 648.
Semenenko, K. N., Naumova, T. N., Gorokhov, L. N., Semenova, G. A., and Novoselova, A. V. (1964b). *Dokl. Akad. Nauk SSSR* **154,** 169.
Semenov, G. A., and Shalkova, E. K. (1969). *Vestn. Leningrad Univ. Fiz. Khim.* **24,** 111.
Semenova, A. I., and Tsiklis, D. S. (1970). *Zh. Fiz. Khim.* **44,** 2505.
Shaffner, R. O. (1971). *Proc. IEEE* **59,** 622.
Shaw, M. (1967). *In* "Fast Breeder Reactors" (P. V. Evans, ed.), p. 35. Pergamon, Oxford.
Shelton, R. A. J. (1968a). *Trans. Inst. Min. Metall.* **77C,** 113.
Shelton, R. A. J. (1968b). *Trans. Inst. Min. Metall.* **77C,** 31.
Shelton, R. A. J., and Holt, G. (1972). *In* Kubaschewski (1972), p. 559.
Shelton, R. A. J., Bhatti, M. A., Copley, D. B., and Normanton, A. S. (1969). *In*

"Reactivity of Solids" (*Proc. Int. Symp. Reactivity Solids, 6th, Schenectady, New York, 1968*) (J. W. Mitchell, R. C. DeVries, R. W. Roberts, and P. Cannon, eds.), p. 255. Wiley (Interscience), New York.

Sholts, V. B., Sidorov, L. N., and Korenev, Y. M. (1970). *Zh. Fiz. Khim.* **44,** 2164.

Shubaev, V. L., Suvorov, A. V., and Semenov, G. A. (1970). *Russ. J. Inorg. Chem.* **15,** 479.

Shukla, R. V., and Srivastava, A. N. (1970). *Indian J. Pure Appl. Phys.* **8,** 497.

Shuler, K. E., and Fenn, J. B. (eds.) (1963). "Ionization in High Temperature Gases." Academic Press, New York.

Sidorov, L. N., Belousov, V. I., and Scholtz, V. B. (1971). *Advan. Mass Spectrom.* **5,** 394.

Siegel, B., and Schieler, B. (1964). "Energetics of Propellant Chemistry." Wiley, New York.

Silver, I. (1962). *In* "Materials Science and Technology for Advanced Applications" (D. R. Mash, ed.), p. 1. Prentice-Hall, Englewood Cliffs, New Jersey.

Silver, L. T. (1972). *Sci. News* **101,** 12.

Silvestri, V. J. (1972). *J. Electrochem. Soc.* **119,** 775.

Simmons, R. F., and Wolfhard, H. G. (1955). *Trans. Faraday Soc.* **51,** 1211.

Simmons, R. F., and Wolfhard, H. G. (1956). *Trans. Faraday Soc.* **52,** 53.

Simons, E. M. (1970). *Reactor Mater.* **13,** 49.

Singer, J. M., and Grumer, J. (1962). Carbon Formation in Very Rich Hydrocarbon Flames. U.S. Bur. Mines Rep. 6007.

Sinha, R. K., and Walker, P. L., Jr. (1972). *Fuel* **51,** 329.

Sittig, M. (1968). "Inorganic Chemical and Metallurgical Process Encyclopedia." Noyes Develop. Corp., New Jersey.

Skell, P. S. (1971). *Int. Congr. Pure Appl. Chemi., 23rd, Boston* **4,** 215. Butterworths, London and Washington, D.C.

Skinner, G. B., and Snyder, A. D. (1964). *In* "Heterogeneous Combustion" (H. G. Wolfhard, I. Glassman, and L. Green, Jr., eds.), p. 345. Academic Press, New York.

Sladek, K. J. (1971). *J. Electrochem. Soc.* **118,** 654.

Slunder, C. J. (1959). *In* "Corrosion and Deposits in Coal- and Oil-fired Boilers and Gas Turbines," p. 152. ASME, New York.

Smith, C. O. (1967). "Nuclear Reactor Materials." Addison-Wesley, Reading, Massachusetts.

Smith, D., and Agnew, J. T. (1956). *Symp. (Int.) Combust., 6th* p. 83. Van Nostrand-Reinhold, Princeton, New Jersey.

Smith, D. L. (1973). *Nucl. Tech.* **20,** 190.

Smith, D. S., and Starkman, E. S. (1971). *Symp. (Int.) Combust., 13th* p. 439. Combust. Inst., Pittsburgh, Pennsylvania.

Smith, H. M., and Turner, A. F. (1965). *Appl. Opt.* **4,** 147.

Smith, S. R., and Gordon, A. S. (1956). *J. Phys. Chem.* **60,** 759.

Smyly, D. S., Townsend, W. P., Zeegers, P. J. T., and Winefordner, J. D. (1971). *Spectrochim. Acta* **26B,** 531.

Snelleman, W. (1968a). *Metrologia* **4,** 117.

Snelleman, W. (1968b). *Metrologia* **4,** 123.

Snelleman, W. (1969). *Flame Emiss. At. Absorption Spectrom.* **1,** 213.

Snyder, L. E. (1972). *Phys. Chem. Ser. 1,* **3,** 193.

Solon, L. R. (1971). *Ind. Res.* **13,** 46.

Sourirajan, S., and Kennedy, G. C. (1962). *Amer. J. Sci.* **260,** 115.

Spalding, D. B., and Stephenson, P. L. (1971). *Proc. Roy. Soc. London* **A324,** 315.

Spear, K. E. (1972). *J. Chem. Ed.* **49,** 81.

Speiser, R., and St. Pierre, G. R. (1964). *In* "The Science and Technology of Selected Refractory Metals" (N. E. Promisel, ed.), p. 289. Pergamon, Oxford.

Spencer, F. E., Jr., Hendrie, J. C., and Bienstock, D. (1973). *Symp. Eng. Aspects Magnetohydrodynam., 13th* Stanford Univ., p. VII.4.1; VII.4.6.

Spengler, G., and Buchner, E. (1969). *Brennstoff. Chem.* **50,** 51.

Spengler, G., and Haupt, G. (1969). *Erdol. Kohle. Erdgas. Petrochem.* **22,** 679.

Speros, D. M., and Caldwell, R. M. (1972). *High Temp. Sci.* **4,** 99.

Spiridonov, V. P., and Malkova, A. S. (1969). *Zh. Strukt. Khim.* **10,** 332.

Spokes, G. N., and Evans, B. E. (1965). *Symp. (Int.) Combust., 10th* p. 639. Combust. Inst., Pittsburgh, Pennsylvania.

Sprague, R. W. (1968). *In* "The Performance of High Temperature Systems" (G. S. Bahn, ed.), Vol. 1, p. 19. Gordon and Breach, New York.

Springer, R. H., and Taylor, R. P. (1971). *Proc. IEEE* **59,** 617.

Stafford, F. E., and Berkowitz, J. (1964). *J. Chem. Phys.* **40,** 2963.

St. Clair, H. W. (1958). *In* "Vacuum Metallurgy" (R. F. Bunshah, ed.), p. 295. Van Nostrand-Reinhold, Princeton, New Jersey.

Steck, S. J., Pressley, G. A., Jr., Lin, S. S., and Stafford, F. E. (1969). *J. Chem. Phys.* **50,** 3196.

Steindler, M. J., Anastasia, L. J., Trevorrow, L. E., and Chilenskas, A. A. (1969). *In* Chiotti (1969), p. 177.

Steinmeyer, R. H., and Kershner, C. J. (1971). *J. Inorg. Nucl. Chem.* **33,** 2847.

Stemprok, M. (1963). *In* "Problems of Postmagmatic Ore Deposition" (J. Kutina, ed.), Vol. I, p. 69. Czech. Acad. Sci., Prague.

Steunenberg, R. K., Pierce, R. D., and Johnson, I. (1969). *In* "Nuclear Metallurgy" (P. Chiotti, ed.), Vol. 15, Reprocessing of Nuclear Fuels, p. 325. USAEC Conf-690801.

Stogryn, D. E., and Hirschfelder, J. O. (1959). *J. Chem. Phys.* **31,** 1531.

Stokes, C. S. (1969). *In* "Chemical Reactions in Electrical Discharges" (R. F. Gould, ed.), Advan. Chem. Ser. 80, p. 390. Amer. Chem. Soc., Washington, D.C.

Stokes, C. S. (1971). *In* "Reactions under Plasma Conditions" (M. Venugopalan, ed.), Vol. II, p. 259. Wiley (Interscience), New York.

Stott, J. V., and Fray, D. J. (1972). *In* Jones (1972), p. 95.

St. Pierre, G. R. (ed.) (1959 and 1961). "Physical Chemistry of Process Metallurgy," Parts 1 and 2. Wiley (Interscience), New York.

Street, J. C., and Thomas, A. (1955). *Fuel* **34,** 4.

Strehlow, R. A. (1968). "Fundamentals of Combustion." Int. Textbook Co., Scranton, Pennsylvania.

Stull, D. R., and Sinke, G. C. (1956). "Thermodynamic Properties of the Elements," Advan. Chem. Ser. 18. Amer. Chem. Soc., Washington, D.C.

Sugden, T. M. (1956). *Trans. Faraday Soc.* **52,** 1465.

Sugden, T. M. (1962). *Annu. Rev. Phys. Chem.* **13,** 369.

Sugden, T. M. (1963). *Progr. Astronaut. Aeron.* **12,** 145.

Sugden, T. M. (1965). *Symp. (Int.) Combust., 10th* p. 539. Combust. Inst., Pittsburgh, Pennsylvania.

Sullivan, C. J. (1957). *Econ. Geol.* **52,** 5.

Summerfield, M. (ed.) (1960). "Solid Propellant Rocket Research." Academic Press, New York.

Sundheim, B. R. (ed.) (1964). "Fused Salts." McGraw-Hill, New York.

Suvorov, A. V., and Krzhizhanovskaya, E. K. (1969a). *Russ. J. Inorg. Chem.* **14,** 468.

Suvorov, A. V., and Krzhizhanovskaya, E. K. (1969b). *Russ. J. Inorg. Chem.* **14,** 434.

Suvorov, A. V., and Malkova, A. S. (1968). *Vestn. Leningrad Univ. Fiz. Khim* **23,** 108.
Suvorov, A. V., and Novikov, G. I. (1968). *Vestn. Leningrad Univ.* **4,** 83.
Suvorov, A. V., and Shubaev, V. L. (1971). *Russ. J. Inorg. Chem.* **16,** 16.
Suvorov, A. V., Malkov, A. S., and Avrorina, V. I. (1969). *Russ. J. Inorg. Chem.* **14,** 720.
Svehla, R. A. (1964). Thermodynamic and Transport Properties for the Hydrogen-oxygen System. NASA SP-3011.
Svirskiy, L. D., and Pirogov, Yu. A. (1971). Processes of Formation of Coatings Deposited by the Method of Gas-flame Spraying. NASA Tech. Tr. TT F-13534.
Swings, P. (1969). *Mem. Soc. Roy. Sci. Liege Collect.* **8,** 6.
Szego, G. C. (1971). The U.S. Energy Problem. NSF-RANN 71-1-2 (PB-207518).
Szekely, J. (ed.) (1972). "Blast Furnace Technology, Science and Practice." Dekker, New York.
Szekely, J., and Evans, J. W. (1972). *In* "Blast Furnace Technology, Science and Practice" (J. Szekely, ed.), p. 35. Dekker, New York.
Takahashi, T., Sugiyama, K., and Itoh, H. (1970). *J. Electrochem. Soc.* **117,** 541.
Takeuchi, S., Tezuka, M., Kurosawa, T., and Eda, S. (1961). *In* "Physical Chemistry of Process Metallurgy" (G. R. St. Pierre, ed.), Part 2, p. 745. Wiley (Interscience), New York.
Tananaev, I. V. (1971). *Izv. Akad. Nauk SSSR, Neorg. Mater.* **7,** 361.
Tanford, C. (1947). *J. Chem. Phys.* **15,** 433.
Tanford, C., and Pease, R. N. (1947). *J. Chem. Phys.* **15,** 431.
Tarasenkov, D. W., and Klyachko-Gurvich, L. L. (1936). *J. Gen. Chem. USSR* **6,** 305 (*Chem. Abstr.* **30,** 7427).
Taylor, M. V., and Petrozzi, P. J. (1968). *In* "The Performance of High Temperature Systems" (G. S. Bahn, ed.), Vol. 1, p. 109. Gordon and Breach, New York.
Tewarson, A., and Palmer, H. B. (1967). *J. Mol. Spectrosc.* **22,** 117.
Tewarson, A., Naegeli, D. W., and Palmer, H. B. (1969). *Symp. (Int.) Combust., 12th* p. 415. Combust. Inst., Pittsburgh, Pennsylvania.
Thiery, P. (1970). "Fireproofing" (Translated by J. H. Goundry). Elsevier, Amsterdam.
Thomas, D. E., and Hayes, E. T. (eds.) (1960). "The Metallurgy of Hafnium." USAEC, U.S. Govt. Printing Office, Washington, D.C.
Thompson, M. A., Marshall, R. S., and Standifer, R. L. (1969). *In* Chiotti (1969), p. 163.
Thorn, R. J. (1966). *Annu. Rev. Phys. Chem.* **17,** 83.
Thorpe, M. L. (1971). *In* "Advances in Extractive Metallurgy and Refining," (*Proc. Symp. Inst. Mining Met., London*), paper 18.
Thorpe, M. L. (1972). *In* Jones (1972), p. 275.
Thrash, R. J., Von Weyssenhoff, H., and Shirk, J. S. (1971). *J. Chem. Phys.* **55,** 4659.
Timmins, R. S., and Ammann, P. R. (1967). *In* "The Application of Plasmas to Chemical Processing" (R. F. Baddour and R. S. Timmins, eds.), p. 99. MIT Press, Cambridge, Massachusetts.
Timms, P. L. (1972). *Advan. Inorg. Chem. Radiochem.* **14,** 121.
Timms, P. L. (1973a). *Accounts Chem. Res.* **6,** 118.
Timms, P. L. (1973b). Atoms and small molecules as useful chemical reagents. *In* High Temperature Studies in Chemistry, Faraday Div. Chem. Soc., paper 6.
Tischer, R. L., and Scheller, K. (1968). *Combust. Flame* **12,** 367.
Tischer, R. L., and Scheller, K. (1969). *Proc. Conf. Western States Sect., Combustion Inst.* (*Chem. Abstr.* 68833r, 1970).
Tischer, R. L., and Scheller, K. (1970). *Combust. Flame* **15,** 199.
Tkachuk, B. V., Bushin, V. V., and Smetankina, N. P. (1966). *Ukr. Khim. Zh.* **32,** 1256.

Trombe, F., and Male, G. (1968). *Rev. Hautes Temp.* **5,** 287.
Tskhvirashvili, D. G., and Vasadze, L. E. (1966). *Soobshch. Akad. Nauk Gruz. SSR* **42,** 653 (*Chem. Abstr.* **65,** 13405e).
Tsuji, T. (1964). *Proc. Jap. Acad.* **40,** 99.
Turkdogan, E. T., Grieveson, P., and Darken, L. S. (1963). *J. Phys. Chem.* **67,** 1647.
Ulrich, G. D. (1971). *Combust. Sci. Technol.* **4,** 47.
Vale, H. J. (1963). *In* "Kinetics, Equilibria and Performance of High Temperature Systems" (G. S. Bahn, ed.), p. 271. Gordon and Breach, New York.
Van der Hurk, J., Hollander, T., and Alkemade, C. T. J. (1973). *J. Quant. Spectrosc. Radiat. Transf.* **13,** 273.
Vanpee, M. (1968). A Study of Metal Combustion in Flowing Systems. U.S. Govt. R and D Rep. AD 676 060.
Vanpee, M., Tromans, R. H., and Burgess, D. (1964). *In* "Heterogeneous Combustion" (H. G. Wolfhard, I. Glassman, and L. Green, Jr., eds.), p. 419. Academic Press, New York.
Vanpee, M., Hinck, E. C., and Seamans, T. F. (1965). *Combust. Flame* **9,** 393.
Vanpee, M., Kineyko, W. R., and Caruso, R. (1970). *Combust. Flame* **14,** 381.
Van Tiggelen, A. (1963). *In* "Ionization in High Temperature Gases" (K. E. Shuler, and J. B. Fenn, eds.), p. 165. Academic Press, New York.
Venugopalan, M. (1971). *In* "Reactions under Plasma Conditions" (M. Venugopalan, ed.), Vol. II, p. 1. Wiley (Interscience), New York.
Veprek, S. (1972). *J. Cryst. Growth* **17,** 101.
Verhaegen, G., Colin, R., Exsteen, G., and Drowart, J. (1965). *Trans. Faraday Soc.* **61,** 1372.
Vickers, T. J., and Winefordner, J. D. (1972). *In* "Analytical Emission Spectroscopy" (E. L. Grove, ed.), Part II, p. 255. Dekker, New York.
Vidal, R., and Poos, A. (1973). *In* "The Steel Industry and the Environment" (J. Szekely, ed.), p. 201. Dekker, New York.
Vidal, B., Dessaux, O., Marteel, J. P., and Goudmand, P. (1969). *C.R. Acad. Sci. Paris Ser. C.* **268,** 574.
Vig, S. K., and Lu, W. K. (1971). *J. Iron Steel Inst. London* **209,** 630.
Viswanathan, R., and Spengler, C. J. (1970). *Corrosion* **26,** 29.
Von Munch, W. (1971). *J. Cryst. Growth* **9,** 144.
Von Wartenberg, H. (1951). *Z. Elektrochem.* **55,** 445.
Vree, P., and Miller, W. J. (1968). *Fire Res. Abstr. Rev.* **10,** 12.
Vucuroic, D., and Miodrag, S. (1969). *Glas. Hem. Drus., Beograd* **34,** 541 (*Chem. Abstr.* **73,** 122606).
Vurzel, F. B., and Polak, L. S. (1970). *Ind. Eng. Chem.* **62,** 8.
Wachtell, R. L., and Jefferys, R. A. (1968). *In* "Fundamentals of Refractory Compounds" (H. H. Hausner and M. G. Bowman, eds.), p. 249. Plenum Press, New York.
Wadsworth, M. E. (1972). *Annu. Rev. Phys. Chem.* **23,** 355.
Wagman, D. D., Evans, W. H., Parker, V. B., Halow, I., Bailey, S. M., and Schumm, R. H. (1969). Nat. Bur. Std. Tech. Note 270-4.
Wagner, C. (1958). *Acta Met.* **6,** 309.
Wagner, C. (1968). *In* "Molecular Processes on Solid Surfaces" (E. Drauglis, R. D. Gretz and R. I. Jaffee, eds.), p. 17. McGraw-Hill, New York.
Wagner, C. (1970a). *In* "Physicochemical Measurements in Metals Research" (R. A. Rapp, ed.), Part 1, p. 1. Wiley (Interscience), New York.
Wagner, C. (1970b). *Corrosion Sci.* **10,** 641.

Wahl, G., and Batzies, P. (1970). *IEEE Conf. Record Thermionic Conversion Specialist Conf., Pap. Annu. Conf.* p. 119.
Walter, L. S., and Giutronich, J. E. (1967). *Solar Energy* **11,** 163.
Ward, C. C. (1973). ASTM Special Tech. Publ. 531, 133.
Ward, R. G. (1962). "An Introduction to the Physical Chemistry of Iron and Steel Making." Arnold, London.
Warren, I. H., and Shimizu, H. (1965). *Can. Mining Met. Bull.* **58,** 551.
Watanabe, H., Nishinaga, T., and Arizumi, T. (1972). *J. Cryst. Growth* **17,** 183.
Waymouth, J. F. (1971a). "Electric Discharge Lamps." MIT Press, Cambridge, Massachusetts.
Waymouth, J. F. (1971b). *Proc. IEEE* **59,** 629.
Webb, W. W., Norton, J. T., and Wagner, C. (1956). *J. Electrochem. Soc.* **103,** 107.
Weber, B., Fusy, J., and Cassuto, A. (1971a). *In* "Recent Developments in Mass Spectroscopy" (K. Ogata and T. Hayakawa, eds.), p. 1319. Univ. Park Press, Baltimore, Maryland.
Weber, M., Trautmann, N., and Herrmann, G. (1971b). *Radiochem. Radioanal. Lett.* **6,** 73.
Wehner, J. F. (1964). *Advan. Chem. Eng.* **5,** 1.
Weinberg, F. (ed.) (1970). "Tools and Techniques in Physical Metallurgy," Vols. 1 and 2. Dekker, New York.
Weller, A. E., and Bagby, F. L. (1967). *In* "High Temperature Materials and Technology" (I. E. Campbell and E. M. Sherwood, eds.), p. 3. Wiley, New York.
Weltner, W., Jr. (1967). *Science* **155,** 155.
Wentink, T., Jr., and Brown, N. B. (1971). *J. Geophys. Res.* **76,** 7006.
Wentorf, R. H., Jr., and Hanneman, R. E. (1974). *Science* **185,** 311.
Westenberg, A. A., and Fristrom, R. M. (1960). *J. Phys. Chem.* **64,** 1393.
Westenberg, A. A., and Fristrom, R. M. (1965). *Symp. (Int.) Combust., 10th* p. 473. Combust. Inst., Pittsburgh, Pennsylvania.
Whatley, M. E., McNeese, L. E., Carter, W. L., Ferris, L. M., and Nicholson, E. L. (1970). *Nucl. Appl. Tech.* **8,** 170.
Wheeler, R. (1968). *J. Phys. Chem.* **72,** 3359.
White, D. E. (1963). *In* "Problems in Postmagmatic Ore Deposition" (J. Kutina, ed.), Vol. I, p. 432. Czech. Acad. Sci., Prague.
White, W. B., Johnson, S. M., and Dantzig, G. B. (1958). *J. Chem. Phys.* **28,** 751.
Whiteway, S. G. (1971). *Can. Met. Quart.* **10,** 185.
Whitman, M. J., Fortescue, P., Hannum, W. H., Leitz, F. J., Jr., and Trauger, D. B. (1967). *In* "Fast Breeder Reactors" (P. V. Evans, ed.), p. 275. Pergamon, Oxford.
Whittingham, G. (1971). *J. Inst. Fuel* **44,** 316.
Wilkins, R. L. (1963). "Theoretical Evaluation of Chemical Propellants." Prentice-Hall, Englewood Cliffs, New Jersey.
Williams, F. A. (1965). "Combustion Theory." Addison-Wesley, Reading, Massachusetts.
Williams, G. J., and Wilkins, R. G. (1973). *Combust. Flame* **21,** 325.
Williams, M. L., Baer, A. D., Ryan, N. W., Isaacson, L. K., and Seader, J. D. (1969). The Chemistry and Mechanics of Combustion with Application to Rocket Engine Systems. Annu. rep., AD-695811 (THEMIS-UTEC-TH-69-073; AFOSR-69-2548 TR; N70-15771).
Willis, J. B. (1970). *In* "Analytical Flame Spectroscopy" (R. Mavrodineanu, ed.), p. 525. Springer-Verlag, New York.

Wilson, R. H., Jr., Conway, J. B., Engelbrecht, A., and Grosse, A. V. (1951). *J. Amer. Chem. Soc.* **73,** 5514.

Wilson, R. W., Penzias, A. A., Fefferts, K. B., Kutner, M., and Thaddeus, P. (1971). *Astrophys. J.* **167,** 197.

Wilson, W. E., Jr. (1965). *Symp. (Int.) Combust., 10th* p. 47. Combust. Inst., Pittsburgh, Pennsylvania.

Wilson, W. E., Jr., O'Donovan, J. T., and Fristrom, R. M. (1969). *Symp. (Int.) Combust., 12th* p. 929. Combust. Inst., Pittsburgh, Pennsylvania.

Wimber, R. T. (1971). High Temperature Oxidation of Iridium. Rep. No. RLO-2228-T1-1.

Winefordner, J. D., and Vickers, T. J. (1964). *Anal. Chem.* **36,** 1947.

Winge, R. K., Fassel, V. A., and Kniseley, R. N. (1971). *Appl. Spectrosc.* **25,** 636.

Wirtz, G. P., and Siebert, D. C. (1971). *Mater. Res. Bull.* **6,** 381.

Wolfhard, H. G., Clark, A. H., and Vanpee, M. (1964a). *In* "Heterogeneous Combustion" (H. G. Wolfhard, I. Glassman, and L. Green, Jr., eds.), p. 327. Academic Press, New York.

Wolfhard, H. G., Glassman, I., and Green, L., Jr., eds. (1964b). "Heterogeneous Combustion." Academic Press, New York.

Wood, G. C. (1970). *Oxid. Metals* **2,** 11.

Wood, L., and Nuckolls, J. (1972). *Environment* **14,** 29.

Woodbridge, D. D. (1961). *In* "Chemical Reactions in the Lower and Upper Atmosphere" (*Symp., Stanford Res. Inst.*), p. 373. Wiley (Interscience), New York.

Woods, W. G., and Bower, J. G. (1970). *Mod. Plast.* **47,** 140.

Woodward, C. (1971). *Spectrosc. Lett.* **4,** 191.

Worrell, W. L. (1971). *Advan. High Temp. Chem.* **4,** 71.

Wyatt, J. R., and Stafford, F. E. (1972). *J. Phys. Chem.* **76,** 1913.

Wynnyckyj, J. R. (1972). *High Temp. Sci.* **4,** 205.

Yamaoka, S., Fukunaga, O., and Saito, S. (1970). *J. Amer. Chem. Soc.* **53,** 179.

Yanagi, T., and Mimura, Y. (1972). *Combust. Flame* **18,** 347.

Yannopoulos, L. N., and Pebler, A. (1971). *J. Appl. Phys.* **42,** 858.

Yannopoulos, L. N., and Pebler, A. (1972). *J. Appl. Phys.* **43,** 2435.

Ybarrondo, L. J., Solbrig, C. W., and Isbin, H. S. (1972). The Calculated Loss-of-coolant Accident: A Review. AIChE Monogr. Ser. No. 7. Amer. Inst. Chem. Eng.

Yoshikawa, S. (1971). *Amer. Sci.* **59,** 463.

Zabetakis, M. G. (1965). Flammability Characteristics of Combustible Gases and Vapors. U.S. Bur. Mines Bull. 627.

Zahradrik, R. L., and Glenn, R. A. (1971). *Fuel* **50,** 77.

Zeegers, P. J. T., Townsend, W. P., and Winefordner, J. D. (1969). *Spectrochim. Acta* **B24,** 243.

Zeegers, P. J. T., and Alkemade, C. T. J. (1970). *Combust. Flame* **15,** 193.

Zeleznik, F. J., and Gordon, S. (1965). Equilibrium Computations for Multicomponent Plasmas. NASA TN D-2806.

Zielinski, A. (1968). *Proc. Magnetohydrodynam. Symp., Warsaw,* **4,** 2211. IAEA, Vienna.

Zollweg, R. J., and Frost, L. S. (1969). *J. Chem. Phys.* **50,** 3280.

Zubler, E. G., and Mosby, F. H. (1959). *Illum. Eng.* **54,** 734.

Zvarova, T. S., and Zvara, I. (1969). *J. Chromatogr.* **44,** 604.

Zvarova, T. S., and Zvara, I. (1970a). ANL-TRANS-843, *Joint Inst. Nucl. Invest. Rep.* **6,** 4911.

Zvarova, T. S., and Zvara, I. (1970b). *J. Chromatogr.* **49,** 290.

Zvezdin, A. G., Ketov, A. N., and Fomin, V. K. (1970). *Zh. Prikl. Khim.* **43,** 2541.

Index

A

B

C

D

E

F

G

H

N

O

P

W

Z

A
B 5
C 6
D 7
E 8
F 9
G 0
H 1
I 2
J 3